Springer
Proceedings in Physics 64

Springer Proceedings in Physics

Managing Editor: H. K. V. Lotsch

44 *Optical Fiber Sensors*
Editors: H. J. Arditty, J. P. Dakin, and R. Th. Kersten

45 *Computer Simulation Studies in Condensed Matter Physics II: New Directions*
Editors: D. P. Landau, K. K. Mon, and H.-B. Schüttler

46 *Cellular Automata and Modeling of Complex Physical Systems*
Editors: P. Manneville, N. Boccara, G. Y. Vichniac, and R. Bidaux

47 *Number Theory and Physics*
Editors: J.-M. Luck, P. Moussa, and M. Waldschmidt

48 *Many-Atom Interactions in Solids*
Editors: R .M. Nieminen, M. J. Puska, and M. J. Manninen

49 *Ultrafast Phenomena in Spectroscopy*
Editors: E. Klose and B. Wilhelmi

50 *Magnetic Properties of Low-Dimensional Systems II: New Developments*
Editors: L. M. Falicov, F. Mejía-Lira, and J. L. Morán-López

51 *The Physics and Chemistry of Organic Superconductors*
Editors: G. Saito and S. Kagoshima

52 *Dynamics and Patterns in Complex Fluids: New Aspects of the Physics–Chemistry Interface*
Editors: A. Onuki and K. Kawasaki

53 *Computer Simulation Studies in Condensed Matter Physics III*
Editors: D. P. Landau, K. K. Mon, and H.-B. Schüttler

54 *Polycrystalline Semiconductors II*
Editors: J. H. Werner and H. P. Strunk

55 *Nonlinear Dynamics and Quantum Phenomena in Optical Systems*
Editors: R. Vilaseca and R. Corbalán

56 *Amorphous and Crystalline Silicon Carbide III, and Other Group IV-IV Materials*
Editors: G. L. Harris, M. G. Spencer, and C. Y.-W. Yang

57 *Evolutionary Trends in the Physical Sciences*
Editors: M. Suzuki and R. Kubo

58 *New Trends in Nuclear Collective Dynamics*
Editors: Y. Abe, H. Horiuchi, and K. Matsuyanagi

59 *Exotic Atoms in Condensed Matter*
Editors: G. Benedek and H. Schneuwly

60 *The Physics and Chemistry of Oxide Superconductors*
Editors: Y. Iye and H. Yasuoka

61 *Surface X-Ray and Neutron Scattering*
Editors: H. Zabel and I. K. Robinson

62 *Surface Science: Lectures on Basic Concepts and Applications*
Editors: F. A. Ponce and M. Cardona

63 *Coherent Raman Spectroscopy: Recent Advances*
Editors: G. Marowsky and V. V. Smirnov

64 *Superconducting Devices and Their Applications*
Editors: H. Koch and H. Lübbig

65 *Present and Future of High-Energy Physics*
Editors. K.-I. Aoki and M. Kobayashi

66 *The Structure and Conformation of Amphiphilic Membranes*
Editors: R. Lipowsky, D. Richter, and K. Kremer

67 *Nonlinearity with Disorder*
Editors: F. Abdullaev, A. R. Bishop, and S. Pnevmatikos

68 *Time-Resolved Vibrational Spectroscopy V*
Editor: H. Takahashi

69 *Evolution of Dynamical Structures in Complex Systems*
Editors: R. Friedrich and A. Wunderlin

Volumes 1–43 are listed on the back inside cover

H. Koch · H. Lübbig (Eds.)

Superconducting Devices and Their Applications

Proceedings
of the 4th International Conference SQUID '91
(Sessions on Superconducting Devices),
Berlin, Fed. Rep. of Germany, June 18–21, 1991

With 356 Figures

Springer-Verlag
Berlin Heidelberg New York
London Paris Tokyo
Hong Kong Barcelona
Budapest

Dr. Hans Koch
Dr. Heinz Lübbig

Physikalisch-Technische Bundesanstalt, Institut Berlin,
Abbestraße 2–12, W-1000 Berlin 10, Fed. Rep. of Germany

ISBN 3-540-55396-7 Springer-Verlag Berlin Heidelberg New York
ISBN 0-387-55396-7 Springer-Verlag New York Berlin Heidelberg

Printed in Germany

Typesetting: Camera ready by authors/editors
54/3140 - 5 4 3 2 1 0 - Printed on acid-free paper

Preface

Summer 1991 was just the right time to organize a conference on electronic quantum effect devices: (i) research and development on high-T_c superconductivity (HTS) had matured to such a degree that device development had become a serious business, (ii) low-T_c superconductive (LTS) device technology had undergone a considerable refinement initiated by a major breakthrough: the commercial availability of complex multichannel SQUID systems for biomagnetic applications, and (iii) another field of rapid and exciting development had evolved concerning non-superconducting quantum effect devices that nontheless use many common concepts known from superconducting devices.

Thus the WE-Heraeus conference entitled the 4th International Conference on Superconducting and Quantum Effect Devices and Their Applications "SQUID '91" held in Berlin, Germany, June 18–21, 1991, tried to follow the tradition of the former conferences SQUID '76, SQUID '80, and SQUID '85 to collect together and summarize the achievements of an important period of research and development of superconducting quantum effect devices. The proceedings of these former conferences consequently served as reference books on the state of the art of this field. It is our intention with this and the companion book – published by the same editors under the title "Single-Electron Tunneling and Mesoscopic Devices" (Springer Series in Electronics and Photonics, Vol. 31) – to achieve the same status again. The publication of a companion book covering the sessions on single-electron tunneling and mesoscopic devices was an obvious need, because this topic dominated a surprisingly large fraction of the programme. Accordingly, the present book concentrates on the traditional content of IC SQUID proceedings: superconducting quantum effect devices and their applications.

This volume covers three main areas: HTS devices, LTS devices, and applications. We deliberately did not consider any papers on HTS thin film preparation or fabrication technology but asked A.I. Braginski for his excellent up-to-date review as an introduction. A kind of extension to this introduction is formed by M.Yu. Kupriyanov's contribution on the physics of HTS junctions and Y. Okabe's survey of the difficulty of fabricating real HTS tunnel junctions. The many papers that follow remind one of the early days of LTS device development in the sense that also now a truly evolutionary process is taking place involving a large variety of competing concepts preceding the survival of the fittest. Remarkable progress can already be seen with the highlight of a working HTS dc SQUID with integrated flux transformer fabricated in a seven-layer technique which already includes all major ingredients of a modern LTS counterpart. Other achievements, e.g., IR detectors and flux-flow transistors, underline as well the fact that the threshold to relevant HTS applications has apparently been crossed.

However, the chapter on LTS devices strongly indicates the long way HTS research and development still has to go. Field sensitivities of LTS SQUID sensors have routinely been pushed into the range of a few $fT/\sqrt{Hz}$ due to the improved understanding of the role of resonances in thin-film coupling structures. This was accompanied by a number of novel concepts for SQUID read-out electronics. Both developments were pushed by the requirements dictated by the specifications of multichannel SQUID systems for biomagnetic applications. Consequently biomagnetic instrumentation became the topic of a specialized session. Other applications are in the field of detection or they exploit mostly the extraordinary rf properties of superconducting devices. The latter are of special value in metrology, as is highlighted by two papers on the status of the Josephson voltage standard and noise thermometers.

Only the combined effort of numerous colleagues, co-workers, friends, institutions, and, of course, the authors of all talks, posters and papers made this successful conference and these proceedings possible. We highly appreciate their support. In addition, we greatly acknowledge the generous sponsorship of the Dr. Wilhelm Heinrich Heraeus and Else Heraeus Stiftung and the very helpful provisions we received from the Japanese-German-Center Berlin and from our institute, the Physikalisch-Technische Bundesanstalt.

Berlin
October 1991

Hans Koch
Heinz Lübbig

Contents

Part I High-T_c Superconducting Devices

Part III Applications

Part IV Summary and Conclusions

Part I

High-T_c Superconducting Devices

Toward a Technology of Electronic Circuits with High-T_c Superconductors

A.I. Braginski

Institut für Schicht und Ionentechnik (ISI),
Forschungszentrum Jülich (KFA), Postfach 1913,
W-5170 Jülich, Fed. Rep. of Germany

Abstract. Material requirements for a thin film high-T_c integrated electronic circuits technology are formulated. On this base, discussed are obstacles and recent progress in substrates, films and superconductor/insulator multilayers, patterning, crossovers, contacts, passive and active elements, and finally, the circuit integration.

1. Introduction

The purpose of this paper is to state the material requirements for a thin-film technology of integrated electronic circuits (ICs) with high-critical-temperature (T_c) superconductors (HTS) and, after identifying the major technical obstacles, to highlight the progress in this field. Barely over four years after the discovery of T_c above 77K in $YBa_2Cu_3O_7$ (YBCO) [1], it is virtually certain that a practical HTS IC technology will be developed. It is, in fact, already becoming a reality, at least in the case of 1:2:3 cuprate material base. Other HTS materials are offering specific advantages of performance, e.g. a higher T_c and stability and might, eventually, replace the 1:2:3. Until present, however, the advantage crucial for the technologist has been that high epitaxial quality, relatively smooth 1:2:3 films and superconductor/insulator multilayers could have been grown at moderately high substrate temperatures of, approximately, 600° to 750°C without any higher-temperature post-annealing ("as-grown" or "in-situ grown" films). Here, we will concentrate our attention on YBCO as a representative of 1:2:3 cuprates.

In contrast to the Nb and NbN-based low-temperature-superconductor (LTS) electronics, any viable technology of HTS electronic circuits must accommodate the high temperatures of film growth and the consequences of the extremely short and anisotropic coherence length, ξ, or the strong electronic anisotropy of HTS in general. In the past, we advocated the use of oriented film growth and epitaxy in electronic circuits based on isotropic but short-ξ LTS materials, especially those also requiring high growth temperatures, i.e. Nb_3Sn and Nb_3Ge [2]. This approach was proposed to minimize the film surface and interface degradation. In the case of HTS, both the anisotropy and interface quality requirements make epitaxial film and multilayer growth mandatory.

We note that more than a decade of efforts did not result in a viable LTS technology of short-ξ Josephson junctions and circuits. However, the prospects for HTS electronics are certainly brighter.

After listing the general requirements which should be met, we will briefly address the following topics: substrates, films and superconductor/insulator (S/I) multilayers, patterning, crossovers, superconductor/normal metal (S/N) contacts, superconductor/superconductor (S/S) contacts between layers (vias), passive and active elements, and finally, the circuit integration. Only selected examples of representative recent results will be mentioned with emphasis on material limitations and fabrication approaches.

2. General Requirements

To implement a technology of HTS superconductive integrated circuits, one must be able to integrate controllable Josephson junctions (JJs) or flux-flow nonlinear devices with passive devices such as superconducting inductors, transmission/delay lines, normal resistors and multilayered superconducting interconnects. To fabricate these, one needs, in order of increasing complexity: epitaxial substrates, patterned films and S/I multilayers of c-axis orientation (for interconnect and all other passive devices), and, possibly, a,b-axis orientation (for sandwich-type Josephson junctions). Furthermore, one needs patterned low-ohmic S/N contacts to layers, S/S superconducting contacts between them (in vias) and insulated cross-

Table 1. Physical and Geometrical Requirements for HTS Films and Multilayers

1. Growth and processing of single phase material at a lowest possible substrate temperature, ideally below T_S = 500°C.
2. Film adhesion to substrate but minimal interdiffusion or chemical reaction at interface(s).
3. Typical film thickness, d, in excess of the London penetration depth, λ_L, i.e d > 200 nm for YBCO.
4. Film thermal compatibility with substrate(s) i.e. comparable thermal expansion coefficients.
5. Minimal film surface roughness << 10 nm, no particles ("boulders" or second phase precipitates).
6. Large film area (A) of uniform thickness and properties, e.g. on a 7.5 to 10 cm diameter wafer.
7. Close match of film lattice parameters with epitaxial substrate.
8. Well defined epitaxial c-axis film orientation; possibility of local orientation control (c- vs a,b-axis normal to plane).
9. Epitaxial film growth on ex-situ processed underlayer.
10. No grain boundaries in epitaxial films (except those controllably nucleated at prescribed locations), low defect number density.
11. No pinholes in thin and very thin nonsuperconducting (insulating, normal- or semiconducting) layers separating superconducting films.

overs with undegraded critical current density (J_c). For JJs, pinhole-free thin barriers (insulating, semi- and normal-conducting) between films are required. For hybrid circuits, semiconducting (SE) films and devices must also be integrated on the chip or wafer and connected to superconductors via S/SE contacts and interfaces. Finally, methods of HTS patterning should be available which permit one to fabricate linewidths down to and well below the 1 micrometer level without any electrical degradation.

The physical and geometrical requirements for thin HTS films and multilayers are listed in Table 1. Obviously, some of these requirements cannot be fully met due to the intrinsic material limitations. Note that the "in-situ" film growth is not a requirement by itself but rather an approach which presently facilitates attaining the required quality of films and multilayers. Recent results on low O_2-pressure post-annealing of amorphous layers at temperatures below 850°C [3] raise hopes that large area "ex-situ" grown films will also be useful, e.g. for passive devices. Epitaxial growth on an ex-situ patterned HTS underlayer is a necessary requirement for the fabrication of multilayer circuits. The pertinent electrical requirements will be given in the following sections of the paper.

3. Film Substrates and Buffer Layers

Czochralski-grown, large-area, single-crystal substrate wafers are desirable technically and economically. Large silicon wafers, used for LTS ICs, are rather unsuitable for HTS, even with an insulating epitaxial buffer layer such as yttria-stabilized zirconia (YSZ) used to prevent the film-substrate reaction during growth. The excessive thermal expansion coefficient mismatch (average $\beta = 3.8\text{x}10^{-6}$/°C for Si vs $16\text{x}10^{-6}$/°C for YBCO in a-b plane, between 20°C and 700°C) results in a too low critical thickness for thermal fracture of only d_c = 50 nm [4]. As the silicon substrate or underlayer will be a virtual must for Si-based semiconductor-superconductor hybrid electronics, the promise resides in commercial silicon-on-sapphire (SOS) wafers. The sapphire average β is approximately $9\text{x}10^{-6}$/°C and a satisfactory d_c = 400 nm was recently determined by Fork for SOS/YSZ/YBCO [5]. Unfortunately, no epitaxial and highly conductive buffers have been found yet. Hence, silicon to YBCO contacts require intermediate noble metal connectors. The same is true for GaAs.

The SOS and the properly buffered sapphire (α-Al_2O_3) might eventually become a preferred choice in microwave passive devices due to the sapphire's very low $\tan\delta < 10^{-6}$ at 77K below 100 GHz [6] and a moderate albeit anisotropic dielectric constant, ϵ = 9-11. However, with thus far investigated sapphire buffers, such as YSZ, MgO and La_2CuO_4 (LSCO), the film crystalline quality still remains lower and the resulting microwave losses (surface resistance, R_s) higher than in YBCO on a well-matched, relatively nonreactive substrate such as $SrTiO_3$ (STO).

Fortunately, one of many rare-earth gallates and aluminates ($LaAlO_3$, $NdGaO_3$, $PrGaO_3$...), which can be Czochralski-grown and offer a selection of close lattice parameter matches, are an acceptable albeit less economical and more lossy alternative: ϵ = 16-25, $\tan\delta < 10^{-3}$ at 77K (100 GHz) and $\beta = 10\text{x}10^{-6}$/°C. At present, wafers of (100) $LaAlO_3$, 50 mm in diameter, are the most popular commercial substrates used for passive and active device prototype development. The twinning of $LaAlO_3$ has not too deleterious consequences for R_s, at least in single-layer devices. Unfortunately, it also enhances the dielectric anisotropy so that more expensive but almost twin-free $NdGaO_3$ (NGO) may win out. The bulk STO epi-substrates, now extensively used in research and active device development, have no bright technological future. Unsuitability for large-area Czochralski growth and the borderline-ferroelectric properties with extremely high $\epsilon = 1900$ and $\tan\delta = 10^{-1}$ at 77K are the main disadvantages.

The epitaxial mechanical/chemical polishing methods developed for semiconductor epi-substrates work satisfactorily for HTS films. A growing number of commercial vendors are developing proprietary recipes for an optimal epitaxial surface finish. One can expect that, eventually, the vendors will be offering large, 75-100 mm in diameter, properly buffered SOS and sapphire substrates for HTS films. For it to occur, further buffer layer development is still required. Typical buffers thus far being explored are YSZ, MgO, LSCO, STO, NGO, $MgAl_2O_4$, Y_2O_3, CeO_2 and PrO_2 (many refs. in [4]). The attained crystalline quality of some buffers can be quite high when characterized by the minimum channeling yield, e.g. $\chi_{min} = 4\ \%$ in YSZ on sapphire [7]. However, the crystalline quality of 1:2:3 on buffers is generally poorer ($\chi_{min} > 10\ \%$). Epitaxy of 1:2:3 on YSZ and also MgO is not without problems. Depending upon growth conditions, two or more matching film orientations may coexist on bulk substrates. The result is the presence of 45° and other grain boundaries in the film: e.g. YSZ [100]//YBCO[100] and YSZ[110]//YBCO[100] [8]. To attain high crystalline quality of HTS films on Si, SOS and sapphire, it might eventually be necessary to fabricate double or multiple buffer layers insuring simultanously a close epitaxial match and a chemical and thermal compatibility.

4. Film Fabrication

The already attained crystalline quality and superconducting properties of epitaxial 1:2:3 HTS films are qualifying the following three "as-grown" film fabrication methods as suitable for the IC technology: pulsed laser deposition (PLD), off- and on-axis variants of high pressure dc- and rf-sputtering, and reactive co-evaporation of "MBE"-type. The metal-organic chemical vapor deposition (MOCVD) is a less developed but attractive fourth alternative. All four methods have been used to produce "as-grown" films with $T_c > 90K$ ($R = 0$), J_c's at 77K in the high $10^6\ A/cm^2$ range. Comparably low microwave losses (surface resistance, R_s) are now also attainable in MOCVD films: $R_s = 400\ \mu ohm$ at 77K and 10 GHz [9]. As $R_s(T)$ is particularly sensitive to the presence of grain boundaries and weak links at other crystalline defects, it is the best figure of merit for epitaxial films.

A description of the four "in situ" growth methods is beyond the scope of this paper, reviews can be found in the literature [10, 11]. Table 2 presents a comparative summary of their demonstrated (projected) capability and problems, along

Table 2. Capabilities and Problems of "In-Situ" Film Growth Methods [12,13,15,9].

Method	PLD	High-press. Sputtering	Reactive Co-evaporation	MOCVD
Growth rate, nm/min	10-100 {2} small A only	2-10 {4}	120 {1}	2-10 {4}
T_s range, °C	650-750	650-750	570-650	740-820
Max. area A_{max}, cm^2	6-9 [13] {4} (100)	20 [15] {2} (100)	9 {2} (100)	20 {2} (100)
Crystal Quality on STO, χ_{min}	1-3	3-5	4 (on MgO)	no data
Surface Morphology, Roughness, nm	"boulders" < 10^4 nm {5} (1-10)	1-10 {2}	smooth {1}	"boulders" < 10^3 nm rough {5}
Best R_s, mohm (scaled to 10GHz) at 77K	0.26	0.16	0.33	0.4 [9]
Reproducibility	v. good {1}	average {3}	good {2}	good {2}
Ease of Control	v. good {1}	v. good {1}	poor {4}	good {2}

Rating on 1 to 5 scale: {1} - very good, {5} - major problem

with subjective ratings. We are not able to predict which will win out as the main technology base. The "ex-situ" coevaporation of amorphous films from BaF_2 and oxide precursors followed by post-annealing at lower O_2 pressures, which was mentioned above, may also find some practical use.

While a dominant majority of available data is limited to single layer c-axis films, each of the methods is in principle suitable for the template growth of a-oriented films [13] and multilayers of either orientation. Basically, they are all adaptable to a major scale-up and largely automatic operation with the attending advantages of quality, reproducibility and cost reduction. Of recently reported progress in film manufacturability, we should mention a demonstration of large-area substrate scanning PLD [14] and double-sided deposition of YBCO on 2" wafers by CVD [15] and off-axis sputtering [16].

5. Multilayers

Epitaxial, ex-situ processed c-axis S/I/S trilayers with well over 100 nm thick insulation of $SrTiO_3$, pinhole-free over patterned areas of the order of 1 mm^2, have been demonstrated [17]. The eventual extension to multiple layers with pinhole-free areas of many cm^2, as required for multilayered interconnects, should not present any fundamental problems. Epitaxy and largely two-dimensional growth of "in-situ" deposited S/I/S multilayers with very thin I-layer was also observed for quite many other perovskite and cubic structure insulators, e.g. MgO. The interdiffusion between S and I perovskites was found minimal, not only when depositing by high-rate PLD but even in low-rate sputtering. The extension to thick insulation (>>100 nm) may be limited by the thermal expansion mismatch leading to microcracks and related microshorts through the I-layer, as seen e.g. in YBCO/$LaAlO_3$ multilayers [18]. Optimization of microshort-free c-axis multilayers with thick low-loss insulators is still an important open issue. Pinhole-free a-axis S/I/S trilayers with very thin barriers (order of 1 nm), which could eventually be used as Josephson tunnel junctions, have not been reported to date.

Epitaxial, in- and ex-situ S/SE/S multilayers of c- and a-axis orientation have been made with SE = $SrTiO_3$:Nb [19], $PrBa_2Cu_3O_7$ (PrBCO) [20,21,22], substituted PrBCO [23] and $La_{1.5}Ba_{1.5}Cu_3O_7$ (LBCO) [24]. On (100) substrates, a-axis multilayers grow with a mosaic domain structure with two orthogonal orientations of c-axis in the plane. To be useful for S/N/S trilayer Josephson junctions, the relatively thin N-layers (<100 nm) must be pinhole-free. An unambiguous proof of it is still outstanding.

6. Patterning

Of issues especially important for the further development of standard patterning methods for HTS we highlight only three: (1) the deterioration of superconducting properties upon wet (by EDTA or diluted H_3PO_4) and ion beam etching (IBE) of small, especially submicron linewidth patterns, (2) the lack of reactive gases suitable for material-selective and high-rate reactive ion etching (RIE) and (3) the need for lift-off processing of "as grown" films.

In 1:2:3, the first problem is caused mainly by oxygen loss at surfaces and defects. It is thus intimately related to the film quality and presence of grain boundaries or other defects in films. In IBE, the oxygen loss occurs upon ion-beam and radiation heating in the low Ar-pressure atmosphere [25]. Consequently, the best remedies are: (a) an effective cooling, e.g. down to liquid nitrogen (LN_2) temperatures, (b) processing in a reactive, O_2-containing atmosphere (RIBE) and (c) deposition of only low average energy densities onto the processed surface, i.e. the use of low duty cycles during etching. In the present practice, the usual threshold for deterioration is at about 1 micrometer linewidth. However, at low etching rates, even with water cooling only, 150-250 nm wide microbridges could recently be fabricated without any substantial damage [26,27]. The use of LN_2-cooling and O_2-RIBE should make higher etching rates practical. Focused ion beam patterning (FIBE) using Ga-ions was also shown to produce submicron structures without damage [28] but this patterning method is not readily accessible.

We are not aware of any effective solution to (2). Lack of selectivity, the resulting difficulty in end-point control and generally low IBE rates make further search for suitable etching gases most desirable. Halide gases were proposed for RIBE [29] but their reaction products are not volatile at processing temperatures and not useful for high-rate RIE.

A promising solution to (3) is the recently reported use of lift-off patterned CaO/Zr masks which can withstand the high temperatures of "as-grown" film deposition without reacting with YBCO and can be lifted-off with water [30]. Standard resist lift-off techniques can be, obviously, applied to "ex-situ" amorphous films grown at or near room temperature, prior to the post-anneal. This is a significant advantage of such films.

A patterning method specific to HTS layers is their selective conversion to insulator e.g. by in-diffusion of a patterned Si template or overlayer. This approach reduces the need for planarization (see below) but is probably unsuitable for submicron linewidth patterns.

7. S/I/S Crossovers

Insulated, ex-situ patterned but epitaxial YBCO/$SrTiO_3$/YBCO line crossovers have been demonstrated in the fabrication of SQUID input coils, initially using shadow mask patterning [31] and eventually with standard photolithography [32]. These simple crossovers were not planarized.

A critical issue is the reduction of J_c on steps of the upper S-layer lines. To maintain the c-axis orientation in the upper YBCO layer and avoid nucleation of J_c-reducing grain boundaries the steps should be less than $\theta = 45°$ steep [33]. This is easy to achieve with shadow masks but difficult with photolithography and IBE. Even below 45°, the J_c over a step can be reduced by up to an order of magnitude [34], since a component of the supercurrent must flow transverse to Cu-O planes. When the upper film is thickner than the step hight, the grain boundaries can be shunted by a c-oriented overlayer even if $\theta > 45°$ [33] but this is not a convienent solution either. Hence, epitaxial planarization of insulator layers might eventually become unavoidable in patterned multilayered interconnects.

8. Contacts

The external S/N contacts and pads on a chip are unproblematic. On freshly prepared or cleaned YBCO c-axis film surfaces, the ex-situ room-temperature deposition of silver and gold results in contact resistances of $R_c = 10^{-6}$ to 10^{-7} ohmcm^2. Optimized annealing in O_2 between 300 and 600°C causes only moderate interdiffusion but reduces R_c by one-two more orders of magnitude. Ultrasonic wire-bonding to metal pads is straightforward when the metal layer is sufficiently thick, say 300 nm. The "in situ" deposition of metallic overlayer can produce even lower R_c's, down to the 10^{-9} and 10^{-10} ohmcm^2 range, in unannealed contacts on surfaces of c- and a-axis films, respectively.

The residual contact resistance points to a reduced S/N interface transmissivity for Cooper pairs which weakens the proximity effect in the normal metal thus reducing the I_cR_n product in SNS Josephson junctions. This is observed especially in the case of metals with long electron mean free path, , (Ag, Au) but also in S/SE contacts between YBCO and epitaxial perovskites such as PrBCO deposited on an ion-milled surface. All the causes of a limited transparency of interfaces are not yet known but interface damage is clearly one of them. The characterization of interface transmissivity along the Cu-O_2 planes (a,b-axis) is still incomplete.

The S/S contacts between two insulated YBCO layers have been demonstrated in vias of prototype planar SQUID input coils and flux transformers [35,32]. The problems still to overcome is the J_c-reduction at (1) the ex-situ patterned via interface and (2) on via step-edges. The temporary remedies are to compensate for J_c-weakening by designing vias of a diameter much larger than the conductor line-width, to either use shadow masks or to ion mill at an angle producing shallow via steps and, after opening the window in the insulator, to "heal" the damaged HTS underlayer surface by a sequence of chemical and/or IB-cleaning and thermal or plasma reoxidation prior to the deposition of the overlayer. However, large via diameters will eventually be unacceptable in high-density multilayer interconnects, e.g. in logic circuits.

9. Passive Devices

The performance of a typical and already demonstrated low-frequency device, the planar SQUID flux transformer with a multiturn input coil [35,32], will be limited by the film and S/I/S/ multilayer homogeneity, the absence of grain boundaries and other defects which contribute to 1/f noise and by effective interlayer insulation over large areas, of the order of 1 cm^2. Another limitation, the J_c reduction in vias and on crossover steps was discussed above. None of these difficulties is fundamental and incremental improvements can be expected. A thin film insulator with high ϵ and tanδ, such as STO, is tolerable.

Relatively low rf-losses (R_s at 77K and f << 100 GHz, $R_s \propto f^2$) already attained in epitaxial, c-axis films at low power levels (Table 2) made it possible to demonstrate prototypes of passive microwave HTS devices with low insertion losses (high Q's), such as resonators, multipole filters, delay lines and antennas [36]. Recently observed significant decrease of residual R_s at T < 50K [37] is hinting at the practicality of lower-T high-Q devices for frequencies up to 100 GHz.

Thus far, mostly coplanar or microstrip configurations have been used. The coplanar structures require only one film level but have the lowest Q. Higher Q microstrip and highest Q stripline structures now require clamping together the "hot" conductor substrate with the HTS or normal-conducting ground plane(s) to obtain a "flip-chip" configuration. This leads to difficult packaging and thermal stability problems. The now substantial substrate thickness (typically 0.5 mm) defines the spacing between conducting planes, h, and thus sets a limit to miniaturization at a given characteristic impedance $Z_o = (\mu/\epsilon)^{1/2}h/w$ where w is the conductor width. Deposition of YBCO on two sides of a 2 inch, 0.5 mm thick $LaAlO_3$ wafer, which was already demonstrated [15,16], may reduce packaging difficulties. An extension of this approach to buffered and very thin SOS wafers (less than, say, 50 microns thick) could bring the microstrip device fabrication technology to a level comparable to that demonstrated in LTS prototypes with down to 20 μm thick substrates [38].

A desired long-term solution will be a multilayered S/I/S film structure with insulator thickness of several microns, i.e. reduced to the lower h limit defined by $Q \propto fh/R_S$. Ideally, the dielectric thick film should not only have low losses but also be isotropic. The dielectric anisotropy, found in perovskites and sapphire, will ultimately limit the directivity of coupled microwave structures and their time-bandwidth product [38]. However, smooth and epitaxial thick films of requisite dielectric properties are still out of sight, a challenge to technologists.

Integration with planar Josephson devices implemented in c-axis films is simple, as shown in the example of a microwave rf-SQUID $\lambda/2$ microstrip YBCO resonator with an integrated step-edge junction [39].

10. Active Devices

The Josephson junction is the essential active element of superconductive ICs. Several types of controllable (engineered) HTS Josephson devices operating at 77K have been demonstrated recently [36] and new data are being reported continuously (e.g. at this Conference). Independent of the type, all these devices exhibit weak-link, SNS-type I-V characteristics and low temperature I_cR_n products, one-two orders of magnitude lower than $\pi\Delta/2e$ = 31 mV where Δ = 20 mV is the assumed gap voltage. Also, it is by now well documented that Josephson current between HTS electrodes is flowing only along the $Cu\text{-}O_2$ planes, i.e. in the direction of the longer ξ (a or b-axis). The main categories of demonstrated devices are (1) grain boundary (GB) junctions artificially nucleated in c-axis films at defined substrate locations, (2) junctions with artificial barriers (N,SE) and (3) microbridges with a weakened/normalized superconductor link. No practical SIS tunnel junctions emerged.

To the first category of demonstrated devices belong GB-junctions on bicrystal substrates [40], bi-epitaxial GB-junctions [41] and, probably, step-edge junctions [42]. Methods of nucleating the GB are schematically shown in Fig. 1. The GB-junctions are true Josephson devices of barrier thickness L comparable to ξ. They exhibit RSJ-type I-V characteristics when weakly coupled ($J_{cGB}/J_c \leq 100$) and their width, w, does not greatly exceed the Josephson penetration depth, λ_J, as shown in Table 3 (where RSJ+ symbolizes the excess current). For a set GB-angle

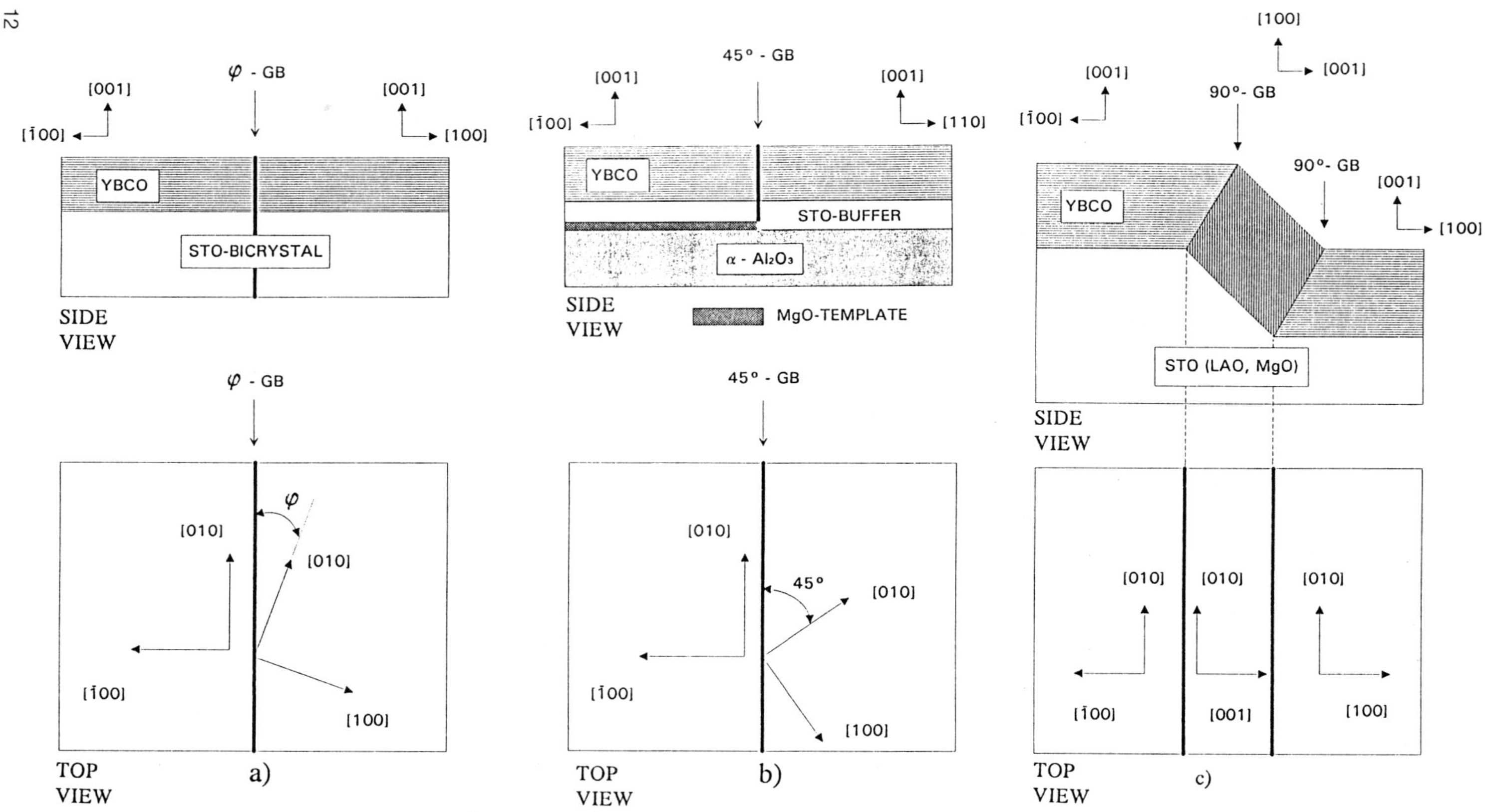

Fig. 1. Configurations of GB-junctions: a) on bicrystal substrate, b) on bi-epitaxial substrate, c) over a step in epi-substrate.

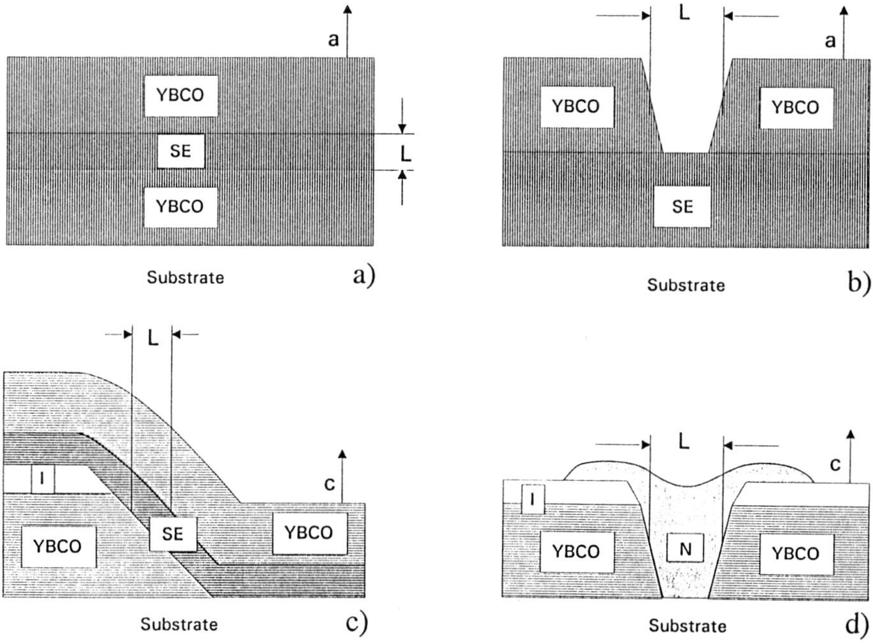

Fig. 2. Configurations of junctions with artificial barriers: a) a-axis sandwich, b) a-axis planar junction, c) edge-junction with interface edge milled into a c-axis film, d) S/N/S junction with interface edges milled into a c-axis film. Parallel lines symbolize the Cu-O planes.

and type, the junction's I_c can only be controlled by the GB-area and by thermal treatment, a disadvantage of bi-epitaxial junctions. However, they appear to be particularly suitable for circuit integration as the single GB-position is defined by a photolithographically patterned boundary of thin epitaxial film template. A buffer layer grows on the template with an orientation rotated by φ with respect to that of the same buffer grown directly on the substrate (Fig. 1b). For the MgO template [41] φ equals 45°.

Artificial barrier Josephson structures are either heteroepitaxial YBCO/SE/YBCO junctions where SE is a semiconducting perovskite [19,24,43,44] or proximity S/N/S junctions with N = Au or Ag [45,46]. Their essential configurations are shown in Fig. 2. Data of Table 3 are a sampling of recently reported results. They tend to agree with the recent theoretical prediction that low carrier density barriers lead to higher I_cR_n values than long- metals [47]. As theoretically expected, the I_c is an exponential function of barrier thickness or coupling length, L. However, some preliminary $I_c(L)$ data for SE-barriers result in abnormally long ξ_N, (Table 3) not yet explained. Such long ξ_N-values, if confirmed, could facilitate designing the S/SE/S junctions to specified I_c-values and tolerances. Edge-junctions are more suitable for integration with the c-axis interconnects than a-axis YBCO/SE/YBCO trilayer sandwich and planar

Table 3. Selected Recent Results on YBCO Josephson Devices

Barrier	Config. (Fig. No)	I_cR_n 4.2K,mV	(I-V) Type	ξ_N nm	Shapiro Steps	$I_c(H)$	SQUID	w/λ_J	Ref.
GB	1a	0.2-8	RSJ FF	-	+	+	dc	<1, >1	[40]
GB	1b	0.04-0.4	RSJ	-	+	+	dc	<1	[41]
GB?	1c	<1-5	RSJ+ FF	-	+	+	rf,dc	<1, >>1	[42]
$SrTiO_3$:Nb	5c	1-3	FF	5	+	-	dc	>>1	[19]
PrBCO	2a	0.1-2.5	RSJ+	10	+	?	--	1,>1	[20]
PrBCO	2c	8 (?)	FF	?	-	+	dc	>>1	[22]
SE-YBCO	2c	0.7	RSJ+	2	+	+	--	1,>1	[44]
LBCO	2b	1	FF	65	-	-	--	>>1	[24]
Au	2d	<0.5	RSJ+ FF	-	-	+	--	>1	[45]
Ag	2d	0.07-0.2	RSJ+	-	+	-	dc	>1	[46]
Ag	2d	0.2	RSJ+	12	+	-	--	>1	[49]

Legend: + measured, - not measured, -- not fabricated

junctions. Etching of edges may lower the pair transmissivity at S/E or S/N interfaces but natural edges can be obtained in films directionally deposited over sharp steps [46].

Microbridges, where the superconductor in the constriction is locally weakened e.g. by in-diffusion of a "poisoning" metal (Al) stripe or ion-implantation are flux-flow (FF) devices, as $L >> \xi, \xi_N$, unless the weakening process converts the link into N with a long $\xi_N = L$. Integration of c-axis microbridges is straightforward but operation at 77K with a sufficient I_c-level control might prove difficult to attain.

Lack of space prevents us from discussing all device types and fabrication approaches in any detail. Referenced are only those most important or not included in [36]. Also, we should bear in mind that only preliminary and incomplete characterization data are available in most cases, limited to the I-V characteristics without and with microwave irradiation (Shapiro steps), and evidence of operation in dc- or rf-SQUIDs. There is still a paucity of firm data on $I_c(H)$ Fraunhofer patterns (those available usually show that I_c in the junctions is not homogeneous), 1/f noise in a junction, reproducibility, I_c tolerances and stability. We note that relatively homogeneous I_c could be inferred from $I_c(H)$ in a YBCO edge junction with a CF_4-treated native oxide barrier not included in Table 3 [48].

At this writing, the most completely characterized, and probably best in overall performance, are the bicrystal GB-junctions. Unfortunately, bi-crystal substrates are not a convenient vehicle for an IC technology. However, properties already attained in bi-crystal substrate junctions permit one to assert that either bi-epitaxial or some other form of GB-junctions should soon mature for use in practical ICs. Similarly, the sum of information now available on various versions of heteroepitaxial YBCO/SE/YBCO edge and planar devices gives one confidence that practical heteroepitaxial JJs will eventually emerge. The future of SIS tunnel junctions and of three-terminal devices, not addressed here, is uncertain.

11. Integration

The demonstration of a functioning dc-SQUID containing bi-epitaxial junctions and a flux transformer completely integrated in a 7-layer YBCO/insulator structure [50] is the first milestone attained on the path toward multilayer circuit integration of active and passive elements.

Among many remaining fabrication challenges, the particularly difficult one is the hybrid semiconductor-HTS integration on the same chip. Another, perhaps avoidable, is the integration of a-axis devices with c-axis interconnect. In both cases, new creative ideas and experiments are needed to attain high transmissivity connections at S/E and S/S (c- to a-axis film boundary) interfaces, respectively. Urgent tasks, which to date received only a limited attention in the literature, are the planarization, passivation, encapsulation and packaging of HTS circuits on chips and boards. Long-term stability tests under various environmental constraints are also due.

12. Conclusions

This introductory overwiew shows that essential ingredients of HTS integrated circuits operating up to 77K are feasible and already demonstrated in the 1:2:3 multilayer film technology. A SQUID magnetometer integrated on a chip is becoming a reality. Large scale integration of complex, multi-junction circuits, including the SQUID electronics, appears fundamentally possible. Major areas on uncertainty, which limit the projected range of applications, include tunnel junctions, three-terminal devices and hybrid semiconductor-superconductor integration.

Acknowledgement

Thanks are due to all colleagues and institutions who provided yet unpublished data and information. Partial support of the German (BMFT) Consortium "First Application of HTS in Micro- and Cryoelectronics" is gratefully acknowledged.

References

[1] M.K. Wu, J.R. Ashburn, C.J. Torng, P.H. Hor, R.L. Meng, L. Gao, Z.J. Huang, Y.Q. Wang, and C.W. Chu, Phys. Rev. Lett. **58**, 908 (1987).

[2] A.I. Braginski, J.R. Gavaler, M.A. Janocko and J. Talvaccio, in Proc. SQUID'85 (H.D. Hahlbohm and H. Lübbig, eds.), W. de Gruyter, Berlin 1985, p. 591.

[3] R. Feenstra, T.B. Lindemer, J.D. Budai and M.D. Galloway, J. Appl. Phys. **69**, 6569 (1991).

[4] D.K. Fork, D.B. Fenner, A. Barrera, J.M. Phillips, T.H. Geballe, G.A.N. Connel and J.B. Boyce, IEEE Trans. Appl. Supercond. **1**, 67 (1991).

[5] D.K. Fork, F.A. Ponce, J.C. Tramontana, N. Newman, J.M. Philips and T.H. Geballe, Appl. Phys. Lett. **58**, 2432 (1991).

[6] J. Talvacchio and G.R. Wagner, SPIE Proc. **1292**,2 (1990) and references therein (also for other substrates).

[7] F. Wang, M. Siegel, R. Smithey, J. Geerk, G. Linker and O. Meyer (KfK), submitted to Proc. ICAM 1991, Strassbourg 1991.

[8] S.M. Garrison, N. Newman, B.F. Cole, K. Char and R.W. Barton, Appl. Phys. Lett. **58**, 2168 (1991).

[9] R. Hiskes, S.A. DiCarolis, J.L. Young, S.S. Laderman, R.D. Jacowitz and R.C. Taber, Appl. Phys. Lett. **59** (1991), in print.

[10] J. Geerk, G. Linker and O. Meyer, Mat. Sci. Repts. **4**, 193 (1989).

[11] R.G. Humphreys, J.S. Satchell, N.G. Chew, J.A. Edwards, S.W. Goodyear, S.E. Blenkinsop, O.D. Dosser and A.G. Gullis, Supercond. Sci. and Technol. **3**, 38 (1990) and references therein.

[12] Compilation from data of German (BMFT) Consortium "First Application of HTS in Micro- and Cryoelectronics"; PLD - J. Schubert (KFA), Sputt. - U. Poppe (KFA), Evap. - H. Kinder (TU Munich), MOCVD - A. Müller (Siemens).

[13] Inam, C.T. Rogers, R. Ramesh, K. Remschnig, L. Farrow, D. Hart, T. Venkatesan and B. Wilkens, Appl. Phys. Lett. **57**, 2484 (1990).

[14] M.F. Davis, J. Wosik, K. Forster, S.C. Deshmukh, H.R. Rampersad, S. Shah, P. Siemsen, J.C. Wolfe and D.J. Economou, J. Appl. Phys. **69**, 7182 (1991).

[15] J.H. Takemoto, C.M. Jackson, H.M. Manasevit, D.C. St. John, J.F. Burch, K.P. Daly and R.W. Simon, Appl. Phys. Lett. **58**, 1109 (1991).

[16] T.T. Braggins, J.R. Gavaler and J. Talvacchio, Proc. ICMC '91, in print

[17] F.C. Wellstood, J.J. Kingston and J. Clarke, Appl. Phys. Lett. **56**, 2336 (1990).

[18] A.E. Lee, J.F. Burch, R.W. Simon, J.A. Luine, R. Hu and S.M. Schwarzbek, IEEE Trans. on Magnetics **27**, 1365 (1991).

[19] D.K. Chin and T. Van Duzer, Appl. Phys. Lett. **58**, 753 (1991).
[20] U. Poppe, P. Prieto, J. Schubert, H. Soltner, K. Urban and Ch. Buchal, Solid State Comm. 71, 569 (1989).
[21] A. Inam, C.T. Rogers, R. Ramesh, K. Remschnig, L. Farrow, D. Hart, T. Venkatesan and B. Wilkens, Appl. Phys. Lett. **57**, 2484 (1990) and references therein.
[22] J. Gao, W.A.M. Aarnink, G.J. Gerritsma and H. Rogalla, Physica C **171**, 126 (1990).
[23] E. Sodtke, C. Andrzejak, D. Guggi and Y. Xu (KFA), submitted to Physica C and unpublished data.
[24] Y. Tarutani, T. Fukazawa, U. Kabasawa, A. Tsukamoto, M. Hiratani and K. Takagi (Hitachi), Appl. Phys. Lett. **58**, 2707 (1991).
[25] R. Lehmann, K. Herrmann, F. Schroeteler, A. van der Hart, J. Schneider, E. Sodtke, H. Soltner, W. Zander and A.I. Braginski, in "High-Temperature Superconductors" (Proc. ICMC '90, H.C. Freyhardt, R. Flükiger and M. Peuckert, eds.), DGM-Verlag, Oberursel, 1991, p. 581.
[26] M. Kawasaki, P. Chaudhari, T. Newman and A. Gupta, Appl. Phys. Lett. **58**, 2555 (1991).
[27] B. David, O. Dössel, J.-P. Krumme and M. Knechtel, in "Supraleitung und Tieftemperaturtechnik", VDI-Verlag 1991, p. 265.
[28] L.R. Harriot, P.A. Polakos and C.E. Rice, Appl. Phys. Lett. **55**, 495 (1989).
[29] S. Matsui, N. Takado, H. Tsuge and K. Asakawa, Appl. Phys. Lett. **52**, 69 (1988).
[30] B. Roas, Appl. Phys. Lett. **59** (1991), in print.
[31] J.J. Kingston, F.C. Wellstood, P. Lerch, A.H. Miklich and J. Clarke, Appl. Phys. Lett. **56**, 189 (1990).
[32] B. Oh, R.H. Koch, W.J. Gallagher, R.P. Robertazzi and W. Eidelloth (IBM), Appl. Phys. Lett. **59**, 123 (1991).
[33] C.L. Jia, B. Kabius, K. Urban, K. Herrmann, G.J. Cui, J. Schubert, W. Zander, A.I. Braginski and C. Heiden, Physica C **175**, 545 (1991).
[34] L. Schultz, G. Sölkner, H. Schmidt, G. Gieres, G. Friedl, M. Römheld, R. Seeböck and H.E. Hoenig, in "Supraleitung und Tieftemperatur-Technik", VDI-Verlag 1991, p. 488.
[35] F.C. Wellstood, J.J. Kingston and J. Clarke, Appl. Phys. Lett. **56**, 2336 (1990); F.C. Wellstood, J.J. Kingston, M.J. Ferrari and J. Clarke in [36], p. 2569.
[36] A representative recent compendium can be found in IEEE Trans. on Magnetics **27**, 2533-3327 (1991).
[37] N. Klein, U. Dähne, U. Poppe, H. Soltner, B. Kabius, K. Urban, J. Schubert, S. Orbach, S. Hensesn, G. Müller, H. Piel, B.P. Gorshunov, G.V. Kozlov, S.P. Lebedev and A.A. Volkov, submitted to Physica C 1991 (Proc. M S-HTSC '91).
[38] R.S. Withers in "Superconducting Devices" (S.T. Ruggiero and D.A. Rudman, eds.), Acad. Press, Boston 1990, p. 227, and refs. therein.
[39] C. Heiden (KFA), presented at the EMRS Workshop on HTS Films, Rome, 1991 (unpublished).

[40] R. Gross, P. Chaudhari, M. Kawasaki and A. Gupta, in [36], p. 3227, and refs. therein.
[41] K. Char, M.S. Colclough, S.M. Garrison, N. Newman and G. Zaharchuk, Appl. Phys. Lett. **59** (1991), in print and this Conference.
[42] K.P. Daly, W.D. Dozier, J.F. Burch, S.B. Coons, R. Hu, C.E. Platt and R.W. Simon, Appl. Phys. Lett. **58**, 543 (1991); G. Cui, Y. Zhang, K. Herrmann, Ch. Buchal, J. Schubert, W. Zander, A.I. Braginski and C. Heiden, Supercond. Sci & Tech. **4**, 130S (1991).
[43] J.B. Barner, C.T. Rogers, A. Inam, R. Ramesh and S. Bersey, Appl. Phys. Lett. **59** (1991), in print.
[44] B.D. Hunt, M.C. Foote and L.J. Bajuk, Appl. Phys. Lett. **59** (1991), in print.
[45] M.G. Forrester, J. Talvacchio, J.R. Gavaler, M. Rooks and J. Lindquist, in [36], p. 3098.
[46] M.S. DiIorio, S. Yoshizumi, K.-Y. Yang, J. Zhang and M. Maung (BTI), Appl. Phys. Lett. **58**, 2552 (1991) and this Conference.
[47] G. Deutscher and R.W. Simon, J. Appl. Phys. **69**, 4137 (1991).
[48] G. Koren, E. Polturak, E. Aharoni, and D. Cohen, this Conference.
[49] M.A.M. Gijs, in "High-T_c Superconductors" (Materials Science Forum, Pouch, Alterovitz and Romanofsky eds.), Trans Tech Publ. Zürich 1992, in print.
[50] K. Char, M.S. Colclough, L. Lee, J.M. Rowell, G. Zaharchuk, J.J. Kingston, F.C. Wellstood and J. Clarke, this Conference.

Physics of Josephson Junctions Made of High-T_c Superconductors

M. Yu. Kupriyanov

Department of Physics, Moscow State University, 119899 Moscow, Russia

Abstract. The main results achieved in the HTS Josephson junction technology are reviewed and the possible reasons leading to the reduction of their characteristic voltage are discussed. It is emphasized that the electron diffraction at the SN boundaries as well as at the inhomogeneities in CuO planes plays a decisive role in HTS weak links and grain boundary junctions.

1. Introduction

So far the possibility of making HTS Josephson junctions with the value of the characteristic voltage as high as $V_c \simeq 20 - 30$ mV at $T = 4.2$ K have been proven experimentally in the monocrystalline break junctions [1]. But till now all attempts to develop a way to fabricate reproducible HTS weak links [2-8] or single grain boundary junctions [9-12] with the same characteristics have never led to a noticeable success.

The goal of this work is to review possible mechanisms leading to the V_c suppression in HTS Josephson junctions and to discuss their relevance to the experimental data.

2. Josephson effect in HTS weak links

The calculations in the BCS based weak link models [13,14] show that their properties depend on two dimensionless parameters which in the "dirty" limit given by the equalities

$$\gamma = \rho_S \xi_S^* / \rho_N \xi_N^*, \quad \gamma_B = R_B / \rho_N \xi_N^* \tag{1}$$

where $\rho_{N,S}$ and $\xi_{N,S}^*$ are the normal-state resistivities of the junction materials, and their coherence lengths respectively, while R_B is the specific resistance of the SN boundary.

Springer Proceedings in Physics, Vol. 64 Superconducting Devices and Their Applications
Editors: H. Koch, H. Lübbig

The characteristic voltage V_c of SNS sandwich-type junctions in the limits of small thickness of the junction $L \ll \xi_N^*$ or large $\gamma_B \gg \max\{(T_c/T)^{1/2}, \ \gamma(1-T/T_c)^{-1/2}\}$ is given by equations (2) and (3) respectively:

$$\frac{eV_c}{2\pi T_c} = \frac{T}{T_c}\max_\phi \sum_{\omega>0} \frac{\Delta^2 G_S \omega^{-1} \sin(\phi)}{\sqrt{\omega^2(1+\gamma_{BM}\omega G_S/2\pi T_c)^2 + \Delta^2\cos^2(\phi/2)}} \tag{2}$$

$$\frac{eV_c}{2\pi T_c} = \frac{2}{\gamma_B}\sum_{\omega>0}\frac{\Delta^2}{\omega^2+\Delta^2}\sqrt{\frac{\pi T^2}{\omega T_c}}\mathrm{sh}^{-1}[\frac{L}{\xi_N^*}\sqrt{\frac{\omega}{\pi T_c}}], \tag{3}$$

where ϕ is the phase difference of the superconductor order parameters, $\omega = \pi T(2n+1)$ are the Matsubara frequencies, $G_S = \omega\sqrt{\omega^2+\Delta^2}$, $\gamma_{BM} = \gamma L/\xi_N^*$, L is the thickness of the N-layer. The specific normal junction resistance is determined by the sum of the interface resistances:

$$R_N = 2\xi_N^* \rho_N \gamma_B. \tag{4}$$

In the SN-N-NS variable thickness bridges the SN interface resistances also make a noticeable contribution to the full junction resistance R_N :

$$R_N = 2\xi_N^* \rho_N \gamma_{BM}^{1/2}, \ \gamma_{BM} = \gamma d_N/\xi_N^*, \tag{5}$$

where d_N is the thickness of the N-film. The temperature dependencies of the parameter V_c in the cases of large $\gamma_{BM} \gg T_c/T$ or small bridge length L

$$L \ll \max\{\xi_N^*, \ \xi_N^*\sqrt{\gamma_{BM}}\min\{1, \ \sqrt{T/(T_c(1+\gamma_{BM}))}\}\}$$

are determined by equations (6) and (7) respectively:

$$\frac{eV_c}{2\pi T_c} = \frac{T}{T_c}\sum_{\omega>0}\frac{\Delta^2}{\omega^2+\Delta^2}(\frac{\pi T_c}{\omega\gamma_{BM}})^{3/2} exp(-\frac{L}{\xi_N^*}\sqrt{\frac{\omega}{\pi T_c}}), \tag{6}$$

$$\frac{eV_c}{2\pi T_c} = \frac{T}{T_c}\max_\phi \sum_{\omega>0}\Delta^2\sin(\phi)[2\Omega_c^2\Omega(\Omega+\Omega_c^2)]^{-1/2} \tag{7}$$

where $\Omega = \sqrt{\omega^2+\Delta^2}$, $\Omega_c = \omega\sqrt{(1+\gamma_{BM}\Omega/\pi T_c)^2 + \Delta^2\cos^2(\phi/2)}$.

Expressions (2)-(7) are valid if the conditions of dirty limit are fulfilled for N- and S-metals as well as the suppression of the superconductivity in the S electrodes of the structures due to both the proximity effect and the supercurrent flow is negligible.

To use those results for HTS weak links one should also suppose that in HTS material the mean-field approximation holds and the weak coupling limit takes place. The first condition is probably satisfied in the HTS due to the smallness of the ratio T_c/E_f [15]. Since the experimentally obtained $\Delta(T)$ dependence is similar to that predicted by the BCS theory the account of strong-coupling effects results only in a modification of the ratio $\Delta(0)/T_c$ and can result in temperature independent corrections to V_c of the order of unity.

Rough estimations [14] of the suppression parameters for the atomically sharp and "clean" (without additional dielectric layers) HTS/N-interface give

$$\gamma_{\perp} \simeq 10^2, \ \gamma_{B\perp} \simeq 10^4, \tag{8}$$

if the C axis of HTS material is perpendicular to the SN boundaries of the junctions and

$$\gamma_{\parallel} \simeq 5, \ \gamma_{B\parallel} \simeq 20, \tag{9}$$

if the C axis is parallel to the boundaries.

From equations (2)-(9) it follows that at $T \simeq 0.5T_c$ and $L \simeq \xi_N^*$ the simple estimations for the parameter V_c are valid for the sandwich-type junctions

$$V_c \simeq 80\gamma_B^{-1} \ \mathrm{mV}, \ V_{c\parallel} \simeq 4 \ \mathrm{mV}, \ V_{c\perp} \simeq 10^{-2} \ \mathrm{mV} \tag{10}$$

and for the variable thickness bridges

$$V_c \simeq 80\gamma_B^{-3/2} \ \mathrm{mV}, \ V_{c\parallel} \simeq 1 \ \mathrm{mV}, \ V_{c\perp} \simeq 5 * 10^{-3} \ \mathrm{mV}. \tag{11}$$

From Eqs. (10) and (11) it follows that the suppression of the parameter V_c in HTS weak links is the result of the finite transparency of HTS/N-metal boundaries. It is also interesting to note that equations (2), (4) and (5), (6) determine the scaling law

$$V_c \propto R_N^{-k}, \tag{12}$$

where $k = 1$ or $k = 3$ for the sandwich type junctions and the variable thickness bridges respectively.

The theoretical predictions (2)-(12) can be easily used for interpretation of recent experimental results.

2.1 HTS/N/LTS sandwiches

These structures are usually used to investigate the quality of the HTS/N interface (N = Ag, Au). Unfortunately the useful information is seriously masked by the suppression of the superconductivity in the vicinity of the N/LTS boundary (LTS = Nb, Pb, Sn) since the parameter γ is greater or nearly equals unity here. In this case the value of the anomalous Green's functions responsible for the critical current have to be proportional to $(T_c^* - T)$ at the interface and at $T \simeq T_c^*/2$ their absolute value is an order of magnitude less than the order parameter of the bulk LTS electrode. Hence in the vicinity of the critical temperature T_c^* of LTS material the $V_c(T)$ dependencies have to be proportional to $(T_c^* - T)$ and the absolute value of parameter V_c be strongly suppressed [16].

These predictions fit well the experimental $V_c(T)$ dependencies [17, 18]. To obtain better agreement with absolute values $V_c \simeq 50 - 500$ mV it is necessary to assume that the order parameter near HTS/N interface is also suppressed ($\Delta \simeq 0.1 - 1$ mV). This fact can be due to the ion cleaning of the HTS surface before the N film deposition leading to the formation of the thin oxygen deficient nonsuperconducting layer on the top of the film [13].

However even in the absence of such a layer the properties of HTS/N/LTS sandwiches seem to be close to those of SINS junctions where the role of I layer is played by this boundary. This is due to high γ_B values intrinsic to HTS/N boundaries. Recently, the scaling law $I_c \propto R_N^{-1}$, which is typical for SINS structures has, been proven experimentally in sandwiches YBCO/Ag/Nb [19].

2.2 Variable thickness bridges

The technology of the variable thickness bridges fabrication [2,3] usually starts with $in-situ$ making the HTS/Ag,Au bilayers. The SNS junction is formed using high resolution lithography or ion milling to make a gap and then bridging this gap by Ag or Au.

Since the HTS films in all experimental realizations of the structures were C-oriented, the $in - situ$ fabricated HTS/N boundary have a small transparency and the properties of the junctions were determined by the quality of the interface between the bilayer edges and the bridging film resulting in SNS sandwich type junctions. The experimental data [2] are fitted well by the theoretical predictions for SNS sandwiches following from equation (2) with $\gamma_B = 20$.

Using the typical value of $2 * 10^{-11}$ Ωcm^2 for the $\rho_N \xi_N^*$ product and experimentally defined $R_N \simeq 4\Omega$ (for the junctions made by means of the electron beam lithography) from (4) one can obtain the area $A \simeq 5 * 10^{-12}$ cm^2 for the real cross-section of the junction. The latter differs from the geometrical assessment by nearly two orders of magnitude. This discrepancy may be due to the oxygen diffusion during the fabrication processes which extends preferably into the ab-planes than along the C-axis.

Thermal treatment of the bridges made by the ion milling method [2] increases the critical current of the junctions and decreases their normal resistance due mainly to the growth of the area A because of the normal metal diffusion process. The comparison of the data with the $V_c(T)$ dependence given by equation (2) shows that this method of fabrication results in worse SN-boundary transparency ($\gamma_B \simeq 250$) and lower V_c.

So the properties of the boundaries play a decisive role in these structures. To make these boundaries more controllable it is necessary to begin the Josephson junction fabrication with the deposition of the SN bilayers based on a-oriented films.

2.3. Edge-type junctions.

The technology of the junction fabrication usually begins with the deposition of the HTS/dielectric bilayers. Then a part of the structure is removed by ion milling to obtain the edge. Before the top electrode deposition the edge is exposed to a plasma discharge processe in CF_4 gas [4,5] or a thin PrBaCuO layer is grown [6].

The investigation of the junctions [4] shows that they reveal the Josephson-like behavior for the temperature $T > 0.5T_c$. A decrease of the temperature leads to intensive growth of the parameter V_c and to the shrinking of the Shapiro steps in the iv-curve. This behavior is typical for SS'S structures where the transition between depairing and the Josephson effects takes place when T passes through the critical temperature of the weak link material.

The experimental data obtained for the junctions with the PrBaCuO interlayer with the thickness $d \simeq 6$ nm [6] are fitted well by the theoretical predictions for SNS junctions with rigid boundary conditions at SN interfaces [20] with $d \simeq 6\xi_N^*$. So we can conclude that the coherence length of the Pr-based compound ($\simeq 1$ nm) is close to that of YBCO material and the suppression of the parameter V_c in these structures is mainly due to the large value of the N-layer thickness.

The surface of the film edge in the above mentioned technique can be tormented during the process of the edge fabrication leading to an irreproducibility of the junction parameters. To circumvent this difficulty a new technique has been proposed [7]. First a sharp step in a substrate is produced. Then the HTS film is deposited across the step. The obtained superconducting electrodes are *in − situ* linked via the deposition of the normal metal film. The process results in a fabrication of the SNS sandwiches with different transparencies of the SN boundaries which can be characterized by the parameters γ_{B1} and γ_{B2}.

The theoretical analysis of the structure in the practically interesting case of the small electrode spacing leads to equation (2) in which the product $0.5\gamma_B$ must be replaced by $\gamma_B^* = \gamma_{B1}\gamma_{B2}/(\gamma_{B1} + \gamma_{B2})$. It is important to note that in the limit $\gamma_{B1} \gg \gamma_{B2}$ the stationary properties of the structure happen to be close to that of the SINS junction where the role of I-layer is played by the boundary with larger γ_B. In this case the characteristic voltage V_c depends only on γ_{B2} as for the symmetric SNS sandwiches. Thus to achieve the estimated V_c values (10) it is sufficient to have only one perfect SN boundary.

2.4. Sandwich-type junctions

Despite of the failure with PrBaCuO interlayers the idea of employing epitaxial trilayer structures for the Josephson junction fabrication is still promising. This has been confirmed by recent experiments with Nb-doped $SrTiO_3$ as the new coupling material which at the doping level between 10^{17} and 10^{20} cm^{-3} is a good normal metal [8]. The experimental data obtained in these junctions with the N-layer thickness $d \simeq 25$ nm are in a reasonable agreement with the predictions of the SNS junction's theory [20] with $d \simeq 6\xi_N^*$. The estimated value of coherence length $\xi_N^* \simeq 5$ nm coincides with the estimations obtained in [8] from $I_c(d)$ dependence.

So to get higher V_c values it is necessary to reduce the interlayer thickness.

3. Single grain boundary junctions

This type of junctions can be fabricated either by pattering the randomly occurring grain boundaries in granular films [9] or by epitaxial growth of the HTS films on the complex substrates (bicrystal [10], step-edge [11], or obtained by "planarization" technology [12]).

The scope of experimental data [9-12] (excess current in iv-curve, temperature dependence of $V_c(T) \propto (T_c - T)^2$ at $T \simeq T_c$, the scaling law (11) with $k \simeq 1$) permit to conclude that the junctions are of the SNS type. The only contradiction comes from the values of the normal resistance ($10^{-7} - 10^{-8}$ Ωcm^2). For SNS junctions this values would imply that the resistivity of the N-metal is about 0.2 to 2 Ωcm assuming a thickness of some 0.5 nm for the N-layer. This value of the resistivity is several orders of magnitude larger than that of the grain material.

The key to the problem can be found if we take into account the wave nature of the charge carriers. The estimations show that their de-

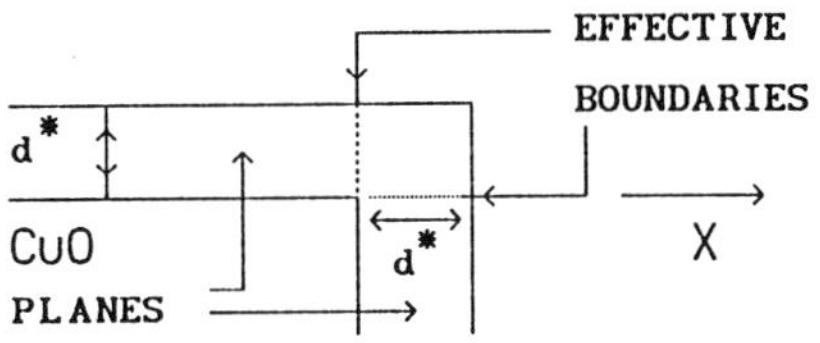

Fig.1. The scheme of the 90^0 grain boundary.

Brogile wavelength λ is comparable with the effective thickness d^* of their localization in the vicinity of CuO planes. So we have the effective waveguide for the carriers and any inhomogeneities of the scale d^* lead to diffraction and as a consequence to an additional resistance.

To calculate this resistance for the 90^0 grain boundary (see Fig.1) we suppose that the quasiparticles are localized within the conductive layer of thickness d^* and the probability of their tunneling to the neighboring layers is negligibly low. Assume also that the wave function equals to zero at the boundaries of the CuO planes and that the parameter $\alpha = (2d^*/\lambda) \simeq 1$.

Solving the Schrodinger equation we get the transparency coefficient

$$D(\theta) = 0, \ \epsilon = \arccos(1/\alpha) > \theta > \pi/2, \qquad (13a)$$

$$D(\theta) \simeq 7.5(\alpha^2 \cos^2(\theta) - 1), \ 0 \leq \theta \leq \epsilon, \qquad (13b)$$

where θ is the angle between the Fermi-momentum of the quasiparticles and the X axis (see Fig.1).

To calculate the grain boundary resistance we solve master equations. Taking into account that $D \ll 1$ and equation (13) for $d^* \simeq 0.5$ nm one can obtain

$$R_N^{-1} = \frac{e^2\alpha}{2\pi d^*} \int_0^{\pi/2} D(\theta) \cos(\theta) d\theta \simeq 5 * 10^3 (\alpha^2 - 1)^{3/2} \ \Omega^{-1}\mathrm{cm}^{-1}. \qquad (14)$$

Expression (14) fits the experimental data if

$$\alpha \simeq 1 + (1 - 5) * 10^{-3} \text{ or } \lambda \simeq 2d^* \simeq 1 \text{ nm}. \qquad (15)$$

Thus the chosen value for λ is closed to that obtained by the other methods [15].

In the framework of the proposed waveguide model it is possible to explain the Josephson properties of the grain boundaries. They look like the SNS sandwich with two effective boundaries formed by the inhomogeneities of the plane waveguide. The diffraction region between these boundaries serve as the weak link material. The calculation of the critical current in this model is in progress now.

4. Conclusion

The analysis of the HTS Josephson junctions have shown that to understand their properties it is necessary to take into account the wave nature of the current carriers in the CuO planes. It is the diffraction of the quasiparticles that determines the specific resistance of the clean HTS/N-metal boundaries and hence the properties of the HTS weak links. The processes in single grain boundary junctions can also be understood in the framework of the waveguide model taking into account this effect.

5. References

[1] B.A. Aminov, A.I. Akimov, N.B. Brandt, Nguen Min Tchu, Ya.G. Ponamarev, M.V.Sudakova, J. Witting, L.M. Fisher, L. Rohta: Physica C **160**, 505 (1989)

[2] M.G. Forrester, J. Talvacchio, J.R. Gavaler, J. Rooks, J. Lindquist: IEEE Trans. Mag. **27** 3098 (1991)

[3] M.S. Wire, R.W. Simon, J.A. Luine, K.P. Daly, S.B. Coons, A.E. Lee, R. Hu, J.F.Burch, C.E.Platt: IEEE Trans.Mag. **27** 3106 (1991)

[4] R.B. Laibowitz, R.H. Koch, A. Gupta, G. Koren, W.J. Gallager, V. Foglietti, B. Oh, J.M. Viggiano: Appl.Phys.Lett. **56** 686 (1990)

[5] G. Koren, E. Aharony, E. Polturak, D. Cohen: Appl.Phys.Lett. **58** 634 (1991)

[6] J. Gao, W.A.W. Aarnink, G.J. Gerritsma, H. Rogalla: Physica C **171** 126 (1990)

[7] M.S. Dilorio, S. Yoshizumi, K-Y. Yang, J. Zhang, M. Maung: To appear in Appl.Phys.Lett. (1991)

[8] D.K. Chin, T. van Duzer: Appl.Phys.Lett. **58** 753 (1991)

[9] S.E. Russek, D.K. Lathrop, B.H. Moeckly, R.A. Buhrman, D.H. Shin, J. Silcox: Appl.Phys.Lett. **57** 1155 (1990)

[10] R. Gross, P. Chaudhari, M. Kawasaki, M.B. Ketchen, A. Gupta: IEEE Tran. Mag. **27** 3227 (1991)

[11] K.P. Daly, W.D. Dozier, J.F. Burch, S.B. Coons, R. Hu, C.E. Platt, R.W. Simon: Appl.Phys.Lett. **58** 543 (1991)

[12] K. Char, M.S. Colclough, S.M. Garrison, N. Newman, G. Zaharchuk: To be published.

[13] M.Yu. Kupriyanov, K.K. Likharev: Uspekhi Fiz. Nauk. [Sov. Phys. - Uspekhi] **160** 49 (1990)
[14] M.Yu. Kupriyanov, K.K. Likharev: IEEE Trans. Magn. **27** 2460 (1991)
[15] L.N. Bulaevskii, V.L. Ginzburg, A.A. Sobyanin: Zh.Eksp.Teor.Fiz. [Sov. Phys. - JETP] **94** 355 (1988)
[16] M.Yu. Kupriyanov, V.F. Lukichev: Fiz.Nizk.Temp. (Sov. J. Low Temp. Phys.) **8** 1045 (1982)
[17] H. Akoh, C. Camerlingo, S. Takada: Appl.Phys.Lett. **56** 1487 (1990)
[18] M. Lee, D. Lew, C.B. Eom, T.H. Geballe, M.R. Beasley: Appl.Phys. Lett. **57** 1152 (1991)
[19] A. Fujimaki, T. Tamaoki, T. Hidaka, M. Yanagase, T. Shiota, Y. Takai, H. Hayakawa: Japan.J.Appl.Phys. **29** L1659 (1990)
[20] K.K. Likharev: Sov.Tech.Phys.Lett. **2** 12 (1976)

Towards Real HTS Tunnel Junctions

Y. Okabe

The University of Tokyo, Research Center for Advanced Science and Technology, 4-6-1, Komaba, Meguro-ko, Tokyo 153, Japan

Abstract. The paper presents a review of works on HTS Josephson tunnel junctions. The ideal junction is not accomplished yet, and the reason will be estimated.

1. Introduction

Lots of works have been done for Josephson tunnel junctions of high temperature superconductors. Some of them are of pure HTS/insulator/HTS structure, some with LTS for counter electrodes, and so on.

These junctions are severely checked about their possibility of the ideal Josephson junction or not by the observations of following effects:

1) DC Josephson effect, that is supercurrent.

2) AC Josephson effect, that is Shapiro steps induced by microwave.

3) Fraunhofer pattern, the critical current dependence on applied magnetic field.

4) Gap structure, with a high resistive region below the gap voltage and a constant normal resistance region above the gap voltage.

All these effects are clearly observed for niobium junctions. However, for HTS junctions there are lot of problems. Some junction can show Josephson characteristics such as 1) to 3) but do not the gap structures, and some junctions show the gap structures without Josephson characteristics. Generally speaking, junctions with thicker tunnel insulators show gap structure without Josephson characteristics. It is also very difficult to get a good coincidence between actual junction area and the area estimated from Fraunhofer patterns.

The paper would review several possible structure of junctions and the reasons of these difficulties.

2. Insulator

Several substrate materials, which are essentially good materials for HTS, are tried for tunnel insulators. Typical materials are MgO, STO, YO and etc. The earlier junctions showed Josephson characteristics without good gap structure[1]. The

Springer Proceedings in Physics, Vol. 64 Superconducting Devices and Their Applications
Editors: H. Koch, H. Lübbig

mismatches between Fraunhofer pattern and the actual junction size were typically one hundreds, and the possibility of pin hole junctions was pretty high. The island structure of insulator materials were observed by SEM[2] or cross-sectional TEM[3] images. This suggests that the superconductors connect directly at several points through holes of insulators.

Recently more precise epitaxial growth of MgO was done on c-axis oriented YBCO[4]. The result is that, films of thickness between 5nm to several 10nm become very smooth, but ones less than 5nm or ones more than several 10nm become island structure. Josephson characteristics were only observed for junctions of island structure, and no Josephson current for 5nm thick insulator. The thickness of 5nm is pretty thick but for even this thickness, superelectrons can penetrate for LTS.

One possible explanation is the degradation of superconductivity. Dislocations release 8% misfit between the MgO and YBCO[5], even though the stress may degrade the superconductivity. The top surface of YBCO beneath the insulator was tested by Raman spectroscopy[6]. The shift shows that the tensile force appears under MgO and YSTO, and the compressive force appears under YO.

Junctions of the structure YBCO/MgO/Pb show rather stable characteristics[7,8]. Supercurrent and the gap structure of Pb are observed clearly. In this case, Pb makes a good insulator of PbO, and it results in high reproducibility. However, current density is rather low compared with estimated one from the actual area, and it again suggests the possibility of pin holes.

All above insulators are epitaxial crystals. However, non crystalline materials or polycrystalline materials are also possible. Fluoriding process on YBCO makes insulator on its surface[9]. However, the thickness of the film is of the order of several 10nm. Chloriding process on YBCO makes thin insulator[2]. The resistivity of this layer checked by N/I/S junctions is proportional to the exponential of the process time. However no further trial for making S/I/S junctions is not yet done.

Reviewing various materials and structure, it seems that some high resistive barrier appears on the YBCO surface just beneath the artificial insulators. Josephson characteristics appear only via pin holes which widely distributed. This results in the big mismatch between the measured area and the actual area. Even under these holes, poor superconductive regions appear, and the IcRn product is mostly less than 1 mV.

3. Proximity Effect

Several types of junctions with normal materials are tried. The most typical one is S/N/S junction[10,11]. However, we will discuss about tunnel junctions, for example, S/N/I/S. As normal material, normal metals such as Au[12] or Ag[13] are used. Normal metals can act for blocking layers for diffusion of upper materials into HTS. For example, Au can protect the diffusion of AlO into YBCO, and enables the structure of YBCO/Au/AlO/Nb. However, the resistance of the junctions is again pretty high. The Au evaporation on ultra-clean surface of YBCO shows also several order higher resistance compared with the theoretical values[14]. Therefore, high resistive region again appear on the YBCO surface under normal metals.

Recently, multi-layer structure of HTS and its relative were reported[15,16]. For example, the structure of YBCO/PrBCO/YBCO is very fascinating combination of similar materials. This structure can be composed by layer by layer epitaxy, and even a monolayer of YBCO or PrBCO can be accomplished. The superconductivity of YBCO surface under PrBCO is very stable[17], and PrBCO is chemically stable. Therefore, if YBCO/PrBCO is used instead of YBCO, the good junction may be accomplished.

Abnormally long proximity effect of the order of several 100 nm is also reported for HoBCO/LBCO/HoBCO[18] or YBCO/LCMO/YBCO[19] structure. This suggests again that CuO relatives of YBCO or BSCCO behave quite differently from the other normal materials, and the use of them might be very effective for the ideal tunnel Josephson junctions.

4. Discussions

The crystal orientation is very important parameter for junction characteristics. As the ab plane conductance is much greater than c-axis conductance, it is very important to make edge junctions. The films with in-plane c-axis orientation are frequently used. But for electronic circuits, c-axis orientation is much important from the point of view of wiring. The edge junctions are very impressive from this reason. Recently lot of works are reported for grain boundary type edge junctions[20]. There characteristics are the most successful ones. Unfortunately most of above mentioned tunnel junctions are layered structure of c-axis oriented films. We should put much efforts to fabricate edge junctions.

5. Conclusion

Many kinds of tunnel junctions of HTS were reported. However, the ideal characteristics have not been observed. The biggest reason will be the degradation of YBCO at the interface between HTS and insulator, which makes high resistive thick layer beneath the insulator. Right now there is a doubt that the most Josephson characteristics appeared by pin hole junctions. Stopping the degradation will be the key technology to accomplish the ideal junction characteristics.

References

[1] A. Inoue, K. Takeuchi, H. Ito, A. Nakayama and Y. Okabe, Jpn. J. Appl. Phys., 26(9) L1443, 1987.

[2] T. Matsui, K. Takeuchi, A. Nakayama, M. Nagano and Y. Okabe, Ext. Abst. ISEC, 222, 1989.

[3] Kobayashi, private communications.

[4] S. Tanaka, H. Nakanishi, T. Matsuura, K. Higaki, H. Itozaki and S. Yazu, IEEE Trans. on Magn., 27 (2), 1607, 1991.

[5] J.H. van der Merwe, C.A.B. Ball, N.H. Fletcher and K.W. Lodge: Epitaxial Growth, ed. J.W.Matthews, Academic Press, New York, Part B, 493-495, 1975.

[6] G. Matsubara, T. Matsui and Y. Okabe, Ext. Abs. FED Workshop, 100, 1991.

[7] M. Gurvitch, J.M. Valles, A.M. Cucolo, R.C. Dynes, J.P. Garno, L.F. Schneemeyer and J.F. Waszczak, Phys. Rev. Lett. 63, 1008, 1989.

[8] I. Iguchi and Z. Wen, IEEE Trans. Magn., 27 (2), 1991.

[9] T. Shiota, K. Takechi, Y. Takai and H. Hayakawa, Advances in Superconductivity, Nagoya, Springer-Verlag, 754, 1989.

[10] H. Akoh, C. Camerlingo and S. Takada, Appl. Phys. Lett., 56, 1487, 1990.

[11] P.M. Mankiewich, D.B. Schwartz, R.E. Howard L.D. Jackel, B.L. Straughn, E.G. Burkhardt and A.H. Dayem, Ext. Abs. FED Workshop, 157, 1988.

[12] A. Nakayama, A. Inoue, K. Takeuchi, H. Ito and Y. Okabe, IEEE Trans. Magn., 25(2), 799, 1989.

[13] Hayakawa, private communications.

[14] J.S. Tsai, I. Takeuchi and S. Ishizaka, Ext. Abs. FED Workshop, 186, 1991.

[15] C.T. Rogers, A. Inam, M.S. Hedge, B. Dutta, X.D.Wu and T. Venkatesan, Appl. Phys. Lett., 55(19), 2032, 1990.

[16] T. Matsushima, Y. Ichikawa, H. Adachi, K. Setsune and K. Wasa, Solid State Comm., 76, 1201, 1990.

[17] Y. Bando, Ext. Abst. FED, 202, 1991.

[18] Y. Tarutani, T. Fukazawa, U. Kabasawa, A. Tsukamoto, M. Hiratani, M. Suga and K. Takagi, Ext. Abst. FED, 190, 1991.

[19] M. Kasai, T. Ohno, Y. Kanke, Y. Kozono, M. Hanazono and Y. Sugita, Jpn. J. Appl. Phys., 29(12), L2219, 1990.

[20] K.P. Daly, W.D. Dozier, J.F. Burch, S.B. Coons, R. Hu, C.E. Platt and R.W. Simon, Appl. Phys,. Lett., 58, 543, 1991.

NS Boundaries and the Proximity Effect in Metal – YBCO Junctions

J. Yoshida, T. Hashimoto, S. Inoue, M. Sagoi, and K. Mizushima

Advanced Research Laboratory, Toshiba R&D Center,
1, Komukai Toshibacho, Saiwaiku, Kawasaki 210, Japan

Abstract. The Josephson characteristics of YBCO/Au–Ag/Pb SNS junctions were investigated both experimentally and theoretically. A photolithography process together with "in–situ" sputtering preparation of YBCO/Au bilayers has enabled the reproducible fabrication of small junctions on YBCO films with various crystal orientations. Most of the fabricated junctions exhibited a dc Josephson current with a clear Fraunhofer pattern dependence on the applied magnetic field. A numerical analysis based on the Usadel equation revealed that a junction model in which a thin insulating layer was assumed to exist on the YBCO film surface could account for the experimentally observed junction characteristics.

1. Introduction

The fabrication of reliable Josephson junctions between high–Tc materials and conventional superconductors is of vital importance not only for the application of these materials to active electronic devices but also for advancing the understanding of high temperature superconductivity itself. One of the promising ways to obtain Josephson junctions on high–Tc materials is to fabricate superconductor – normal metal – superconductor (SNS') structures using high–Tc oxide/noble metal bilayers expecting that the proximity effect works between these materials. Several groups have reported the successful fabrication of SNS' structures and have demonstrated the Josephson effect between Pb or Nb and YBCO[1–5]. However, there still remain many problems to be clarified with regard to the SN boundaries between YBCO and noble metals. The most puzzling problem is the anomalously large normal resistance values (Rn) which were always observed for the SNS' junctions. Also, contradictory results were reported concerning the YBCO anisotropy effect on the junction characteristics.

The present paper describes the recent progress in our research on SNS' junctions on YBCO thin films, highlighting a possible interface structure of actual SN boundaries.

Springer Proceedings in Physics, Vol. 64 Superconducting Devices and Their Applications
Editors: H. Koch, H. Lübbig

2. Fabrication of SNS' junctions

YBCO thin films with various crystal orientations including (001), (100), (110), and (103) were prepared on $SrTiO_3$ (100) and (110) substrates by multi–target rf magnetron sputtering using Y, Cu, and Ba_2CuO_3 targets. A Au layer was formed on the YBCO film by sputtering in the same vacuum chamber without exposing the YBCO surface to air. The substrate temperature during the Au deposition was kept between 50 ℃ and 150 ℃. This "in–situ" preparation of YBCO/Au bilayers was indispensable to obtain reproducible junction characteristics. The nominal thickness of the Au film was 25 to 100 nm.

Small junctions with dimensions ranging from 5X5 μm to 100X 100 μm were fabricated on the YBCO/Au bilayers with a photolithography process. Fig.1 shows a schematic representation of the fabricated junction. The fabrication process started with dry etching the YBCO/Au bilayer using an Ar ion milling technique. After etching, the junction area was defined with a negative resist, and a 600–nm thick resist film was made to remain so as to work as an interlayer isolation dielectric. Preceding the counter electrode formation through a metal mask, an Ar plasma cleaning procedure was provided to eliminate any possible resist residue in the junction area. 100–nm thick Ag and subsequently a 1 to 1.5 μm thick Pb or PbIn alloy was deposited to provide a counter electrode by thermal evaporation. The insertion of the Ag layer was necessary in order to prevent a solid phase reaction between Pb and Au. Thus, the N–layer of the fabricated junction was composed of a Au/Ag bilayer with its thickness ranging from 125 to 200 nm.

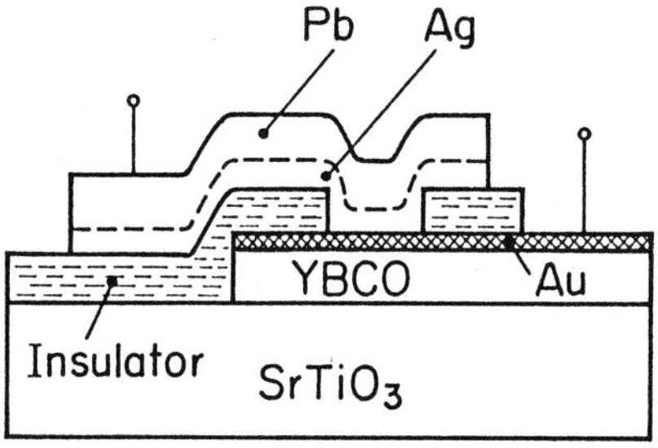

Fig.1 SNS' junction structure fabricated on YBCO thin films

3. Josephson Effect in SNS' Junctions

Fig.2 shows an example of the current–voltage (I–V) characteristic observed at 4.2 K for the 10X10 μm junction fabricated on (103) oriented YBCO thin films. Almost all junctions with a different junction area and a different YBCO film orientation, except those on

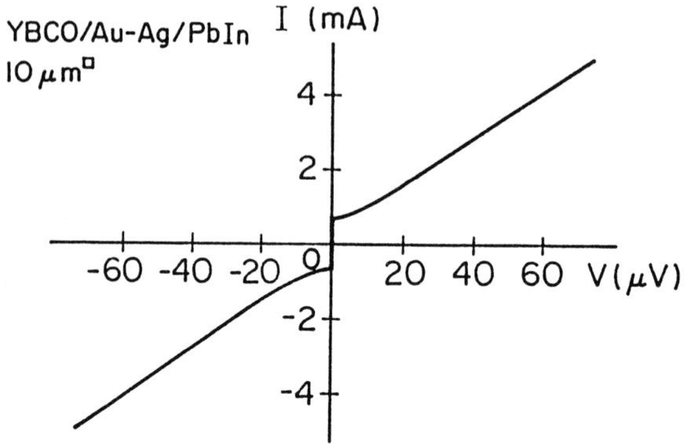

Fig.2 Current–voltage characteristic observed for SNS' junction at 4.2 K

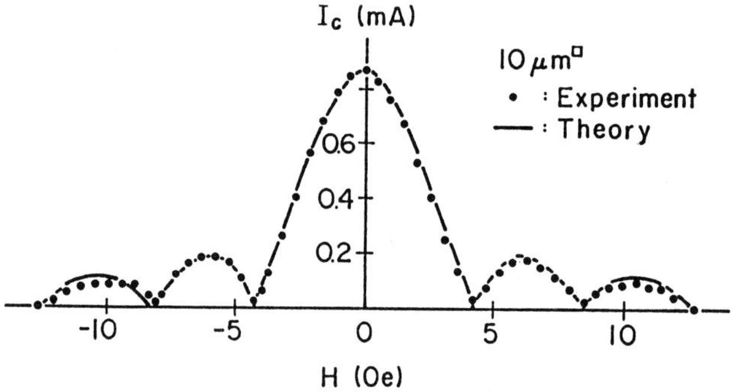

Fig.3 Magnetic field dependence of critical current for 10X10 μm junction

(001) oriented films, exhibited a similar I–V characteristic. The most sensitive test of whether the fabricated structures are indeed homogeneous Josephson junctions is to examine the magnetic field dependence of the critical current. A typical result is shown in Fig.3. The solid line in the figure shows the theoretical magnetic field dependence of the critical current Ic calculated by assuming that the magnetic penetration depth of the junction was equal to 490 nm. The agreement between theory and experiment was excellent. Also, the magnetic penetration depth deduced from the experiment well coincided with the value estimated from the fabricated junction structure (300–nm thick YBCO, 150–nm Au/Ag, and 1.5–μm PbIn). The Fraunhofer pattern dependence deviated from an ideal one for junctions with higher critical current densities, which was reasonably attributed to the self–field effect, i.e., the Josephson penetration depth became small compared to the junction dimension.

Fig.4 summarizes the critical current density Jc versus normalized normal resistance Rn relation for the SNS' junctions for which the

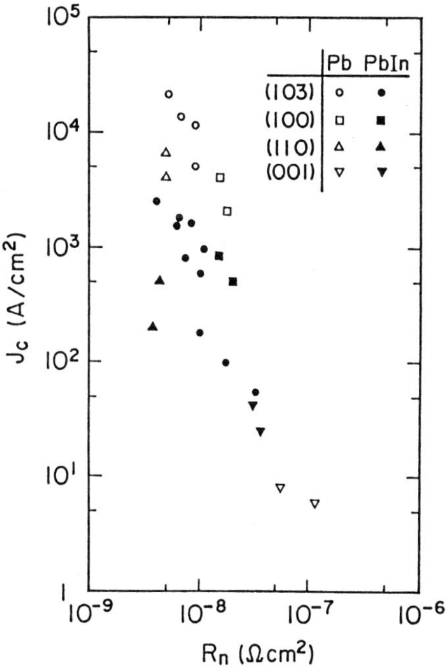

Fig.4 Critical current density versus normal resistance observed for SNS' junctions on YBCO with various crystal orientations. Open symbols denote results obtained for junctions with a Pb electrode and black symbols correspond to those with a PbIn electrode.

Josephson property was observed. A number of junctions fabricated on the YBCO films with (103), (110), or (100) orientation exhibited critical current densities exceeding 10^3 A/cm^2 with the normal resistance around $1X10^{-8}$ Ωcm^2. No systematic dependence on YBCO film orientation was observed among these junctions. On the other hand, the junctions fabricated on the (001) oriented YBCO films often showed no evidence of a superconducting current, but occasionally showed Jc values two orders less than those of others while the Rn values were slightly higher. These results coincided with the report of Akoh et al. [3], who clearly observed an anisotropic superconducting current in the junctions fabricated on (001) and (103) oriented films. However, contradictory results have also been reported by many groups[3,6,7]. The reason for the discrepancy in the anisotropy effect is not well understood at present. One possibility is that the crystal orientation on the very surface of the YBCO film may deviate from that of the inside. We often observed a (100) oriented surface layer on (001) oriented films through RHEED measurement. Such a surface layer could not be well resolved by X–

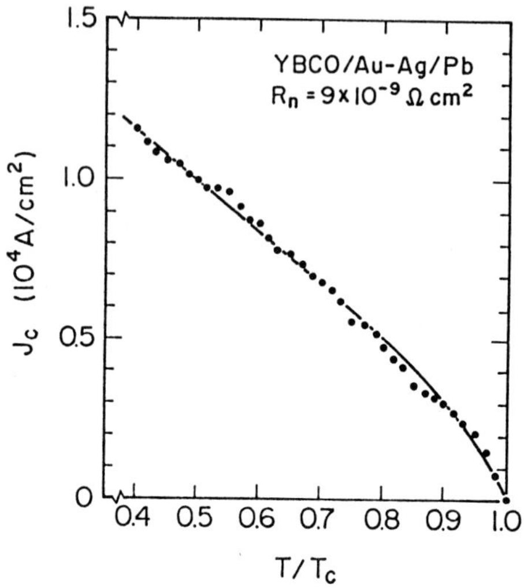

Fig.5 Temperature dependence of critical current density for junction with high IcRn product value

ray diffraction. Further experiments paying close attention to the surface structure are required to give an entire picture of the anisotropy effect.

An apparent difference in Jc between Pb and PbIn electrode junctions seen in Fig.4 can be attributed to the difference in the coherence length between these materials. The short coherence length in PbIn compared to that in Pb results in a large suppression of the order parameter at the interface between Ag and these materials, thereby reducing the critical current in the SNS' junctions.

An IcRn product value around 100 μV was obtained for the junctions fabricated on the (103) oriented YBCO films with a Pb counter electrode. Fig.5 shows an example of the temperature dependence of Jc for these junctions with high IcRn. The temperature dependence was rather weak, i.e., almost linear in the temperature range well below the critical temperature of Pb, while Rn was independent of temperature.

4. Numerical Analysis

Almost all the SNS' junctions fabricated on the YBCO thin films showed Rn values of the order 10^{-8} -10^{-9} Ωcm^2 . These values are comparable to the contact resistance value reported for SN junctions fabricated on a cleaved YBCO surface[6,7], and are more than 100 times larger than that expected for the noble metal layer alone. A plausible explanation is that the high resistance comes from a normal conduction layer on the YBCO surface, or perhaps even from a barrier layer in which a tunneling mechanism dominates electron conduction. These layers may be formed by the segregation of the constituent atoms on the surface in the course of the YBCO film preparation, or, more intrinsically, by contact to noble metals which breaks the translational symmetry of the lattice, resulting in a modification of the electronic structure within a few unit cell length. A numerical analysis was carried out to investigate these possibilities.

The Usadel equation[8] and the self–consistent gap equation were numerically solved in accordance with the recipe given by Kupriyanov[9–11] under appropriate boundary conditions. This numerical model enabled a quantitative prediction of the Josephson characteristics for SNS' junctions with an arbitrary structure, though the validity of the model is restricted to the dirty limit case. The effect of a possible finite transparency of the SN boundaries has been incorporated into the model[10]. Table I summarizes the physical parameters utilized in the present work. It should be noted that the free electron model was assumed to be valid for all these materials including YBCO.

Table I Physical parameters utilized for calculation

Layer	N(0) ($eV^{-1}cm^{-3}$)	v_F (cm/s)	m*	ρ (Ωcm)	ℓ (nm)	D (cm^2/s)	Tc (K)
Pb	$1X10^{22}$	$1.8X10^{8}$	1.0	$1X10^{-6}$	52	312	7.2
Au/Ag	$8X10^{21}$	$1.4X10^{8}$	1.0	$1X10^{-6}$	83.6	390	0
YBCO	$1X10^{22}$	$2.8X10^{7}$	2.5	$1X10^{-4}$	3.2	3	90.0

Two plausible interface structures were investigated, i.e., one was that an insulating layer (or the low transparency of the interface) played an important role in determining the junction characteristics (SINS' model), and another was that the very surface of the YBCO became nonsuperconductive and this normal conduction layer governed the junction properties (SN'NS' model). The calculated results for these two models are summarized in Fig.6, in which the noble metal layer was assumed to be 100–nm thick. The Jc versus Rn relation for the SINS' model is represented by a straight line in the figure. The

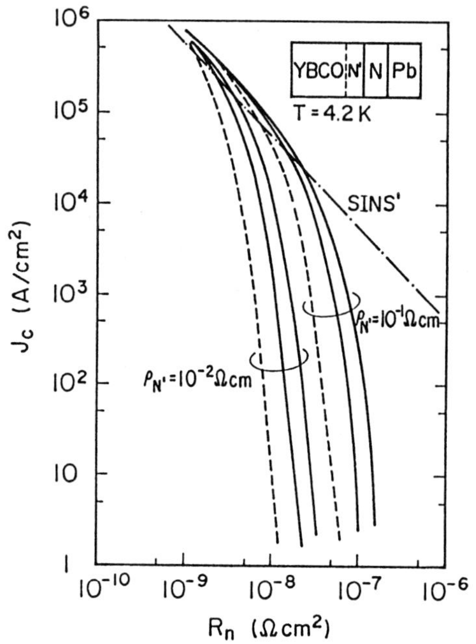

Fig.6 Theoretically calculated critical current density versus normal resistance relation

IcRn product was 0.56 mV, independent of Rn, which was 5 times larger than the experimentally observed maximum value. The situation is rather complicated for the SN'NS' model, because some ambiguity remains for the material parameters of the nonsuperconductive YBCO layer. We varied the material parameters under the assumption that the free electron picture was applicable even for the nonsuperconductive YBCO layer. The independent parameters under this assumption were the density of state at the Fermi surface, the electron effective mass, and the electron mean free path. A wide variety of junction characteristics resulted depending on the choice of these parameters. Two groups of curves shown in Fig.6 represent the cases that the resistivity of the nonsuperconductive YBCO layer was fixed to be either 10^{-2} Ωcm or 10^{-1} Ωcm while the layer thickness was varied. Seemingly, the SN'NS' model with the 10^{-2} Ωcm N' layer well reproduces the experimentally observed Jc versus Rn relation. However, one should note that , in the region where Jc strongly depends on Rn in Fig.6, the N' layer is far thicker than the coherence length of this layer. This results in a strong dependence of Jc on temperature, which is not what was observed experimentally.

The only way to solve the discrepacy between theory and experiment described above is to assume that the actual junction structure is of the SINS' type, and that the Jc values become low

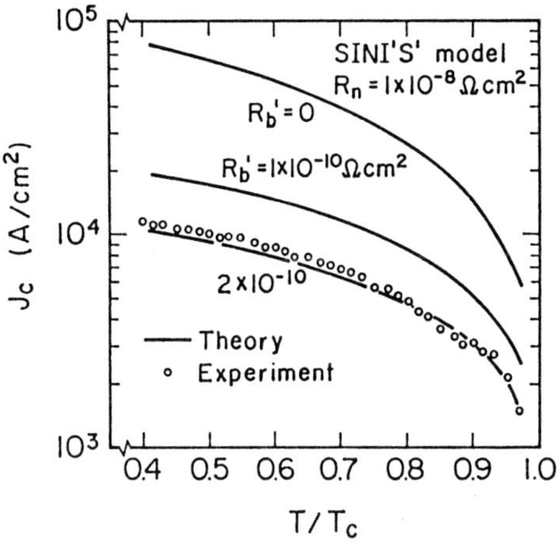

Fig.7 Theoretical fitting of experimentally observed Jc versus T relation with SINS' model including interface resistance Rb' at NS' boundary

because of some physics which is not introduced in the present model. Two possibilities exist as the mechanism to reduce the critical current density. One is the temperature independent pair breaking mechanism at the interface which was duscussed by Golubov in his analysis for grain boundary junctions[12]. Another possibility is electron reflection at the noble metal–Pb interface[13]. In fact, an excellent fitting to the experimentally observed temperature dependence of Jc was possible when an interface resistance Rb' around $2X10^{-10}\,\Omega\,cm^2$ was assumed at the noble metal–Pb interface as shown in Fig.7. However, it should be emphasized that this agreement does not necessarily deny the possibility of gap reduction at the YBCO–noble metal interface, because, as well known in the SIS junction theory, the Jc value is not sensitive to the YBCO gap value at the interface unless the superconductivity on the YBCO side becomes weaker than that on the noble metal side which is induced by the proximity to Pb. A wide scattering of the experimental IcRn values shown in Fig.5 may be attributed to the uncontrolled gap reduction in YBCO at the interface.

5. Conclusion

Experimental and theoretical studies were carried out for SNS' junctions fabricated on YBCO thin films to investigate the SN boundary properties and the possible proximity effect between high–Tc materials and noble metals. The fabricated junctions exhibited well defined Josephson characteristics including an ideal Fraunhofer pattern dependence of the critical current on the applied magnetic field. An anisotropy effect was confirmed between junctions fabricated on (001) oriented YBCO films and those on other orientation films including (103), (110), and (100). A numerical simulation based on the Usadel equation clarified that a junction model in which an insulating layer was assumed to exist on the YBCO surface could account for the experimentally observed junction characteristics.

Acknowledgement

This work was performed under the management of FED (The R&D Association for Future Electron Devices) as a part of the R&D of Basic Technology for Future Industries supported by NEDO (New Energy and Industrial Technology Development Organization).

References

[1] L. H. Greene, J. B. Barner, W. L. Feldman, L. A. Farrow, P. F. Miceli, R. Ramesh, B. J. Wilkens, B. B. Bagley, J. M. Tarascon, J. H. Wernick, M. Giroud and J. M. Rowell, Physica C162–164, 1069 (1989)

[2] H. Akoh, F. Shinoki, M. Takahashi and S. Takada, Jpn. J. Appl. Phys., 27, L519 (1988)

[3] H. Akoh, C. Camerlingo and S. Takada, Appl. Phys. Lett., 56, 788 (1990)

[4] A. Fujimaki, T. Tamaki, T. Hidaka, M. Yanagase, T. Shiota, Y. Takai and H. Hayakawa, Jpn. J. Appl. Phys., 29, L1659 (1990)

[5] M. Lee, D. Lew, C. B. Eom, T. H. Geball and M. R. Beasley, Appl. Phys. Lett., 57, 1152 (1990)

[6] T. W. Jing, Z. Z. Wang and N. P. Ong, Appl. Phys. Lett, 55, 1912 (1989)

[7] J. S. Tsai, I. Takeuchi, H. Tsuge, N. Matsukura and S. Miura, Physica, B165–166, 1627 (1990)

[8] K. D. Usadel, Phys. Rev. Lett., 25,507 (1970)

[9] M. Yu. Kupriyanov and V. F. Lukichev, Sov. J. Low Temp. Phys., 7, 137 (1981)

[10] M. Yu. Kupriyanov and V. F. Lukichev, Sov. J. Low Temp. Phys., 8, 526 (1982)

[11] M. Yu. Kupriyanov and V. F. Lukichev, Sov. Phys. JETP,67, 1163 (1988)

[12] A. A. Golubov and M. Yu. Kupriyanov, Phys. Lett., A154, 181 (1991)

[13] H. W. Lean and J. R. Waldram, J. Phys. Cond. Matt., 1, 1285 (1989)

In-Situ $YBa_2Cu_3O_7/SrTiO_3/YBa_2Cu_3O_7$ a-b Plane Josephson Edge Junctions

E. Aharoni, G. Koren, E. Polturak, D. Cohen, and E. Iskevitch

Physics Department and the Crown Center for Superconductivity, Technion-Israel Institute of Technology, Haifa 32000, Israel

Abstract. YBCO/$SrTiO_3$/YBCO thin film edge junctions were prepared *in-situ* and characterized. The epitaxial growth of $SrTiO_3$ on YBCO led to a sharp and well defined junction edge with a very high yield. Typical junctions showed critical currents up to 83 K, with $I_c \propto (1-T/T_c)^2$ temperature dependence. Sharp Shapiro steps were observed under microwave radiation at temperatures up to 82 K. A typical diffraction pattern was found in the voltage response of the junctions to transverse magnetic field.

One of the obstacles facing the attempts of fabricating all high T_c Josephson junctions is the difficulty to obtain a well defined thin barrier at the high deposition temperatures involved. One of the promising candidates for a barrier material is $SrTiO_3$, which has a relatively high resistivity at low temperatures [1], a small thermal mismatch with YBCO, and no interdiffusion into it, as was shown by TEM micrographs of YBCO/$SrTiO_3$ multilayers [2]. Ultrathin films of $SrTiO_3$ were also observed to grow uniformly and free of pinholes on $EuBa_2Cu_3O_7$ [3].

Recently, YBCO-YBCO edge junctions were prepared *in situ* using excimer laser deposition through metallic masks, with a barrier formed by a plasma discharge process [4,5]. They showed ac Josephson effect and SQUID operation at temperatures above 80 K. Here, we report preliminary results on edge junctions with a thin $SrTiO_3$ layer as the barrier.

The fabrication procedures of the junctions were described previously [4,5]. A schematic diagram of the junction cross section is given in Fig. 1. All the layers were deposited under 0.2 Torr O_2 and substrate temperature of 730 °C. Following the deposition of the base electrode and the insulating layer (with thicknesses of about 0.5 μm each), a barrier was formed by covering the edge of the two layers with a thin $SrTiO_3$ film

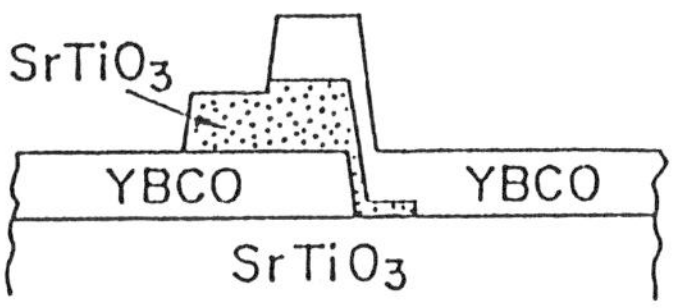

Fig. 1: A schematic cross section of the edge junction.

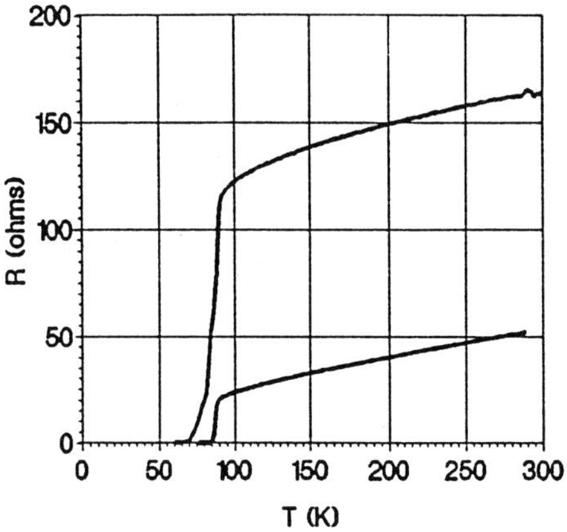

Fig. 2: R vs. T of two junctions: a) With a barrier formed by 100 laser pulses, W = 18 μm, and acid etching patterning. b) 200 laser pulses barrier, W = 20 μm, and ion-milling patterning.

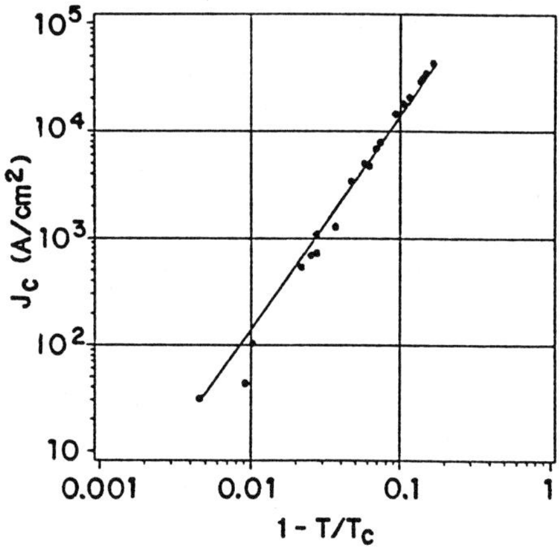

Fig. 3: Critical current density of the 100 pulses junction vs. (1 − T/T$_c$). The solid line, with a slope of 2, is a fit to the measured data.

(100 laser pulses). Finally, an YBCO counterelectrode was deposited. It was found that the resulting junction had a sharp and well defined edge. For patterning we used a deep UV photoresist (PMMA) and a dilute phosphoric acid for the etching. The $SrTiO_3$ layer protected the base YBCO film during the etching, and as a result the edge junction was adjacent to the base YBCO pad. Because of this close proximity, a strong focusing of the applied magnetic field into the junction is expected.

Figures 2 and 3 show the resistance R and critical current density J_c of an 18 μm wide junction as a function of temperature. In Fig. 3 one can see that the critical current densities are relatively high, hence that the junction will function as a strong link. The critical current obeyed the $(1 - T/T_c)^2$ temperature dependence near T_c which is typical of SNS junctions but also of SIS junctions provided that the depression of the order parameter of the superconducting electrodes at the interface with the barriers is taken into account [6]. A typical value of R_N was 30 mΩ at 77 K, which led to an I_cR_N product of only 0.1 mV. Figure 4a is an oscillogram of the I-V characteristic of the junction at 78 K under 10.7 GHz microwave radiation. Many Shapiro steps are clearly observed as well as a suppression of I_c. The ac Josephson effect persisted up to 82 K.

Figure 5 shows the voltage response of the junction to a transverse magnetic field, $H_\perp$, at 81.5 K. At this temperature the junction is still large since the Josephson penetration length, $\lambda_J \sim 4\,\mu$m, is smaller than

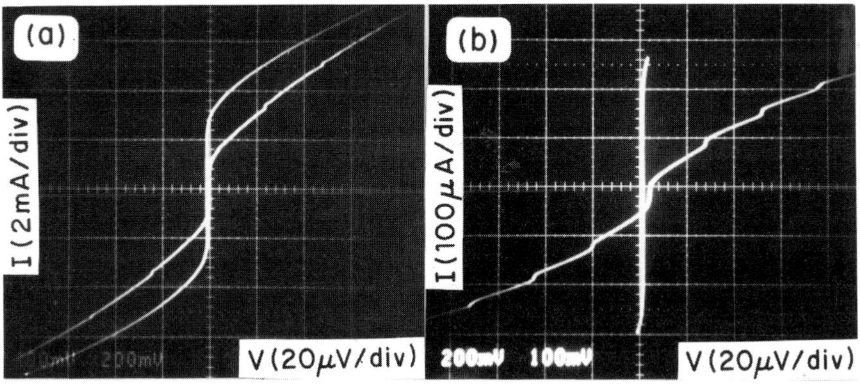

Fig. 4: I-V curves with and without microwave radiation. a) 100 laser pulses barrier, T=78 K, patterned with acid etching. b) 200 laser pulses barrier, T=75 K, patterned by ion-milling.

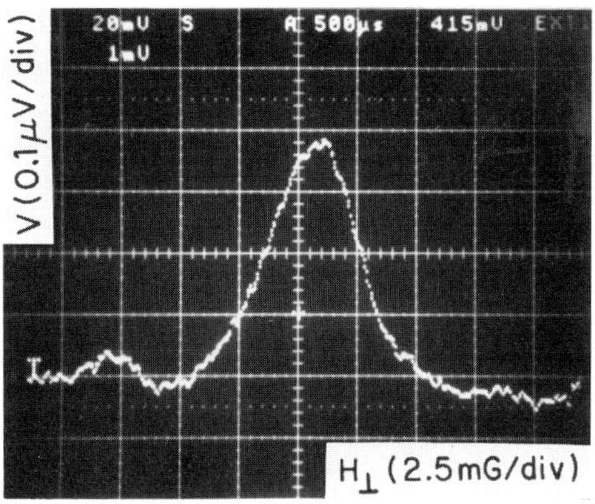

Fig. 5: Voltage response of the junction to a transverse magnetic field, at 81.5 K.

the junction width W (18 μm). Therefore, the response is characteristic of a junction with a nonuniform current density distribution across it. The observed magnetic period in Fig. 5 is $\Delta H_p \approx 12.5$ mGauss, and using the two fluid model relation we find that $\lambda_L(81.4\,K) \approx 2700$ Å. With these values, the junction obeys the relation $f\Delta H_p 2\lambda_L(T) = \Phi_0$ where f is the focusing factor (~ 100). Similar results were obtained previously in junctions with a barrier produced by plasma discharge [5].

In order to produce weaker links with lower critical current densities, barriers with higher R_N should be produced. To this end a similar junction with a barrier formed by 200 laser pulses, has been prepared. The thicker $SrTiO_3$ layer protects the base electrode more efficiently during the acid etching. Therefore, one cannot rely on a possible etching undercut, for the separation of the junction from the YBCO pad. Ion milling was used instead for the patterning. A 20 μm wide microbridge was patterned across the junction using 0.3 mA/cm^2 (3000 V) Ar beam for 6 hours. The resistance of this junction vs. T is also given in Fig. 2. The film deterioration results from heating degradation during the milling process. Similar deterioration of films was also reported by others [1]. Nevertheless, Shapiro steps did appear under microwave radiation at temperatures up to 61 K (see Fig. 4b). Modulation of the voltage across the junction in magnetic field has been observed with a period of $\Delta H_p \approx 2.5$ Gauss at temperatures up to 70 K. This period is two orders of magnitude higher than that of the former junction. This indicates that the focusing factor, here, is of the order of unity. This results from the fact that the junction here was about 50 μm far away from each pad.

In conclusion, an all high T_c *a-b* plane Josephson edge junction, with YBCO/$SrTiO_3$/YBCO structure has been fabricated and characterized. The ac Josephson effect and voltage modulation versus magnetic field have been observed at temperatures above 80 K. The YBCO pad adjacent to the junction has led to a high magnetic focusing, which disappeared when the junction was far from the two superconducting electrodes. The parameters affecting the barrier formation and the ion-milling process are yet to be optimized.

References

1. F. C. Wellstood, J. J. Kingston, and J. Clarke, Appl. Phys. Lett. **56**, 2336 (1990)
2. M. E. Tidjani, R. Gronsky, J. J. Kingston, F. C. Wellstood, and J. Clarke, Appl. Phys. Lett. **58**, 765 (1991)
3. O. Michikami and M. Asahi, Jap. J. Appl. Phys. **30**,446 (1991)
4. G. Koren, E. Aharoni, E. Polturak, and D. Cohen, Appl. Phys. Lett. **58**, 634 (1991)
5. G. Koren, E. Polturak, E. Aharoni, and D. Cohen, Submitted to Appl. Phys. Lett. (1991)
6. R. Gross, P. Chaudhari, D. Dimoss, A. Gupta, and G. Koren, Phys. Rev. Lett. **64**, 228 (1990)

Fine Structure in the Tunneling Characteristis of Bi-Sr-Ca-Cu-O:Pb

B.A. Aminov[1], *A.A. Bush*[2], *L.I. Leonyuk*[2], *T.E. Oskina*[1,2], *M.V. Pedyash*[2], *D.K. Petrov*[2], *Ya.G. Ponomarev*[2], *H.T. Rakhimov*[2], *K. Sethupathi*[2], *and M.V. Sudakova*[2]

[1]Physical Technical Institute, Aini str., 299/1, 734063 Dushanbe, Tadjikistan
[2]Physical Faculty, Moscow State University, 119899 Moscow, Russia

Abstract. Tunneling effects in poly- and monocrystalline samples of Bi-Sr-Ca-Cu-O :Pb compounds (2:2:1:2- and 2:2:2:3- phases) have been investigated. A reproducible vertical rise of current on I(V) - characteristics of break junctions at gap voltages have been found. A clearly defined " knee "- structure has been observed at voltages $V \cong 2\Delta / e$. Equidistant oscillations of significant amplitude on dI/dV-characteristics of Bi-Sr-Ca-Cu-O: Pb- junctions with a period $\Delta V = (2.5 \pm 0.2)$mV (2:2:1:2- phase) have been registered in a wide range of temperatures.

1. Introduction

Tunneling spectroscopy is one of the basic methods for the determination of fundamental superconducting parameters. To obtain I(V) - characteristics with a well-defined gap feature a perfect superconducting surface must be available. In the case of HTSC materials such a requirement becomes exceptionally hard, which is the result of very short coherence length ξ and mean free path l of the quasi-particle exitations. One way of overcoming these problems is a creation of a break junction in the superconducting sample at helium temperatures. In this report the I(V)-characteristics of break junctions based on Bi-Sr-Ca-Cu-O:Pb mono- and polycrystalline samples are discussed. The method of preparation of contacts is described in detail in [1].

2. Results

From the X - ray analysis it was found that monocrystalline samples of Bi-Sr-Ca-Cu-O: Pb (LIL - series) have the structure of the 2:2:1:2 - phase and polycrystalline samples (BUCH - and OS- series) consist approximately 90% of the 2:2:2:3- phase and near 5% each of the 2:2:0:1 - and the 2:2:1:2 - phases. These studies also indicated the presence of an "intergrowth" of these phases similar to that described in [2]. These phases have periodical structure of layers of 2:2:0:1-, 2:2:1:2- and 2:2:2:3- phases along the c- axis.

A good reproducible vertical rise of current in the current-voltage characteristics of break junctions at gap voltages was found at T=4.2 K in polycrystalline samples of Bi-Sr-Ca-Cu-O:Pb of BUSH- and OS- series (Fig.1).

Besides a vertical current rise there are several typical peculiarities on the I(V)- characteristics of the investigated samples: a clearly defined "knee"- structure

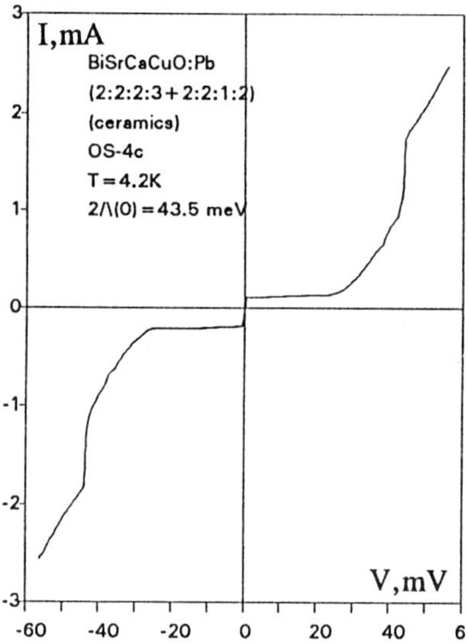

Fig.1.I(V)-characteristic of OS-4 sample at T = 4.2 K.

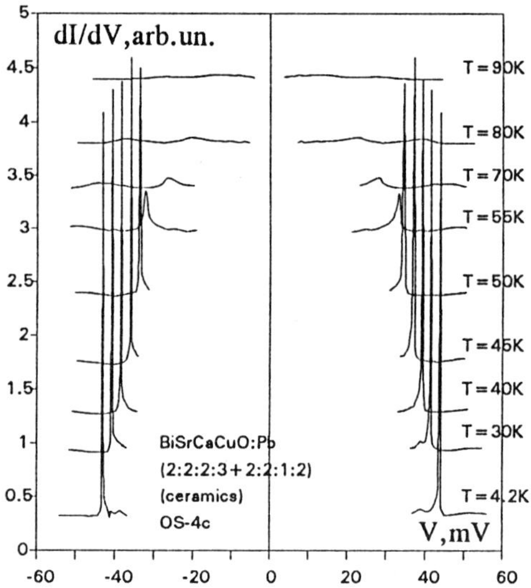

Fig.2.dI/dV-characteristics of OS-4 sample at different T.

at voltages, comparable with $V \approx 2\Delta / e$ (Fig.1) , and a considerable excess current at innergap voltages. The I (V) - dependence shown in Fig.1 is qualitatively similar to current-voltage characteristics of niobium contacts whose surface is "infected" by oxygen [3,4]. It is customary to believe that these contacts have S-N-I-S- or S-N-I-N-S- structure [5]. It is important to note, that the presence of a "knee"-structure on I(V)-characteristics is also an attribute property for tunneling structures of S-I-S_2-S_1 and S_1-S_2-I-S_2-S_1 type, and the proximity effect plays an important role in these structures [6,7] as well as in the case of S-N-I-N-S- structures [5]. It is possible that in the investigated break junctions for OS- series and BUSH- series symmetrical tunneling structures of either S-N-I-N-S- type or S_1-S_2-I-S_2-S_1- type are formed.

As temperature increases the peaks on dI/dV- characteristics corresponding to the gap value $V_g = \pm 2\Delta/e$, shift towards lower voltages (Fig.2, temperature range (4.2 , 90)K). This is the result of the decrease of the gap parameter Δ . In this case, as temperature increases we can see a nonmonotonous decrease of the peak amplitude in the dI/dV- characteristics: a very rapid decrease of the amplitude is observed in a short temperature range (45K , 55K), which is close to the region, where the critical temperatures T_c for different samples of the 2:2:1:2- phase are grouped. The latter is in favour of the existence of S_1-S_2-I-S_2-S_1 - structures in the contact region, where S_1: superconducting phase of 2:2:2:3 and S_2 : thin superconducting layer of the 2:2:1:2- phase. The temperature dependence of Δ determined from the peak positions of the dynamic conductance for the sample OS- 4(Fig.1), is shown in Fig.3 (curve 2). For $T \geq 60$ K in the I(V)- characteristics for the sample OS-4 additional gap structures with small amplitudes were observed which gave the second branch of 2Δ (T) in Fig.3 (curve 1) and this agree well with the calculations made by using the Thouless formula [8] (solid line inFig.3) and in

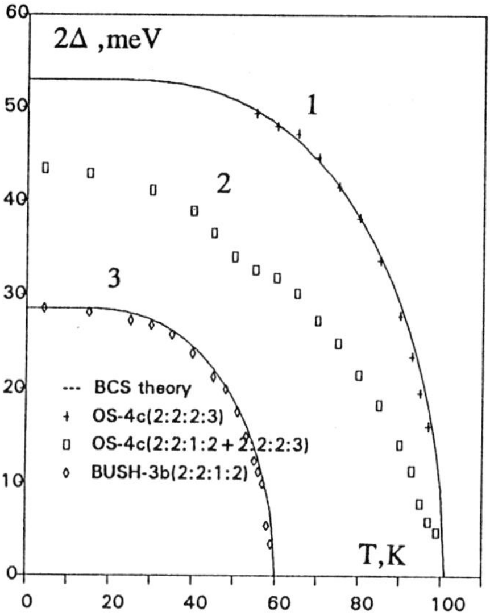

Fig.3. 2Δ(T)- dependence for 1: 2223-phase, 2 :2212+2223, 3: 2212-phase.

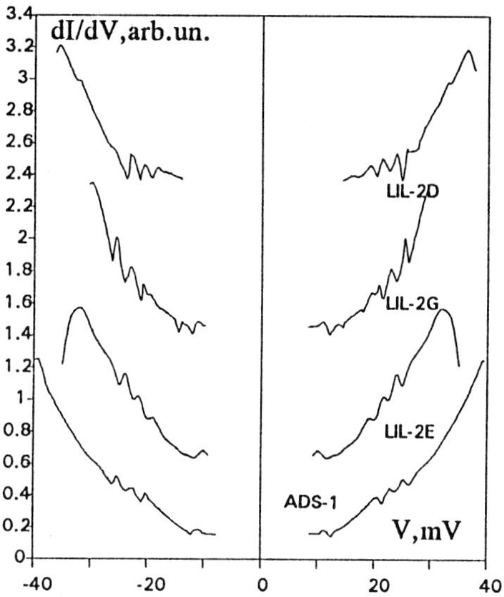

Fig.4. dI/dV- characteristics for different samples (2212-phase) at T = 4.2K.

our point of view the latter coincides to the pure 2:2:2:3- phase. Curve 2 in Fig.3, which coincides with the basic gap structure in the I (V) - characteristics of the sample OS-4 (Fig.1,2), may be related to a thin layer (probably monolayer) of 2:2:1:2-phase . It is possible that the gap parameter in this phase was increased due to the influence of the proximity effect in the structure S_1-S_2-I-S_2-S_1. Curve 3 in Fig.3 was obtained for the pure 2:2:1:2 - phase. One can see in curve 2 a nonmonotonous temperature dependence of 2Δ in the temperature region, where the gap of the pure 2:2:1:2- phase rapidly decreases (curve 3). Perhaps the existence of intergrowth, which we discussed earlier may lead to the formation of S_1-S_2-I-S_2-S_1 - type structures (S_1: 2:2:2:3-phase, S_2: 2:2:1:2-phase) in the process of breaking of the sample and also to the temperature dependence of 2Δ , typical for bilayer superconductors with different critical temperatures [6,9]. In this case there is a possibility of a "mending" influence of the "strong" superconductor (2223) on the thin layer of the "weak" (2212) superconductor. This influence could smooth fluctuations of the gap parameter in the surface layer which in turn could lead to the increase of the quasiparticle lifetime τ , to the reduction of the smearing of the peaks in the density of states and to the vertical rise of current in the I(V)- characteristics (Fig.1).

Another peculiarity, which was observed in the I(V)-characteristics for most of the investigated contacts was the step-like behaviour of the essential part of the excess current for $V \leq 2\Delta/e$. In the dI(V)/dV- characteristics this structure is manifested with equidistant oscillations (Fig. 4). The basic peculiarities of these oscillations are: 1) the oscillations are observed in monocrystalline as well as in polycrystalline samples; 2) the oscillations have the equidistant character and in some cases propagate beyond the gap region; 3) for 2:2:1:2-phase samples the period of oscillations is reproducible with good accuracy $\Delta V=(2.5\pm 0.3)$mV and does not depend on the macrostructure (mono- or polycrystalline) and geometry of the contact (Fig.4, T=4.2K; dI(V)/dV for different 2:2:1:2-phase samples); 4) when the temperature changes the position of the oscillation maxima remains unchanged ; 5) as the temperature increases, the amplitude of the oscillations decreases so that they completely disappear at $T \geq T_c$ .The reason for the appearance of this resonance structure is not yet clear. One of the possible ways of explaining it may be the formation of bound states in the surface region, which are caused by the formation of S-N-I- or S_1- S_2-I- structures (de Gennes- oscillations)[10]. As the temperature increases the gap parameter decreases and these oscillations should move towards the lower bias region, which is not happening in our case. Recently a number of theoretical papers were puplished [11-12], in which the appearance of complicated additional structure in I(V)- characteristics of tunneling contacts is attributed to the quasi two-dimensional character of conductance in HTSC. In some cases [11] the calculated dynamic conductance for HTSC tunneling contacts have characteristics qualitatively similar to that observed in our case. But according to this model additional structure in the I(V) - characteristics also should move with the increase of temperature. Perhaps, the most suitable model is the one proposed by Hahn [12], according to which the quasiparticle exitation spectrum in HTSC is quantized because of the layered character of the structure.

The authors are most grateful to Prof. Dr. H. Piel for his support in this work.

References

1. Aminov B.A.,Akimov A.I.,Brandt N.B.,Nguen Minh Thu, Ponomarev Ya.G., Sudakova M.V.,Wittig J.,Fisher L.M.,Roshta L.: Physica C **160**, p.505 (1989).
2. Bush A.A.: Sverhprovodimost, **3**, 9, p.1 (1990).
3. Shen L.Y.L., in Superconductivity in d- and f- Band Metals, ed. D.H.Douglas (AIP, Conf.Proc., N4, N.Y., 1972).
4. Broom R.F.,Raider S.I., Oosenberg A.: IEEE Trans.E.D.,**ED-27**,p.1998 (1980).
5. Golubov A.A., Kupriyanov M.Yu.: JETF **96**, p.1420 (1989).
6. Gilabert A.,Romagnan J.P.,Guyon E.: Sol. State Comm. **9**, p.1295 (1971).
7. McMillan W.L.: Phys. Rev. **175**, N2, p.537 (1968).
8. Thouless D.J.: Phys. Rev. **117**, p.1256 (1960).
9. Vbra J., Woods S.B.: Phys. Rev. B **3**, N7, p.2243 (1971).
10. Wolf E.L.,Arnold G.B.:Proximity electron tunneling spectroscopy ,N.-H.,(1982).
11. Tachiki M., Takahashi S.: Physica B, **160**, p.80 (1990).
12. Hahn A.: Physica B, **165&166**, p.1065 (1990).

Suppression of Magnetic Flux Noise in $YBa_2Cu_3O_{7-x}$ Flux Transformers

M.J. Ferrari[1], *F.C. Wellstood*[2], *J.J. Kingston*[1], *M. Johnson*[3], *and J. Clarke*[1]

[1]Department of Physics, University of California, and Center for Advanced Materials, Materials Sciences Division, Lawrence Berkeley Laboratory, Berkeley, CA 94720, USA

[2]Center for Superconductivity Research, Department of Physics, University of Maryland, College Park, MD 20742-4111, USA

[3]Bell Communications Research, 331 Newman Springs Road, Red Bank, NJ 07701-7030, USA

Abstract. We have constructed hybrid magnetometers by coupling a flux transformer of $YBa_2Cu_3O_{7-x}$ to a Nb-PbIn SQUID. The low frequency (1/f) noise is dominated by fluctuations in the supercurrent circulating in the transformer, driven by vortex motion. The application of a small static magnetic field induces a supercurrent that reversibly suppresses the noise, in quantitative agreement with a model of thermally activated vortex hopping between symmetrical pairs of pinning sites. A persistent current can be used to reduce the low frequency noise power of high-T_c flux transformers by at least one order of magnitude.

1. Introduction

Although the SQUID is a sensitive detector of magnetic flux, its small size implies a relatively poor magnetic field sensitivity, a deficiency generally overcome by the use of a flux transformer.[1] The need for a flux transformer is particularly acute in the case of SQUIDs fabricated from high transition temperature (T_c) superconductors such as $YBa_2Cu_3O_{7-x}$ (YBCO) since these devices operate at higher temperatures than conventional SQUIDs, and therefore have smaller inductances. To our knowledge, complete YBCO thin-film magnetometers consisting of a grain boundary dc SQUID and a multilayer flux transformer have been successfully operated by two groups, Miklich *et al.*[2] and Oh *et al.*,[3] who reported field sensitivities of 0.6 pT $Hz^{-1/2}$ and 12 pT $Hz^{-1/2}$, respectively, at 10 Hz and 77 K. The sensitivity in both cases was limited by the 1/f noise of the SQUID (f is frequency).

High-T_c magnetometers are attractive for applications such as magnetocardiography and magnetoencephalography because they require much less thermal shielding than their low-T_c counterparts, allowing pickup loops to be brought closer to the subject and thereby improving spatial resolution. At present, however, the 1/f noise of high-T_c SQUIDs is too high for the more demanding of these applications. One might seek means to reduce this low frequency noise in order to produce a sensitive all-high-T_c magnetometer. Alternatively, one might construct a hybrid magnetometer combining a low-T_c SQUID, with its low 1/f noise, and a high-T_c flux transformer, which provides high resolution; this is the approach we describe in this paper. In either case, noise from the flux transformer itself will become significant, and it is important to understand its origin and to find ways to reduce it.

2. Noise Mechanisms

We have investigated the noise in YBCO flux transformers by coupling them to a Nb-PbIn SQUID, as shown schematically in Fig. 1. Wellstood *et al.*[4] have described the design and fabrication of the flux transformers. The motion of vortices in a transformer causes variations in the flux Φ through the SQUID to which its input coil is coupled. Consider an ensemble of vortices, the position r of each one of which fluctuates with an average spectral density $<S_r(f,T,I)>$, which may depend on the temperature T and the time-averaged circulating current I in the transformer. If there are n_v uncorrelated vortices per unit area of the transformer, the power spectrum of the flux noise in the SQUID is

$$S_\Phi(f,T,I) = n_v \int_A dA <S_r(f,T,I)> \left(\frac{\partial\Phi}{\partial r}\right)^2 , \quad (1)$$

where the integral is taken over the area A of the transformer. When $<S_r(f,T,I)>$ is independent of position, Eq. 1 becomes

$$S_\Phi(f,T,I) = \eta n_v \Phi_0^2 <S_r(f,T,I)> , \quad (2)$$

where Φ_0 is the flux quantum and η is a dimensionless geometrical parameter.

Evaluating the parameter η in Eq. (2) requires knowledge of the transfer function $\partial\Phi/\partial r$, which we now discuss for two coupling mechanisms, one of which involves currents circulating in the transformer (indirect noise) and one of which does not (direct noise). We first consider indirect noise. The magnetic field lines produced by a vortex close around the transformer line in which it is pinned, preferentially around the nearer edge. If the vortex were displaced the entire distance w from one edge of the line to the other, the flux which it applies to the transformer would change by one flux quantum. A small displacement δr across the line produces, in a linear approximation, an applied flux change $\Phi_0\delta r/w$. Because the transformer is a closed superconducting circuit, a current $\delta I_i = \Phi_0\delta r/w(L_i+L_p)$ flows to oppose this flux, where $L_i \approx 75$ nH and $L_p \approx 20$ nH are the estimated inductances of the input coil and pickup loop, respectively. This current produces a flux in the SQUID, and we immediately obtain

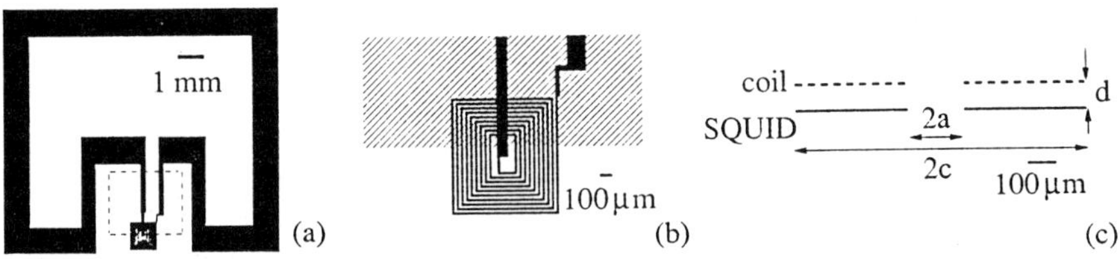

Fig. 1. YBCO flux transformer. (a) Pickup loop formed by wide lines, input coil centered on bottom edge, insulating $SrTiO_3$ layer indicated by dashed box. (b) Enlargement of input coil: $SrTiO_3$ is shaded region, crossunder is vertical strip connected to center of coil. (c) Cross section of input coil coupled to SQUID.

$$\frac{\partial \Phi}{\partial r} = \frac{\Phi_0 \alpha \sqrt{LL_i}}{w(L_i+L_p)} , \tag{3}$$

where L = 0.44 nH is the SQUID inductance and $\alpha \approx 0.5$ is the coefficient of inductive coupling between the input coil and the SQUID. Thus for the case of indirect noise,

$$\eta^{(in)} = \frac{\alpha^2 LL_i}{(L_i+L_p)^2} \left(\frac{\ell_i}{w_i} + \frac{\ell_{cr}}{w_{cr}} + \frac{\ell_p}{w_p} \right) \approx 1.2 , \tag{4}$$

where ℓ_i, ℓ_{cr}, and ℓ_p are the lengths of the input coil, crossunder, and pickup loop, respectively, and w_i, w_{cr}, and w_p are their widths. Most of the indirect noise is due to the input coil because of the large ratio $\ell_i/w_i \approx 1200$, as compared to $\ell_{cr}/w_{cr} \approx 30$ and $\ell_p/w_p \approx 40$.

Vortex motion also produces flux noise in the SQUID even in the absence of a closed superconducting path around the transformer, by a mechanism we call direct noise. Consider an unpatterned YBCO film in place of the transformer input coil in our apparatus. A vortex in this film couples a fraction of its flux directly into the SQUID, and this fraction varies with position. The relevant coordinate r is now the radial distance of the vortex from the center of the SQUID, not position across the transformer line as in the case of indirect noise; the approximate radial symmetry of the SQUID implies that azimuthal vortex motion produces little noise.

The direct-noise transfer function $\partial\Phi/\partial r$ is difficult to calculate exactly for our experimental geometry because of the distortion of the vortex field by the body of the SQUID and by the lines of the transformer. Approximate calculations of η have been made by Ferrari *et al.*,[5] and are summarized in Table I. Note that the indirect noise is only about a factor of three less than the direct noise from an unpatterned film. On the other hand, the direct noise from the input coil is much less, because most of the field lines from a vortex in a given turn close around that turn and do not couple into the SQUID. We are therefore justified in neglecting correlations between the direct and indirect contributions.

TABLE I. Estimates of flux noise parameter η defined in Eq. (2). Values for an unpatterned film, and for the input coil and crossunder of a flux transformer, are direct noise. Direct noise from pickup loop is negligible because of its distance from the SQUID. Total indirect noise from all transformer components is given by Eq. (4). Noise also shown as a fraction of $\eta^{(u)}$, the value for an unpatterned film.

Noise source	η	$\eta / \eta^{(u)}$
Unpatterned film	3.9	1
Input coil	0.0015	0.00038
Crossunder	0.11	0.03
Indirect	1.2	0.31

3. Noise Measurements

We measured the low-frequency noise produced by two similarly fabricated[4] YBCO flux transformers, T1 and T2. Each transformer was supported on a thermally isolated hot stage[6] with its 10-turn input coil approximately 100 μm from a Nb-PbIn SQUID maintained at 4.2 K; the temperature of the transformer could be raised above its transition temperature. Mu-metal and superconducting Pb shields surround the apparatus, which was cooled in a field B < 1 μT. The measured noise power spectra scaled approximately as 1/f, as shown in Fig. 2(a). Figure 3(a) indicates that the magnitude of the noise and its temperature dependence were similar for both transformers. The transition temperatures of T1 and T2 were 77 K and 59 K, respectively; however, above 62 K other Nb components of the apparatus became normal, preventing us from making meaningful noise measurements.

To test the mechanism of noise generation, we made a cut through the pickup loop of T2 which prevented the flow of screening current. Figure 3(b) illustrates the order-of-magnitude decrease observed in the noise power. According to Table I, this is consistent with noise in the intact transformer being

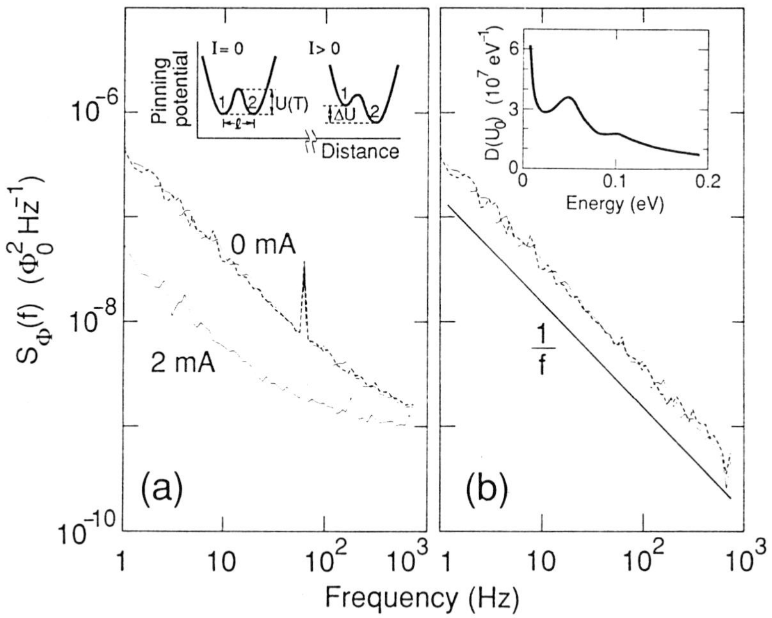

Fig. 2. (a) Total noise power vs. frequency for transformer T1 at 39 K. Upper solid curve: initial spectrum (I = 0); lower solid curve: application of I = 2 mA; dashed curve: subsequent return to I = 0. Spike is 60 Hz pickup. Flattening at high frequencies caused by Johnson noise from normal metal in apparatus. Inset: Schematic of pinning potential for single hopping process with I = 0 and with I > 0. Vortex hops distance ℓ between pinning sites 1 and 2. (b) Spectra for I = 0 with spectrum for I = 2 mA subtracted. Line indicates 1/f scaling. Inset: Distribution of activation energies from zero-current noise.

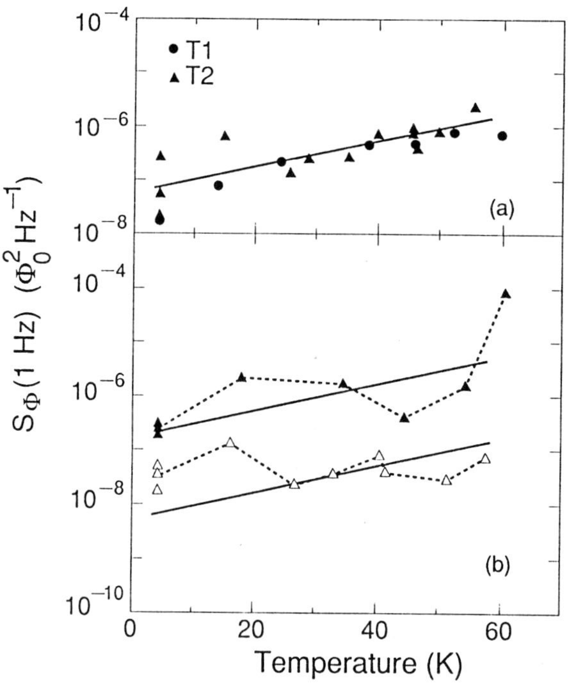

Fig. 3. Total noise power vs. temperature in ambient field. (a) Both transformers. (b) Transformer T2 after being scribed open (open symbols) and after SQUID has been moved beneath a 1-mm-wide line of pickup loop (solid symbols). For purposes of comparison, linear least-squares fit made to T2 data in (a) is replotted in (b) multiplied by 0.09 (lower line) and by 3 (upper line), indicating noise expected from the configurations represented by the open and solid symbols, respectively, according to Table I.

dominated by indirect noise, while in the cut transformer only direct noise from the crossunder is significant. Next we moved the 1 mm strip of part of the pickup loop over the SQUID. Figure 3(b) shows that the noise from this wide portion of the film ("unpatterned film" geometry) is larger than that from the intact transformer by approximately a factor of three, as Table I predicts.

4. Suppression of Noise by a Supercurrent

A Nb magnetic field coil operated in a persistent current mode allowed us to apply a small static magnetic field ($< 10^{-5}$ T) to the flux transformer, inducing a circulating current I. This current reversibly suppressed the noise power at low frequencies by an order of magnitude, as shown in Fig. 2. We have shown[7] that only indirect noise is significantly suppressed, so the direct noise background is subtracted from the data plotted in Fig. 4.

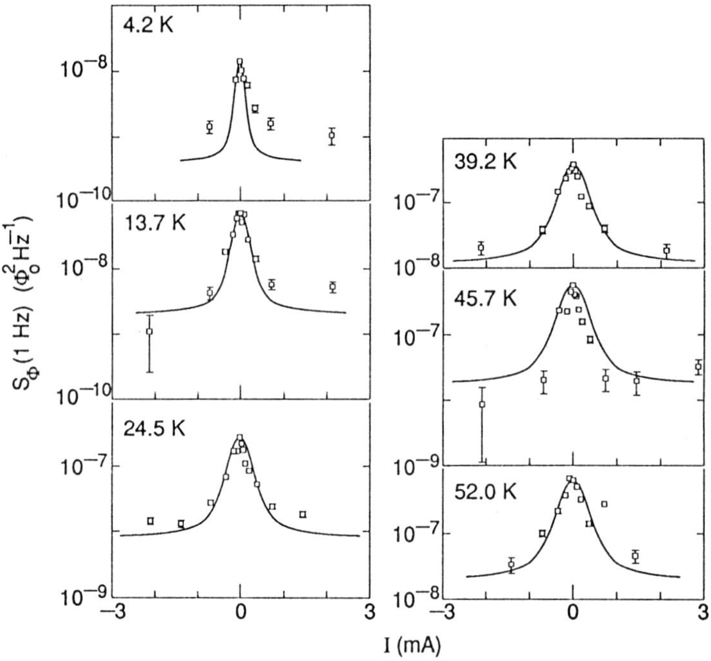

Fig. 4. Noise power S_Φ(1 Hz) vs. current in transformer T1 for six temperatures. Points are experimental data, from which the least noise measured at each temperature has been subtracted to remove background. Curves are prediction of Eqs. (1) and (7) with indirect noise transfer function from Eq. (3).

The reduction of indirect noise by current can be explained by our model[6,7] for 1/f noise in YBCO, based on the work of Dutta *et al.*[8] This model postulates an ensemble of thermally activated bistable processes such as the one depicted in the inset to Fig. 2(a). A vortex hops between two pinning sites separated by a distance ℓ. The temperature-dependent activation energy for hopping out of either site is U(T); we define $U_0 = U(0)$. Our model depends on two-level systems which are symmetric in the absence of the current, as would be the case, for example, if similar defects were present at each pinning site. The current, which we assume to be uniformly distributed across the line, exerts a Lorentz force $F = I\Phi_0/w$ on each vortex. The inset to Fig. 2(a) shows that this force introduces a misalignment $\Delta U = F\ell \cos\theta$ between the minima of the pinning potential, where θ is the angle between the Lorentz force and the vortex trajectory. Increasing current decreases the probability that the vortex will be activated out of site 2, reducing the indirect noise. Direct noise will not be significantly affected because it originates primarily in the crossunder, where the noisiest (radial) processes involve vortices moving perpendicular to the Lorentz force. The lifetimes of states 1 and 2 are

$$\tau_1(T,I) = 2\tau_0 \exp[U_0\beta - \delta] \quad (5a)$$
$$\tau_2(T,I) = 2\tau_0 \exp[U_0\beta + \delta] \ , \quad (5b)$$

where the attempt time $2\tau_0$ and $\beta(T) = U(T)/U_0 k_B T$ are assumed to be the same for all processes in the ensemble, and $\delta = I\Phi_0 \ell \cos\theta/2wk_BT$. For consistency with Ref. 6, we choose $U(T)/U_0 = 1-(T/T_c)^4$ and $\tau_0 = 10^{-11}$ s; our results are relatively insensitive to these parameters.

The spectral density for a single hopping process is given by[9]

$$S_r(f,T,I) = \frac{4\ (\ell \cos\theta)^2}{(\tau_1+\tau_2)[(\tau_1^{-1}+\tau_2^{-1})^2 + (2\pi f)^2]} \ . \tag{6}$$

To solve Eq. (1) we need the spectral density averaged over angle and energy in the ensemble:

$$\langle S_r(f,T,I)\rangle = \frac{1}{\pi n_v A} \int_0^{\pi} d\theta \int_0^{\infty} dU_0\ D(U_0)\ S_r(f,T,I) \ , \tag{7}$$

where $D(U_0)\,dU_0$ is the number of processes with zero-temperature activation energies between U_0 and U_0+dU_0. Two functions remain to be specified, $D(U_0)$ and the hopping distance ℓ. We have shown[7] that for $I = 0$, Eqs. (1), (3), and (7) can be inverted to yield $D(U_0)$ from $S_\Phi(f,T,0)$, as shown in the inset to Fig. 2(b). We assume that the minima in the inset to Fig. 2(a) are parabolic, so the hopping distance is given by $\ell = (8U_0/k_0)^{1/2}$, where the spring constant k_0 is the only significant adjustable parameter in our model.

We obtain the best fit to our data with $k_0 = 6 \times 10^{-5}$ N/m, as shown in Fig. 4. This gives a hopping distance for the processes contributing to the measured noise which increases from 13 nm at 4.2 K to 53 nm at 52 K.[7] Integrating $D(U_0)$ gives a number density of vortices n_v which can be expressed as an effective field $B_{eff} = n_v\Phi_0 \approx 40\ \mu T$.[7] Accounting for processes outside the experimental energy range may increase B_{eff}, but it is already much greater than the field B in which the transformer was cooled, implying that there is another mechanism for generating vortices in the film, such as the freezing-in of vortex-antivortex pairs as the film is cooled below T_c.

5. Discussion

The dominant source of noise in our hybrid magnetometer is the fluctuation of the screening current in the YBCO flux transformer, caused by vortex motion in the input coil. Inducing a dc current in the transformer significantly reduces its noise; for example, the field sensitivity of the magnetometer [Fig. 2(a)] improves from 0.22 pT $Hz^{-1/2}$ at 10 Hz to 0.08 pT $Hz^{-1/2}$ when we induce a current of 2 mA. Nevertheless, this latter figure is almost an order of magnitude above the limit set by the noise in our Nb-PbIn SQUID. Further improvements could be effected by designing a transformer with a narrower crossunder, thereby reducing the direct noise which dominates for $I > 0.7$ mA, or by fabricating the transformer from higher-quality YBCO films, the noise in which has been observed[10] to approach that in our SQUID. The optimization of future magnetometers with YBCO flux transformers will presumably rely on such intrinsically quiet films, as well as a geometry chosen to minimize both direct and indirect noise, and the suppression of indirect noise by a supercurrent. We note that the existence of a large number of mobile vortices despite the small ambient field ($B \ll B_{eff}$), which we have observed in YBCO, could also explain 1/f flux noise in low-T_c SQUIDs.[11]

Acknowledgements

We have benefited greatly from stimulating discussions with J. R. Clem. This research was supported by the California Competitive Technology Program and by the Director, Office of Energy Research, Office of Basic Energy Sciences, Materials Sciences Division of the U. S. Department of Energy under contract number DE-AC03-76SF00098.

References

1. J. E. Zimmerman and N. V. Frederick, *Appl. Phys. Lett.* **19,** 16 (1971); F. Wellstood, C. Heiden, and J. Clarke, *Rev. Sci. Instrum.* **55,** 952 (1984).
2. A. H. Miklich, J. J. Kingston, F. C. Wellstood, J. Clarke, M. S. Colclough, K. Char, and G. Zaharchuk, *Appl. Phys. Lett.* (submitted).
3. B. Oh, R. H. Koch, W. J. Gallagher, R. P. Robertazzi, and W. Eidelloth, *Appl. Phys. Lett.* (submitted).
4. F. C. Wellstood, J. J. Kingston, M. J. Ferrari, and J. Clarke, *Appl. Phys. Lett.* **57,** 1930 (1990).
5. M. J. Ferrari, J. J. Kingston, F. C. Wellstood, and J. Clarke, *Appl. Phys. Lett.* **58,** 1106 (1991).
6. M. J. Ferrari, M. Johnson, F. C. Wellstood, J. Clarke, D. Mitzi, P. A. Rosenthal, C. B. Eom, T. H. Geballe, A. Kapitulnik, and M. R. Beasley, *Phys. Rev. Lett.* **64,** 72 (1990).
7. M. J. Ferrari, F. C. Wellstood, J. J. Kingston, and J. Clarke, *Phys. Rev. Lett.* (submitted).
8. P. Dutta, P. Dimon, and P. M. Horn, *Phys. Rev. Lett.* **43,** 646 (1979).
9. S. Machlup, *J. Appl. Phys.* **25,** 341 (1954).
10. M. J. Ferrari, M. Johnson, F. C. Wellstood, J. Clarke, A. Inam, X. D. Wu, L. Nazar, and T. Venkatesan, *Nature* **341,** 723 (1989).
11. J. Clarke, W. M. Goubau, and M. B. Ketchen, *J. Low Temp. Phys.* **25,** 99 (1976).

Anomalous Properties of Weak-Link-Containing Superconductors: Flicker Noise

S.A. Sergeenkov

Laboratory of Neutron Physics, Joint Institute for Nuclear Research, 141980 Dubna, Russia

Abstract. Low-frequency magnetic noise spectral density is calculated within the superconductive glass model. The model predicts the existence of both white noise and flicker-like noise $\omega^{-\alpha}$ with α strongly dependent on applied magnetic field and temperature. It is shown, in particular, that α increases with field from 0.5 to 1, and that in the critical region $1/\sqrt{\omega}$-law should prevail in the noise spectrum.

1. Introduction

Probably the most intriguing, perplexing, and still unresolved problem in SQUIDology is the problem of noises (see, e.g.,[1] and references therein). As was indicated by Donaldson [2], all high-T_c RF SQUIDs suffer from excessive $1/\omega$ noise contributions at low frequencies generated by a dense network of superconducting loops interrupted by weak Josephson junctions. According to Gross et al. [3], temperature and field dependencies of the flux noise power spectrum in DC SQUIDs strongly correlate with the similar behavior of the critical current (magnetization) in weak-link-containing systems. Gabovich et al. [4] have observed a flicker noise in ceramic superconductor $BaPb_{0.75}Bi_{0.25}O_3$ in the range $20Hz \le \omega \le 500Hz$ with the noise spectral density $S_m(\omega)$ well-fitted by the law $\omega^{-\alpha}$, where α increased with applied field.

On the other hand, as was shown in our previous paper [5], the superconductive glass (SG) model enables us to describe quite satisfactory experimentally observed anomalies of critical currents and current-voltage characteristics of weak-link-containing superconductors.

In this paper we consider via the SG model the low frequency noise properties of granular superconductors due to the long-time magnetization fluctuations of the sample.

2. The SG Model

As is known [5] the Hamiltonian of the SG model in a pseudospin representation has a form

$$\mathcal{H} = -\mathrm{Re} \sum_{i,j} J_{ij} S_i^* S_j \quad , \tag{1}$$

where

$$J_{ij} = J(T)\exp(iA_{ij}), \; S_i = \exp(if_i),$$

$$A_{ij} = \pi H(x_i + x_j)(y_i - y_j)/\phi_0, \quad \phi_0 = hc/2e. \tag{2}$$

The model (1) describes the interaction between superconductive grains via Josephson junctions on a 2-D disordered lattice in a frustrating magnetic field **H** = (0,0,H).
According to the fluctuation-dissipation theorem one can calculate the enhancement of excess magnetic noise density within the SG model by the Kubo formula :

$$S_m(\omega) = \int_0^\infty dt \cos(\omega t) \overline{M(t)M(0)} \tag{3}$$

where

$$M(t) = M(H)D(t), \; M(H) = -M_0(H/H_0)(1+H^2/H_0^2)^{-3/2},$$

$$H_0 = \phi_0/2S, \; M_0 = SNJ/\phi_0. \tag{4}$$

Here $S^{1/2}$ is an average grain size. The correlator $D(t) = D_{ii}(t)$ is defined as follows :

$$D_{ij}(t) = \langle S_i(t) S_j^*(0) \rangle \tag{5}$$

The bar denotes the Gauss-like averaging over random grain coordinates (x_i, y_j) with a mean square deviation S/π [5].
In the mode-coupling approximation one obtains the self-consistent equation in correlator D(t) [5] :

$$d^2D/dt^2 + \Omega^2 D(t) + \int_0^t d\tau K(t-\tau)[dD(\tau)/d\tau] = 0 \tag{6}$$

The Laplace transform of the memory kernel K(t) in a pair

spin approximation has the form [5]:

$$K_{ij}(z)=\Omega^2 f(T,H)D_{ij}(z) \ , \qquad (7)$$

where

$$f(T,H)=J^2(T,H)/T^2, J(T,H)=J(T)/(1+H^2/H_o^2)^{1/2}.$$

Here $\Omega = 2e^2RT/h^2N$ is a characteristic frequency of the system (1), N is the number of superconducting grains.

Let us consider the solution of the system (6)-(7) for equilibrium relaxation in the form :

$$D(t) = L + (1 - L)\phi(t) \ . \qquad (8)$$

Here we have introduced the nonergodicity parameter L(T,H) which is defined via correlator D(t) as [5] :

$$L(T,H) = \lim_{t\to\infty} D(t) \ . \qquad (9)$$

Taking into account eq.(9) from eqs.(6) and (7) one gets the equation for L :

$$L^3 - L^2 + L - (1 - T^2/T_c^2(H)) = 0 \ . \qquad (10)$$

Using as Ansatz a power-like relaxation law :

$$\phi(t) = (1 + t/\tau)^{-n} \ , \qquad (11)$$

for the exponent n(T,H) one obtains near $T_c(H)$:

$$n = 1/2 + L/(1 - L) \qquad (12)$$

3. Flicker Noise in the SG Model

To obtain the magnetic noise density owing to the magnetization fluctuations within the SG model we use the power-law Ansatz (11). Taking into account eqs.(4),(8),(11),(12) from eq.(3) one gets finally:

$$S_m(\omega) = S_0(H)[L\delta(\omega)+(1-L)\tau\Gamma(\alpha)(\omega\tau)^{-\alpha}] \ , \qquad (13)$$

where

$$\alpha(T,H) = 1-n(T,H), \ S_0(H) = M^2(H). \qquad (14)$$

Thus, according to eq.(13), the magnetic fluctuations due

to redistribution of the magnetic flux between Josephson network loops lead to emergence of both the white noise ~ $\delta(\omega)$ and the flicker noise ~ $\omega^{-\alpha}$. Near $T_c(H)$, as it follows from eq.(10), the order parameter $L \sim 1-[T/T_c(H)]^2$. In turn, the phase boundary $T_c(H)$ is determined by the equation $f(T_c,H)=1$ (see eq.(7)), so that near $T_c(H)$ the white noise contribution is strongly depressed in comparison with the flicker noise one. Moreover, in view of eqs.(12) and (14), the power exponent α will increase with field H from $\alpha=0.5$ (at low fields $H/H_0 \ll 1$) to $\alpha=1$ (at higher fields $H/H_0 \geq 1$, when the SG state is established throughout the sample).It is important to stress that $\alpha(T_c,H)=0.5$, i.e. in the critical region $1/\sqrt{\omega}$ noise should prevail in the fluctuation spectrum.

References

1. M.J. Ferrari, Proceedings of this Conference.
2. G.B. Donaldson, Cryogenics **28**, 668 (1988).
3. R. Gross, P. Chaudhari, M. Kawasaki, M. Ketchen, and A. Gupta, Physica C **170**, 315 (1990).
4. A.M. Gabovich, D.P. Moiseev, V.M. Postnikov, V.A. Kulikov, and L.V. Matveets, Physica B **165&166**, 1165 (1990).
5. S.A. Sergeenkov, Physica C **167**, 339 (1990).

Exotic Manifestations of Intrinsic High-T_c Interferometer Loops

G. Jung, J. Konopka, and B. Solek*

Instytut Fizyki PAN, PL-02-668 Warszawa, Poland
*Also at: Department of Electrical Engineering, University of Duisburg, Fed. Rep. of Germany

Abstract: Experimental evidence for intrinsic quantum interferometer loops in YBaCuO thin films was obtained from the microwave emission, magnetoresistance quantization, and low frequency noise measurements. The average loop size distribution peak is around 10 μm^2.

1. Introduction

Intrinsic Josephson junctions, frequently arranged into spontaneous quantum interferometer loops, strongly influence the properties of high-Tc superconductors (HTS). In this paper we present examples of the "exotic" manifestations of intrinsic interferometers in YBaCuO thin films. In particular, we determine the structure and size of intrinsic loops from the microwave emission properties and from magnetoresistance quantization. Finally, magnetic field dependent random telegraph signals can be used for the independent evaluation of intrinsic loop geometry.

2. Microwave emission

Microwave power emitted from HTS films demonstrates sharp excess "lines" when recorded as a function of bias current. We have ascribed such excess CSL (current stable line) emissions to the synchronous radiation of intrinsic Josephson junctions, spontaneously arranged into clusters of multiple parallel connected interferometers and synchronized by activated flux jumps across the entire cluster [1]. Intrinsic interferometer loop dimensions can be thus estimated from the periodicity of the power vs. magnetic field patterns. In order to determine the actual pattern periodicities Fourier power spectra of the recorded data were calculated using standard FFT procedures, enhanced by the Hanning numerical filter which eliminated spurious ripples due to the data windowing.

Fourier spectrum of the data recorded at a current bias set for a pronounced CSL emission shows multiple peaks in the low and high field range, while the spectrum corresponding to an arbitrary bias, no CSL line, is completely smooth (see Fig.1). The peaks in the low

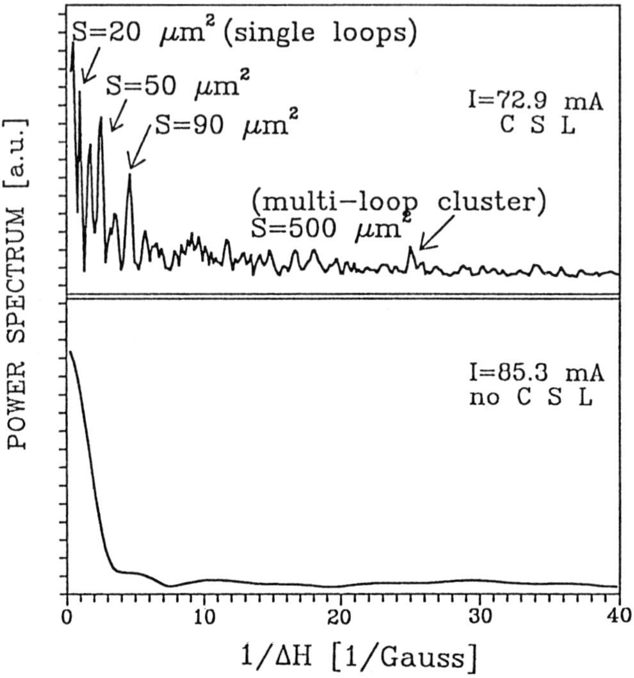

Figure 1. Fourier power spectra of the microwave power vs. magnetic field characteristics recorded at 4.2 K.

$1/\Delta H$ range can be attributed to flux quantization in an elementary single loop and those at large $1/\Delta H$ to the change of the quantum state of the entire radiating cluster. The typical single interferometer size ranges from 15 to 100 μm^2 and that of an entire multi-loop cluster from 500 to 2500 μm^2. However, rich multiple periodicities in the Fourier power spectrum suggest that the radiating cluster has more complicated structure than initially assumed. In particular, it may comprise serial-parallel interferometers [2] arranged in the ladder or chess board like networks.

3. Magnetoresistance of the radiating films

Spectral features and high levels of the detected emissions indicate that intrinsic junctions radiate synchronously. The locking mechanism relies on uniform flow of flux lines across the junction cluster, and thus the resistance of the film biased at the CSL position is dominated by the flux-flow. The change of a single loop quantum state, say from n to n+1, results in the change of number of flux quanta contained in a moving flux line. This causes the in-

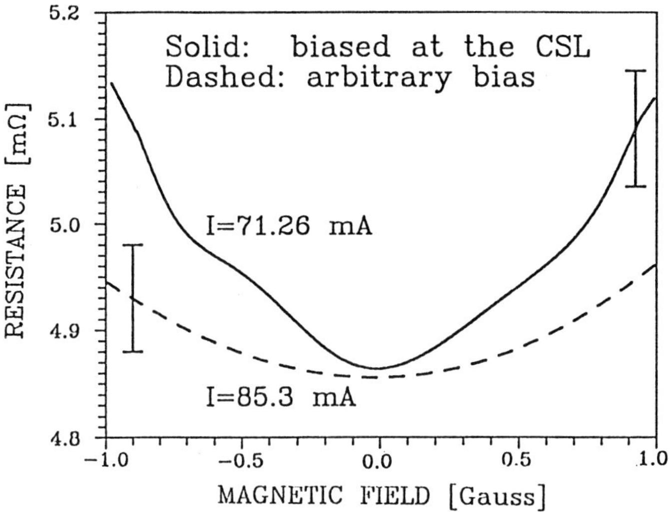

Figure 2. Magnetoresistance of the radiating (solid line) and non-radiating (dashed line) YBaCuO thin film.

crease in the slope of flux-flow resistivity ρ_f vs. magnetic field:

$$\Delta(d\rho_f/dB)=(n+1)\Phi_0^2/\eta c^2-n\Phi_0^2/\eta c^2=\Phi_0^2/\eta c^2.$$

Therefore, the change of the quantum state of an elementary interferometer leads to the quantization in the slope of sample resistance vs. magnetic field characteristics [3]. An example of such behaviour is shown in Fig.2. The plot recorded for the bias current corresponding to a CSL demonstrates clear quantum effects. The single loop area estimated from the periodicity of the slope variations is consistent with those obtained from the r.f power vs. magnetic field patterns. Resistivity of the film biased outside CSL (dashed line in Fig.2) demonstrates no quantum effects.

4. Random telegraph noise signals (RTS)

Single two-level fluctuators (TLF) producing RTS noise are seen as source of the 1/f noise in solids [4]. We have observed the RTS voltage in current biased YBaCuO films which can be attributed either to the charge traps in the barrier of intrinsic junction or to the flux jumps between two pinning centers. In the latter case the flux noise is converted into the voltage noise by self-detection within an intrinsic dc SQUID loop.

The life times of the TLF two states are exponentially distributed and depend strongly on the bias current, see Fig.3. We can

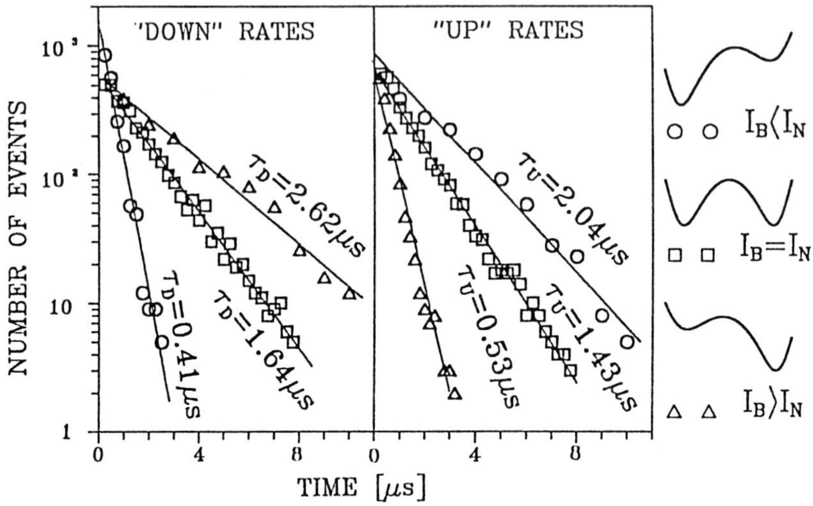

Figure 3. Distribution of the life times in the RTS of the low frequency voltage noise observed at 4.2 K and proposed variations of the energy structure of the TLF with changing bias current.

attribute changes in the energy structure of the TLF to the influence of the Lorentz force. In Fig.3 I_N denotes the bias current causing a symmetrical RTS. Small deviations ΔI from I_N result in symmetrical changes of the RTS life times. Assuming that the observed switching is thermally activated and separating the TLF barrier energy into current independent part E_0 and $\Delta E(I)$ we have [5]:

$$\Delta E(I) = \frac{kT}{2} \ln(\tau_D/\tau_U) \ .$$

Expanding $\Delta E(I)$ in series and linearizing around E_0, and assuming that energy change is due to a flux jump $\Delta E(I) = (\Delta \Phi)^2/2L$, we can evaluate the effective flux change from the measured data [5]:

$$\frac{\partial \Delta E(I)}{\partial I} = \Delta \Phi = \frac{kT}{2} \ln(\tau_d/\tau_u) \frac{1}{\Delta I} = (7.2 \pm 0.8) \mathrm{x} 10^{-4} \ \Phi_0 \ .$$

The loop size can be now estimated from the transfer function of a dc SQUID, $\Delta\Phi/\Delta V = 2L/R$. The transfer function of an intrinsic SQUID is obviously not known, but in the first approximation we can substitute for the shunt resistance R the doubled differential resistance of the film measured at I_N. The intrinsic loop can be treated as a flat. Then if A is linear dimension of a loop:

$$L = 1.25 \mu_0 A = \frac{\Delta V}{\Delta \Phi} \frac{1}{2R} = 1.57 \pm 0.6 \ \mathrm{pH}.$$

The loop area estimated from this relation is of the order of 1 μm^2, i.e., an order of magnitude smaller than obtained from the microwave emissions and magnetoresistance quantization. This can be explained either by assuming that TLF loop is smaller than the loops responsible for the microwave emissions, or by non optimal β parameters of the intrinsic SQUID. In the latter case the transfer function is lower than R/2L and the loop size must be estimated higher.

References

1. G.Jung, J.Konopka and S.Vitale, J.Appl.Phys. 68, 3029, 1990
2. S.J.Lewandowski, Phys.Rev.B43, 7776, 1991
3. G.Jung, J.Konopka, and B.Solek, Physica C, in print
4. M.B.Weissman, Rev.Mod.Phys. 53 (1988) 537
5. G.Jung, M.Bonaldi, S.Vitale, and J.Konopka, Physica C, in print

Percolation and Josephson Effects in High-T_c Polycrystalline Thin Films

Yu.Ya. Divin and V.N. Gubankov

Institute of Radioengineering & Electronics of the Academy of Sciences, Marx Avenue 18, 103907 Moscow, Russia

Abstract. The electrical properties and Josephson behaviour of high-T_c polycrystalline thin films are shown to be consistent with the model of a 2D array of intergrain Josephson junctions with an exponentially wide distribution of parameters. In these films the electrical currents flow through the critical subnetwork, with characteristic percolation length L being much larger than the grain size a. Single-Josephson-junction behavior is realized in polycrystalline high-T_c bridges with the dimensions w,l in the limits $\xi << a < w,l < L$, and high responsive radiation detectors and SQUID's based on these bridges are demonstrated.

1. Introduction

Polycrystalline high-T_c superconductors are of particular interest both for the physics of inhomogeneous superconductors and for application in electronics. In contrast to conventional superconductors the high-T_c materials can be considered as an array of weakly coupled, strongly superconducting grains [1,2]. A number of electromagnetic properties of bulk high-T_c superconductors was discussed within the simplified version of this model, considering identical intergrain Josephson-junctions and grains [1,2].

But before the high-T_c discovery a model of a Josephson array with exponentially wide distribution of parameters was suggested to analyze critical behaviour of bulk inhomogeneous superconductors [3]. In this paper we use this approach to describe the large-scale percolation in high-T_c thin films at different temperatures, illustrate this percolation by a direct probing experiment and show the applicability of this approach to realize Josephson devices with high responsivity to electromagnetic fields.

2. The Model

The model in question is a regular array of grains with the same values of conductance G_g and size a, connected by intergrain junctions with extremely different conductances $G_i = G_o \exp \zeta_i$, where ζ_i is a random variable ($0 \leq \zeta_i \leq \zeta_o$, $\zeta_o >> 1$), and $G_g >> G_i$. The qualitative picture of electrical transport in this array is presented in Fig. 1, where the critical subnetwork relevant for conduction is shown by solid lines [4].

Springer Proceedings in Physics, Vol. 64 **Superconducting Devices and Their Applications**
Editors: H. Koch, H. Lübbig

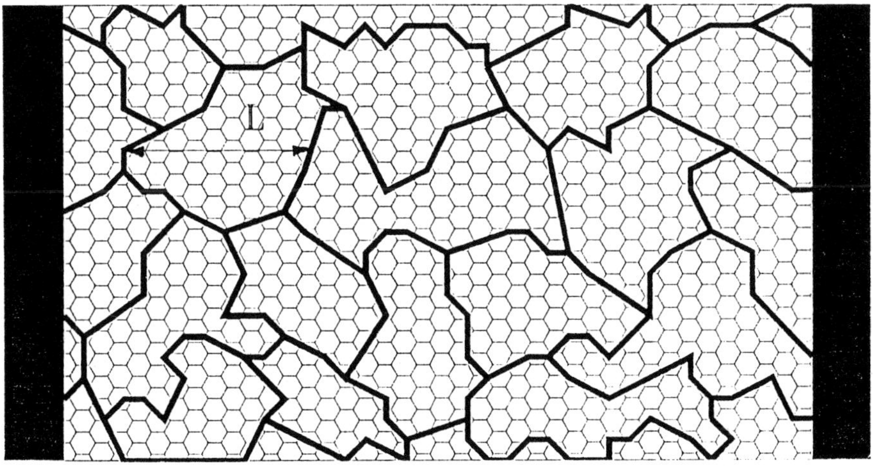

Fig. 1. Schematic presentation of critical subnetwork (solid lines) in the array of intergrain Josephson junctions with exponentially wide distribution of parameters.

The smallest conductance in each current path between the nodes of the subnetwork is of the order of the critical value $G_c = G_0 \exp \zeta_c$, which is defined so that the fraction of intergrain conductances with $G_i \geq G_c$ is equal to the threshold probability x_c for bond percolation on this array [4].

The characteristic size of critical subnetwork or the percolation length is about $L \simeq a\,\zeta_c^{1.3}$. The conductance $G_\square$ of the array with $w = l >> L$ will be $G_\square \simeq G_c = G_0 \exp \zeta_c$. When the dimensions w,l of the array are decreased so that $a < w,l < L$, the exponential decrease in conductance is expected due to the change of percolation threshold ζ_c [4], and single-junction behaviour should be expected for the array containing large number of junctions.

Considering the intergrain junctions as tunnel junctions with critical currents $I_{Ji}(T) = [\pi\Delta(T)/2e] \times G_{ni} \times \mathrm{th}[\Delta(T)/2kT]$ and using the above-mentioned considerations, the behaviour of the array can be discussed at different temperatures. The results are summarized in Table 1.

Table 1. Summary of parameters of a 2D array, consisting of extremely different Josephson tunnel junctions.

Temperature range	Parameter of junction	Percolation length L	Parameter of array (w,l > > L)
$T>T_c$	$G_{ni} = G_{no}\exp \zeta_i$	$L_n \simeq a\zeta^{1.3}$	$G_{n\square} \simeq G_{no}\exp \zeta_c$
$T<<T_c$	$I_{Ji} = I_{Jo}\exp \zeta_i$	$L_s \simeq a\zeta^{1.3}$	$j_c \simeq (I_{Jo}/L_s a) \exp \zeta_c$
$T\lesssim T_c$	$G_i = G_o\exp \gamma_i(T)$	$L_{fl} \simeq a\gamma^{1.3}$	$G_\square \simeq G_o\exp \gamma_c(T)$
	$\gamma_i(T) = \hbar I_{Ji}(T)/ekT$	$L_{fl} >> L_s \simeq L_n$	

3. Laser Probing of Current Distribution in High-Tc Films

The main difference between the model discussed in the previous chapter and the model with identical intergrain junctions [2] is in the characteristic scale of current distributions. In a homogeneous array this scale is equal to the grain size a, but in our model the percolation length L is much larger than the grain size, especially at the fluctuation region $T \lesssim T_c$, where $\hbar I_{Ji}/ekT > 1$ (see Table 1).

To study the spatial current distribution in high-T_c polycristalline thin films we used the focused laser beam for local heating of the film and the resulting change of voltage ΔV on the film was measured as a function of the beam position [5,6]. The laser-probe data for polycrystalline $YBa_2Cu_3O_{7-x}$ films ($a \approx 1$ μm) at $T < T_c$ are shown in Fig. 2. The characteristic percolation length is about 100 μm in this case. In normal and superconducting states the percolation length L was shown to be about 15 μm for these films [5].

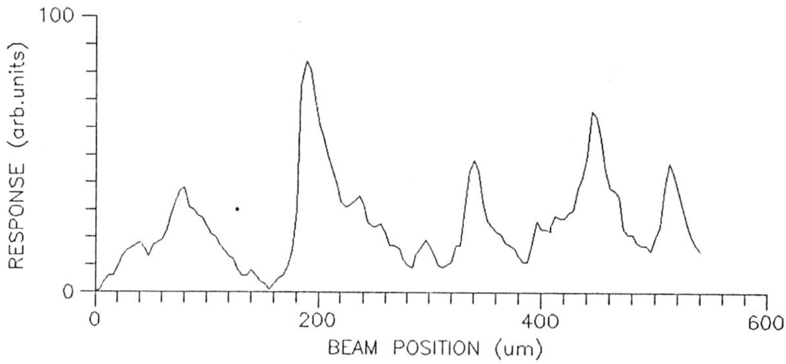

Fig. 2. Response ΔV of polycrystalline $YBa_2Cu_3O_{7-x}$ thin film to the focused He-Ne laser beam, as the beam was scanned across the width of the film. T = 84 K.

4. Josephson Behavior of High-T_c Thin Film Bridges

The decrease of the dimension w,l of high-T_c thin films below percolation length L results in very pronounced size effects in their electrical and Josephson behavior [6,7].

Thin-film bridges show a single-Josephson-junction behaviour when their dimensions are in the limits $a << w$, $l < L$ and they contain a large amount of intergrain Josephson junctions [7]. The response of one of such $YBa_2Cu_3O_{7-x}$ bridge with $w \times l = 12 \times 9$ μm^2 to millimeter-wave radiation with power $\lesssim 10^{-10}$ W is shown in Fig. 3, where selective response is clearly seen near $V = \hbar f/2e$. The voltage dependence of the response is consistent with the prediction of the RSJ model. The bridges with $a < w,l < L$ are very sensitive to radiation both in wide-band (NEP $< 10^{-12}$ $W/Hz^{1/2}$) and selective (NEP $< 10^{-11}$ $W/Hz^{1/2}$) regimes.

Also macroscopic quantum interference in SQUIDs using polycrystalline $YBa_2Cu_3O_{7-x}$ thin-film bridges was only observed when bridge dimensions were

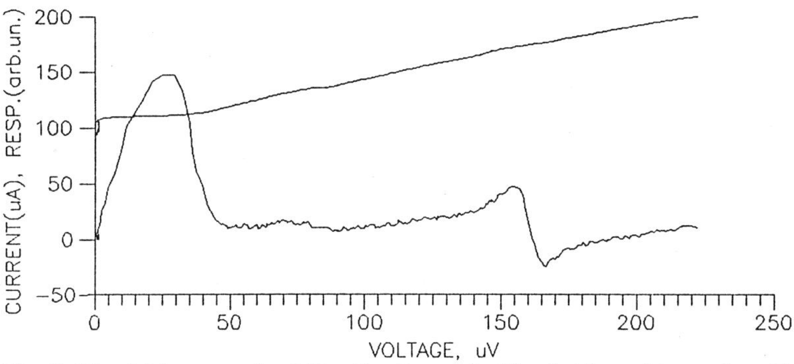

Fig. 3. The I-V curve of a $YBa_2Cu_3O_{7-x}$ thin film bridge with $w \times l = 12 \times 9\ \mu m^2 < L \simeq 15 \times 15\ \mu m^2$ and its response $\Delta V(V)$ to electromagnetic radiation with f = 78 GHz. T = 24 K.

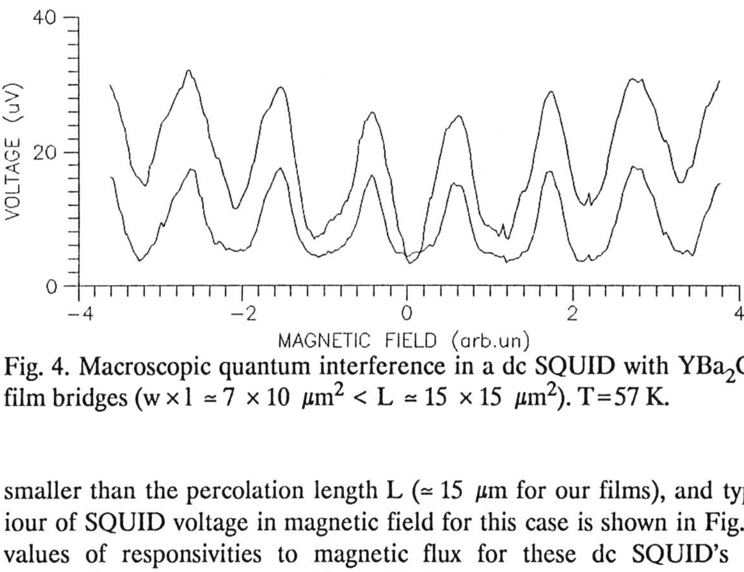

Fig. 4. Macroscopic quantum interference in a dc SQUID with $YBa_2Cu_3O_{7-x}$ thin film bridges ($w \times l \simeq 7 \times 10\ \mu m^2 < L \simeq 15 \times 15\ \mu m^2$). T=57 K.

smaller than the percolation length L ($\simeq$ 15 μm for our films), and typical behaviour of SQUID voltage in magnetic field for this case is shown in Fig. 4. The best values of responsivities to magnetic flux for these dc SQUID's were about 100 $\mu V/\Phi_0$ at 60 K and 10 $\mu V/\Phi_0$ at 77 K.

Acknowledgement

Discussions and assistance provided by I.M. Kotelyanskii, A.S. Afanasyev, A.V. Andreev and P.M. Shadrin are gratefully acknowledged.

References

[1] J.R. Clem, Physica C, **153-155**, 50 (1988)
[2] M. Trinkham, C.J. Lobb, Solid State Physics, **42**, 91 (1989)

[3] L.B. Ioffe, A.I. Larkin, ZhETF, **81**, 707 (1981) (in Russian)
[4] B.I. Shklovskii, A.L. Efros, Electronic properties of doped semiconductors. M., Nauka, 1979 (in Russian)
[5] Yu.Ya. Divin, F.Ya. Nad', V.Ya. Pokrovski, P.M. Shadrin, IEEE Trans. on Magnetics, **27**, N2 (1991).
[6] A.S. Afanasyev, V.N. Gubankov, Yu.Ya. Divin, P.M. Shadrin, JETP Letters, **49**, 776 (1989).
[7] V.N. Gubankov, Yu.Ya. Divin, I.M. Kotelyanskii, V.B. Kravchenko, IEEE Trans. on Magnetics, **27**, N2 (1991).

Observation of Magneto-Thermoelectric Effects in High-T_c Superconducting $Tl_2Ba_2CaCu_2O_x$ Single Crystals

G. Yu. Logvenov, A.E. Koshelev, V.A. Larkin, V.V. Ryazanov, and K.Ya. Soifer

Institute of Solid State Physics, Academy of Science,
142432 Chernogolovka, Moscow District, Russia

Abstract. Anomalous peaks on the temperature dependences of the magnetic moment and the transverse Nernst voltage were observed in $Tl_2Ba_2CaCu_2O_x$ single crystals, when the perpendicularly directed temperature gradient and transport current were applied. The results obtained are interpreted within the context of fluctuation vortex dynamics near T_c.

1. Introduction

The layered anisotropic structure is an important feature responsible for unusual properties of High T_c superconductors (HTS). Quasi-two-dimensionality of the new superconductors is exhibited by the fluctuation effects near the superconduction transition temperature T_c. The characteristic resistive behavior was observed in Ref. [1] in accordance with the model of the free vortex-antivortex pair dynamics at a temperature higher than the one of the Berezinskii-Kosterlitz-Thoules (BKT) topological phase transition [2].

In this paper we report the observation of new thermoelectric effects in HTS single crystals for the case when the perpendicularly directed temperature gradient and transport current are applied in the crystal base plane in the absence of an external magnetic field.

2. Experimental

The investigations were carried out with single crystals $Tl_2Ba_2CaCu_2O_x$ with T_{c0} in the vicinity of $T_c = 105K$. For the measurements of the sample magnetic moment M as a function of the temperature a non-commercial $SQUID$-magnetometer was

Springer Proceedings in Physics, Vol. 64 Superconducting Devices and Their Applications
Editors: H. Koch, H. Lübbig

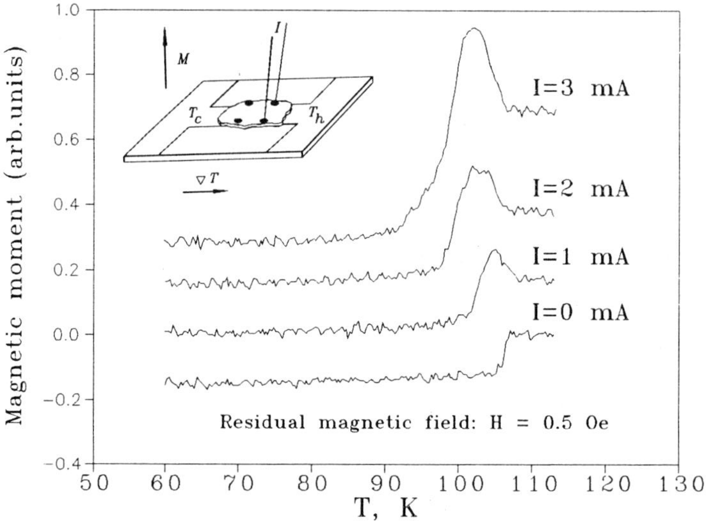

Figure 1: The temperature dependence of the generated magnetic moment $M(T)$ for different currents ($\nabla T \approx 1K/cm$)

used . The measurements were performed at the residual magnetic field value $0.5Oe$. In addition the dc transverse component of the thermoelectric voltage was measured. The experimental setups are shown in the inserts of the Figs. The experimental dependence $M(T)$ plotted in Fig.1. The lowest curve shows the superconducting magnetic transition in the residual field without any transport current. The switching on of the transverse current yields an unusual peak of magnetic moment near T_c that increases with the current. The sign effect of the peak of magnetic moment is determined by the sign of the current. A similar thermoelectric effect takes place for the resistive behavior as can be seen from the temperature dependence of the transverse voltage to the temperature gradient (see Fig.2). The experiments were performed with a single-phase sample at various temperature gradients and transverse currents (the values of ΔT shown in Fig.2 correspond to the temperature difference between the holder positions where the temperature was measured). The following behavior of the voltage peaks observed should be noted. 1) The peak arises in the superconducting state in the vicinity of T_{c0}. 2) The peak decreases with the decreasing of the current and/or the temperature gradient. 3) The peak

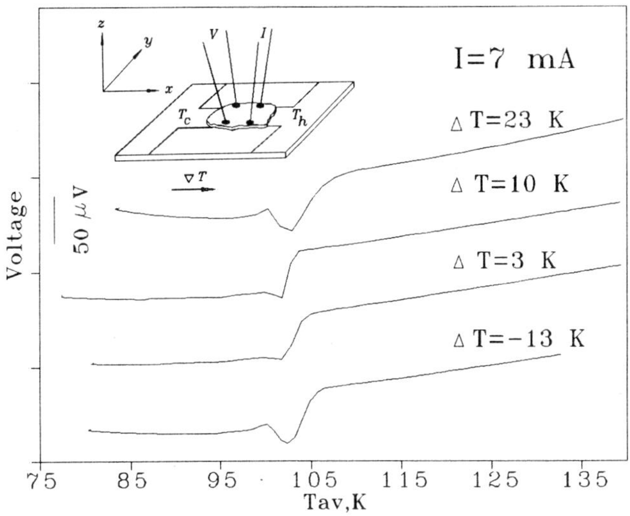

Figure 2: The dependence of the transverse voltage from the average sample temperature for different ΔT and the same current $I = 7mA$

vanishes at zero current or at zero temperature gradient. 4) The peak inversion without drastic change in its form takes place when the current direction is inverted. 5) The sign of the effect is the same when the temperature gradient is inverted . 6) An external magnetic field along the c-axis ($20 - 200Oe$) suppresses the effect remarkably independently of the sign.

As it was proposed in Ref.[4] and was firstly observed on In/InO_x-film the transport current induces a vortex-antivortex imbalance in the presence of a transverse temperature gradient and creates the magnetic induction B. By means of a simple model (see Ref.[3]) we can obtain a relation between the peaks observed and a resistive superconducting transition broadening. The expression for the average induction B due to the vortex plasma imbalance in quasi-2D-superconductors in the presence of the mutually perpendicular temperature gradient and transport current is given by

$$B = \frac{1}{12}\frac{4\pi}{\mathrm{c}}\frac{\partial \ln \varrho_{\mathrm{f}}}{\partial T}[I \cdot \nabla \mathrm{T}]\frac{\mathrm{L}}{\mathrm{d}} \quad . \tag{1}$$

Here c is the light velocity, ϱ_{f} is the flux flow resistivity, I is the value of the transport current , ∇T is the temperature gradient, L is the sample dimension along the temperature gradient and d is the its thickness The voltage peak arises at a temperature higher then T_{BKT} when the average magnetic induction B arises in the sample. The peak is related with an additional contribution of the transverse voltage

$$V_N = -\frac{1}{3}\frac{\pi}{c^2}\frac{\partial \ln \varrho_f}{\partial T}\frac{S_\varphi}{\eta}(\nabla \mathrm{T})^2 \mathrm{I}\frac{\mathrm{LW}}{\mathrm{d}} \quad . \tag{2}$$

Where S_φ is the transport entropy, η is the vortex viscosity, W is the dimension along the current direction. The proposed model [3] explains all experimental results.

References

[1] D. H. Kim, A. M. Goldman, J. H. Kang, and R. T. Kampwirth, Phys. Rev. **B 40** (1989) 8834.

[2] J. M. Kosterlitz and D. J. Thouless, J. Phys. **C 6** (1973) 1181.

[3] A. E. Koshelev, G. Yu. Logvenov, V. A. Larkin, V. V. Ryazanov, K. Ya. Soifer, Phsica C.**177** (1991) 129.

[4] J. C. Garland and D. J. Van Harlingen, Phys. Rev. Lett. **55** (1985) 2047.
H. J. Lee, D. A. Rudman, and J. C. Garland, Phys. Rev. Lett **55** (1985) 2051.

Enhancement of the Superconducting Transition Temperature of Y-Ba-Cu-O Thin Films by Silver Over-Layer Deposition

S.S. Tinchev and A. Baranyak

Forschungsgesellschaft für Informationstechnik, Postfach 1147,
W-3202 Bad Salzdetfurth, Fed. Rep. of Germany

Abstract.Deposition of a silver over-layer on the laser ablated Y-Ba-Cu-O thin films was found to enhance the superconducting transition temperature and to decrease the superconducting transition width. The Tc of the Y-Ba-Cu-O/Ag is increased over that of Y-Ba-Cu-O films typically by 0.5-1.5 K and the transition width is usually twice smaller. The importance of the HTc film surface properties is pointed out.

Shortly after the discovery of Y-Ba-Cu-O superconductors, the effect of silver was investigated widely. First Ag doping was found to have positive effect on the ceramics, reducing its porosity and increasing the critical current density. Silver is now very often used for making low-resistance contacts to Y-Ba-Cu-O and other HTc superconductors. A much less moisture-induced degradation was also observed in bulk samples containing silver, or in layered structures with silver. Therefore a knowledge of the behaviour of silver on the surface of Y-Ba-Cu-O is necessary.

In our earlier work [1] we reported on our preliminary observation that deposition of silver over-layer without annealing caused an enhancement of the superconducting transition temperature of Y-Ba-Cu-O laser ablated thin films. We used mainly an eddy-current technique for this investigation although the observation was also confirmed by the four-point DC resistance measurement. In this letter we present more details for the observed phenomenon. The experimental data were obtained by means of the DC four-point technique with different contact positions.

In our first experiments we deposited silver over-layers by thermal evaporation at room temperature typically 300 nm thick on the whole Y-Ba-Cu-O surface. Then all four contacts were placed as usually over the silver layer (Fig.1a) by means of a silver paint. A typical result is shown in Fig. 2. For comparison there is also shown the R(T) dependence for the same film before silver deposition. One can note enhancement of the critical temperature as well as a reduction of the transition width by a factor of two. Below Tc one can see the remainder of the silver resistance. The typical enhancement of Tc because of silver deposition was rather small (about 0.5-1.5 K). To exclude any possible inaccuracies, for example owing to the temperature measurement of the sample, we carried out two types of other experiments using different locations of the contacts on the sample surface.

In both cases the silver over-layer was deposited only over a part of the Y-Ba-Cu-O film (Fig. 1b, c). On the remained silver-free surface of the Y-Ba-Cu-O

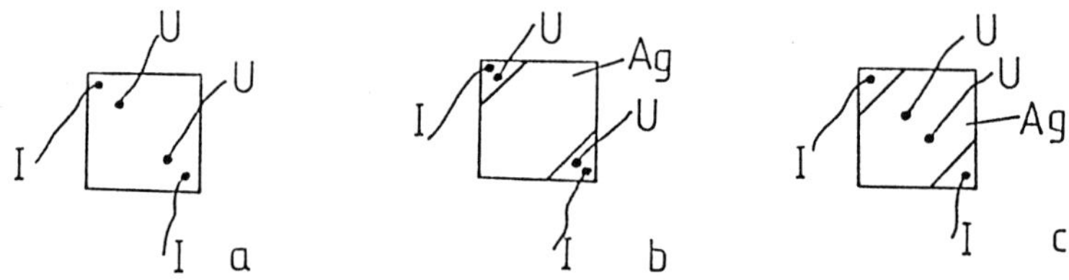

Fig.1: Different positions of the four contacts for the DC resistance measurement: whole YBa2Cu3O7 film surface is covered by silver (a), the four contacts placed on the silver free YBa2Cu3O7 (b) and only voltage contacts on the silver (c).

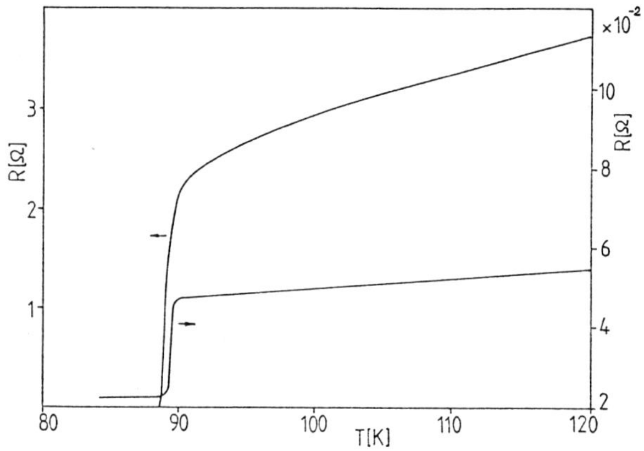

Fig.2: Superconducting transition of the YBa2Cu3O7 thin film before (left curve) and after deposition of 300 nm silver over-layer (right curve) recorded by four-point DC resistance measurement.

film, different contacts for the four-point resistance measurement have been placed. In the first measurements all four contacts were placed there (Fig. 1b). The recorded curve is shown on Fig. 3b. One can see that the superconducting transition consists of two parts having different slope. This can be very well recognized from the derivative dR/dT depicted also in Fig. 3b, where two well separated peaks exist. The first one coincides exactly with the peak before Ag deposition as one can see in Fig.3a. The second peak is obviously due to the enhanced Tc of the silver covered Y-Ba-Cu-O film. To prove this suggestion in the last experiment the voltage contacts were moved over to the silver over-layer as is shown in Fig.1c. In this case only one peak of the derivative is observed exactly at the same temperature, where the second peak was placed (Fig. 3b).

Therefore there is no doubt that the transition temperature is increased because of Ag deposition, and in Fig.1b (all four-contacts placed outside Ag) we measured serially connected superconductors Y-Ba-Cu-O and Y-Ba-Cu-O/Ag with different Tc.

In conclusion, we have demonstrated that the deposition of a silver over-layer on the surface of Y-Ba-Cu-O thin films enhances Tc and reduces the

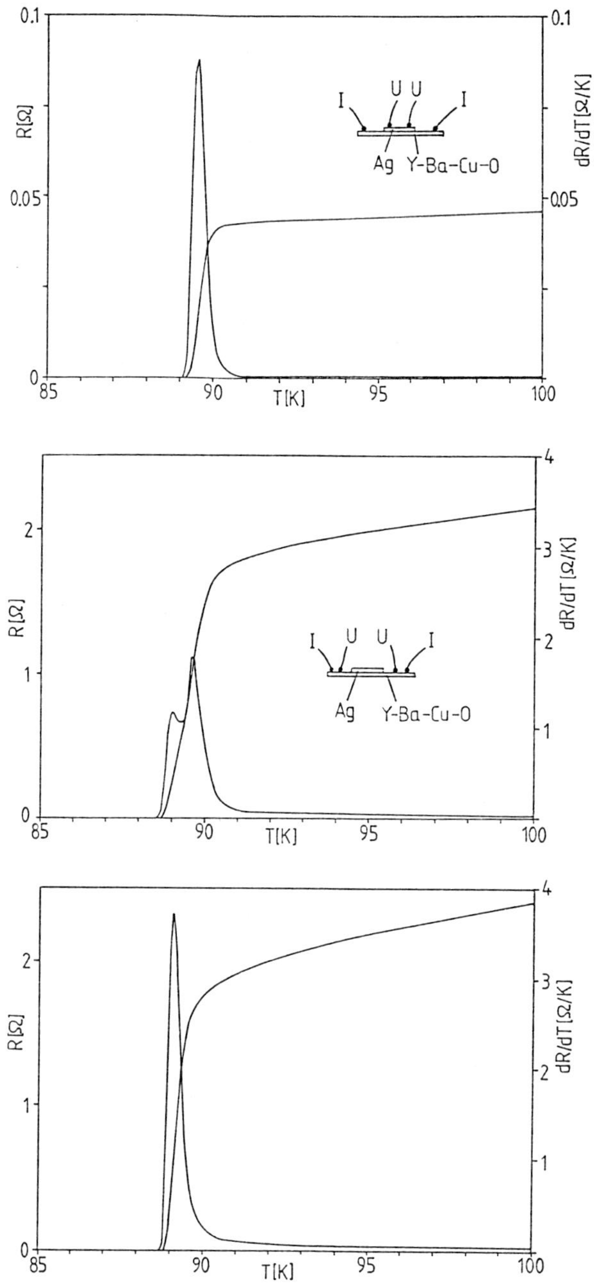

Fig.3: R(T) and (dR/dT)(T) dependence for the YBa2Cu3O7 film before silver deposition (a) and after that (b and c) for different position of the voltage contacts, see inserts.

superconductor transition width. Similar behavior was observed in [2], where the effect of an aditional silver and gold layer on the Y-Ba-Cu-O films made by electron beam evaporation was studied. An increase of Tc of 1 K over that without Ag was observed. The transition into the superconducting state is also sharper for the sample with silver layer than that without it.

In Y-Ba-Cu-O/Ag ceramic composites containing 10 or 30 wt.% Ag such a slightly higher Tc (91 K) than that for the pure Y-Ba-Cu-O sample (90.5 K) was also observed [3]. Therefore one can think that such an improvement of Tc is general effect of silver on Y-Ba-Cu-O. Our explanation of this fact [1] is that the Ag atoms saturate the Y-Ba-Cu-O dangling bonds and by this improve the order of the film surface region, where actually the supercurrent flows. This may be a possible explanation of the observed Tc enhancement of gold doped Y-Ba-Cu-O polycrystalline samples [4, 5], but other explanations can not be excluded. Another possibility is that the adsorbed oxygen on the metal surface causes full oxidation of the Y-Ba-Cu-O surface. It is well known that the Tc of Y-Ba-Cu-O becomes higher with increasing oxygen content. The existence of a non-superconducting insulating layer on the Y-Ba-Cu-O surface can also explain the observed phenomenon. From the low temperature superconductors it is known that if such an oxide layer exist on the surface of a thin superconducting film, a double layer with strong electrical field can arise. It leads to changes of the superconducting transition temperature of about a tenths of a °K [6].

The observed phenomenon can improve our understanding of the physics of the HTc superconductivity, pointing out the importance of the HTc film surface properties, and it seems that it can be interesting for some applications (for example in HTc bolometers, where the reduced transition width is useful).

Acknowledgement.Prof. J.H. Hinken is acknowledged for his continuous support.

References

1. S.S. Tinchev, Electrical and magnetic properties of the Y-Ba-Cu-O interfaces, to be published.
2. C.Chang, Effect of silver and gold on the YBaCuO superconducting thin films with the use of Ag/Cu/BaO/Y2O3 and Au/Cu/BaO/Y2O3 structures, Appl. Phys. Lett., 52, No.11 (1988), 924-926.
3. J. Jung, M.A.-K. Mohamed, S.C. Cheng, J.P. Frank, Flux motion, proximity effect, and critical current density in YBa2Cu3O7-x/silver composites, Phys. Rev.B, 42, No.10(1990),6118-6195.
4. M. Z. Cieplak, G. Xiao, C.L. Chien, J.K. Stalik and J.J. Phyne,Unexpected effect of gold on the structure, superconductivity, and normal state of YBa2Cu3O7, Appl. Phys. Lett., 57, No.9(1990), 934-936.
5. M.Z. Cieplak, G. Xiao, C.L. Chien, A. Bakhshai, D. Artymowicz, W. Bryden, J.K. Stalik and J.J. Phyne, Incorporation of gold into YBa2Cu3O7: Structure and Tc enhancement, Phys. Rev. B, 42, No.10(1990), 6200-6208.
6. W. Rühl, Die Übergangstemperatur der Supraleitung verschiedener Metalle unter dem Einfluß sehr dünner Oxidschichten, Z.Phys.,186(1965),190.

High-Temperature Ceramic RF-SQUIDs

B.V. Vasiliev

Joint Institute for Nuclear Research, Low Temperature Physics Department, 101000 Moscow, Dubna, Russia

Abstract. It has been proved that there are intergranular contacts of an order of coherent length in size in the high-T_c ceramics. This fact enables a successful development of one-contact rf-SQUIDs operating at liquid nitrogen temperature. Signal and noise characteristics of these yttrium ceramic SQUIDs are given. Some aspects of these SQUIDs application for magnetic cardiography and sensitive galvanometry are reported.

1. Intergranular contacts in high-Tc ceramics

Easy practical use of high-T_c SQUIDs mainly under field conditions, in clinics and so on, has focused the attention of a number of research laboratories on the problem of SQUID fabrication and at present the SQUIDs are developed in many laboratories. Attempts of building HTSC-SQUIDs using diverse methods were done. A successful way was found to fabricate one-contact rf-SQUIDs from high-T_c ceramics with the weak link in the intergranular contacts. The results of parameter investigations of independent high-T_c ceramic bridges determined the present concept of the HTSC-SQUIDs development [1]. The measurements were carried out on bridges made from $Y_1Ba_2Cu_3O_{7-\delta}$ ceramics which were prepared by different methods and had diverse properties.

These measurements demonstrated a very interesting fact: although the critical current of these contacts and their normal resistance were changing from one contact to another in a range of more than two orders the product of the critical current and the normal resistance, the so called characteristic voltage of the bridge, remained the same and equal approximately to $20\mu V$ with a precision of ten per cent.

The simple speculation permits us to draw a conclusion from this experiment that a constancy of the characteristic voltage of the HTSC-contact is due to the fact that between the ceramic grains bridges must exist with nearly a constant size of one crystalline lattice, i.e. of an order of coherent length. It seems a very difficult task to prepare artificially bridges of a coherent length size - for it is a very small value - from HTCS material. The fact that the nature had created the bridges with such a size between the grains makes us to prefer this way of building SQUIDs as opposed to that i.e. of using the weak links in intergranular ceramic contacts.

We prefer one-contact rf-SQUIDs, which are easier to prepare from the ceramics and which are naturally more stable mechanically and rather reliable in operation.

Thus, we can state at the moment that it is more convenient to develop and apply only the ceramic rf-SQUIDs based on the intergranular contacts. The correctness of the chosen way is proven by practical results.

2. Ceramic rf-SQUIDs

After the procedure of optimization this one-contact high-T_c rf-SQUID shows the main characteristics very close to those of the conventional low-temperature rf-SQUIDs. In figures 1 and 2 volt-current and volt-oersted characteristics of the SQUID at liquid nitrogen temperature are shown with the inductance approximately $2 \cdot 10^{-10} H$.

In order to prepare such a high-T_c SQUID (fig.3a) first there were drilled two 1 mm holes in the ceramics pellet and then a saw-cut was made with a little bridge of 50x50x50 mm size. This construction is mechanically stable, and being protected from moisture can stand several cooling/heating cycles. This SQUID is sensitive to the magnetic flux introduced into one of its holes. But high T_c two-hole SQUIDs cannot be used in magnetic measurements due to the lack of thin superconductive wire.

Fig.1

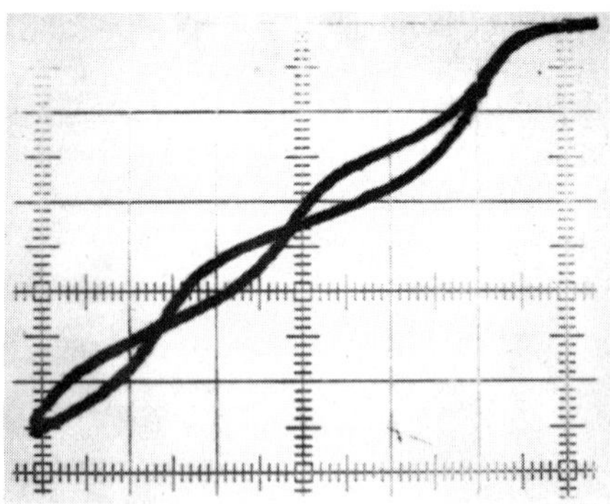

Fig.2

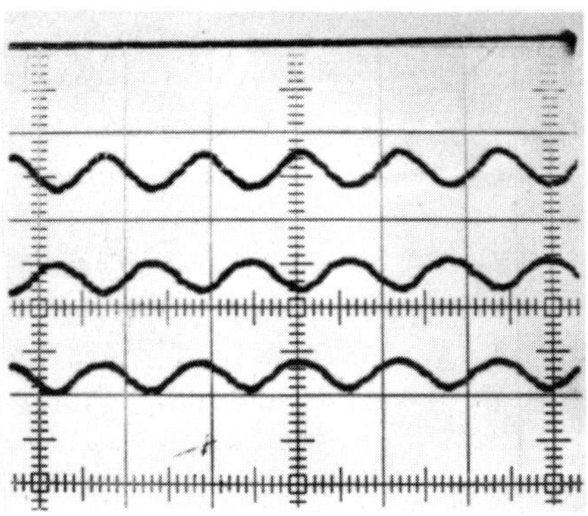

Fig.1. Volt-current characteristic of the high T_c SQUID with $Ls \simeq 2 \cdot 10^{-10} H$ at $78K$ in 1 kHz bandwidth.

Fig.2. Volt-magnetic field characteristic of the high T_c SQUID with $Ls \simeq 2 \cdot 10^{-10} H$ at $78K$ in a bandwidth of 10 kHz.

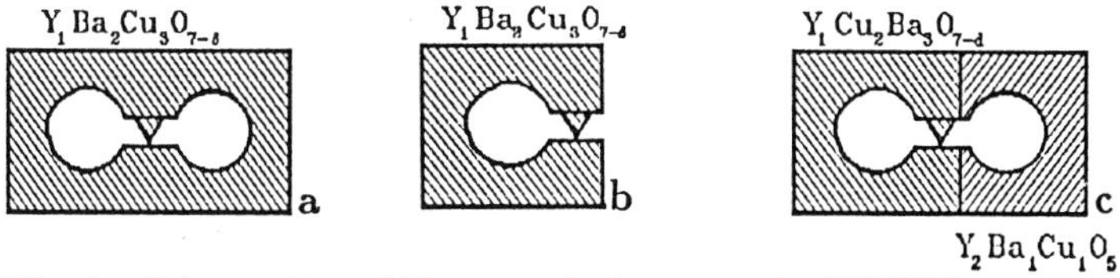

Fig.3a.Schematic of the two-hole ceramic SQUID.

Fig.3b.Schematic of the one-hole ceramic SQUID.

Fig.3c.Schematic of the one-hole ceramic SQUID with a reinforcer made of the dielectric plate.

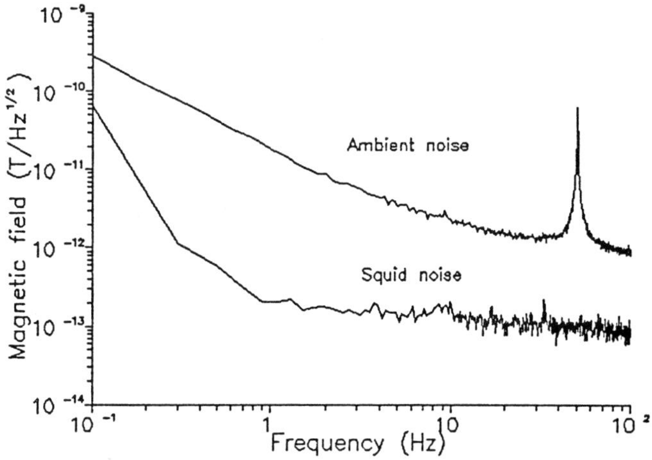

Fig.4. Upper curve is the spectral noise density of the Earth's magnetic field in Dubna measured by the one-hole high T_c SQUID. Lower curve is a one-hole high T_c SQUID self sensitivity measured like a magnetic field noise in a ferromagnetic shield.

One-hole high-T_c SQUIDs (fig.3b) can be used practically, but the problem of their mechanical stability has to be solved. One of the ways of solving the problem is sintering the two-layer pellet, one layer of which is formed from standard yttrium based superconducting ceramics and another from dielectric green phase. The procedure is the same: first the two holes were drilled in the pellet - one in the superconducting layer another in the dielectric layer. Then a cut was made connecting the two holes with the bridge in the superconducting part. This SQUID (fig.3c), single-inductive and sensitive to the external field, owes a rather good mechanical stability and reliability.

The self sensitivity of an one-inductive high T_c SQUID in the white noise region lies on the level of $10^{-13}T/Hz^{1/2}$ (fig.4). The $1/f$ type noise begins from higher frequency levels than for low- temperature SQUIDs - approximately from $1Hz$. Being so sensitive this SQUID can be used in geophysical investigations since geomagnetic field noise and the induced fields are considerably higher than its sensitivity (fig.4).

3. SQUID-Gradiometer for Magnetic Cardiogram Measurement

The possibility of magnetocardiogram measurement with a single-hole SQUID was demonstrated. But the sensitivity of such a SQUID depends on the level of the external noise (fig.4), and the quality of the cardiograms are too poor to be used in medical diagnostics. To improve the quality a technique of spatial discrimination must be used. But the usage of input flux transformers is impossible here, as well as in low-temperature magnetic cardiography, for the lack of thin high T_c superconductive low-noise wire. Good results are obtained by the spatial discrimination made by the extraction of the signals of two single-hole high T_c SQUIDs [2]. When the magnetocardiogram is being taken one SQUID is placed in the nitrogen cryostat as near as possible over the human chest and the other a little further. The baseline used in the experiments was 9 cm. The extraction of the signals was made by means of electronics in the magnetometer's output. The calibration of the system has shown that along the Z axis a balancing up to 60 dB can be attained, but along the transversal axes the balancing

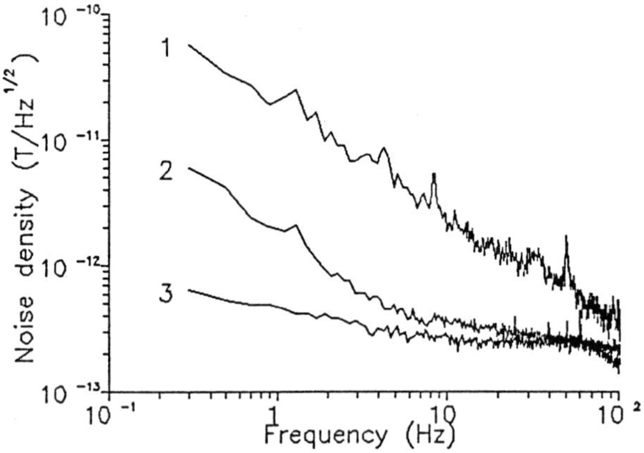

Fig.5. 1 is the spectral noise density of the Earth's magnetic field measured in Dubna by one channel of the gradiometer.
2 is the spectral noise density of the Earth's magnetic field measured in Dubna by two high T_c SQUID-gradiometers.
3 is the self sensitivity of the gradiometer (noise in a ferromagnetic shield).

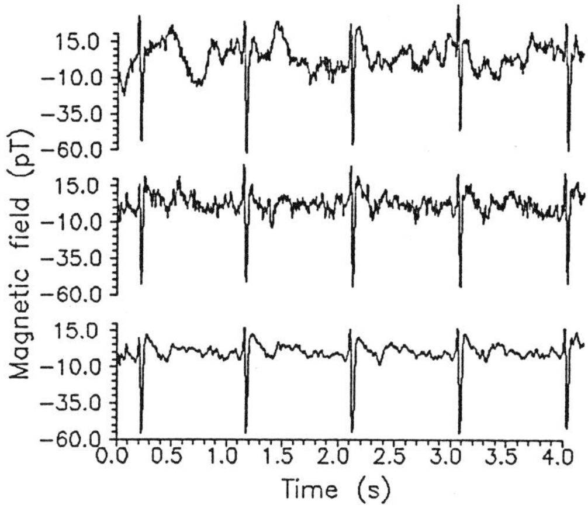

Fig.6. Human magnetocardiogram taken simultaneously:
Above - taken by one channel of the gradiometer.
In the middle - taken by the whole gradiometer without averaging in a bandwidth of $60Hz$.
Below - taken by the whole gradiometer in a bandwidth of $60Hz$ after computer averaging by four periods.

was approximately 1.5 orders worse, and the external field suppression obtained in the experiment was approximately 30 dB. The spectral noise density of this gradiometer measured under different conditions (fig.5) shows that although using this gradiometer one can suppress external homogeneous field noise by more than an order, the self noise of the gradiometer is of an order lower.

The results of the magnetocardiogram measurement by this gradiometer are shown in fig.6. Time averaging by four heart cycles (lower curve) reveals also the presence of the P- and T-peake besides QRS-peake. But the necessary level of self-noise of the gradiometer can be reached only in a shielded room. The measurements show that inside the shield with a 9 cm baseline gradiometer a sensitivity of the magnetic field gradient of up to $3 \cdot 10^{-12} T/(m \cdot Hz^{1/2})$ can be obtained, which apparently is good enough in order to obtain high-quality magnetocardiograms.

4. Galvanometer Based on a HTSC-SQUID

The lack of thin superconducting wire restricts the using of high-T_c SQUIDs as magnetometers. At the same time there are some areas of high-T_c SQUID applications were this obstacle can be overpassed.

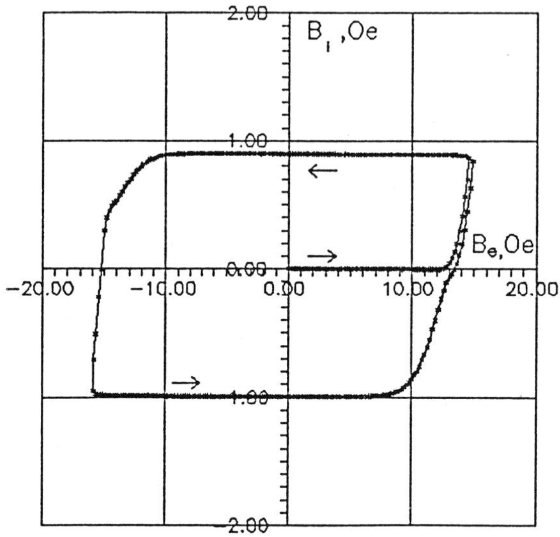

Fig.7. The dependence of the magnetic field intensity inside the HTSC-shield at liquid nitrogen temperature on the applied external field.

Using high-T_c SQUIDs in electric measurements, one can obtain their high energy resolution of up to $10^{-28} J/Hz$, by shielding them from the external magnetic fields and by coupling the measured signal by standard wire coils. Then the high-T_c SQUID can be used to measure the direct and the low-frequency alternative currents induced by the external sources.The best shape for the superconducting shield, used for the galvanometric measurements appears to be a cylinder with a bottom 150 mm high, with the inner diameter 15 mm and a 3 mm thick wall made of $Y_1Ba_2Cu_3O_{7-\delta}$ ceramics with a 10% addition of Ag. This shield screens the external magnetic field up to 12 Oersted and gives the possibility to get almost completely rid of external disturbances. To prevent the degradation of the ceramic the shield must be protected from moisture by applying to it a thin layer of varnish. Figure 7 illustrates the dependence of the field inside the HTSC-shield on the applied external field.

The measurement characteristics of the SQUID-galvanometer are mainly determined by the coupling coil with a current passing through it to the SQUID. The coil for experiments was wrapped on glass frames 0.3 mm in diameter by 0.02 mm isolated copper wire and consisted of two parts of 810 and 420 turns respectively. Its active resistance at liquid nitrogen temperature was about 16Ω, the inductance was about 90μH. The two parts were placed into different SQUID holes and were switched in series. The coefficient of transformation of the galvanometer

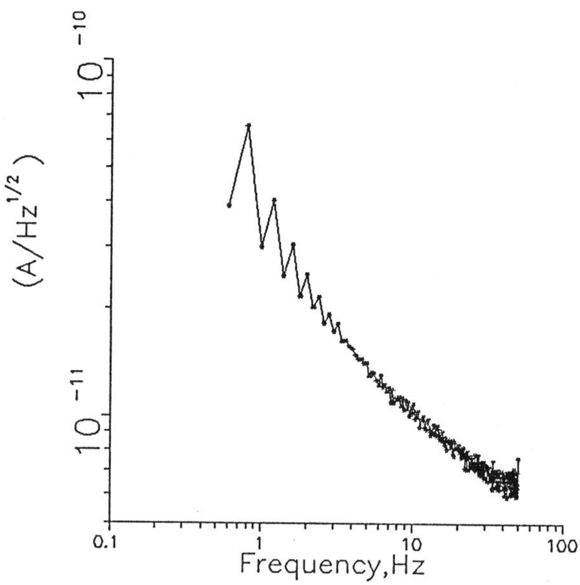

Fig.8. The sensitivity of the galvanometer based on the high-T_c SQUID operating at liquid nitrogen temperature.

30 nA/ϕ_o was obtained with the abovementioned parameters of the coil. In fig.8 the spectral noise density of the galvanometer is shown in current units. It can be seen that the spectral noise density in the white noise region in current units is about $10^{-11} A/Hz^{1/2}$. The sensitivity at lower frequency and on direct current is limited by the high $1/f$ noise. For Direct current measurement for 10 *sec* one can estimate a sensitivity on the level of 0.5 nA.

The above given data make possible the evaluation of the energy sensitivity of the galvanometer $\varepsilon_g \cong 10^{-21} J/Hz$ which gives us hope that practical application of such a galvanometer will be possible as the obtained energy resolution is several orders of magnitude better than that of "warm" electronic devices. Taking into account the high potential sensitivity of HTSC-SQUIDs, one can expect further improvement of the HTSC-galvanometer parameters.

5. Conclusions

The use of liquid nitrogen as a cooling medium is based on two main properties of it: it's not expensive and easy to operate with as it has more evaporation heat than liquid helium and a very small diffusion through the walls of the cryostat. Due to this fact high T_c SQUIDs are more practical under field conditions, in clinics and so on. Evidently, the HTSC-SQUIDs of the present state are good enough to perform a great part of magnetic cardiography investigations and in the future they may take the role of the low-temperature ones in this field.

We can say it now for sure that high T_c SQUIDs of the contemporary state-of-art may successfully solve some of the medical and current measurement problems and their application will be widespread in the near future.

References

1. Vasiliev B.V., Uchaikin S.V., Journ.of Supercond. No.3,1991.
2. Vasiliev B.V., Journ.of Supercond. No.4. 1991.

A Compact High-T_c RF-SQUID System with Reduced Tank Circuit Damping

Yi Zhang and C. Heiden

Institut für Schicht und Ionentechnik (ISI),
Forschungszentrum Jülich (KFA), Postfach 1913,
W-5170 Jülich, Fed. Rep. of Germany

One requirement for good rf-SQUID operation is a coupling between SQUID-ring and tank circuit with $k^2Q \approx 1$. For typical bulk SQUID structures of high-T_c materials, however $k^2Q < 1$ usually is found experimentally, when using a copper coil at 77 K for the tank circuit and a resonance frequency near 20 MHz. We therefore investigated the operation of high-T_c rf-SQUIDs where the damping of the tank circuit was reduced by a positive feedback circuit in order to achieve $k^2Q \geq 1$. The reduction of damping leads to noise reduction. The SQUID can be operated without external rf-oscillator solely driven by the tank circuit noise. This elimination of the oscillator leads to a simplified compact "SQUID-STICK".

1. Experimental Details

For the investigations we used two different types of SQUID structures: The first one is made of a rectangular $Bi_2Sr_2Ca_2Cu_3O_{10-\delta}$ pellet with one or two holes of 0.8 mm diameter. The weak link is a microbridge that is formed by mechanical erosion.

The critical current of the weak link can be adjusted by squeezing the pellet. For this purpose it is inserted into a special frame [1]. Dispersive ($\beta < 1$) as well as dissipative operation ($\beta > 1$) thus can be obtained for one and the same SQUID. It is found, that at 4.2 K the maximum value of the transfer function $\partial V / \partial \phi$ varies very little for different values of β ($\beta \geq 1$), whereas at 77 K it decreases with increasing β ($\beta \geq 1$). (see fig. 1)

The second SQUID structure is of the washer type, made of a thin $Y_1Ba_2Cu_3$-$O_{7-\delta}$ film deposited by laser ablation [2]. The weak link is a step edge junction in this case [3].

The tank circuit coil is made of thin (0.06 mm diam.) copper wire wound a) in form of a thin solenoid that fits into a hole of the bulk SQUID, or b) in pancake shape laid over the thin film SQUID.

Different circuits to reduce the damping of the tank circuit were tested. A simple version is shown in fig. 2. It allows one to choose different Q-values by adjusting its dc-working point. Stable operation with Q-values up to 200 were possible without danger of wild oscillations.

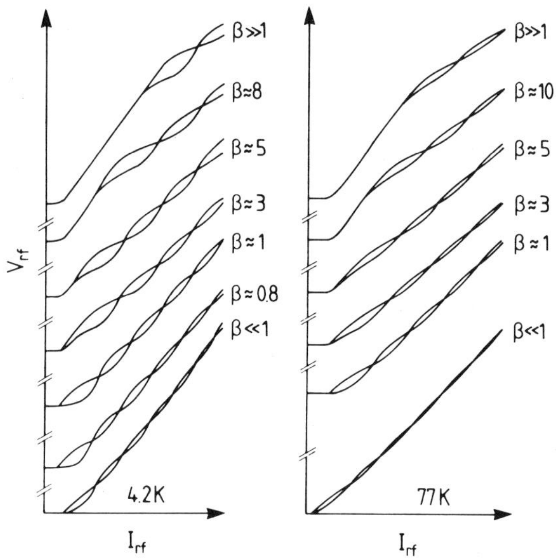

Fig. 1 Tank circuit rf-voltage V_{rf} vs driving rf-current I_{rf} for one and the same bulk rf-SQUID, obtained at 4.2 K and 77 K for various β-values.

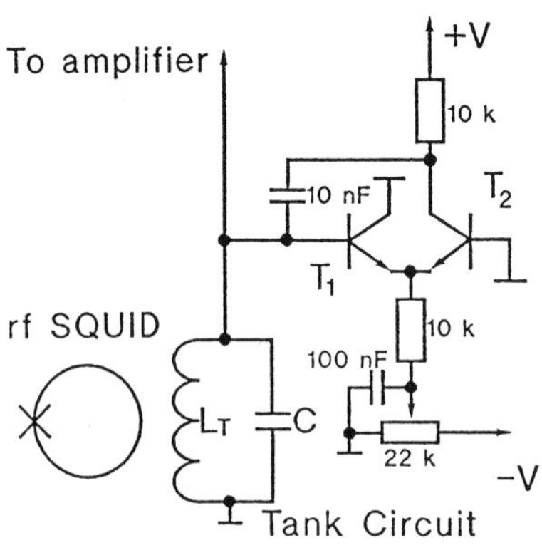

Fig. 2 Simple circuit to reduce the damping of the tank circuit.

2. Results

2.1 Voltage-Flux Transfer Function for Reduced Tank Circuit Damping

Increasing the effective Q of the tank circuit by reducing its damping increases also the value of the voltage-flux transfer function $\partial V/\partial \phi$.

This is shown by two examples in fig. 3 for a bulk SQUID and in fig. 4 for a thin film SQUID. In the former case $\partial V/\partial \phi$ increases by about a factor of 1.5 whereas in the latter case the enhancement is by a factor 4. This is due to the fact that without positive feedback the quality factor Q of the pancake coil is less than that of the solenoid for the bulk SQUID.

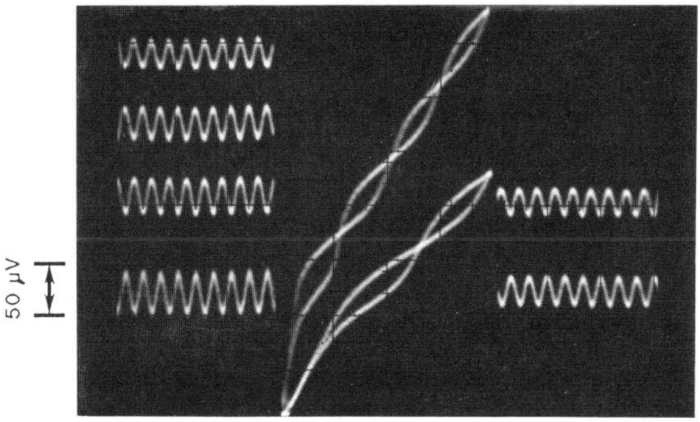

Fig. 3 Tank circuit I-V characteristics of a bulk SQUID with positive feedback (upper curves) and without (lower traces) for $\beta \approx 6$ at 77 K. The corresponding flux voltage characteristics are shown on the left and right side respectively.

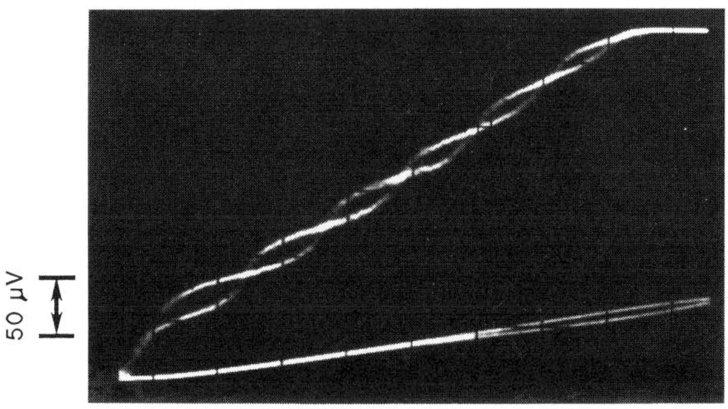

Fig. 4 Tank circuit I-V characteristics of a thin film SQUID with (upper curves) and without (lower curves) positive feedback for $\beta \approx 4$ at 65 K.

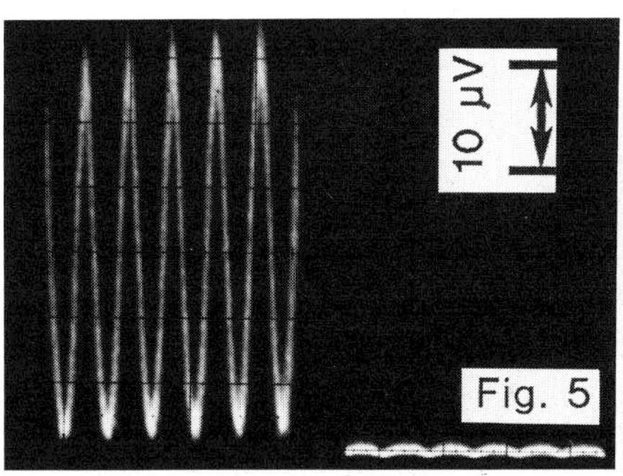

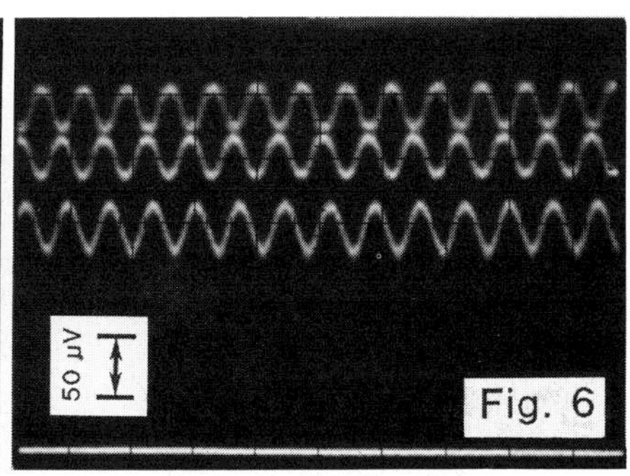

Fig. 5 Flux voltage characteristics of a noise driven bulk SQUID in the dispersive mode at 77 K with positive feedback (left) and without feedback (right).

Fig. 6 Flux voltage characteristics of a noise driven bulk SQUID in the dissipative mode at 77 K. Bottom trace: without feedback. Other curves: feedback increasing with level.

2.2 Noise Driven Compact rf-SQUID

For sufficient strong positive feedback the amplitude of the tank circuit noise current is sufficient to operate the SQUID, both in the dispersive and in the dissipative mode. This is shown in fig. 5 for $\beta<1$, and in fig. 6 for $\beta>1$ for three different feedback levels. The rf-oscillator together with its shielding thus can be eliminated, which leads to a rather compact design of the complete read-out electronics including the circuits for flux-locked loop operation. Without power supply it fits in a cylinder of 25 mm diam. and 9 cm length.

3. Discussion

The flux noise power spectrum S_ϕ is determined by the voltage noise power spectrum S_V at the output of the SQUID system and the transfer function $\partial V/\partial\phi$ by $S_\phi = S_V/(\partial V/\partial\phi)^2$. As shown, positive feedback increases $\partial V/\partial\phi$, but it also increases the noise temperature of the tank circuit and thereby S_V. In our first results the ratio of S_V and $(\partial V/\partial\phi)^2$ however is found to decrease with reduced damping of the tank circuit. At $k^2Q\approx1$ minimum equivalent flux noise was observed. Fig. 7 shows the equivalent flux noise for a bulk SQUID for three different operating conditions, fig. 8 noise data for a thin film SQUID. It should be emphasized that the noise shown is the total noise consisting of contributions from the SQUID-ring, intrinsic thermal noise of the tank circuit itself, and noise of the associated electronic circuit. No effort was made here to separate the individual contributions.

Except for some interference peaks the frequency spectrum in fig. 8 is almost white down to 0.5 Hz (curve I) which suggests, that the dominant noise stems from

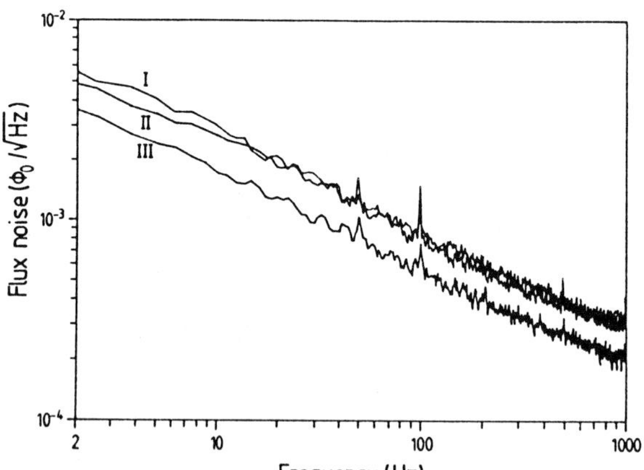

Fig. 7 Equivalent flux noise spectrum of a bulk SQUID at 77 K with and without positive feedback ($\beta\approx1$).
I) With oscillator
II) Without osc. but with reduced damping
III) With osc. and reduced damping

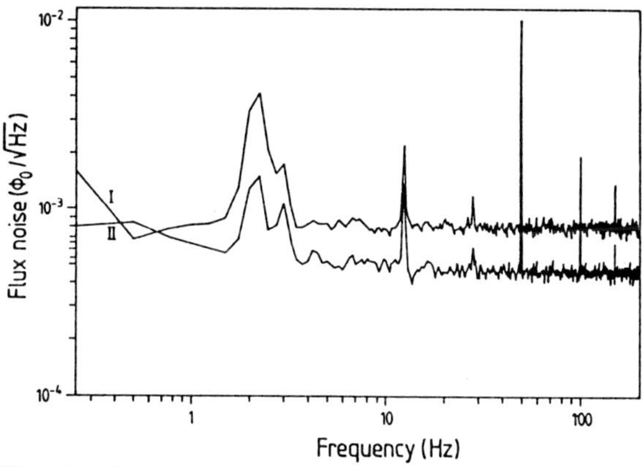

Fig. 8 Equivalent flux noise spectrum of a thin film SQUID at 65 K with and without positive feedback ($\beta \approx 4$)
I) With oscillator
II) With osc. and reduced damping

the electronics. This implies that the intrinsic thin film SQUID 1/f-noise is remarkably low! Lowest values of the equivalent flux noise were found for SQUID operation with external rf-oscillator and positive feedback.

References

[1] Yi Zhang, Yunhui Xu, and C. Heiden, Appl. Phys. Lett. 56, 1579(1990).
[2] G. Cui, Y. Zhang, K. Hermann, C. Buchal, J. Schubert, A.I. Braginski, and C. Heiden, Supercon. Sci. and Tech. 4,S130(1991).
[3] C.L. Jia, B. Kabius, K. Urban, K. Herrmann, G.J. Cui, J. Schubert, W. Zander, A.I. Braginski, and C. Heiden, Physica C 175,545(1991).

High-T_c RF-Biased SQUID Magnetometers

V.I. Shnyrkov, G.M. Tsoi, D.A. Konotop, V.P. Timofeev, S.S. Khvostov, and V.V. Shapovalenko

Institute for Low Temperature Physics & Engineering, 310164 Kharkov, Ukraine

Abstract. The basic results are given for a development of three-component HTSC SQUID-magnetometers. The sensitivity of the RF SQUID-based sensor pumped at 24 MHz was found to be 2×10^{-28} J/Hz (2×10^{-13} T/Hz$^{1/2}$). The cryogenic module consists of a fiberglas cryostat 8.6 l in volume allowing a continuous operation for 30 days. The interferometers used demonstrated stable characteristics for 1 year. With the pumping frequency of 360 MHz the SQUID was sensitive to 3×10^{-29} J/Hz for frequencies higher than 700 Hz.

A three-component magnetometer has been designed on the basis of RF SQUIDs with polycrystal YBaCuO ceramic interferometers (Fig.1). The interferometers have been choosen for the following considerartions. As earlier measurements [1-3] have demonstrated, the HTSC RF SQUIDs made of bulk ceramics show a stable operation under the earth magnetic field and are highly sensitive. On the other hand, only a few bulk DC SQUIDs fabricated by the same technique operated in a stable mode. This has to do with the interferometer inductance L. For a film interferometer with a linear size d [4] the inductance is as follows

$$L \simeq 1.25\,\mu_0 d, \qquad \mu_0 = 4\pi\times10^{-9}\ \mathrm{H/cm} \tag{1}$$

and, respectively,

$$L \simeq \mu_0\,[S/h]\ K \tag{2}$$

for a bulk interferometer of a height h and area S. Here, K < 1 is a function weakly dependent on S/h. The ratio between the inductances in Eqs. (1) and (2) seems to approximate h/d; therefore, if the signals are supplied directly to the quanti-

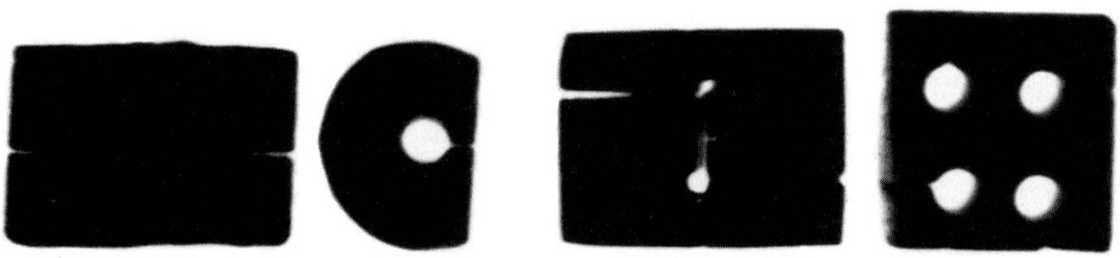

a) b)

Fig.1. Constructions of polycrystalline ceramic one-contact interferometers. (a) $L \approx 2\times10^{-10}$ H; $S \approx 5\times10^{-3}$ cm^3, (b) $L \approx 6\times10^{-11}$ H; $S \approx 2\times10^{-2}$ cm^3.

Springer Proceedings in Physics, Vol. 64 Superconducting Devices and Their Applications
Editors: H. Koch, H. Lübbig

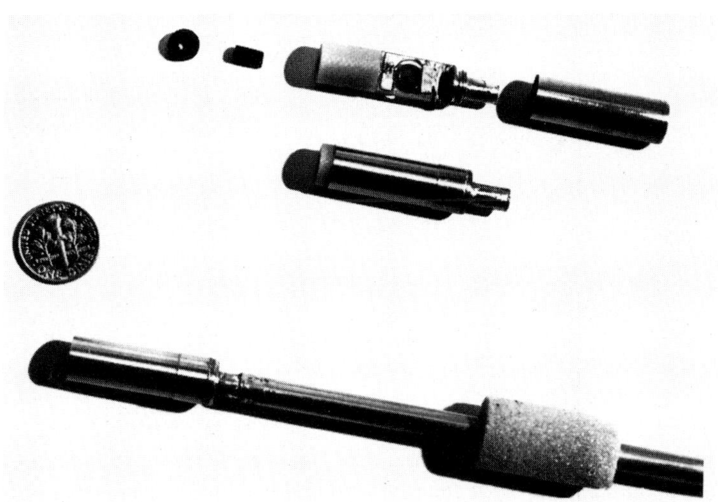

Fig.2. External view (in the centre) and basic components (top from left to right: flux focuser, interferometer, tank circuit on the holding plate, shield) of the polycrystalline YBaCuO ceramic HTSC SQUID. An alternative way of anchoring the module to the rigid coaxial of the cryogenic probe is shown at the bottom.

zation loop the field-period should be 1.5-2 orders of magnitude less for the bulk construction compared with the thin film one. Inductances connected in parallel (Fig.1b) allow that the total interferometer inductance can be made less than the fluctuation inductance $L_F = (\Phi_0/2\pi)^2/(k_B T) \simeq 10^{-10}$ H at 77 K. A weak link was initially prepared as a small bridge of $0.1 \times 0.1 \times 0.1$ mm^3 whose Josephson properties were formed by means of high-voltage electric or laser pulses with a simultaneous control of the current-voltage characteristic. The yield of SQUIDs with a satisfactory sensitivity ($\delta\epsilon \simeq 2\times10^{-28}$ J/Hz) was not large ($\leq$ 10 per cent) if a ceramic was used having a high critical current density $j_c(77K) \geq 200$ A/cm^2.

A study of the "ageing" processes of interferometers has revealed that a degradation of their characteristics results basically from the atmospheric moisture condensed during thermocycling. That is why the interferometers together with the resonant tank circuit of the three-component magnetometer had been placed into a hermetically sealed fiberglass capsule 9.2 mm across and 12 mm long filled with inert gas to promote the heat transfer (Fig.2). A magnetic flux focuser made of bulk ceramics YBaCuO had also been placed into the capsule.

Such a design makes possible a one-year stable operation of this system with a number of thermocycles of about 100.

Fig.3 shows the spectral noise density S_ϕ for the magnetometer sensor pumped at 24 MHz. The value of S_ϕ in the white noise region (f > 100 Hz) is three times as large as the theoretical one derived from the RSJ model, the amplifier and resonant tank circuit noise being taken into account. The intrinsic noise of HTC ceramics (Fig.3) studied in earlier work [5] is essentially of the 1/f type and defines, in

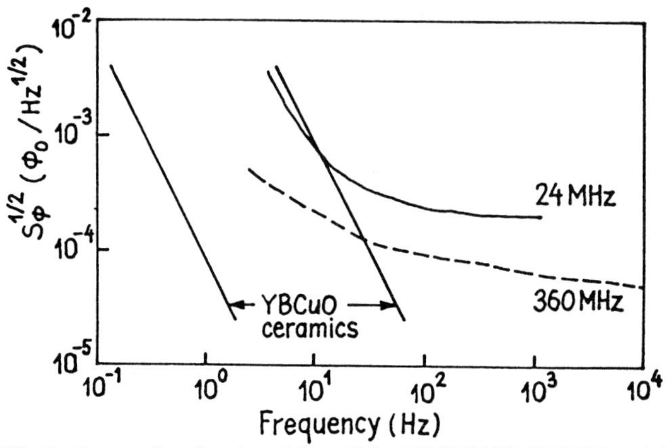

Fig.3. Spectral noise densities of the HTSC RF SQUIDs with $L \approx 2\times10^{-10}$ H at 77 K (solid line: pumping frequency of 24 MHz; dotted line: 360 MHz). Intrinsic noise density of HTSC ceramics at 77 K in the magnetic fields ($\leq 10^{-5}$ T): solid straight lines.

practice, the S_ϕ behaviour within a frequency range f < 10 Hz. This noise caused by a thermoactivated excitation of metastable current states in HTSC materials is almost independent on the value of the magnetic field over a range from 10^{-10} T to 10^{-5} T. The spread in the values of spectral noise density observed from interferometer to interferometer is defined by the intrinsic noise of the Josephson contacts. At critical currents, I_c = 2 - 15 μA, the spectral noise density was found to be proportional to the critical current. This effect can be attributed to increased percolation in the high-I_c contacts.

As is evident from Fig.3, if one succeeds in designing adequate Josephson contacts and interferometers from YBaCuO ceramics ($j_c \simeq 200$ A/cm^2) with $L < L_T$ using, for example, the construction of the type shown in Fig.1b, then the value $S_\phi^{1/2}$ of about 5×10^{-5} $\Phi_0/Hz^{1/2}$ can be achieved in the white range and the onset of the 1/f spectrum could be observed below a frequency of 2-5 Hz.

We have attempted to improve the SQUID sensitivity by increasing the pumping frequency up to 360 MHz. The interferometer construction used was similar to that shown in Fig.1a. The spectral noise density at high frequencies was reduced to $S_\phi^{1/2} \simeq 5\times10^{-5}$ $\Phi_0/Hz^{1/2}$ exceeding the theoretical limit by some factor at lower frequencies (Fig.3). A noise of the 1/f type should dominate in this case a portion of which is inherent to the contact itself. The investigations made on the UHF SQUID are reported in more detail in [6].

A cryogenic module of the three-component SQUID-magnetometer is assembled on the basis of a fiberglass cryostat of 8.6 l in volume (Fig.4). The evaporation rate of LN_2 was about 0.17 l per day at the neck 100 mm in diameter. Consumption of the cryogen was 0.25 l per day if the cryogenic probe for the three-component magnetometer had been used. The other two modifications of the cryostat used could, respectively, provide a continuous operation for 12 days (with a volume of 4 l and a diameter of the neck of 50 mm) and 6 days (with a volume of 1.6 l and the same neck diameter).

Fig.4. External view of measuring instruments **for** three-component SQUID-magnetometers with liquid nitrogen cooling.

The cut-off frequency of the cryostat in the SQUID locations was found to be 25 MHz at 2 dB. It is known [7] that this parameter can be correlated with the intrinsic noise of the superinsulation of the cryostat. For frequencies higher than 20 Hz, the sensitivity of one magnetometer channel fitted with a flux focuser was as high as $\delta B_{min} \simeq 2\times10^{-13}\ T/Hz^{1/2}$. However, we believe that using the interferometer design shown in Fig.1b and HTSC material with $j_c \geq 1000\ A/cm^2$ permits an improvement of the sensitivity for the frequencies higher than 3 - 5 Hz up to the value of $2\times10^{-14}\ T/Hz^{1/2}$ without utilizing an input antenna.

The authors thank the Scientific Council in HTSC Problems of USSR Academy of Sciences for financial support.

References

[1] B.I. Verkin, I.M. Dmitrenko, V.M. Dmitriev, A.G. Kozyr, G.M. Tsoi, and V.I. Shnyrkov, Fiz. Nizk. Temp. **13**, 651(1987) - Sov. J. Low Temp. Phys.

[2] B.I. Verkin, I.M. Dmitrenko, V.M. Dmitriev, V.V. Kartsovnik, A.G. Kozyr, V.A. Pavlyuk, G.M. Tsoi, and V.I. Shnyrkov, Fiz. Nizk. Temp. **14**, 34(1988) - Sov. J. Low Temp. Phys.

[3] V.I. Shnyrkov, G.M. Tsoi, A.G. Kozyr, and V.N. Glyantsev, Fiz. Nizk. Temp. **14**, 770(1988) - Sov. J. Low Temp. Phys.

[4] M.B. Ketchen, IEEE Trans. Magn. **MAG-17**, 387(1981)

[5] V.N. Glyantsev, I.M. Dmitrenko, V.V. Borzenets, and V.I. Shnyrkov, Fiz. Nizk. Temp. **15**, 1001(1989) - Sov. J. Low Temp. Phys.; V.I. Shnyrkov, G.M. Tsoi, V.N. Polushkin, V.N. Glyantsev, and A.G. Kozyr, in: Weak Superconductivity, ed. S. Benacka and M. Kedro (Nova Science Publ.-New York, 1990) p. 131.

[6] V.I. Shnyrkov, V.P. Timofeev, S.S. Khvostov, and G.M. Tsoi, Mod. Phys. Lett. (to be published)

[7] V.V. Shapovalenko, private communication

Performance of BSCCO Thick Film RF-SQUIDs

N. Khare, S. Chaudhry, A.K. Gupta, and V.S. Tomar

National Physical Laboratory, Dr. K.S. Krishnan Road,
New Delhi-110012, India

Abstract. RF SQUIDs have been fabricated by manually scribing a hole shunted with a microbridge in a BSCCO thick film. The V-Φ modulations are observed up to 96.5 K. In the white noise region, the spectral density of flux noise is $5 \times 10^{-4}\ \Phi_0/\sqrt{Hz}$ at 77 K. The SQUID has been operated successfully in flux-locked-loop mode (FLLM). The stability of the SQUID output and observation of hysteresis in external flux measurements are also reported.

1. Introduction

The weak links showing Josephson behaviour are naturally present at the grain boundaries of high T_c superconductors. The rf SQUID behaviour in two hole as well as in one hole geometries in bulk YBCO superconductors using these weak links in a microbridge have already been observed at 77 K by several workers [1,2]. Recently rf SQUID behaviour both in thin and thick films of YBCO [3,4] and in thick films of BSCCO [5] have been reported.

In this paper we present the operation of BSCCO thick film rf SQUID under flux-locked-loop-mode (FLLM) at 77 K. The successful flux locking of the SQUID up to an external flux equivalent to several flux quanta demonstrates its suitability for practical applications. Stability of SQUID output under FLLM and hysteresis in the measurement of external flux are also reported.

2. Experimental

The thick films of BSCCO have been grown on MgO (100) substrates employing screen printing technique with the starting composition of 1112. The details of the preparation technique for the films are described elsewhere [5]. T_c (offset) of the film is 104 K. The films are highly oriented and consist of predominantly high T_c 2223 phase as shown by XRD studies. The rf SQUID geometry basically consists of a hole (diameter 0.4 mm) shunted with a microbridge ($0.15 \times 0.12\ mm^2$) carved manually on the thick film. A ten turn copper spiral is glued on the top of the patterned film. The spiral forms a part of the tank circuit with resonance frequency 20.5 MHz and Q being 32. A twenty turn copper coil (radius 4 mm), used for applying the magnetic field and feedback signal is glued underneath the substrate such that centres of the coil and of the hole of the SQUID coincide.

Springer Proceedings in Physics, Vol. 64 Superconducting Devices and Their Applications
Editors: H. Koch, H. Lübbig

A commercial rf SQUID electronics with suitable modifications has been used to observe the rf SQUID behaviour. The built-in rf oscillator has been disabled and an external rf oscillator (HP 3325B) is connected. An external af oscillator (Wavetek model 166) is connected to the field coil for applying ac magnetic field. The reflected rf signal is amplified using a low noise preamplifier and detected finally by a diode detector. The overall gain of the amplifier has been fixed at 63 dB. A cylinder consisting of three layers of μ-metal is used to shield the rf SQUID.

Flux noise density of the SQUID has been studied under open-loop-mode (OLM) using a lock-in amplifier (SR 530) along with a dynamic signal analyser (HP 35660A) [6]. The amplitude of the ac modulation is adjusted so that the coil current, I(coil) < $I(\Phi_0/2)$, where $I(\Phi_0)$ being the coil current needed to change the flux in the SQUID by a single quantum. The detector output is connected to the lock-in amplifier and its output is connected to that of the dynamic signal analyser (HP 35600 A) for recording the voltage noise spectrum, $S_V(f)$. Flux noise density, $\sqrt{S_\Phi}(f)$, is estimated by

$$\sqrt{S_\Phi(f)} = [S_V(f)/(\delta V/\delta\Phi)^2]^{1/2}.$$

where $(\delta V/\delta\Phi)$ being the transfer function of the SQUID.

In order to operate the SQUID in FLLM, an external lock-in amplifier (SR 530) has been used. The detector output is fed to the input of the lock-in amplifier. An ac modulation signal of 1 kHz frequency is applied to the coil of the tank ciruit. The amplitude of the ac signal has been adjusted such that I(coil) < $I(\Phi_0)/2$. Signal from the same modulation source has been fed to the reference point of the lock-in amplifier. The output of the lock-in amplifier is coupled to a feedback coil through a resistor. In FLLM, the current in the feedback coil produces an opposing magnetic field on the SQIUD so as to cancel the effect of the external applied magnetic field. For an ideal flux locked case the lock-in output is expected to vary linearly with the externally applied magnetic field.

3. Results and Discussion

Fig. 1 shows the V-Φ characteristics of the rf SQUID at different temperatures. At 77 K, peak-to-peak voltage at the detector output is 5 mV (which corresponds to a magnetic flux responsivity of 7 $\mu V/\Phi_0$ at the input of preamplifier in the detection circuit). With the increase in the operating temperature, the peak-to-peak voltage of the oscillations in V-Φ curve decreases. However it is possible to see the modulation up to 96.5 K, while beyond this temperature the modulation almost disappears.

Fig. 2 shows the flux noise spectrum of the SQUID at 77 K which has been recorded under open loop condition. In the white noise region (f > 10 Hz), the spectral density of the flux noise, $\sqrt{S_\Phi}$ is $5 \times 10^{-4}\ \Phi_0/\sqrt{Hz}$.

Below 10 Hz, $\sqrt{S_\Phi}$ increases due to the occurence of 1/f noise which can be attributed to thermally activated flux hopping. The inductance of the SQUID loop, L_{SQ} is 3.9×10^{-10} H. The energy sensitivity of the SQUID in the white noise region can

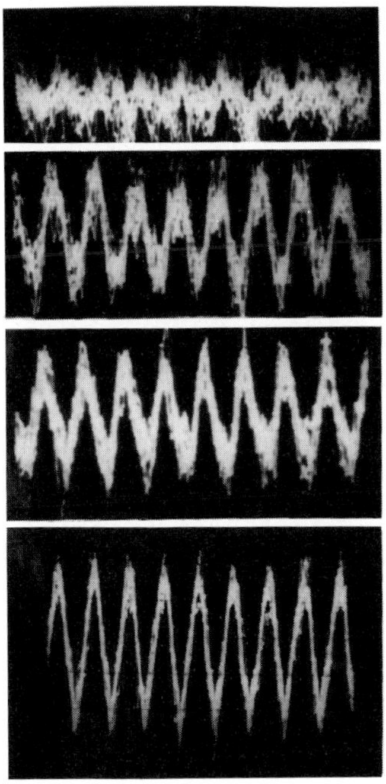

Fig. 1 V-Φ modulations at various operating temperatures.

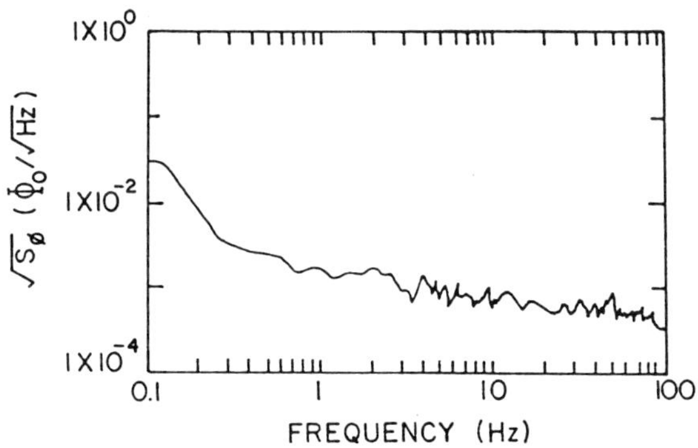

Fig. 2 Flux noise spectrum at 77 K.

be estimated using the formula, $\epsilon = S_\Phi/2L_{SQ}$. The present BSCCO thick film rf SQUID has an energy sensitivity 1.3×10^{-27} J/Hz at 77 K.

Fig. 3 shows variation in the look-in output with the applied magnetic flux for the SQUID operated in FLLM. A linear increase in lock-in output with the increase in the magnetic flux equivalent to several flux quanta confirms the SQUID operation in FLLM.

Fig. 4 shows the variation of the SQUID output with the increase and decrease of flux cycles corresponding to a maximum applied flux of Φ_o, $3\Phi_o$ and $10\Phi_o$. Regular curves show the SQUID output for increasing, while dotted curves correspond to decreasing flux values. When the variation of the flux is Φ_o, no hysteresis in the

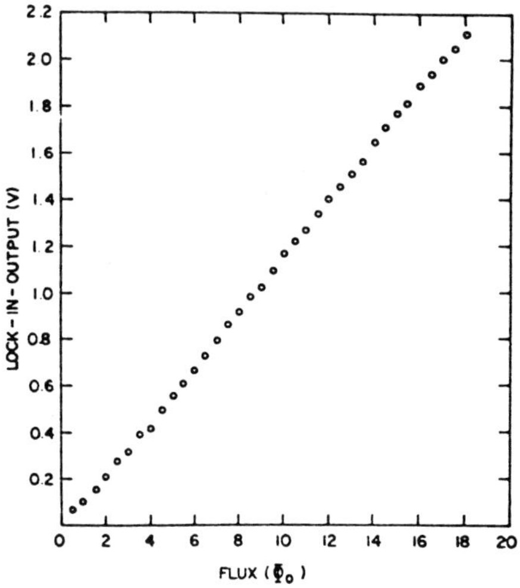

Fig. 3 Variation of lock-in output with the change in magnetic flux for FLLM.

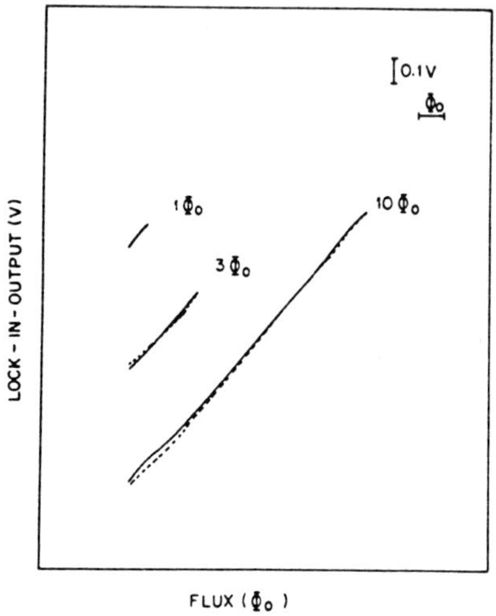

Fig. 4 Hysteresis observed in SQUID for FLLM

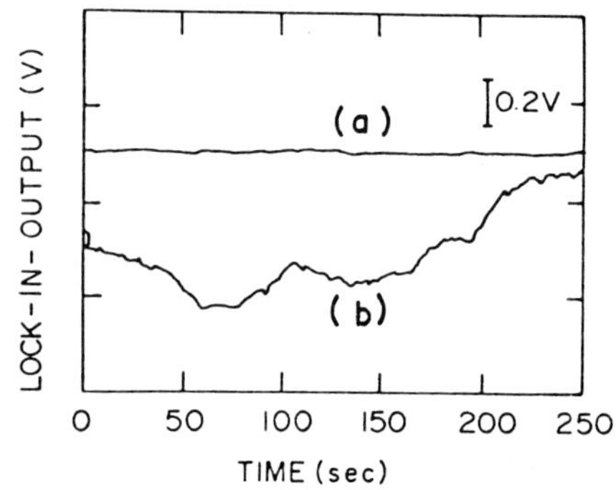

Fig. 5 SQUID output as a function of time for (a) FLLM and (b) OLM.

SQUID output is observed. However for larger flux variations, a small hysteresis has been observed.

The stability of the SQUID output in FLLM is also tested. Fig. 5 shows the comparison of SQUID output for FLLM and OLM (open loop mode) cases for fixed magnetic field conditions. For FLLM, the SQUID output is quite stable. Similarly a comparison of the voltage noise spectra (in the lower frequency region) for FLLM and OLM shows that S_V is smaller in the FLLM condition.

4. Conclusion

The operation of BSCCO thick film rf SQUID under flux locked condition has been demonstrated at 77 K. In this mode the stability of the SQUID improves significantly. The flux noise density of the SQUID in the white noise region (f > 10 Hz) is $5 \times 10^{-4}\ \Phi_0/\sqrt{Hz}$ at 77 K. The thick film rf SQUID shows a small hysteresis when the change in external flux exceeds Φ_0.

References

[1] S. Harrop, C. M. Muirhead, M.S. Colclough and C.E. Gough, Physica C **153 - 155**, 1411 (1988)

[2] A.G. Likhachev, V.N. Polushkin, S.V. Uchaikin and B.V. Vasiliev, Supercond. Sci. Technol. **3**, 148 (1990).

[3] S.S. Tinchev, Supercond. Sci. Technol. **3**, 500 (1900).

[4] N. Khare, A.K. Gupta, S. Chaudhry and V.S. Tomar, Supercond. Sci. Technol. **4**, 107 (1991).

[5] N. Khare, A.K. Gupta, S. Chaudhry and V.S. Tomar, Solid State Commun. **76**, 929 (1990).

[6] M.N. Keene, S.P. Harrop, T.J.Jackson, C.M. Muirhead and C.E. Gough, Supercond. Sci. Technol. **3**, 263 (1990).

Two-Loop Y-Ba-Cu-O RF-SQUID Magnetometer

S.S. Tinchev and J.H. Hinken

Forschungsgesellschaft für Informationstechnik, Postfach 1147,
W-3202 Bad Salzdetfurth, Fed. Rep. of Germany

Abstract: HT_c RF SQUID magnetometers have been made from Y-Ba-Cu-O thin films. The small, i.e. 0.3 μm long region of the microbridge was irradiated with 100 keV oxygen ions at an ion dose as low as 1 x 10^{13} O^+/cm^2. SQUID function was observed up to 86 K. The white noise of the SQUID was 2.5 x 10^{-4} $\Phi_0/\sqrt{Hz}$ at 83 K and the 1/f corner frequency was lower than 5 Hz. A SQUID magnetometer with a directly coupled pick-up coil was also made.

1. Introduction

Thin-film planar superconducting quantum interference devices (SQUIDs) have relatively poor intrinsic sensitivity to magnetic fields because of their small pick-up areas. The need for an enhanced sensitivity to magnetic fields is even more important for thin-film high T_c SQUIDs operated near 77 K, where noise considerations require the inductance and hence the pick-up area to be smaller than in low T_c versions.

The field sensitivity can be improved in principle by making use of: a superconducting flux transformer [1], a multiloop SQUID, where pick-up loops are connected in parallel and form directly the SQUID inductance [2], or by a flux focusser [3].

We report here on RF SQUIDs made from Y-Ba-Cu-O thin films, where the SQUID inductance consists of two square pick-up loops which are connected in parallel. Similar Nb-SQUIDs with inductive shunt have been presented in literature [4]. Because of using only two loops the HT_c SQUID magnetometer can be made on a single substrate from a simple HT_c monolayer. With this geometry the magnetic field applied from both coils has the same direction improving the magnetic field response of the SQUID.

2. SQUID Fabrication

The high T_c SQUIDs were made in the same manner as described in [5]. Here only a brief description of the fabrication process will be given.

The films used in the SQUID fabrication were Y-Ba-Cu-O films deposited on $SrTiO_3$ substrates by laser ablation. A Siemens XP2020 Excimer laser was used for the thin film fabrication. The films were 200 - 270 nm thick having a zero resistance T_c of 87 - 89 K.

Springer Proceedings in Physics, Vol. 64 **Superconducting Devices and Their Applications**
Editors: H. Koch, H. Lübbig

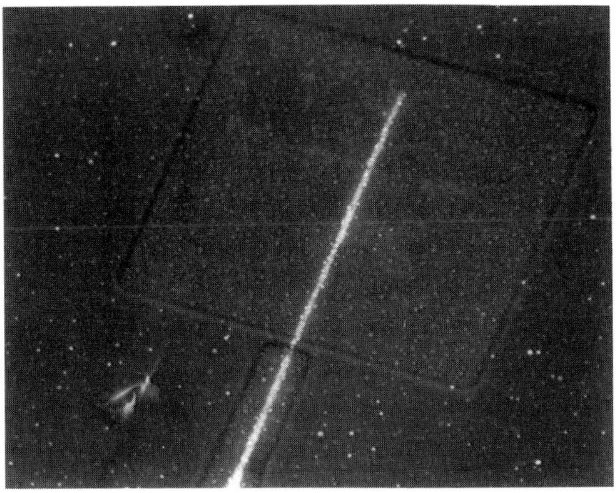

Fig. 1 Scanning electron micrograph of the HT_c RF SQUID.
The SQUID hole is 100 µm x 100 µm in size. The microbridge is about 0.3 µm in length and about 5 µm in width.

The films were patterned by standard photolithography. We used wet chemical etching in EDTA. On a substrate with 10 mm x 10 mm in size usually 16 RF SQUIDS with single-loop or one two-loop RF SQUID are placed. In both cases the SQUIDs have a square hole of 100 µm x 100 µm with a single bridge which in this fabrication step is 10 µm long and about 5 µm wide. Next the bridge is coated with PMMA resist, and a slit across the microbridge is definded by electron beam using a Philips XL30 Scanning Electron Microscope. After the resist development the final fabrication step is to implant 100 keV oxygen ions into the slit. During the implantation the PMMA film is used as a mask.

A scanning electron micrograph of the SQUID is shown in fig. 1. The white line across the bridge is the implanted region which is the microbridge. The implantation dose of 1 x 10^{13} O^+ ions/cm^2 was chosen because it reduces the T_c of the implanted region typically by 15 - 20 K. Thus at LN_2 temperature the bridge operates as an S-N-S (or S-S'-S) microbridge.

3. SQUID Properties

The SQUIDs were tested in the traditional configuration: A ten turn spiral copper coil was pressed onto the SQUID under test. The coil and a cooled capacitor form a tank circuit, resonating at 19 MHz. The voltage across the tank circuit is amplified and detected by a commercial SHE 330 SQUID electronics.

The SQUIDs having the implanted region about 0.3 µm in length showed a good SQUID behavior above 80 K. Fig. 2 shows a SQUID signal characteristic (SQUID pattern - SQUID output voltage vs. external magnetic flux) recorded at a temperature of 83 K. The peak-to-peak amplitude of the triangular pattern was about

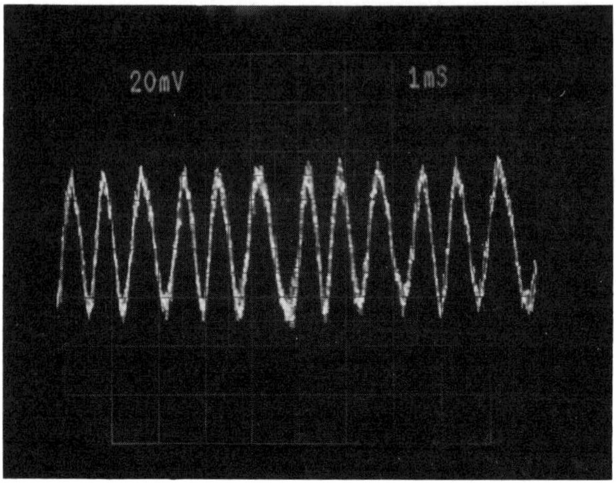

Fig. 2 SQUID pattern on the first plateau recorded at a temperature of 83 K.

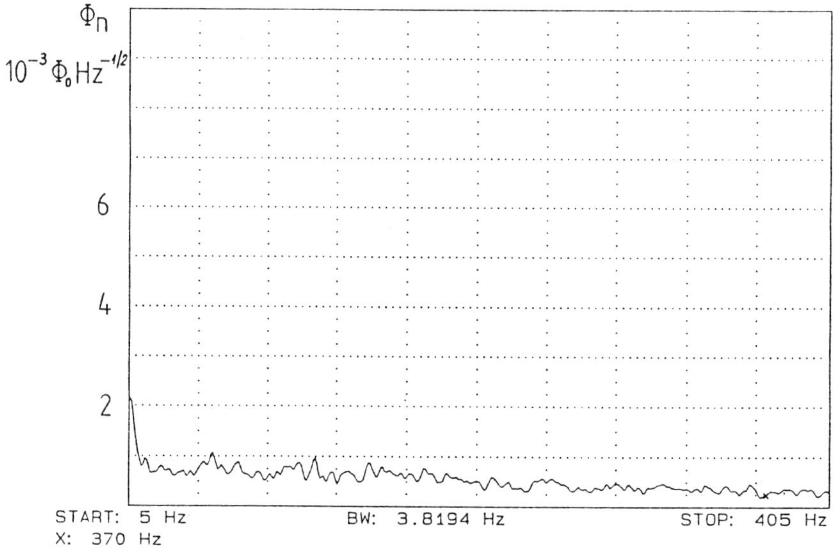

Fig. 3 Measured equivalent flux noise spectral density Φ_n for a single RF SQUID.

50 mV p-p. The same SQUID electronics with a 4.2 K Niobium SQUID showed a p-p amplitude which is only slightly higher, namely 80 mV.

The SQUIDs fabricated in this way usually operated in a temperature range of about 3 - 5 K. Although this range seems to be small, practically it was no problem to have a stable SQUID operation at around 77 K. The maximum operating temperature of one of the SQUIDs was 86 K. The fabrication yield of our SQUIDs was fairly satisfactory: About 70% of these fabricated SQUIDs showed good SQUID behavior.

The noise properties of the SQUIDs were quite good. Fig. 3 shows the measured equivalent flux noise spectral density of a single SQUID. Even in conditions of inadequate shielding, the white noise level was about $2.5 \times 10^{-4}\Phi_0 \cdot Hz^{-1/2}$. We can not observe clearly the crossover to 1/f noise and we assume that it lies below 5 Hz.

4. Two-loop RF SQUID Magnetometer

The design of such a SQUID is demonstrated in Fig. 4a. Actually, the SQUID itself has the same dimensions like a single-loop SQUID. The only difference is the parallel connected large pick-up loop. Although, it can be expected that the field sensitivity of such a directly coupled loop will be smaller than that of a flux transformer with a multiturn input coil, the advantage of this SQUID is that only a single film process is required for the entire magnetometer.

Now work is in progress to measure the SQUID parameters, especially the magnetic field sensitivity and its improvement compared to the bare SQUID.

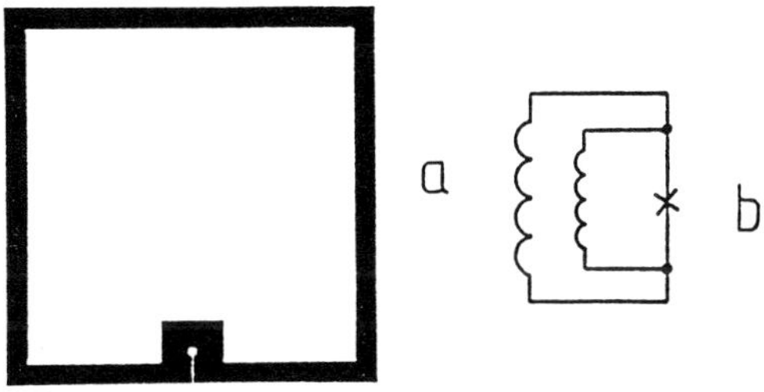

Fig. 4 RF SQUID directly coupled to the pick-up loop (a) and its equivalent circuit (b).

References

[1] F.C. Wellstood, J.J. Kingston, M.J. Ferrari, J. Clarke, Superconducting thin-film flux transformers of $YBa_2Cu_3O_{7-x}$, Appl. Phys. Lett., 57, No. 18 (1990), 1930-1932.
[2] J. Zimmerman, Sensivity enhancement of superconducting quantum interference devices through the use of fractional-turn loops, J. Appl. Phys.,42 (1971), 4483-4487.
[3] A.N. Matlashov, V.P. Koshelets, P.V. Kalashnikov, Yu. E. Zhuravlev, V. Yu. Slobodchikov, S.A. Kovtonyuk, L.V. Filippenko, High sensitive magnetometers and gradiometers based on DC SQUIDs with flux focusser, IEEE Trans. Magn., 27, No. 2 (1991), 2963-2966.
[4] H. Koch, An inductively shunted DC-SQUID, in Proc. 10th Int. Cryog. Eng. Conf.(ICEC 10), H. Collan, R. Berglund, M. Krusius, eds., Butterworth (1984), 834.
[5] S.S. Tinchev, Investigation of RF SQUID made from epitaxial YBCO films, Supercond. Sci. Technol., 3 (1990), 500 - 503.

YBCO/PBCO/YBCO Edge Junctions and DC-SQUIDs

H. Rogalla

University of Twente, Faculty of Applied Physics, Low Temperature Group, P.O. Box 217, NL-7500 AE Enschede, The Netherlands

One of the first demonstrated low power applications of high-T_c superconductivity was the DC-SQUID making use of natural grain boundaries as Josephson junctions[1/3]. That was (and still is) a difficulty in high current applications turned out to be a benefit in the early stages of the research in high-T_c superconducting electronics: grain boundaries in polycrystalline thin films of $YBa_2Cu_3O_{7-x}$ (in short YBCO) work as Josephson junctions with superconductor/normal metal/superconductor (SNS) like behavior[4]. Based on these early investigations two main directions are currently followed in the Josephson junction and SQUID research, (i) the preparation of artificial grain boundaries on bicrystals[5] and on edges in the substrate[6], and (ii) the development of SNS-like junctions with thin film sandwiches of YBCO/normal metal[7] or semiconductor/YBCO[8/10]. Up to now the best performance at 77 K has been demonstrated for a DC-SQUID with artificial grain boundary junctions on $SrTiO_3$ bicrystals[11]. It offers relative low noise and a periodic voltage modulation by an external magnetic field at temperatures up to 87 K. However, bicrystal fabrication, control of the critical current and the normal resistance, and reproducibility pose a serious problem.

These difficulties can be overcome by making use of epitaxial multilayer structures forming a SNS sandwich. The N-layer could be realized by a normal metal or a semiconductor. For epitaxial sandwich structures the barrier (N) material has to match the lattice constant of the base and top layer in order to enforce epitaxial growth. This drastically reduces the number of possible barrier materials: one of the few ones left is $PrBa_2Cu_3O_{7-x}$ (in short PBCO) which has the same structure and nearly the same lattice constants as YBCO. First attempts to make planar SNS-type junctions from *c*-axis oriented sandwiches made clear that the barrier has to lie between the CuO-planes of the adjacent superconducting YBCO electrodes. One possiblility for a realization may be the growth of *a*-axis oriented layers[12]. Another one, which we chose, to cut through a *c*-axis oriented layer under a flat angle and thus open the CuO-planes for a subsequent deposition of PBCO barrier and YBCO counter electrode, will be described in this contribution. An essential ingredient is the flat angle ($<35^\circ$) for cutting with an Ar ion beam so that epitaxy is even maintained at the edge[13]. The fabrication turned out to be very reliable and controllable and the junctions show good Josephson behavior. In the following sections the preparation and properties of YBCO/PBCO/YBCO edge junctions and DC-SQUIDs will be described and at the end first results on an integrated YBCO multiturn coupling coil will be presented.

I. YBCO/PBCO/YBCO Josephson Edge Junctions

I-1. YBCO/PBCO/YBCO multilayer structures.

For practical thin film applications of high-T_c superconductivity, like SQUIDs and other superconducting electronic devices, a single layer

Springer Proceedings in Physics, Vol. 64 Superconducting Devices and Their Applications
Editors: H. Koch, H. Lübbig

technique is mostly not sufficient. In order to realise cross-overs of electrical connections and active elements, e.g. Josephson junctions, a multilayer technique is indispensible. This means that stacks of superconductor layers (e.g. YBCO), insulating layers (e.g. MgO, ZrO_2) and normal- or semiconducting layers (e.g. Ag, PBCO) have to be fabricated, each with good electrical, structural and surface quality.
For the realization of the edge-type Josephson junctions high-quality trilayers of YBCO/PBCO/YBCO had to be prepared.

Both YBCO layers and the PBCO layer in the YBCO/PBCO/YBCO sandwich structure were deposited in situ by a modified off-axis rf-sputtering technique. Details of this technique have been described in a previous publication[14]. Stoichiometric sintered YBCO and PBCO targets with a diameter of 50 mm have been used. To avoid contamination of the interfaces of the multilayer structure, we applied a multi-target arrangement in our sputter system; so the substrates could be turned under different targets and various materials could be deposited sequentially without breaking the vacuum. The layers of PBCO and those of YBCO were prepared using the same deposition procedures. As substrates we used Y-stabilized ZrO_2 (100). The sputter gas is a mixture of argon and oxygen (typically 50% oxygen and 50% argon). Typical sputter pressures are between 5×10^{-2} and 3×10^{-1} mbar. During deposition the substrates were heated to 640-700 °C. The incident rf sputter power density was between 1-5 W/cm^2. After film deposition an anneal step at 400°C in one bar oxygen was carried out. This procedure was earlier found to consistently produce YBCO thin films showing a zero resistance temperature $T_{c,zero}$ around 90 K and a critical current density $J_c > 10^6$ A/cm^2 at 77 K. The $T_{c,zero}$ and J_c(77 K) of the YBCO layer in YBCO/PBCO bilayers are not degraded by the presence of the PBCO layer and are comparable to results on an YBCO single layer with the same film thickness[15]. Therefore it can be concluded that the good superconducting properties of YBCO layers are preserved throughout the whole fabrication procedure of the YBCO/PBCO/YBCO sandwich.

Since the composition and quality of interfaces will strongly affect the performance of the high-T_c superconducting devices[16], the microstructure and interfaces of the YBCO/PBCO/YBCO multilayer have been thoroughly investigated, using several analysis techniques[15].

X-ray diffraction measurements indicated that these multilayers are highly oriented with the *c*-axis perpendicular to the substrate, as in the case of individual YBCO and PBCO layers. Channeling analysis using 2.0 MeV He-ions aligned parallel to the (001) direction of the film confirmed this maintenance of epitaxy throughout the YBCO/PBCO/YBCO trilayer. However, the analysis also indicated that misfit dislocations may be present on the YBCO/PBCO interface due to very small differences in the lattice constants of YBCO and PBCO.

The microstructure of these multilayers has been studied by high resolution transmission electron microscopy (HRTEM) with atomic resolution. Fig.1 shows a cross-sectional bright-field TEM image of the complete trilayer on a (100) ZrO_2 substrate. Stacking fault defects are visible in this image as horizontal fringes. Close to the substrate, only few of these defects are observed. Further away from the substrate/film interface, their density increases. The YBCO/PBCO interfaces are not visible in Fig.1, since their contrast is largely obscured due to the higher defect density in that area. The HRTEM image in Fig.2 reveals the atomic stacking at the interface region between the substrate and the first YBCO layer. This image was made with the electron beam parallel to the (010) orientation of YBCO, as is clear from the inset. The focus condition of the microscope was chosen such that the Y, Ba and Cu atoms are properly imaged as white dots. The stacking sequence of YBCO is indicated. In the atomic layers just above the substrate, a perfect and defect free YBCO stacking with c-axis texture is revealed, in agreement with the bright-field observations. Looking in more detail at the

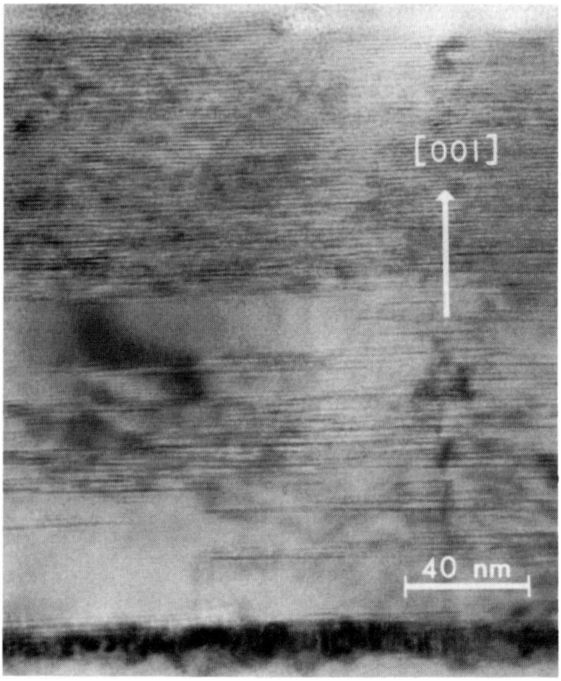

Fig.1. Cross-sectional TEM bright-field image of the complete YBCO/PBCO/YBCO trilayer.

interface area in Fig.2, it is found however, that no real coherent match is realized between film and substrate, and that the interface is not perfectly sharp at an atomic scale. Interface steps are present as can be observed in the bottom-right area of Fig.2. Also local small orientation differences are revealed in the substrate area, which have surprisingly no effect on the YBCO epitaxy. This can be seen in the bottom-left area of Fig.2. The substrate diffraction spots, which were marked by "S" in the inset of Fig.2, are also slightly misaligned with respect to the YBCO spots.

Indeed, only the alignment of the *c*-axis of the film perpendicular to the substrate is clearly maintained. Therefore, it seems that the film quality is not critically dependent on the substrate conditions. Effects due to possible reactions between film and substrate are improbable.

From Fig.2, the defect-free layer of the first grown film was estimated around 30 nm. This corresponds to the thickness range of the YBCO thin films, which are often made by our preparation technique[14,17]: it illustrates the excellent quality of our ultra thin YBCO films.

A combination of Rutherford Backscattering Spectroscopy (RBS), Scanning Auger Microscopy (SAM) and secondary ion mass spectroscopy (SIMS) studies on diffusion at the interfaces indicated that in the YBCO/PBCO/YBCO trilayer a pronounced interdiffusion has not occurred; the interfaces are very sharp and well defined. The maximum interdiffusion depth between YBCO and PBCO layers was estimated to be less than 7 nm, being the experimental resolution of RBS and SAM. Thus we can conclude that even if any interdiffusion takes place, it will be limited to a few nanometers.

In addition to these promising properties, the trilayers are very smooth at the surface and only a very limited amount of outgrowths can be

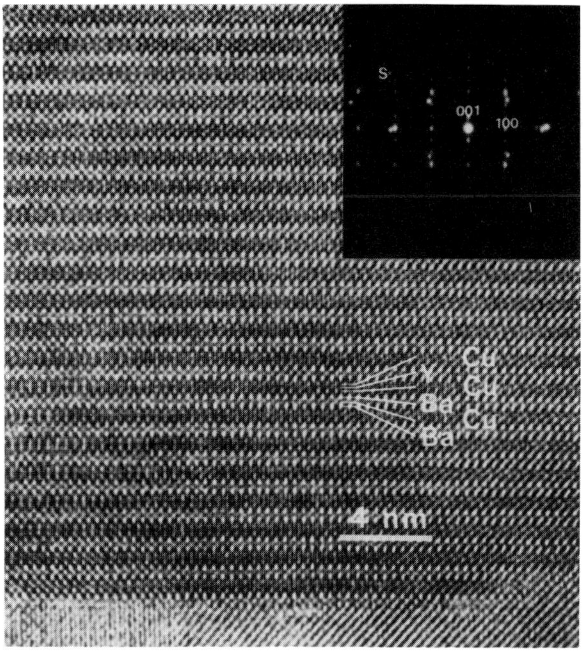

Fig.2. HRTEM image of the interface of the substrate and the first YBCO layer. The corresponding electron diffraction pattern characterizes the epitaxy of the film.

observed. Thus the trilayers are well applicable for even complicated multilevel structures.

I-2. Fabrication of edge junctions

The hetero-epitaxially grown multilayers that we described in the previous section, were used for the fabrication of our high T_c Josephson junctions, using PBCO as a barrier[5]. Such a PBCO barrier has to (i) survive the high temperature deposition process for the top electrode, (ii) be sufficiently thin (a few nm, but without pinholes) in order to have a usefully high Josephson tunnel current density and (iii) allow the epitaxial growth of the top YBCO layer.

In order to fabricate YBCO/PBCO/YBCO junctions in the ab-plane we used a two step process. In the first step, an YBCO layer with a thickness of about 200 nm and a PBCO top layer have been grown in situ. This PBCO layer will prevent Josephson coupling in the *c*-axis direction between the bottom YBCO layer and the top one, which was deposited later. Therefore a relatively thick (> 250 nm) PBCO layer was used. Then a part of this bilayer was covered by a photoresist stencil and the unprotected area was etched by an Argon ion beam with an incident angle of 45-65 degree in order to fabricate a smooth edge. Our ion beam parameters were typically 3.5 kV acceleration voltage and 30 $\mu A/cm^2$ beam current density at the sample. After removing the photoresist stencil a thin PBCO barrier layer and an YBCO top layer were deposited. As the edge may be contaminated

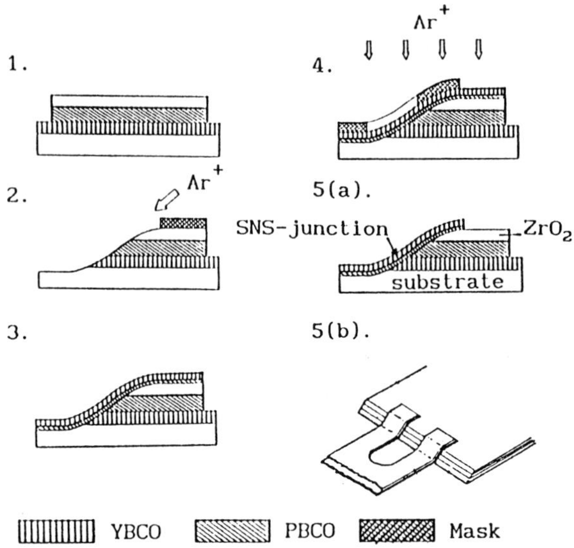

Fig.3. Fabrication sequence for the DC-SQUIDs: 1) In situ growth of the YBCO bottom layer, PBCO and insulating ZrO_2. 2) Fabrication of the edge in the base electrode by Ar ion beam etching. 3) Deposition of the PBCO barrier and YBCO top electrode. 4) Patterning of the sample into a DC-SQUID by lithographic methods and ion milling. Deposition of four silver electrodes by sputtering (not shown). 5) Schematic view of a finished DC-SQUID. (a) the cross-section and (b) the perspective view.

between the two processes we cleaned it with the Ar ion beam before depositing the barrier layer. The critical temperature of the final junction structure was mostly found to be slightly degraded, to 80 K.

After completing the complete multilayer structure, a junction or a DC-SQUID was defined across the edge using photo lithographic patterning and ion beam etching. The resulting edge junction cross section is shown in Fig.3-(5). Finally the lateral dimension of the edge was reduced to obtain junctions with a width ranging between 4 and 1000 μm. The height of these junctions is typically between 30 and 100 nm.

Normally we used a thick (> 250 nm) PBCO layer to insulate the base electrode from the counter electrode, as mentioned previously. A rough estimate of the resistivity of PBCO in *c*-axis direction at low temperatures in our structures yields a lower limit of 10^4 Ω·cm. Recent results reported by the *Bellcore* group on all *a*-axis oriented junctions demonstrated however, that *a*-axis PBCO layers as thick as 150 nm are thin enough to be used as a barrier[12]. Although PBCO is an anisotropic material and its normal state resistivity, like that of other $ReBa_2Cu_3O_x$ compounds (Re: rare earth element), is supposed to be lower in the *c*-axis direction by some orders of magnitude than in the *a*-axis direction, some junctions with a $SrTiO_3$ or ZrO_2 layer instead of the PBCO layer were fabricated to be sure that the *c*-axis PBCO layer indeed insulates the YBCO base layer from the YBCO counter electrode. The normal state resistance and the shape of the *I-V* curves of the resulting junctions are almost the same as those of the junctions which used a PBCO layer as an insulator[18]. From this we conclude that a 250 nm PBCO layer is sufficient to insulate the two electrodes.

I-3. Characteristics of the edge junctions.

To gain insight in the possibility to fabricate really controllable and uniform junctions, it is important to study their scaling behavior and their Josephson properties. Therefore the dependence of the critical current I_c, the normal resistance R_n and their product I_cR_n on the barrier thickness L and junction area A, as well as the junction response to an applied magnetic field and irradiation with microwaves, have been studied extensively[18].

Fig.4 shows the dependence of the critical current and the inverse resistance of the junction area, defined as $A=w\cdot d/\sin\vartheta$ (w is the junction width, d is the thickness of the base electrode, and ϑ is the angle of the junction edge), for junctions with the same barrier thickness. The width of the smallest junctions may be somewhat overestimated due to the relatively larger influence of the damaged edges by ion etching. Anyhow, these results indicate that the junctions show good uniformity in the scaling behavior of the critical current and device conductance with the area of the junction, upto quite large areas.

A strong dependence of the critical current density J_c on the barrier thickness has been observed. Fig.5 shows $\ln(J_c)$ as a function of L at 4.2 K. It indicates that, to a good approximation, J_c decreases exponentially as L increases. Such exponential scaling is also predicted by the theory for large superconductor/normal-metal/superconductor (SNS) junctions[19]. From the equation $\ln J_c \approx \ln J_{c0} - L/\xi_n$ (J_{c0} is the critical current density of the electrodes) we have estimated the effective normal metal coherence length ξ_n for the PBCO layer at 4.2 K to have a value of 5 to 8 nm. For very thin barriers ($L \leq 6$ nm) the values of J_c are found to be higher than predicted by the exponential dependence. There are several possible reasons. Firstly, when $L \leq \xi_n$, the Josephson coupling between the base and the counter electrode is stronger than for $L \gg \xi_n$ and described by another $J_c(L)$ dependence[19]. Secondly, a junction with a very thin barrier is more sensitive to nonuniformities or the existence of small pinholes in the barrier. Thirdly, the intrinsic nature of the PBCO layer might play a role.

The dependence of the critical current density on the R_nA product has been investigated as well. The data shown in Fig. 6 are distributed along

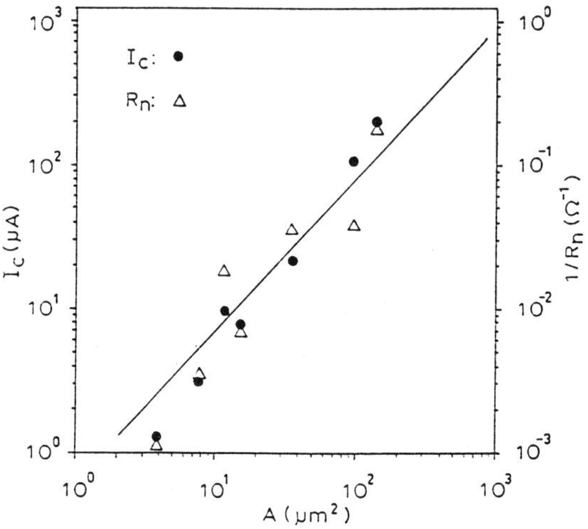

Fig.4. Scaling of the critical current I_c and resistance R_n with the junction area A for junctions with the same barrier thickness.

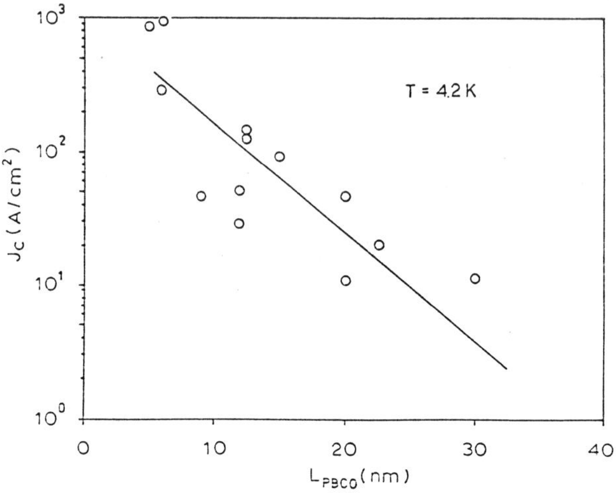

Fig.5. Dependence of the critical current density J_c on the barrier thickness L.

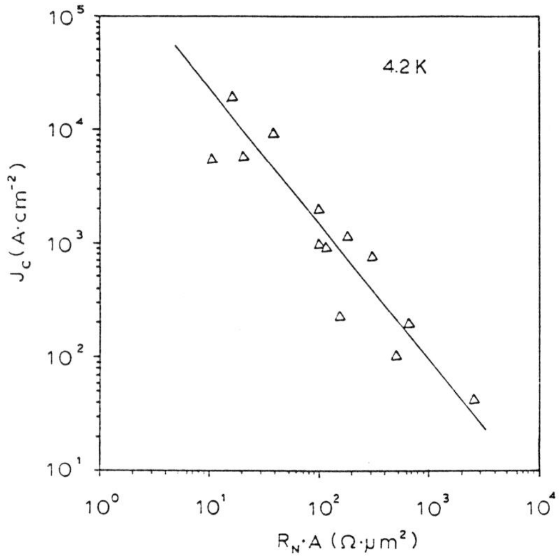

Fig.6. Critical current density J_c as a function of the product R_nA at T = 4.2 K.

the line $J_c \sim (R_nA)^{-m}$ with m=1.2. For SNS structures with $L \ll \xi_n$ one would expect $J_c \sim (R_nA)^{-2}$, whereas for superconductor/insulator/normal-metal/superconductor (SINS) structures $J_c \sim (R_nA)^{-1}$ is expected[16,19]. The observed dependence differs from the prediction for SNS structures and more closely resembles the dependence of SINS structures. On the other hand, we found for the dependence of the critical current on temperature $I_c(T) \sim (1-T/T_c)^2$. This result is in good agreement with the SNS model[19]. This discrepancy between different properties of the junction may be attributed to the interface resistance between the base electrode and the PBCO barrier, and the influence of the specific resistance of PBCO, which may be different for different barrier thicknesses, $L \gg \xi_n$ or $L \le \xi_n$.

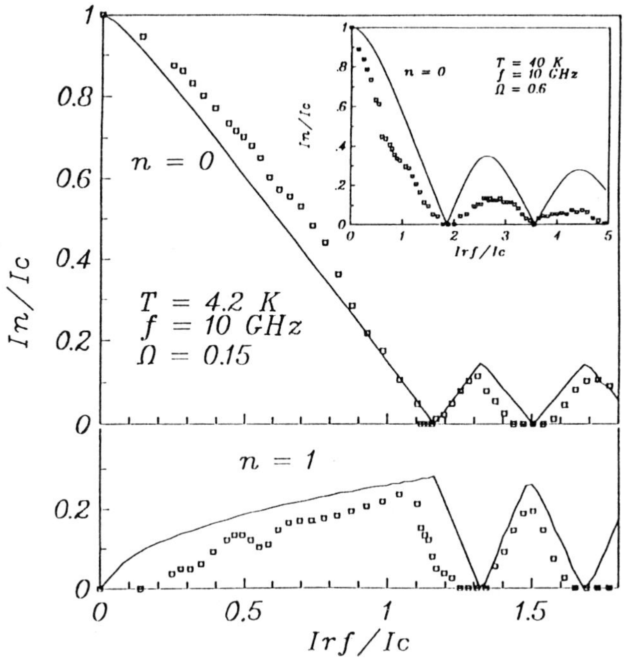

Fig.7. Magnitude of the current steps as a function of microwave current.

In addition, the I_cR_n products have been found to scale approximately linearly with the barrier thickness, ranging from 0.8 to 5 mV. The highest value obtained was 8 mV. The damage of the edge surface due to the ex-situ process could detrimentally impact the I_cR_n product[16].

However, the presented results clearly demonstrate good uniformity of the critical parameters of the YBCO/PBCO/YBCO edge junctions. The fabrication process is indeed controllable and well reproducible.

For a clear observation of Shapiro steps in the I-V curve of the junctions under microwave irradiation, the large overlap area which, the first versions of the junctions had[9], needed to be reduced. The large shunting capacitance of the junctions due to such a large overlap was a problem in the microwave measurements.

With reduced overlap area, clear Shapiro steps were observed at voltages $V = n\Phi_o f$ corresponding to the Josephson frequency (Φ_o is the flux quantum) in the presence of microwave irradiation with a frequency f = 10 GHz, between 4.2 and 60 K. Fig.7 shows the magnitude of the measured current steps as a function of the normalized microwave current. At higher temperatures the Shapiro steps become rounded but the critical current I_c still exhibits an oscillatory dependence, clearly revealing Josephson behavior. To examine the applicability of the RSJ model, we simulated the I-V characteristics for different microwave amplitudes by solving the appropriate differential equation using a standard fourth-order Runge-Kutta method. The measured data are in reasonable agreement with the calculated values.

The critical current depends very strongly on the magnetic field. Many junctions showed diffraction patterns $I_c(H)$ approximating the expected Fraunhofer behavior, with a strong central peak and periodic modulation of I_c with more than 95% up to 80 K. Fig.8 shows the $I_c(H)$ modulation at 60 K for one of these junctions. The field was applied parallel to the film and

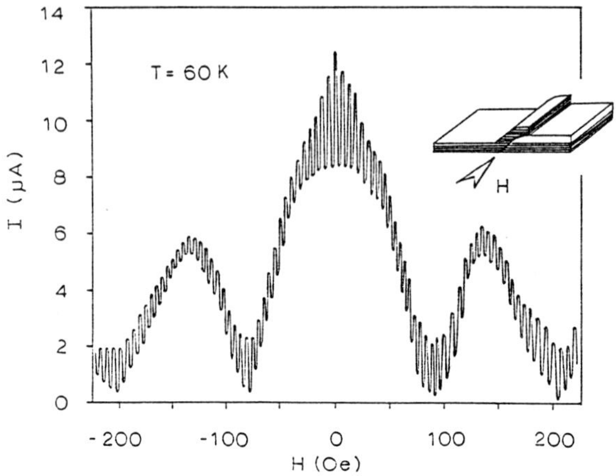

Fig.8. Modulation of the critical current vs magnetic field directed as shown in the inset.

the junction, as shown in the inset. For $J_c = 5\times10^2$ A/cm^2 at 60 K the corresponding Josephson penetration depth $\lambda_J \approx 15$ μm is larger than the width of the junction (w = 10 μm), which means that the junction can be treated as "small"[20]. For this "in plane" direction of the magnetic field, the period H_0 of the $I_c(H)$ dependence is found to be between 90 and 100 Oe which approaches the calculated value given by $H_0 = \phi_0/(2\lambda_L d/\sin\vartheta)$. A characteristic feature for many of the studied junctions is the presence of additional periodical structures for all directions of the applied field. For the junction presented in Fig. 8, the small period H_0^s is 7 Oe at 60 K, increasing gradually to H_0^s = 8 Oe at 4.2 K. It is of the same order of what one would expect if the field is applied perpendicular to the junction $H_0^{\perp} \approx 5$ Oe.
The increase of the period H_0^s at lower temperatures correlates with the change of $H_0^{\perp}(T)$ due to the London penetration depth $\lambda_L(T) = \lambda_L(0)/[1-(T/T_c)^4]^{1/2}$. A possible explanation of the observed "interference" picture may be the complex distribution of both the induced Meissner current and Josephson current over the junction area. It would lead to an induced magnetic field at the surface of the superconducting electrodes and a complication of the $I_c(H)$ dependence[20,21].

II. DC-SQUIDs made with YBCO/PBCO/YBCO edge Josephson junctions

One of the first and - in the initial stage - simpliest application of the edge junctions are DC-SQUIDs. Such DC-SQUIDs were fabricated from YBCO/PBCO/YBCO trilayers with a wide edge junction using standard photolithography and Ar-ion beam etching. For the SQUIDs described here edge junctions with a typical barrier thickness of 20 nm were applied. The size of the SQUID-loop is in all cases 5 x 150 μm^2.

In Fig.9 the IV-characteristic of such a SQUID, measured at 4.2 K, is presented together with a simulation based on the RSJ model. The width of the junctions is 4 μm.The best fit between RSJ model and measured data is obtained for I_0=1.7μA, R=40Ω, Γ=0.104 and β_c=0.83. From this the capacitance of a single junction is calculated to be approximately 50fF. This rather high value is due to a large overlap of the electrodes. Presently this overlap is reduced to 3 μm by photolithography and Ar-ion

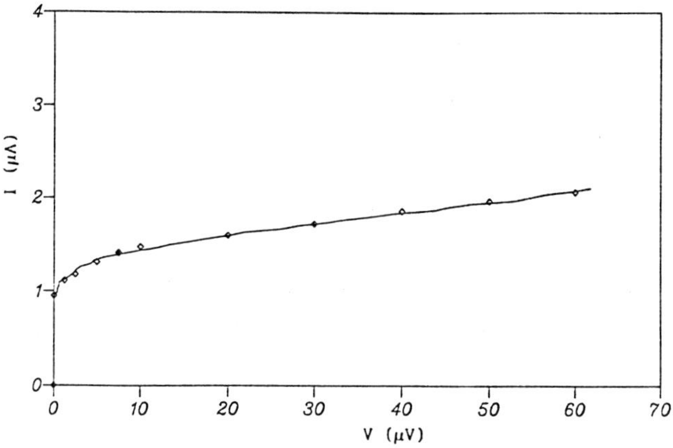

Fig.9. IV-curve of YBCO/PBCO/YBCO SQUID and RSJ-simulation at 4.2 K.

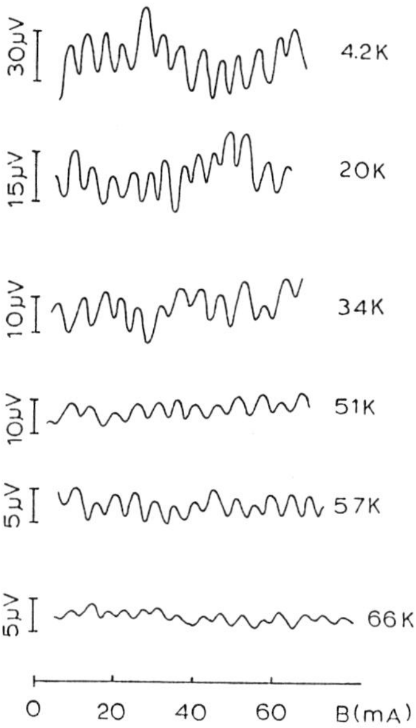

Fig.10. Voltage vs magnetic field for one of the DC-SQUIDs at different temperatures. The magnetic field axis 20 mA corresponds approximately to 7 flux quanta in the SQUID-loop.

etching resulting in much smaller capacitances. We have found that also for higher voltages the junctions can very well be described by the RSJ-model.

Flux-voltage characteristics for one of our DC-SQUIDs are shown in Fig.10 for the temperature range from 4.2 K to 66 K. The height of the 15 μm wide junctions is about 120 nm and the critical current density at 4.2 K was approximately 150 A/cm^2. The normal state resistance of about 20 Ω is almost independent of temperature. The V-ϕ curves are well periodic up to an applied flux of several tens of flux quanta. Even above 66 K some modulation could be observed, but the signal was too small and

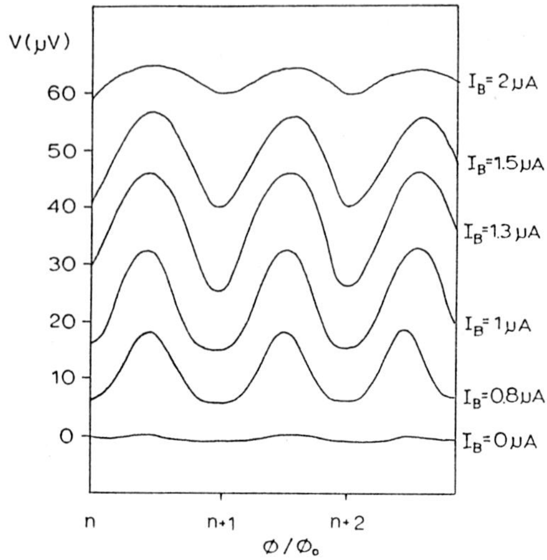

Fig.11. A set of typical V_{mod} vs. Φ characteristics of this SQUID for various bias currents measured at 4.2 K.

noisy for practical applications. It should be mentioned however, that these measurements were carried out in an unshielded environment. The modulation period corresponds to approximately one flux quantum in the SQUID loop and does not change between 4.2 K and 66 K. The total modulation depth $m = \Delta I_c/I_c$ is roughly 20% of the critical current and the modulation voltage at 4.2 K and 66 K are typically 25 μV and 2 μV, respectively.

Fig.11 shows a set of typical V_{mod} vs. Φ characteristics of a SQUID for various bias currents measured at 4.2 K in a well shielded environment. The curves are very smooth in the usual bias regime $I_B \cong I_c - 2.5I_c$. The maximum voltage modulation at $I_B \cong 1.3$ μA was approximately 25 μV which corresponds to approximately 35% of the I_cR_n-product.

The noise power spectrum of this particular SQUID was measured at 4.2 K and is depicted in Fig.12. It shows an increased noise level at frequencies below about 40 Hz (at 40 Hz the flux noise was determined to be 1.8×10^{-4} $\Phi_0/Hz^{1/2}$) having a 1/f frequency dependence of $S_\phi(f) \cong 2\times10^{-8}/f$ Φ_0^2/Hz. This noise is probably caused by vortex motion. Real time observations on our oscilloscope screen revealed random jumping between two different base line voltage levels (telegraph noise). The remaining variations around these two base lines are one order of magnitude lower than the base line jumps. Compared to the noise of classical DC-SQUIDs the noise of high T_c SQUIDs is still high, but it is low enough to make these high T_c SQUIDs already useful for any applications. Furthermore it is likely that by a better design of the DC-SQUIDs vortex jumps may be avoided and hence the resulting telegraph noise be reduced.

No significant changes in the voltage modulation data of our SQUIDs were seen even after 6-8 cryogenic/room temperature cycles, showing the relatively robust nature of these edge junctions. This statement also holds for the electrical stability: it is possible to increase the bias current of the SQUIDs up to 1 mA at 4.2 K without any damage to the junctions.

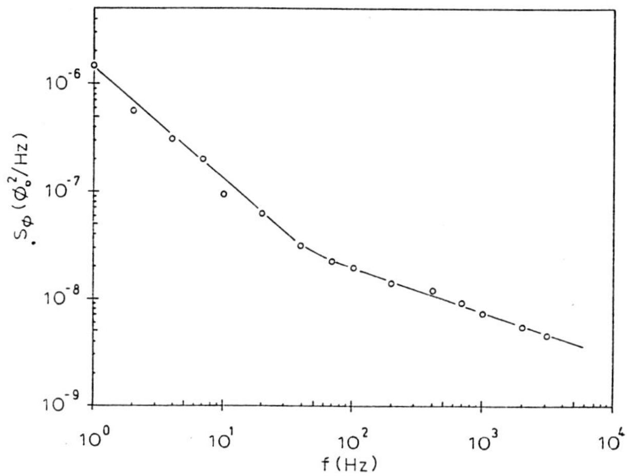

Fig.12. The noise spectrum of one of the SQUIDs measured at 4.2K.

III. High Tc multiturn coils

A necessary ingredient of a practical DC-SQUID magnetometer is a fluxtransformer which consists of a large pick-up coil and a multiturn input-coil.

An 8-turn input-coil was fabricated by laserablation using $SrTiO_3$ as an insulating layer between an YBCO base-stripline and 8 YBCO windings. Two key problems in the fabrication of such a multiturn coil are the insulating cross-over and the superconducting window-contact through the insulating layer. In order to grow $SrTiO_3$ and YBCO epitaxial on top of the edge of an YBCO stripline, the edge should be very smooth and have an angle of 30° or less with the substrate surface[13]. Although high quality multi-turn coils were produced by standard photolithography and wet chemical etching using a bromine-solution[22] or EDTA[23] we found that a more suitable way to produce the smooth edges may be Ar-ion beam bombardment under an angle. For this an AZ 1518-photoresist layer is baked in an oven at 90°C for a short time (10 minutes). After structuring the photoresist the film is etched using an Ar-ion source. The angle between the substrate normal and the Ar ion beam is 45°. Because of simultaneous etching of YBCO and photoresist very smooth edges with an angle less than 35° can be obtained on both sides of the YBCO stripline after only one etch-run. Also for the fabrication of the window contact this procedure is useful. To provide a good superconducting contact the surface of the window has to be cleaned by Ar ion milling before deposition of the YBCO toplayer.

In this way a first 8-turn superconducting coil has been made. The thickness of the YBCO-layers is 250 nm and the thickness of the $SrTiO_3$ layer is 80 nm. Four point measurements yielded a critical temperature of 27 K and a critical current density of 400 A/cm^2 at 4.2 K. After intentionally disconnecting the coil at the center by wet chemical etching, it did not become superconducting anymore, proving that no shorts are present in the cross-overs.

Although the superconducting properties of this first coil were not good enough for practical applications we expect further improvement in the near future. Presently cross-overs can be fabricated with critical temperatures of the YBCO layers higher than 87 K, $j_c(4.2K) > 5\cdot10^5\ A/cm^2$, and resistivity of the $SrTiO_3$ of $6\cdot10^5\ \Omega cm$ at 77 K. Window contacts can be fabricated with critical temperatures up to 84 K and $j_c(77K)$ of $9\cdot10^2$ A/cm^2.

IV. Conclusions

In conclusion, we have demonstrated the feasibility to prepare DC-SQUIDs with multiturn coupling coil incorporating YBCO/PBCO/YBCO edge junctions. In order to realise such devices, high quality trilayers of YBCO/PBCO/YBCO have been prepared and thoroughly studied by TEM, X-ray analysis, He-ion channeling, RBS, SAM and SIMS. These analysis techniques indicate excellent epitaxy between the films and the substrate, which is maintained throughout the whole trilayer. The orientation of the c-axis perpendicular to the substrate is fairly perfect. No signs of a pronounced lattice mismatch between YBCO and PBCO could be found. The maximum thickness of a diffusion layer between YBCO and PBCO is less than the experimental resolution limits of the RBS and SAM measurements. In addition the superconducting properties of the individual YBCO thin films were found to be preserved in the multilayer structures.

YBCO/PBCO/YBCO edge junctions fabricated using the trilayer techniques are very uniform and controllable for different barrier thicknesses and junction areas. Periodic modulation of the critical current in applied magnetic field and microwave power are clear evidences of the Josephson behavior of the YBCO/PBCO/YBCO multilayer structures.

In the next step DC-SQUIDs were fabricated which make use of the YBCO/PBCO/YBCO junctions. These SQUIDs show very good modulation behavior up to 66 K. This temperature is not a real limitation; higher operating temperatures (e.g. 77 K) could be reached with a change in the structuring process.

Finally, a first multiturn input coil was fabricated, which demonstrates the feasibility of crossings and superconductor/superconductor contacts between different layers.

Acknowledgements

The author is greatly endebted to J. Gao, D. Veldhuis, W.A.M. Aarnink, R.P.J. Ijsselsteijn, J.W.M. Hilgenkamp, A.J.H.M. Rijnders, B.B.G. Klopman, D. Terpstra, Yu. Boguslavkij, E.C.M.C. Reuvekamp, J. Flokstra and G.J. Gerritsma. This work was supported by the Dutch National Program on High Tc superconductivity.

References

1. A.Th.A.M. de Waele, R.T.M. Smokers, R.W. van der Heiden, K. Kadowaki, Y.K. Huang, M van Sprang and A.A. Menovsky, Phys.Rev.B **35**, 8858 (1987).
2. R.H. Koch, C.P. Umbach, G.J. Clark, P. Chaudhari and R.B. Laibowitz, Appl.Phys.Lett.**51**, 617 (1987).
3. B. Häuser, M. Diegel and H. Rogalla, Appl.Phys.Lett.**52**, 846 (1988).
4. B. Häuser, B.B.G. Klopman, G.J. Gerritsma, J. Gao and H. Rogalla, Appl.Phys.Lett.**54**, 1368 (1989).
5. D. Dimos, P. Chaudhari, J. Mannhart and F.K. LeGoues, Phys.Rev.Lett.**61**, 219 (1988).
6. G.J. Cui, K. Herrmann, Y. Zhang, C.L. Jia, Ch. Buchal, J. Schubert, W. Zander, A.I. Braginski and Ch. Heiden, in *Nonlinear Superconductive Electronics and Josephson Devices* Plenum Press (1991).

7. M.S. DiIorio, S. Yoshizumi, K. Yang, J. Zang, M. Maing. Appl.Phys.Lett.**58**, 2552 (1991).
8. D.K. Chin, T. Van Duzer. Appl.Phys.Lett.**58**, 753 (1991).
9. J. Gao, W.A.M. Aarnink, G.J. Gerritsma, H. Rogalla. Physica C **171**, 126 (1990).
10. R.B. Laibowitz, R.H. Koch, A. Gupta, G. Koren, W.J. Gallagher, V. Foglietti, B. Oh and J.M. Viggiano, Appl.Phys.Lett.**56**, 686 (1990)
11. R. Gross, P. Chaudhari, M. Kawasaki, M.B. Ketchen and A. Gupta. Appl.Phys.Lett **57**, 727 (1990).
12. J.B. Barner, C.T. Rogers, A. Inam, R. Ramesh, and S. Bersey, Appl.Phys.Lett.**59**, 742 (1991).
13. C.L. Jia, B. Kabius, K. Urban, K. Herrman, G.J. Cui, J.S. Schubert, W. Zander, A.I. Braginski and Ch. Heiden. Physica C **171**, 545 (1991).
14 J. Gao, B. Häuser, and H. Rogalla, J.Appl.Phys.**67**, 2512 (1990).
15. J. Gao, W.A.M. Aarnink, G.J. Gerritsma, A.J.H.M. Rijnders, H. Rogalla, F. Hakkens, W. Coene and M.A.M. Gijs, Physica C **177**, 384 (1991).
16. M.Yu. Kupriyanov and K.K. Likharev, IEEE Trans.Magn.**27**, 2460 (1991).
17. J. Gao, W.A.M. Aarnink, G.J. Gerritsma and H. Rogalla, Appl. Surface Science **46**, 74 (1990).
18. J. Gao, Yu. Boguslavkij, B.B.G. Klopman, D. Terpstra, G.J. Gerritsma and H. Rogalla, To be published in Appl.Phys.Lett.**59** (1991).
19. K.K. Likharev, Rev.Mod.Phys.**51**, 116 (1979).
20. A. Barone and C. Paterno, Physics and Application of the Josephson Effect, John Wiley Cons., New York, 1982.
21. S.L. Miller, K.R. Biagi, J.R. Clem and D.K. Finnemore, Phys.Rev.B **31**, 2684 (1985).
22. F.C. Wellstood, J.J. Kingston, M. Ferrari and J. Clarke. IEEE Trans.Magn.**27**, 2569 (1991).
23. W. Eidelloth, W.J. Gallagher, R.P. Robertazzi, R.H. Koch, B. Oh, R.L. Sandstrom. To be published.

Reproducible High-T_c SNS Josephson Junctions and DC-SQUIDs

M.S. DiIorio, S. Yoshizumi, K.-Y. Yang, J. Zhang, M. Maung, and N. Fan

Biomagnetic Technologies, Inc., 9727 Pacific Heights Blvd., San Diego, CA 92121, USA

Abstract. Reproducible high-T_c $YBa_2Cu_3O_7/Ag/YBa_2Cu_3O_7$ Josephson junctions have been fabricated using an entirely in-situ step-edge technique. Josephson junctions have been produced with I_cR_n products up to 4.6 mV at 4.2 K and normal resistances up to 5 Ω. The expected modulation of the critical current in a magnetic field is observed in single junctions as well as dc SQUIDs. The devices exhibit a well-defined supercurrent from 4.2 K to 87 K. The lowest measured flux noise sensitivity of a dc SQUID at 4.2 K is 1.4 x 10^{-6} $\Phi_o/\sqrt{Hz}$ at 2.5 kHz leading to an energy sensitivity of 2.8 x 10^{-31} J/Hz. At 10 Hz the best measured flux noise at 4.2 K is 2.6 x 10^{-5} $\Phi_o/\sqrt{Hz}$ leading to an energy sensitivity of 4.8 x 10^{-29} J/Hz.

1. Introduction

SQUID applications utilizing high temperature superconductors require the development of a practical Josephson junction. The junction must be fabricated in a reproducible fashion and be rugged enough to withstand packaging, handling, and thermal cycling. Recently, a number of techniques have been developed to produce superconductor - normal metal - superconductor (SNS) Josephson junctions that hold promise for device applications [1-6]. The key element common to all techniques is the contact of the normal metal to an a-axis exposed surface of the superconducting film, which allows use of the longer a-axis superconducting coherence length. Generally, this has been accomplished by contacting the edge of a film grown with its c-axis normal to the substrate. A potential drawback of many of these techniques is that the surface of the high-T_c film edge is exposed during the lithographic processing needed to produce the film edge, and any degradation of this superconducting surface can hinder process reproducibility.

To circumvent these difficulties we have utilized a step-edge technique to fabricate SNS Josephson junctions of $YBa_2Cu_3O_7/Ag/YBa_2Cu_3O_7$ using an entirely in-situ process. The precisely defined small dimensions generated by the use of the step-edge permit us to obtain the requisite Josephson coupling between two exposed a-axis superconducting film edges. The coupling is through the normal metal, Ag, which possesses a relatively large normal metal coherence length (100-200 nm at 4.2 K and 20-60 nm at 77 K) [7]. The junction fabrication process requires the deposition of a single high-T_c film and eliminates the need for processing before the deposition of the normal metal.

Springer Proceedings in Physics, Vol. 64 **Superconducting Devices and Their Applications**
Editors: H. Koch, H. Lübbig

2. Fabrication

The junction fabrication process begins by first producing a sharp step in a (100) $LaAlO_3$ substrate, which is then used to produce a narrow gap in the high-T_c superconductor deposited across the step. The normal metal is deposited on top of the superconductor, without breaking vacuum, and joins the two superconducting "banks." Fig. 1 shows a schematic diagram of the device structure. The step height primarily determines the strength of the Josephson coupling and hence is chosen so that the gap between the superconducting banks is on the order of the normal metal coherence length. We believe that the deposition of the film across the step is such that the a-axis edges are left exposed near the step-edge, allowing the device structure to make use of the longer a-axis superconducting coherence length. X-ray diffraction measurements confirm that the films are c-axis oriented perpendicular to the plane of the substrate with a typical rocking curve FWHM of 0.5°. The final devices are completed using standard photolithography followed by ion-beam etching and the process is intrinsically compatible with routine production methods. Details of our fabrication process can be found in a separate publication [5].

Typically, 140-280 nm steps in $LaAlO_3$ substrates are produced using ion-beam etching. A Nb etch mask is used to improve the quality of the step as the inherent reproducibility of the process stems from the precise fabrication of the step-edge.

The c-axis oriented $YBa_2Cu_3O_7$ films are deposited on using off-axis single-target magnetron sputtering. Both dc and rf sputtering have proven successful in producing high-T_c devices. Depending on the process, the substrate temperature during deposition is 710-750°C, the argon partial pressure is 100-170 mTorr, and the oxygen partial pressure is 20-40 mTorr. The deposition rate is 2-4 Å/min for dc sputtered films and 12 Å/min for rf sputtered films. The 50-150 nm thick films used to fabricate the junctions have a zero resistance transition temperature of 85-89 K. Films deposited in a similar fashion for different purposes have critical current densities reaching 5×10^6 A/cm^2 at 77 K.

The normal metal is deposited, without breaking vacuum, from a second magnetron source shortly after the superconductor deposition in order to ensure a clean interface between the superconductor and the normal metal. The silver is deposited at an angle to help promote adequate step-coverage. Note that along the step the thickness of the silver may be less than the nominal thickness of 80-600 nm. Devices fabricated *without* the silver layer always resulted in open circuits or

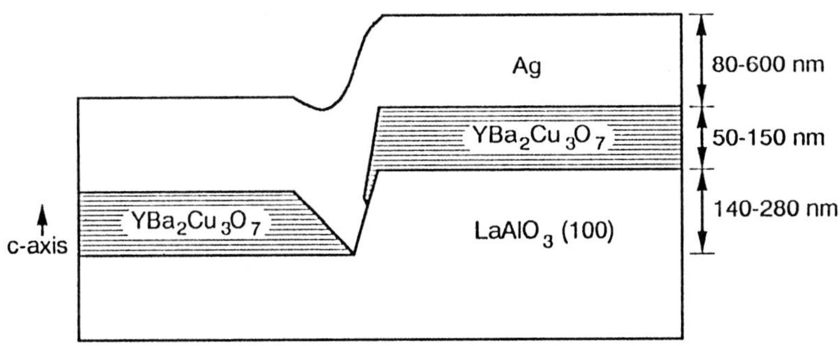

Fig. 1 Schematic diagram of the SNS Josephson junction structure.

high resistances (>300 Ω), clearly indicating that the Josephson coupling takes place through the normal metal.

Following the film deposition, the superconductor-normal metal bilayer is processed using standard photolithography and ion-beam etching. The widths of most junctions are presently conservatively patterned to 5 μm; a 20 μm-wide junction is patterned for comparison. The dc SQUIDs consist of two 5 μm-wide junctions with typical SQUID loop areas of either 25 μm^2 or 100 μm^2. Note that the entire process is readily capable of producing large numbers of devices, with the main variable being the well-controlled step-edge geometry.

3. Measurements

To date over 90 Josephson junctions and dc SQUIDs patterned and measured have demonstrated a well-defined supercurrent. The first 56 devices fabricated on eight different substrates from four entirely separate deposition runs all showed a supercurrent from 4.2 K (the lowest temperature investigated) to above 77 K. Typically the highest temperature at which the device exhibited a critical current was 83 - 87 K. Subsequently, a few junctions have not proven operational. We believe that this is due to our experimentation with the process, as a number of key parameters such as the step height have been varied in order to investigate their impact on junction quality.

A current-voltage characteristic at 4.2 K for a 4 μm-wide single junction fabricated with our thinnest silver layer (nominally 80 nm) is shown in Fig. 2. The critical current, I_c, is 2.2 mA and the normal resistance, R_n is 2.2 Ω leading to our highest I_cR_n product of 4.6 mV. Note that the I_cR_n product at 76 K is 0.6 mV. The noise rounding at higher temperatures stems mostly from external rf noise impacting the junction via the measurement probe; devices measured in an rf shielded room show much sharper current-voltage characteristics.

The critical current as a function of temperature for the junction is shown in the inset of Fig. 2. The fit to the data indicates that near T_c, $I_c \propto (1-T/T_c)^n$ where n =

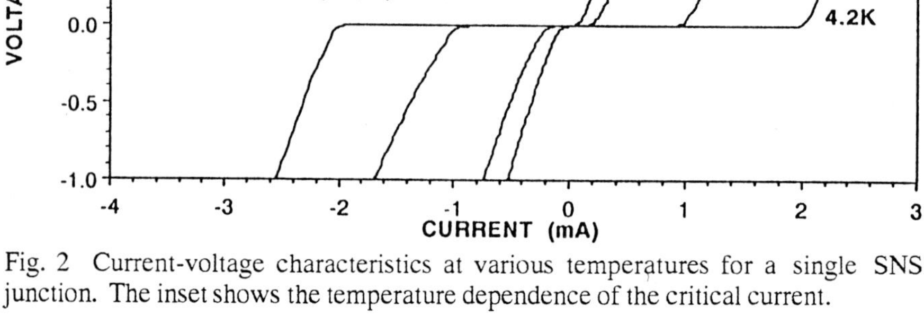

Fig. 2 Current-voltage characteristics at various temperatures for a single SNS junction. The inset shows the temperature dependence of the critical current.

1.0 ± 0.1. This does not follow the behavior expected for SNS devices [8] although it must be emphasized that the present high-T_c structure may have fundamental differences stemming from the nature of the SN interface. In contrast, most of the low resistance (R_n < 60 mΩ) devices we have fabricated yield n = 2, although the small output voltage and substantial noise rounding for most of these devices makes it difficult to accurately determine the critical current near T_c. Compared to the highest resistance devices such as shown in Fig. 2, the current-voltage characteristics of the low resistance devices show the "positive curvature" associated with the simple RSJ model.

Recently, a substantial number of single junctions and dc SQUIDs have been fabricated with I_cR_n products at 4.2 K ranging from 0.8-1.8 mV. The I_cR_n products are somewhat larger than expected from recent calculations for SNS devices containing silver as the normal metal [9,10]. For the devices with the highest I_cR_n products, we have not yet conclusively ruled out the possibility of some of the superconductor being continuous across the step, perhaps providing a parallel grain-boundary junction. The reproducibility of the junctions and the temperature dependence of the junction resistance, however, make this unlikely. Importantly, these high I_cR_n devices all clearly show Josephson behavior and successfully operate as dc SQUIDs.

Without knowing the precise thickness of silver along the step region, it is difficult to quantitatively compare the normal resistances of devices made with different silver thicknesses. A large group of 5-μm-wide junctions with various step heights were fabricated with the larger 300-600 nm nominal silver thicknesses. The resistances were quite low, ranging from 1.3 - 30 mΩ and leading to I_cR_n products of 5 - 50 μV. A key factor here is the quality of the interface between the $YBa_2Cu_3O_7$ and the Ag, which can strongly influence the junction resistance.

The two contributions to the normal resistance stem from the boundary resistance due to the SN interface as well as the resistance of the silver in the bridge region linking the two superconducting banks. The resistivity of our silver has been measured to be about 0.9 μΩ-cm at 4.2 K, which would contribute only about 0.9 mΩ of resistance to a 5-μm-wide junction with a bridge length of 100 nm and a silver thickness of 200 nm. The best contact resistivities reported for a YBaCuO/Ag interface have been 1×10^{-10} Ω-cm^2 for a c-axis oriented film and $<10^{-12}$ Ω-cm^2 for an a-axis oriented film [11]. The resistance contribution from the two such interfaces contained in a 5-μm-wide SNS device, assuming only 100 nm (roughly the thickness of the YBaCuO film) along the a-axis exposed film edge is contributing to the contact area, would then range from 0.4 - 40 mΩ. Some of our lowest resistance devices then may well have a very clean interface (the junction critical current densities are $1\text{-}3 \times 10^5$ A/cm^2). Note that under various process conditions the contact resistivity could readily be higher than the numbers above, and a value of 1×10^{-9} Ω-cm^2 could yield a resistance of 0.4 Ω.

Reproducibility has not been a deliberate initial goal but early results have been encouraging. For example, one substrate containing three microbridges and four dc SQUIDs had a variation in critical current, scaled for the deliberate differences in junction width, of ±17 %. The variation in the normal resistance, also scaled for junction width, was ±15 %. Further improvement is expected from refinements to the process. Many junctions have been thermally cycled from room temperature to liquid helium temperature 3-4 times with minimal or no change in the key device characteristics.

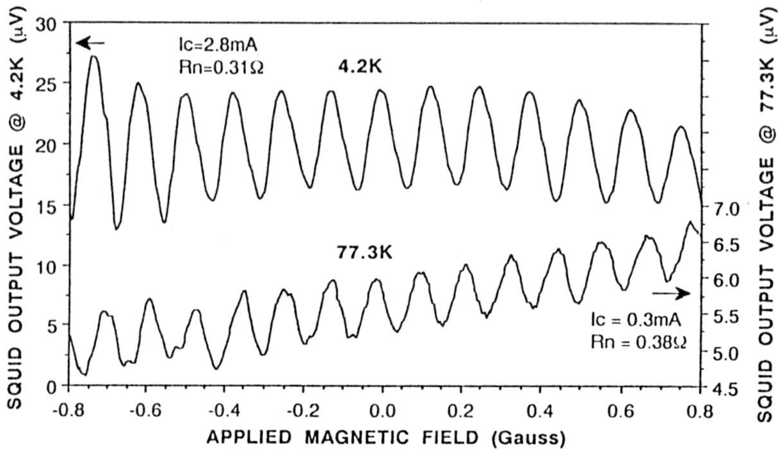

Fig. 3. (a) The output voltage of a dc SQUID at 4.2 K as a function of applied magnetic field. (b) The same SQUID at 77.3 K. The two junctions are each 3 μm wide and the SQUID is biased just above its zero field critical current.

The behavior at 4.2 K and 77.3 K of a dc SQUID in a magnetic field produced by an external coil is presented in Fig. 3. The period of the modulation of the output voltage is in excellent agreement with that expected for this SQUID, which has a geometrical loop area of 130 μm^2, when the increase in effective area due to the penetration of the field into the film (the perpendicular penetration depth $\lambda_\perp \sim 0.6\ \mu m$ at 4.2 K for this film) is taken into account. Following Tesche and Clarke [12] the inductance can be calculated from the depth of the critical current modulation, assuming the resistively-shunted junction model is applicable to these junctions. The calculated inductance value of 40 pH is reasonably close to the value of 26 pH determined from the geometry [13]. Note that at 77.3 K the period of the modulation is slightly smaller, due to the increase in the loop area stemming from an increase in the perpendicular penetration depth ($\lambda_\perp \sim 1\ \mu m$ at 77.3 K).

The highest temperature at which we have observed modulation of the SQUID output voltage in a magnetic field is 85.7 K. The largest voltage modulation we have seen at 4.2 K is 60 μV, while the largest at 77 K is 2 μV. The period of the modulation for SQUIDs with different loop areas is consistent when $\lambda_\perp$ and "flux-focusing" due to the presence of superconducting film in the SQUID loop are taken into account.

The modulation of the critical current of a *single* junction with applied magnetic field exhibits a clear periodic pattern where the period is typically within 30% of that expected from the area of the junction. The modulation, however, is smaller than expected and generally limited to about 25% of the maximum critical current.

Both single junctions and dc SQUIDs show a clear ac Josephson effect in response to microwave radiation. As many as 16 voltage steps have been observed at 4.2 K and steps have been observed at temperatures up to 80 K.

The spectral density of the flux noise in a number of these high-T_c SQUIDs was measured at 4.2 K using a conventional Nb-based Biomagnetic Technologies

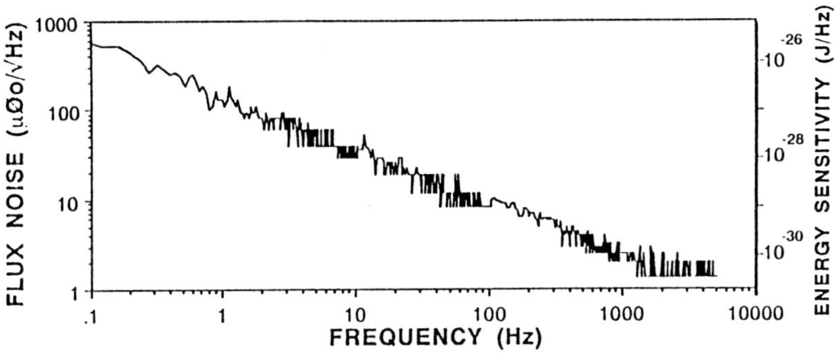

Fig. 4 Flux noise sensitivity of dc SQUID as a function of frequency at 4.2 K. Right-hand axis shows energy sensitivity.

dc SQUID. The high-T_c SQUID was voltage-biased in an open-loop configuration and the BTi SQUID was used as a current amplifier. A field was applied with an external coil. A plot of the flux noise, $(S_\Phi/Hz)^{1/2}$, for a SQUID with an inductance, L, of 15 pH is shown in Fig. 4. The inductance was determined by the depth of the critical current modulation as well as the geometrical calculation, both of which gave similar results. The critical current of this SQUID is 0.4 mA and the normal resistance is 14 mΩ. In the white noise region (2.5 kHz), the flux noise is $1.4 \times 10^{-6}\ \Phi_0/\sqrt{Hz}$ corresponding to an energy sensitivity of $\varepsilon = S_\Phi/2L = 2.8 \times 10^{-31}$ J/Hz. This is comparable to the lowest value of energy sensitivity reported for a high-T_c SQUID [14], although it is important to note that the inductance is smaller than desired for a practical SQUID, and the temperature of the measurement is 4.2 K.

Note that if the usual SQUID theory [12] is valid for our SQUIDs, then we expect the spectral density of the thermal flux noise $S_\Phi \approx 16\ k_BTL^2/R$, where R is the resistance for each junction and k_B is Boltzmann's constant. For the SQUID of Fig. 4, the calculated value of S_Φ is $1.7 \times 10^{-12}\ \Phi_0^2/Hz$, which is close to the measured value of $2.0 \times 10^{-12}\ \Phi_0^2/Hz$. We have not yet made measurements at higher temperatures or on the SQUIDs with the high normal resistances.

The noise spectrum is close to 1/f as the slope of the S_Φ data extracted from Fig. 4 is 1.1 ± 0.1. The lowest value of flux noise at 4.2 K measured on any of our dc SQUIDs at 10 Hz is $2.6 \times 10^{-5}\ \Phi_0/\sqrt{Hz}$ corresponding to $\varepsilon = 4.8 \times 10^{-29}$ J/Hz.

In summary, we have developed a new process to reproducibly fabricate $YBa_2Cu_3O_7/Ag/YBa_2Cu_3O_7$ Josephson junctions and dc SQUIDs. The process appears reliable and produces robust devices with I_cR_n products up to 4.6 mV, including junction resistances up to 5 Ω. The devices operate over a very wide temperature range, namely from 4.2 K to over 87 K, and both the ac and the dc Josephson effect have been observed from 4.2 K to above 80 K. The noise of various dc SQUIDs has been measured at 4.2 K and has been shown to be very low. The properties of the junctions are still evolving rapidly as key process parameters are optimized and there is good reason to expect further improvement.

This work has been supported in part by the Department of the Air Force Wright Laboratory (F33615-89-C-5655) and DARPA (N00014-88-C-0468).

4. References

[1] J. Gao, W. A. M. Aarnink, G. J. Gerritsma, D. Veldhuis, and H. Rogalla, IEEE Trans. Magn. **27**, 3062 (1991).
[2] R. B. Laibowitz, R. H. Koch, A. Gupta, G. Koren, W. J. Gallagher, V. Foglietti, B. Oh, and J. M. Viggiano, Appl. Phys. Lett. **56**, 686 (1990)
[3] G. Koren, E. Aharoni, E. Polturak, and D. Cohen, Appl. Phys. Lett. **58**, 634 (1991).
[4] D. K. Chin and T. Van Duzer, Appl. Phys. Lett. **58**, 753 (1991).
[5] M. S. DiIorio, S. Yoshizumi, K-Y. Yang, J. Zhang, and M. Maung, Appl. Phys. Lett. **58**, 2552 (1991).
[6] C. T. Rodgers, A. Inam, M. S. Hegde, B. Dutta, X. D. Wu, and T. Venkatesan, Appl. Phys. Lett. **55**, 2032 (1989).
[7] K. K. Likharev, Rev. Mod. Phys. **51**, 101 (1979).
[8] K. K. Likharev, Sov. Tech. Phys. Lett. **2**, 12 (1976).
[9] M. Yu. Kupriyanov and K. K. Likharev, IEEE Trans. Magn. **27**, 2460 (1991).
[10] G. Deutscher and R. W. Simon, J. Appl. Phys. **69**, (1991).
[11] M. Lee, D. Lew, C. B. Eom, T.H. Geballe, and M. R. Beasley, Appl. Phys. Lett. **57**, 1152 (1990).
[12] C. D. Tesche and J. Clarke, J. Low Temp. Phys. **29**, 301 (1977).
[13] J. M. Jaycox and M. B. Ketchen, IEEE Trans. Magn. **17**, 400 (1981).
[14] M. Kawasaki, P. Chaudhari, T.H. Newman, and A. Gupta, Appl. Phys. Lett. **58**, 2555 (1991).

High-T_c Epitaxial Junctions and DC-SQUIDs

Shiguang Wang, Xianghui Zeng, Yuandong Dai, Yifei Hu, Huaming Jiang, Rangjiao Liu, and Jiazhang Li

Department of Physics, Peking University, Beijing 100871, P.R. of China

Abstract. By using the multilevel deposition technique and Ar ion beam milling, we have fabricated YBaCuO epitaxial edge junctions and dc SQUIDs with PrBaCuO buffer layer. The dc and ac Josephson effects have been observed at 77 K. The intrinsic energy resolution for the SQUID was found to be $8*10^{-29}$ J/Hz at 100 Hz at liquid nitrogen temperature.

1. Introduction

Among the potential applications of high Tc superconductivity, the electronics have been considered as the important ones. The devices fabricated from poly-crystalline films have a high noise, except the Tl-based thin film SQUID[1,2], and are difficult to reproduce. Great effort was payed to the fabrication of artificial weak links from epitaxial high Tc films. Some progress has been achieved at IBM and Bell Lab [3-5]. Especially the GBJ dc SQUIDs demostrated a good performance at 77 K. Besides the improvement of device performance, these techniques are of advantage to yield controllable device characteristics and to incorporate into more complex stuctures. We will present our results on the fabrication and the investigation of the epitaxial edge junctions (EEJ) and dc SQUIDs. Our devices exhibited Josephson behaviour and quantum interference effect at liquid nitrogen temperature.

2. Experimental Details

The YBaCuO base and counter electrodes were deposited by dc magnetron sputtering with in situ heat treatment, which becomes a conventional technique for the preparation of high Tc epitaxial thin films [6]. Briefly, a one-to-ten mixture of oxygen and argon was used as the working gas, with total pressure of 26 Pa. The (100) $SrTiO_3$ substrate was mounted onto a substrate holder, which was kept at 750°C during deposition, aligned off-axis to the plasma ring in order to avoid the bombardment by negative ions. A dc current of 160 mA (via a 120 V dc bias) was applied to a sintered stochiometrical $YBa_2Cu_3O_{7-x}$ target of 60 mm in diameter. This yielded a deposition rate of 3 nm/min. A 10 minutes *in situ* thermal annealling was carried out at 400°C in 1 atm oxygen after the deposition. The films had a smooth and shiny surface. XRD and RHEED patterns indicated that the films were grown epitaxially with c-axis normal to the substrates. The zero

Springer Proceedings in Physics, Vol. 64 **Superconducting Devices and Their Applications**
Editors: H. Koch, H. Lübbig

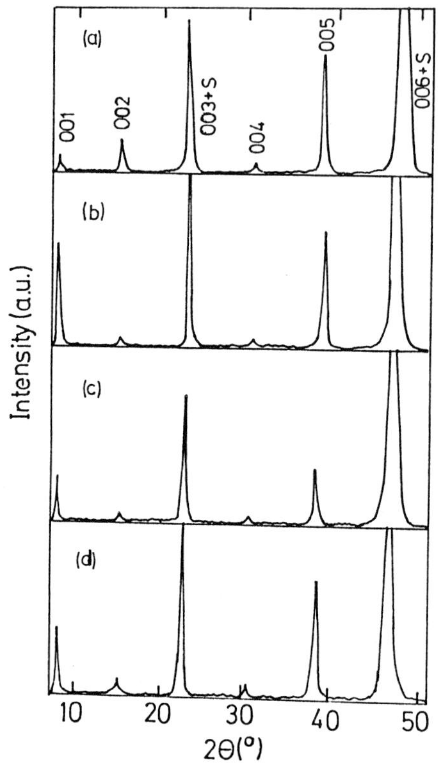

Fig. 1. X-ray diffraction patterns for: YBaCuO (a); PrBaCuO (b); YBaCuO/PrBaCuO bilayer (c); YBaCuO/PrBaCuO/YBaCuO trilayer (d) on $SrTiO_3$ (100) substrates. One sees only the peaks related to c-axis (00l) in the patterns.

resistance temperature was up to 88K and the critical current density $1.2*10^6$ A/cm^2 at 77 K.

$PrBa_2Cu_3O_{7-x}$ was selected as the insulation between base and counter electrodes, since it possesses the same crystal structure as $YBa_2Cu_3O_{7-x}$, and can be etched much easier than $SrTiO_3$ in the ion beam milling process. The PrBaCuO layer was deposited by rf sputtering with the similar condition as used for YBaCuO electrodes. Figure 1 gives the X-ray diffraction patterns of YBaCuO and PrBaCuO on $SrTiO_3$ with different sequences. It shows that they grew epitaxially on each other. The RBS results indicated no interdiffusion between them, but a minor diffusion between the bottom layer and the substrate. The epitaxial growth prevented the interdiffusion, thus provided a clear interface between the elecrodes and the buffer layer.

For the fabrication of the epitaxial edge junctions, the base electrode and the insulation were deposited in a vacuum chamber with three targets. The dc or rf power was applied to the desired target, under which the substrate holder could be aligned without breaking vacuum. The base YBaCuO and the PrBaCuO layers had a thickness of 150 nm and 200 nm respectively. Half plane of the bilayered film

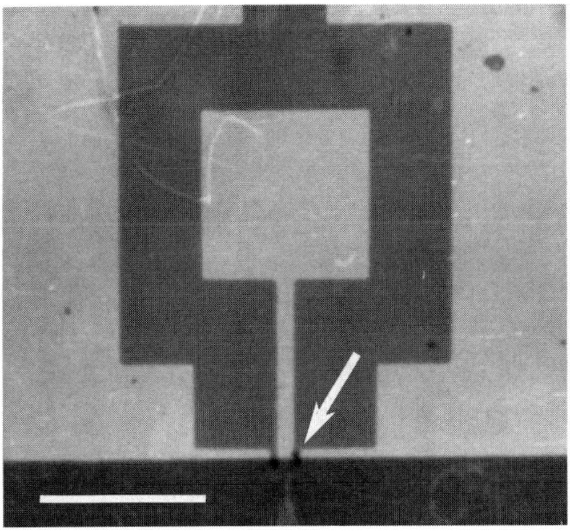

Fig.2. Micrograph of an EEJ dc SQUID with the square-washer superconductive loop. The arrow indicates the location of the junctions, and the white bar is 100μm.

was removed by Ar ion beam milling with photoresist as a mask. Subsequently the YBaCuO counter electrode of 200 nm thickness was deposited and ion milling was used to define the junction pattern. The junctions size was 2-5 μm (wide) by 0.15-0.2 μm (thick). After the patterning the devices were annealed at 400°C. This step restored the superconductivity of electrodes, especially the base, which might be degraded during the depositoion of the counter or the ion milling process. Four silver pads were thermally evaporated for the electrical measurement.

The dc SQUIDs were fabricated with the same procedure as that for EEJs. Figure 2 shows the micrograph of a SQUID. Instead of inserting the junctions into the two arms of the SQUID loop in our old design[7], we modified the pattern to have a square-washer, which was merely formed by the counter electrode, while the two junctions were defined in the edge of two electrodes. The junctions were so closely located that their physical properties were likely to be the same. The area enclosed by the superconductive loop was 100 μm x 100 μm, of which the inductance was about 200 pH. This pattern, moreover, makes it easier for the coupling of the superconducting flux transformer, which is under investigation in our laboratory.

3. Results and Discussion

The junctions possessed a zero resistance temperature of 80 K typically. The critical current at 77K varied from 50 μA to 1 mA for different junctions. Figure 3 is the current voltage characteristics of an edge junctions without and with microwave irradiation. The I-V curves are bridge-typed with no hysteresis at

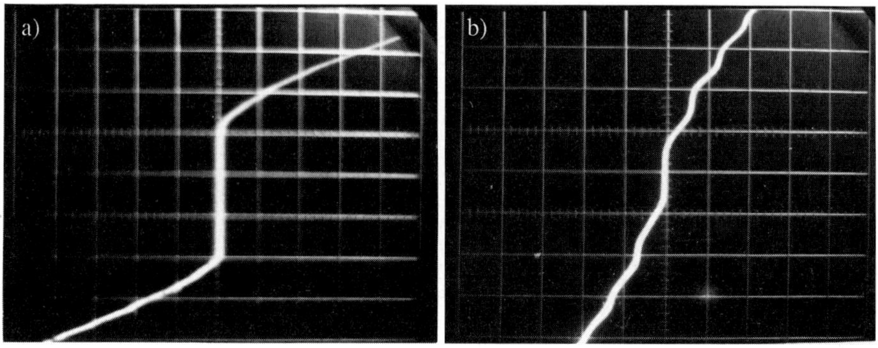

Fig.3. Current-voltage characteristics of an edge junction without (a) and with (b) microwave irradiation at 68 K. X=35μV/div, (a) Y=250μA/div, (b) Y=125μA/div.

liquid nitrogen temperature. The microwave irradiation leads to clear Shapiro steps in the I-V curves. The voltages where the steps occur fit the Josephson relation $V_n=n(h/2e)f$. The heights of the zeroth microwave induced steps varied as a function of microwave amplitude in accordance with the prediction of the resistively shunted junction model[8].

In contrast, the microbridges of the same width fabricated from a single epitaxial YBaCuO slab had a critical current about two orders of magnidute higher than the edge junctions, and we have never observed the microwave induced steps of the bridges unless the film was of poor quality. This implies that the weak links be localized in the boundary between the two electrodes. The barrier was native, which might be due to the oxygen or composition deficiency created by the ion milling process. Although our fabrication process was rudimentary in comparison with that of low Tc devices, half of our samples (16 out of 30) manifested Josephson behaviour at liquid nitrogen temperature. Since the coherence length is short in c-axis (0.3nm) and relatively long in a-b plane (3nm), and the the isolating layer is rather thick (200nm), the weak links occured most likely in the edge perpendicular to the a-b plane of two electrodes.

The dc characteristics of the SQUIDs were similar to the single junctions. The quantum interference effect could be observed at 77K for the samples with I_c<1mA. The investigation was performed in the flux-modulated mode, in which a magnetic flux was applied to the loop through a coil of 1mm in outer diameter attached to the SQUID. The output was coupled to a preamplifier via a tank circuit with a cooled transformer. The noise measurment was carried out in a flux-locked loop. A sinusoidal signal of 50 kHz was used to produce a modulating flux of amplitude about $\pm\Phi_0/4$. A commercial lock-in amplifier was employed to demodulate the preamplified signal. The dc output was integrated to generate a feed-back flux to the SQUID. The feed-back signal was plotted versus time by an X-Y recorder. We determined the total noise in the dc-1Hz band by the root mean square of the fluctuation of the fead-back signal in 100 seconds. The noise spectrum was detected by an HP 35665A dynamic signal analyzer. Since our feed-back system had a rather low slew rate, the study was limited to the frequency

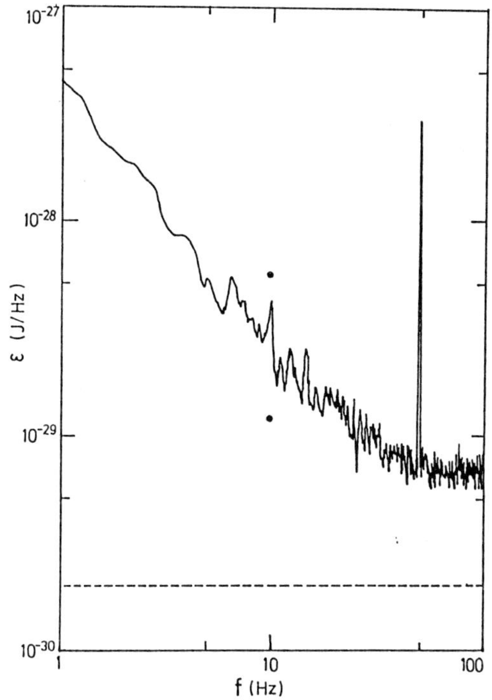

Fig.4. Measured energy resolution of a dc SQUID operated in the flux-lock mode at 63 K. The dashed line indicates the noise level of the BTI rf SQUID, and the dots are the results of grain-boundary-junctions dc SQUID from ref. [4].

below 100 Hz. Scaled with the output level corresponding to one quantum flux, we found the noise was $7.5*10^{-4}\Phi_0/\sqrt{\text{Hz}}$ at 65K , which is equivelant to an energy resolution of $5.6*10^{-27}$ J /Hz. Figure 4 is the intrinsic energy resolution measured from 1 Hz to 100Hz. It indicates that the $1/f$ noise plays a predominant role in this band, and the white noise becomes important arround 100 Hz, in which the noise power density is about $8*10^{-29}$ J/Hz. In comparison with results reported for various high Tc SQUIDs[1-5], the $1/f$ noise of our devices falls in the best cluster, however the white noise is about an order of magnitude higher. Nevertheless the latter is comparable with the commercial rf SQUIDs operated at 4.2 K.

4. Summary

We have developed the multilevel deposition technique of epitaxial YBaCuO/ PrBacuO/YBaCuO films. With proper thickness of PrBaCuO, we obtained a good insulating crossover. By using this technique epitaxial edge junctions and dc SQUIDs have been fabricated. The single junctions manifested dc and ac Josephson effects at 77K. The dc SQUIDs operated in liquid nitrogen temperature have achieved an energy resolution $5.6*10^{-27}$ J/Hz in dc to 1Hz band and $8*10^{-29}$ J/Hz around 100 Hz.

Acknowledgement

The authors did benefit form the discussion with Profs. Z.Z. Gan and D.L. Yin. The work is suported by the FEUYT, SEDC, CHINA and the National R&D Center for Superconductivity.

References

[1] R.H. Koch, W.J. Gallagher, B. Bumble, and W.Y. Lee, Appl. Phys. Lett. **54**, 951(1989).
[2] X.H. Zeng, S.G. Wang, Y.D. Dai, R.J. Liu, Z.Z. Wang, J.Z. Li, S.L. Yan, and L. Chen, Cryogenics **30**, 915(1990).
[3] R.B. Laibowitz, R.H. Koch, A. Gupta, G. Koren, W.J. Gallagher, V. Foglietti, B. Oh, and J.M. Viggiano, Appl. Phys. Lett. **56**, 686 (1990).
[4] R. Gross, P. Chaudhari, M. Kawasaki, M.B. Ketchen, and A. Gupta, Appl. Phys. Lett. **57**, 727 (1990).
[5] C.T. Rogers, A. Inam, M.S. Hegde, B. Dutta, X.D. Wu, and T. Venkatesan, Appl. Phys. Lett. **55**, 2032 (1989).
[6] G.C. Xiong, and S.Z. Wang, Appl. Phys. Lett. **55**, 902(1989).
[7] S.G. Wang, G.J. Cui, Y.D. Dai, H.M. Jiang, J.Z. Li, X.H. Zeng, Z.L. Bao, S.Z. Wang, C.Y. Li, and D.L. Yin, IEEE Tran Magn. **MAG25**, 893(1989).
[8] Y.D. Dai, Y.F. Hu, H.M. Jiang, R.J. Liu, and S.G. Wang, to appear in J. Appl. Phys.

Multilayer $YBa_2CU_3O_x$ Structure – An Approach to SQUID Applications

M. Schilling, T. Bade, and U. Merkt

Institut für Angewandte Physik, Universität Hamburg,
Jungiusstr. 11, W-2000 Hamburg 36, Fed. Rep. of Germany

Abstract. Edge junctions formed by a grain boundary or a $PrBa_2Cu_3O_x$ layer between two $YBa_2Cu_3O_x$ layers are prepared on $SrTiO_3$ and MgO substrates by pulsed laser deposition using a KrF excimer laser. The initial layer of $YBa_2Cu_3O_x$ on the substrate is followed by an insulating layer of MgO. By photolithography and argon ion beam etching an edge is defined in these layers. After that a $PrBa_2Cu_3O_x$ barrier and a second $YBa_2Cu_3O_x$ layer are grown and patterned to form edge junctions with the initial $YBa_2Cu_3O_x$ layer. The junctions are characterized by current-voltage measurements. Weak links with low critical current densities $j_c(4.2K)$ in the 10^4 A/cm^2 regime as well as interconnections with high densities $j_c(4.2K) = 3 \cdot 10^6$ A/cm^2 can be obtained. We also discuss the influence of various preparation steps on the junction properties.

1. Introduction

Superconducting Quantum Interference Devices (SQUIDs) from copper-oxide high-temperature superconductors are one major application of these new materials. Despite of their short coherence lengths many successful realizations have been reported, using natural [1] or artificial grain boundaries [2]. Different approaches use non-superconducting materials as barriers between the superconducting electrodes to form weak-link junctions at the edge of a $YBa_2Cu_3O_x$ film [3] - [5]. This promises the possibility to accurately control all junction properties. Here we study junctions that are patterned from laser deposited multilayer $YBa_2Cu_3O_x$ structures by conventional photolithography and argon ion beam etching (IBE). We investigate the influence of insulating MgO layers and semiconducting $PrBa_2Cu_3O_x$ barriers on the resulting weak-link junctions.

2. Junction preparation

In Fig. 1 the sequence of preparation steps is shown schematically. First a $YBa_2Cu_3O_x$ film is grown on the heated substrate (785 °C) by laser ablation of

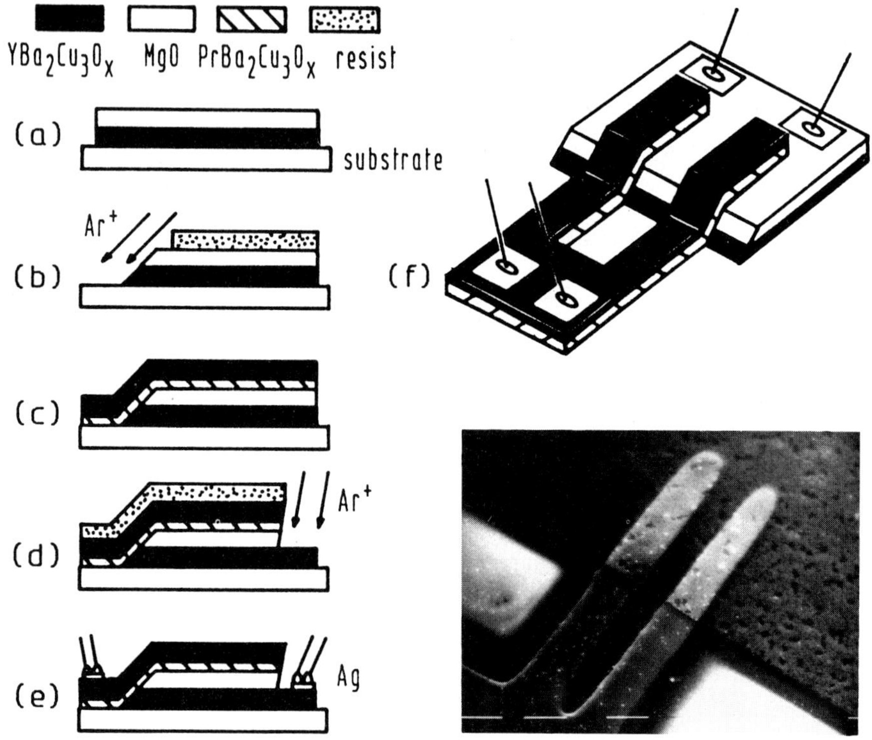

Figure 1. Preparation process for multilayer $YBa_2Cu_3O_x$-$PrBa_2Cu_3O_x$-$YBa_2Cu_3O_x$ weak-link junctions. The micrograph shows a DC-SQUID. The marker in this picture is 10 μm long.

a polycrystalline $YBa_2Cu_3O_x$ target in an oxygen atmosphere of 0.4 mbar [6]. On top of this 200 nm thick film a polycrystalline MgO film is deposited as an insulator between the two $YBa_2Cu_3O_x$ electrodes. The 100 nm thick MgO film is deposited exsitu in a different deposition chamber at a relatively low substrate temperature of 350 °C and $2 \cdot 10^{-5}$ mbar residual pressure. A degradation of the underlying $YBa_2Cu_3O_x$ layer by the MgO deposition was not observed and optimum critical temperatures $T_{C0} = 88$ K were obtained. In the first step this bottom electrode for the junctions is formed by photolithography and argon IBE under an angle of incidence of 60° as shown in Fig. 1(b). This etch process is done in a separate unit. The etch parameters are 5 kV acceleration voltage and 25 μA/cm² current density. Cleaning of the surface is accomplished by acetone rinsing and remaining spots of photoresist are removed in a barrel reactor in an oxygen plasma. For the laser deposition of the $PrBa_2Cu_3O_x$ layer and the $YBa_2Cu_3O_x$ top layer shown in Fig. 1(c) the sample is again mounted in the deposition chamber. In the following step these layers are patterned by photolithography and argon IBE employing the already mentioned parameters. However, we now use an angle of incidence of 30°. In this process the different plasma colors on the film surface during the IBE are used as etch stop indicators. For the contact pads a 100 nm thick silver layer is deposited by thermal evaporation through a metal mask. Wires are soldered with indium resulting in contact resistances less than 10 Ω. On a particular chip we prepare four single-junctions and three DC-SQUIDs as well as two microbridges for the separate characterization of both $YBa_2Cu_3O_x$ films.

3. Current-voltage measurements

For the characterization of the resulting structures current-voltage measurements are performed and resistance versus temperature curves are recorded. Usually we obtain critical temperatures T_{C0} above 82 K for both $YBa_2Cu_3O_x$ layers, despite of the many processing steps and the high argon ion energy. Usually critical currents in both layers lie above 10^5 A/cm^2 at 4.2 K, measured with a criterion of 10 μV/cm and optimum values exceed 10^6 A/cm^2. In Fig. 2 current-voltage curves of a DC-SQUID with a lower critical

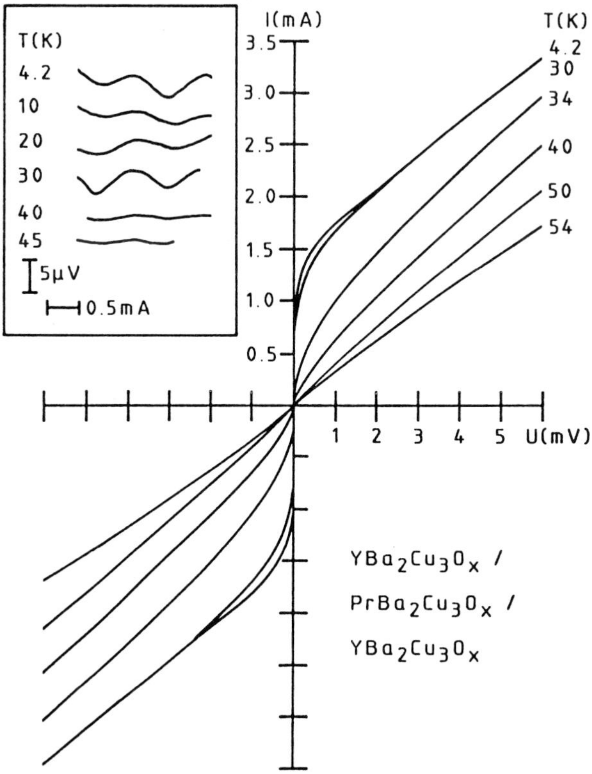

Figure 2. Current-voltage curves of a DC-SQUID for various temperatures. The inset shows the flux modulation of the SQUID versus the current in a modulation coil.

temperature T_{C0} = 60 K are shown for different temperatures. In the inset the voltage modulation of the SQUID is reproduced against the current flowing through an external modulation coil. Traces for different temperatures are given and flux modulation could be detected up to 45 K. The critical current of our junctions depends on the barrier thickness. Much larger $PrBa_2Cu_3O_x$ layer thicknesses are required compared to the values (6nm for I_C = 0.8 mA) of the Twente-group [4]. We started with this value of 6 nm and deposited a barrier layer using our usual deposition rate of 0.27 nm per laser pulse. This junction exhibited nearly no deminution of the critical current in respect to the value measured on the microbridges on both films. We had to raise the $PrBa_2Cu_3O_x$ thickness to 16 nm to get a critical current of the SQUID below 1 mA, while the microbridge of the same cross section exhibited a critical current of 160 mA. The comparatively large thickness of the barrier layer may have two reasons. First the carrier density in the particular $PrBa_2Cu_3O_x$ layer at the operation temperature of the SQUID determines the coherence length and the critical current through the barrier in case of our superconductor-normal metal-superconductor (SNS) junctions [7]. Our $PrBa_2Cu_3O_x$ had a specific resistance of 0.6 Ωcm at 77 K. A second reason is the higher surface roughness of the laser deposited films which may require thicker films to fill up holes in the film in order to avoid short circuits between adjacent layers [8]. We think that the current-voltage curves of our junctions are not influenced by short circuits neither in the vertical nor in the lateral direction, because the resulting heating effects would give an extremely steep voltage increase above the critical current. The insulating MgO layer ensures that there are no vertical shorts between the $YBa_2Cu_3O_x$ layers and the current-voltage curves in fact are linear up to a voltage of about 100 mV.

4. Interconnections with high critical currents

We also investigated structures without a $PrBa_2Cu_3O_x$ barrier and without a MgO insulation layer. The critical currents of such junctions were as high as the values in the microbridges of the two $YBa_2Cu_3O_x$ layers, namely above 10^6 A/cm^2 at 77 K. So these structures could advantageously be used for lateral interconnections with high critical currents between $YBa_2Cu_3O_x$ layers. For example they could be employed as contacts to magnetometer coils on top of SQUID washers.

5. Conclusions

Multilayer $YBa_2Cu_3O_x$ structures are a promising route to SQUID applications. The advantage over other preparation methods is the possibility to control the critical current of the junctions by their geometry and by the thickness and specific resistance of the barrier material. Without a barrier layer and without a vertical insulation layer lateral interconnections with critical currents exceeding 10^6 A/cm^2 can be prepared between two $YBa_2Cu_3O_x$ layers.

For further improvement of the SQUIDs still smoother films are required. Such work is under progress.

We acknowledge financial support from the Bundesministerium für Forschung und Technologie, Federal Republic of Germany.

References

[1] R.H. Koch, C.P. Umbach, G.J. Clark, P. Chaudhari and R.B. Laibowitz, Appl. Phys. Lett. **51**, 200 (1987).

[2] R. Gross, P. Chaudhari, M. Kawasaki, M. Ketchen and A. Gupta, Physica C **170**, 315 (1990).

[3] R.B. Laibowitz, R.H. Koch, A. Gupta, G. Koren, W.J. Gallagher, V. Foglietti, B. Oh and J.M. Viggiano, Appl. Phys. Lett. **56**, 686 (1990).

[4] J. Gao, W.A.M. Aarnink, G.J. Gerritsma, D. Veldhuis and H. Rogalla, IEEE Trans. Magn. **27**, 3062 (1991).

[5] D.K. Chin and T. Van Duzer, Appl. Phys. Lett. **58**, 753 (1991).

[6] M. Schilling, K.D. Laue and U. Merkt, J. Less Common Metals **164 & 165**, 400 (1990).

[7] A.W. Kleinsasser and T.N. Jackson, Phys. Rev. B **42**, 8716 (1990).

[8] M. Schilling, F. Goerke and U. Merkt, Proceedings of the ICAM 1991, E-MRS spring meeting, Strasbourg, to be published.

DC-SQUID with Step Edge Junctions on (100) $SrTiO_3$

M. Siegel[1], *F. Schmidl*[1], *E. Heinz*[1], *K. Zach*[1], *J. Fuchs*[2], *E.-B. Kley*[2], *and P. Seidel*[1]

[1]University Jena, Institute of Solid State Physics,
Helmholtzweg 5, O-6900 Jena, Fed. Rep. of Germany
[2]University Jena, Institute of Applied Physics,
Beutenbergstr., O-6900 Jena, Fed. Rep. of Germany

Abstract. We have fabricated dc SQUIDs based on $YBa_2Cu_3O_{7-x}$ thin films with step edge junctions on steps on (100) $SrTiO_3$ substrates. The steps were etched by ion milling, the $YBa_2Cu_3O_{7-x}$ films were inhibit layer patterned and deposited by laser ablation. The SQUIDs show perfect periodic flux-voltage characteristics in the temperature range between 4.2 K and 63 K. The voltage-flux transfer function was 3.3 mV/Φ_0 and 12 $\mu V/\Phi_0$ at 4.2 K and 63 K, respectively.

1. Introduction

Superconducting quantum interference devices (SQUIDs) based on the naturally occurring weak links in polycrystalline films of YBaCuO, BiSrCaCuO and TlBaCaCuO have been made by a number of groups [1-3]. Multilevel, all high T_C thin film SQUIDs based on edge junctions similar to those used in some low T_C SQUIDs were presented by Laibowitz et. al. [4]. Using a single grain boundary between two epitaxial films it was shown that low noise dc SQUIDs can be made [5]. Wellstood et al. [6] have presented an integrated dc SQUID with a flux transformer based on bi-epitaxial weak links.

In this report we describe a fabrication process of dc SQUIDs with weak links formed at step edges on (100) $SrTiO_3$ substrates and discuss the properties of the SQUIDs in the temperature range from 4.2 K to 63 K.

2. Experimental

The $YBa_2Cu_3O_{7-x}$ thin films were deposited by in-situ laser ablation at an oxygen pressure of 0.3 mbar, and a substrate temperature of 800 °C using a KrF excimer laser operating at a repetition rate of 10 Hz. The X-ray diffraction pattern of the films on (100) $SrTiO_3$ substrates reveal clear (001) lines and show c-axis textured growth. The films exhibit a critical temperature of 89 K (R=0) and a critical current density higher than 10^6 A/cm^2 at 77 K.

Recently, step edge junctions have been fabricated by several groups [7, 8]. The fabrication process of step edge junctions requires etching a sharp step into the (100) $SrTiO_3$ substrate prior to the thin film growth. We have etched the steps by ion milling using Ar^+ ions of 650 eV and an ion current density of 250 $\mu A/cm^2$. Height and angle of the steps were 200 nm and 60°, respectively. According to [8] the

Springer Proceedings in Physics, Vol. 64 Superconducting Devices and Their Applications
Editors: H. Koch, H. Lübbig

$YBa_2Cu_3O_{7-x}$ film thickness was also approximately 200 nm to obtain weak link behaviour of the step edge junctions.

The junctions and the SQUID layout were patterned by an inhibit patterning process. Before depositing the $YBa_2Cu_3O_{7-x}$ thin film, a 20 nm thick layer of TiO_2 or SiO_2 was prepared on the substrate and patterned by a lift-off process. Because $YBa_2Cu_3O_{7-x}$ thin films become not superconducting on both materials, this layer is serving as an inhibit mask. On (100) $SrTiO_3$ substrates without any steps we have found the superconducting properties of the $YBa_2Cu_3O_{7-x}$ films to be not degraded by the patterning process down to structures of 3 μm lateral width.

Based on inhibit layer patterning we have fabricated a rectangular SQUID washer geometry with a SQUID loop of 100 pH inductance and two step edge junctions having a lateral width of 3 μm.

3. Results and Discussion

In Figure 1 we present the current-voltage characteristic of a dc SQUID with two identical step edge junctions. It is comparable to those calculated within the RSJ model including an excess current. $I_C R_N$ products of up to 400 μV have been determined. In contrast to our results the current-voltage characteristics of step edge junctions of other groups show not a RSJ-like but a flux flow behaviour [7, 8]. Figure 2 shows the temperature dependence of the critical Josephson current of the SQUID.

Figure 3 shows the flux-voltage characteristic of one of our dc SQUIDs at temperatures of 4.2 K and 63 K. The flux modulation was periodical over many periods up to a temperature of 63 K. At higher temperatures we have observed phase jumps in the flux-voltage curves that we attribute to vortex jumps in the SQUID loop.

Figure 4 demonstrates the good periodical dependence of $dV/d\Phi$ on Φ, measured using a lock in modulation technique. Figure 5 shows the dependence of voltage-flux transfer function on bias current at 4.2 K. We have obtained a sharp maximum in that bias current range, where the voltage-flux characteristic deviates extremely from sinusoidal behaviour.

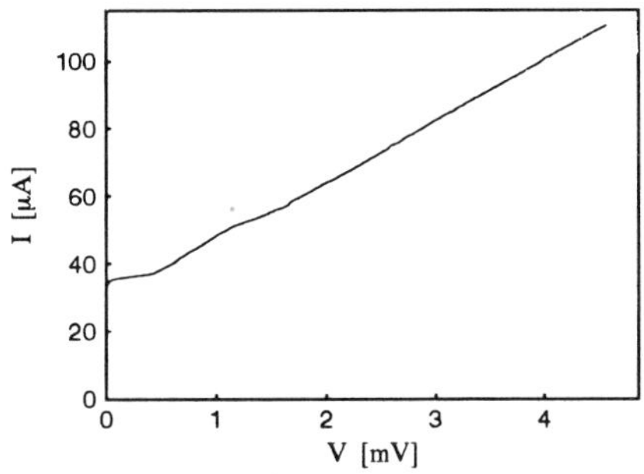

Fig. 1 Current-voltage characteristic of a dc SQUID at 4.2K.

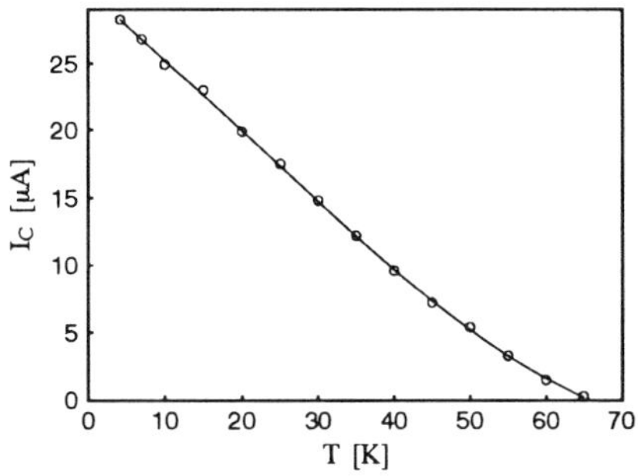

Fig. 2 Dependence of the critical current on temperature.

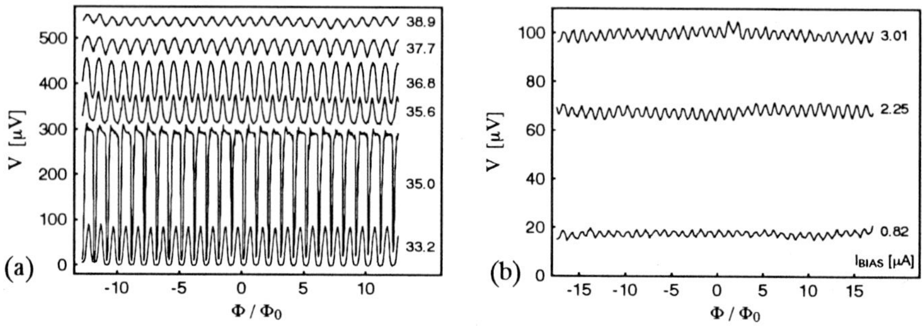

Fig. 3 Voltage-flux characteristics of a dc SQUID at different bias currents (I_{BIAS} [μA]) and temperatures of 4.2 K (a) and 63 K (b).

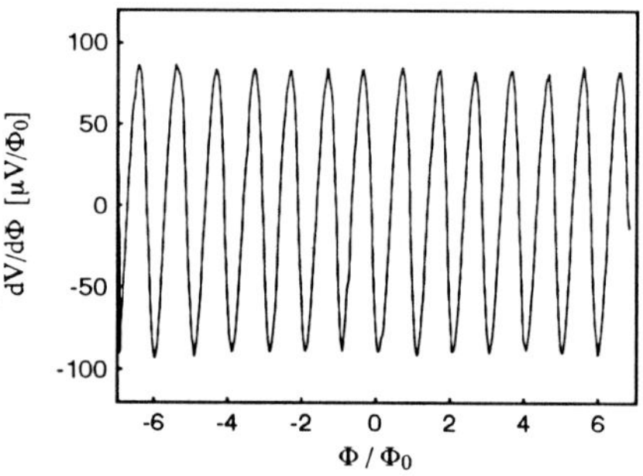

Fig. 4 dV/dΦ versus external magnetic flux at 4.2 K.

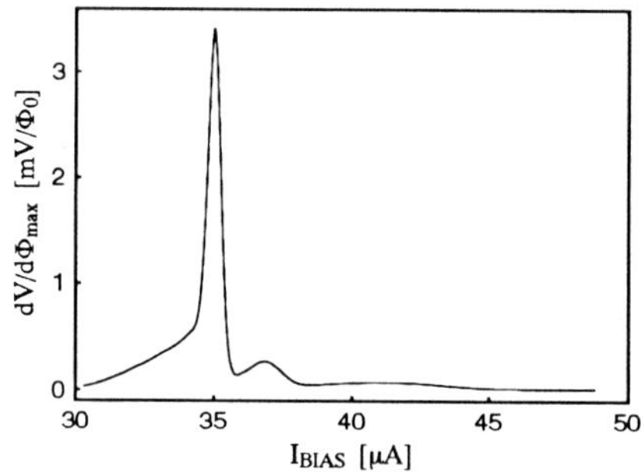

Fig. 5 Dependence of the voltage-flux transfer function on bias current at 4.2 K.

4. Conclusions

We have fabricated dc SQUIDs with step edge junctions on (100) $SrTiO_3$ substrates. The junctions exhibit RSJ-like current-voltage characteristics. Voltage modulation by applied magnetic flux has been observed up to 63 K.

Acknowledgements

We would like to thank Mr. G. Luge and Mrs. W. Gräf for technical assistance. This work was supported by BMFT contract No. 13 N 5924.

References

[1] R.H. Koch, C.P. Umbach, G.J. Clark, P. Chaudhari, and R.B. Laibowitz, Appl. Phys. Lett. **51**, 200 (1987)

[2] I.S. Gergis, J.A. Titus, P.H. Kobrin, and A.B. Harker, Appl. Phys. Lett. **53**, 2226 (1988)

[3] R.H. Koch, W.J. Gallagher, B. Bumble, and W. Lee, Appl. Phys. Lett. **54**, 951 (1989)

[4] R.B. Laibowitz, R.H. Koch, A. Gupta, G. Koren, W.J. Gallagher, V. Foglietti, B. Oh, and J.M. Viggiano, Appl. Phys. Lett. **56**, 686 (1990)

[5] R. Gross, P. Chaudhari, M. Kawasaki, M. Ketchen, and A. Gupta, Appl. Phys. Lett. **57**, 727 (1990)

[6] F.C. Wellstood, J.J. Kingston, A.M. Miklich, M.J. Ferrari, J. Clarke, M.S. Colclough, K. Char, G. Zacharchuk, E-MRS, ICAM '91, Straßbourg, May 27-31, 1991, France

[7] R.W. Simon, J.B. Bulman, J.F. Burch, S.B. Coons, K.P. Daly, W.D. Dozier, R. Hu, A.E. Lee, J.A. Luine, C.E. Platt, S.M. Schwarzbek, M.S. Wire, M.J. Zani, Proc. ASC '90, IEEE Trans. Mag. **27**, 3209 (1991)

[8] G.J. Cui, K. Herrmann, Y. Zhang, C.L. Jia, Ch. Buchal, J. Schubert, W. Zander, A.I. Braginski, Ch. Heiden, to appear in 'Nonlinear Superconductive Electronics and Josephson devices', ed. by N.F. Pedersen, M. Russo, A. Davidson, G. Costabile, A. Oosenburg, Plenum Press, N.Y. (1991)

$YBa_2Cu_3O_x$ DC-SQUIDs on Y-ZrO_2 Bicrystals

Z.G. Ivanov[1], *P.A. Nilsson*[1], *D. Winkler*[1], *G. Brorsson*[1], *T. Claeson*[1], *E.A. Stepantsov*[2], *and A.Ya. Tzalenchuk*[2]

[1]Department of Physics, Chalmers University of Technology, S-412 96 Göteborg, Sweden
[2]Institute of Crystallography, USSR Academy of Sciences, 59 Leninskii pr., 117333 Moscow, Russia

Abstract. We have studied dc SQUIDs fabricated by laser deposition of $YBa_2Cu_3O_x$ thin films on yttria stabilized zirconia bicrystal substrates. The SQUID properties were studied as a function of the misorientation angle θ between the two halves of the bicrystal. The best performance was obtained for $\theta = 32°$. The modulation depth in the voltage-flux curve in this case was 12 μV at 77 K. No hysteresis was observed when the direction of the magnetic field was reversed.

1. Introduction

It has been demonstrated that an artificially created grain boundary in an epitaxial $YBa_2Cu_3O_x$ (YBCO) film shows a Josephson behavior. The pioneering investigations in this field [1] showed that the symmetric boundary between two $SrTiO_3$ crystals sintered together could be transferred to an epitaxial film during the growth process, hence forming a weak link. The properties of the latter depend on the misorientation angle θ of the boundary. We have studied dc SQUIDs formed in YBCO thin films deposited on yttria stabilized zirconia (YSZ) bicrystal substrates with non-symmetric tilt boundaries for different θ.

2. SQUID fabrication

Bicrystal substrates were made by joining together two single crystals of YSZ with different orientations [2] by a solid phase intergrowing method [3]. The interfaces were smooth on an atomic scale. High quality YBCO thin films were grown on the bicrystals by pulsed laser deposition. The films were highly c-axis oriented with FWHM for the [005] peak of 0.25°. No traces of other orientations could be seen. The critical temperature of the films was 90 K and the critical current $j_c \approx 2\text{x}10^6$ A/cm^2 at 77 K (more details in [4]).

A SQUID loop with dimensions 50 x 5 μm^2 and two weak links, 4 μm wide, were patterned by using photo resist masks and Ar ion beam milling (Fig. 1a). No changes were observed in the superconducting properties of the YBCO film after the patterning. The loop inductance was 45 pH as calculated from the dimensions.

Springer Proceedings in Physics, Vol. 64 Superconducting Devices and Their Applications
Editors: H. Koch, H. Lübbig

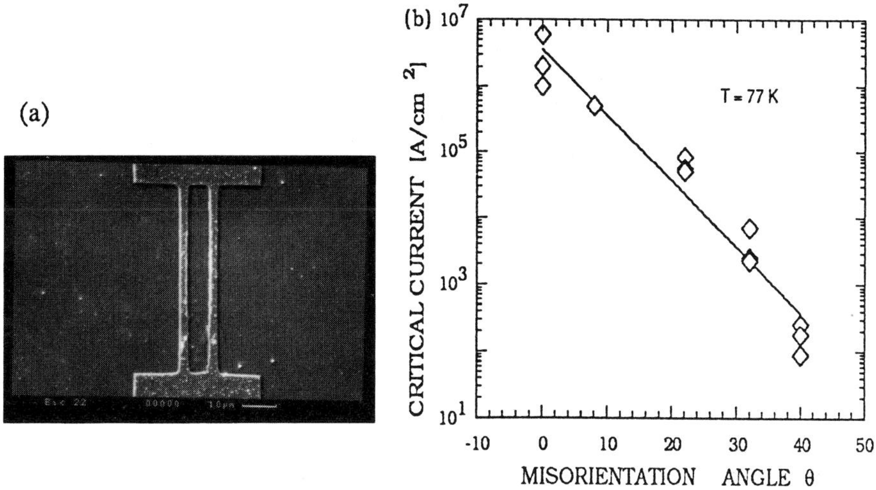

Fig.1 (a) SEM micrograph of the central part of the chip. The SQUID loop is 50 x 5 μm^2 and the YBCO film is 0.25 μm thick; (b) j_c vs. θ dependence for weak links on YSZ bicrystal substrates measured at 77 K.

3. Results and discussion

TEM studies showed that the changes in the film orientation at the boundary takes place in a region less than 10 Å [5]. This sharp transition between the two orientations of the YBCO film assures good Josephson coupling. The critical current density, j_c, of the boundary was studied as a function of θ between two halves of the bicrystal. A summary of the results is presented in Fig. 1b. The j_c of the boundary decreases exponentially with increasing θ up to 45°. For low θ, j_c was high and the weak links showed a linear increase in the dynamic resistance with bias voltage. For high θ, j_c was significantly decreased at the boundary and the weak links showed behavior well described by the resistively shunted junction model (RSJ). Microwave response was observed in both cases [5]. The j_c dependence on θ provides information of how to choose the proper angle in order to design a SQUID with the best performance for a certain operation temperature.

The magnetic field responses of the SQUIDs were investigated for θ = 22– 45°. For θ = 22°, I_c was of the order of 1 mA and the screening parameter $\beta=2LI_c/\Phi_0>>1$ at T = 77 K. This SQUID was far off the optimal performance and only a weak magnetic field response was observed at 77 K. For θ equal to 32, 40 and 45°, I_c was between 40-100 μA. β was of the order of 1 and good magnetic field responses were observed for all these SQUIDs. A SQUID with θ = 45°, which is the maximum possible tilt, had a depressed $T_c(R=0)$ of 72 K. It gave a modulation of 50 μV (peak-to-peak) at 4.2 K and 1 μV at 70 K. For θ = 40°, T_c was 84 K. A good performance was obtained at 77 K.

The best results were obtained for the SQUID with θ = 32°. The voltage responses vs. external magnetic field applied normal to the film surface were measured for the SQUID and for the single weak links. In both cases the devices

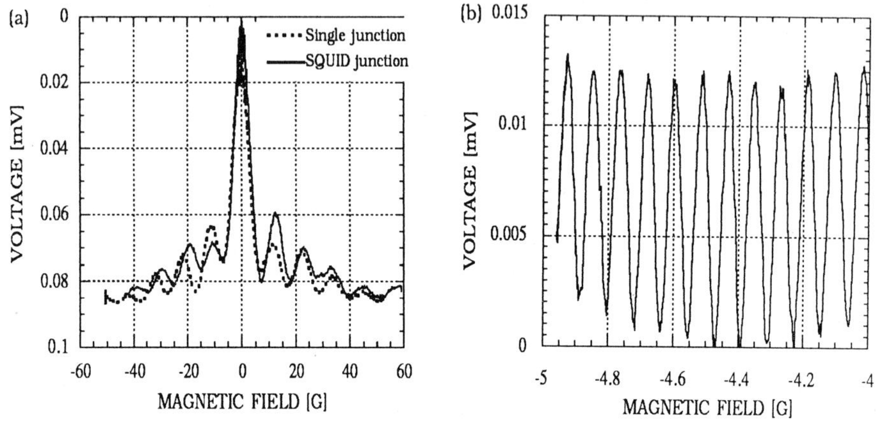

Fig. 2 (a) Voltage modulation vs. magnetic field for a current biased dc SQUID (solid curve) and weak link (dotted curve). The field is applied perpendicular to the SQUID loop and the modulation period corresponds to the area of the weak links; (b) Voltage modulation caused by magnetic field threading the area of the SQUID loop.

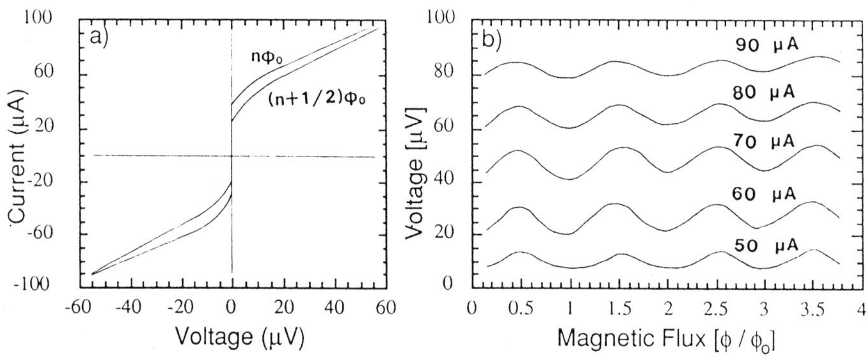

Fig 3. Flux modulation of the I-V characteristics (a), and voltage-flux characteristics at different bias currents (b) for a SQUID measured at 77 K. The V(Φ) curves are displaced along the voltage axis.

were current biased at $I_{bias} = 1.5I_c$ and Fraunhofer like dependences were observed (Fig. 2a). The fine modulation period for the SQUID corresponds to the magnetic flux quantum penetrating the area of the loop (Fig. 2b). The voltage-flux characteristics of the SQUID were periodic and the dimensionless parameter $\delta\Phi/\Delta\Phi_e$ was determined to be 2×10^{-3} for $\Delta\Phi_e = 50\Phi_o$.

Typical current vs. voltage curves with external magnetic field applied normal to the film surface and measured at 77 K are shown in Fig. 3a. The top curve corresponds to a flux equal to an integer number of flux quanta in the loop, the bottom to half integer. The modulation depth of the voltage-flux curve was not critical to the bias point (Fig. 3b) and a periodic behavior was observed for a wide range of I_{bias}. At 77 K, the maximum value of 12 μV was measured at $I_{bias} \approx 1.5I_c$.

A similar chip but with 14 SQUIDs was designed for measuring the noise properties. All devices on the chip were working. Preliminary measurements have been performed on a SQUID with L = 70 pH and a noise level of $1.5\times10^{-9}\ \Phi_0^2/\mathrm{Hz}$ was measured at 10 Hz and at a temperature of 85 K. These measurements were made by H.K. Olsson and R.H. Koch and will be published later.

4. Conclusions

We have demonstrated the possibility to fabricate high quality dc SQUIDs on yttria stabilized zirconia bicrystal substrates. These give nonhysteretic and large magnetic field response at 77 K. Preliminary noise measurements give $1.5\times10^{-9}\ \Phi_0^2/\mathrm{Hz}$ at 10 Hz and at 85 K for a 70 pH loop inductance. The artificial grain boundary junctions have I-V curves in close match to the RSJ model and respond to a magnetic field similarly as for small low T_c junctions with a uniform current distribution.

Acknowledgements

Discussions with K.K. Likharev, R.I. Sheckter, S. Zarembinski were very fruitful. Help from H.K. Olsson and R.H. Koch and microscopy work by J.A. Alarco is gratefully acknowledged. This work was supported by the Swedish Board for Technical Development and the Swedish Industrial Superconductivity Initiative.

References

[1] D. Dimos, P. Chaudhari, J. Mannhart, Phys. Rev. B41, 4038 (1990).

[2] E.A. Stepantsov and A.Ya. Tzalenchuk, Physica Scripta 44, 102 (1991).

[3] E.A. Stepantsov, A.s. USSR no. 1116100 (1982), Bulleten' Izobreteniy 36, (1984) 77.

[4] J.A. Alarco, G. Brorsson, T. Claeson, Z.G. Ivanov, M. Löfgren, and P.Å.Nilsson, to be published in the proceedings of ICAM '91, Symposium A1, Strasbourg, 22-31 May, 1991.

[5] Z.G. Ivanov, P.-Å. Nilsson, D. Winkler, J.A. Alarco, G. Brorsson, T. Claeson, E.A. Stepantsov, and A.Ya. Tzalenchuk, to be published in the proceedings of ISEC'91, Glasgow, 25-27 June, 1991.

YBCO Thin Films SQUIDs Fabricated on Bicrystal Substrates

S.I. Krasnosvobotsev[1], M.Yu. Kuprijanov[2], A.G. Maresov[2], V.G. Pirogov[2], O.V. Snigirev[2], and I.I. Vengrus[2]

[1]Lebedev Institute of Physics, Academy of Science, 117924 Moscow, Russia
[2]Department of Physics, Moscow State University, 119899 Moscow GSP, Russia

Abstract. Properties of the dc SQUIDs made of YBCO thin film deposited on $SrTiO_3$ and ZrO_2 bicrystal substrates and the properties of the fabricated Josephson junctions have been studied. It is found that the characteristics of the junctions are similar to those of the SNS structures. The magnetic field sensitivity has achieved a value of $\delta B \sim 10^{-12}$ T/Hz$^{1/2}$ at frequencies higher than 25 Hz and T = 4.2 K.

1. Introduction

One of the possible ways to create HTS SQUIDs is using a bicrystal substrate of suitable misorientation angle to fabricate the HTS Josephson junctions [1].

The goal of this work is to fabricate single-level dc SQUIDs made on bicrystal ZrO_2 and $SrTiO_3$ substrates and studying their characteristics as well as the nature of the Josephson junctions.

2. Device fabrication

YBCO thin films were deposited by a laser ablation process [2] on ZrO_2 and $SrTiO_3$ bicrystal substrates with a misorientation angle in the basal plane close to 35-40° and 18-20° respectively.

The deposited films had usually a mirror-like surface. The X-ray diffraction pattern of the films typically showed only (001) type reflexes corresponding to a well oriented thin film with the c-axis normal to the substrate.

The films had a thickness d = 0.3-0.5 μm and a resistance in the normal state at 100 K $\rho \simeq 130$ $\mu\Omega\times$cm for ZrO_2 and $\rho \simeq 300$ $\mu\Omega\times$cm for $SrTiO_3$. The critical temperature T_c of the films for both types of substrates lies in the range T_c = 82-90 K.

Springer Proceedings in Physics, Vol. 64 Superconducting Devices and Their Applications
Editors: H. Koch, H. Lübbig

The Josephson junctions and the interferometer loops were fabricated by the laser ablation process [3]. Dayem type microbriges across the boundary had a width W in the limits 10-20 μm for different samples. Loop areas of 1600 μm^2 and 2500 μm^2 were used for dc interferometers. The density of the critical current j_c was about 100-200 A/cm^2 for all junctions.

3. The parameters of the junctions.

The I-V curves of the junctions made on $SrTiO_3$ substrates were close to those followed from the *RSJ* model [4], while the ones of the junctions made on ZrO_2 had a "humped" form, which is typical for inhomogeneous Josephson structures. This difference in the spatial homogeneity of the junctions was also confirmed by the dependences of the critical current I_c on the magnetic field H. For the junctions on ZrO_2 we have not found any $I_c(H)$ variations, while for the those on $SrTiO_3$ we have obtained usual diffraction-like dependences with a period corresponding to the geometrical cross-section of the junctions.

Absolute values of the characteristic voltage V_c were usually less than 2.5 mV at $T = 4.2$ K. The temperature dependences of the parameter V_c (see fig.1) were in a reasonable agreement with the theoretical prediction [5] for the *SNS* sandwich type junctions with paramagnetic impurities in the N-layer (spin-flip time τ_s equals to $0.5\hbar/\pi T_c$).

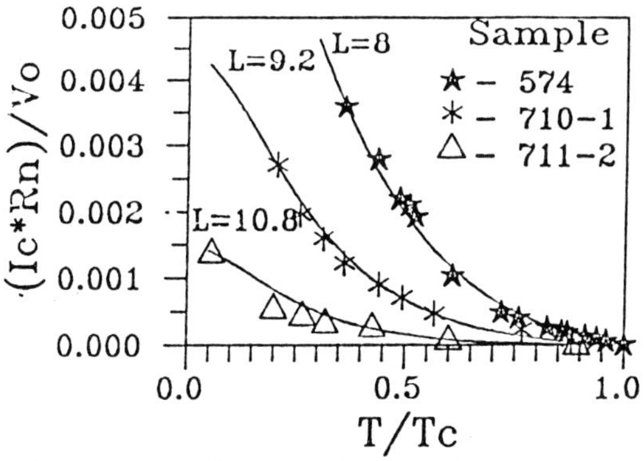

Figure 1. Temperature dependence of the characteristic voltage for the junctions made on $SrTiO_3$ substrates (710-1, 711-2) and ZrO_2 substrate (574), $L = d/\xi^*_N$ is the normalized thickness of the N layer ($V_o = 2\pi k_B T_c/e$).

4. Interferometer characteristics.

Typical modulation voltages V of the current-biased dc interferometers versus applied magnetic flux Φ_e at 4.2 K are presented in fig.2. The peak-to-peak magnitude of V-Φ_e curves of the samples on the $SrTiO_3$ substrates were several times greater than those on ZrO_2. In both cases the oscillation periods were determined by the geometrical loop of the interferometers and hysteresis of the V-Φ_e curves was practically absent.

The noise measurements have been made in a flux-locked loop and the flux-to-voltage factor was determined from the magnitude of the minimal jump of the SQUID output voltage. It was found that $S_\Phi \sim 10^{-3}$ $\Phi_0/Hz^{1/2}$ (at frequencies f > 25 Hz) at 77.3 K for samples fabricated on both types of substrates. However the samples on $SrTiO_3$ usually had better noise characteristics at low temperatures. For example, the best sample 673-1 demonstrated noise level $S_\Phi \simeq 10^{-6}$ $\Phi_0/Hz^{1/2}$. This value corresponds to the field sensitivity close to 10^{-12} $T/Hz^{1/2}$ at T = 4.2 K.

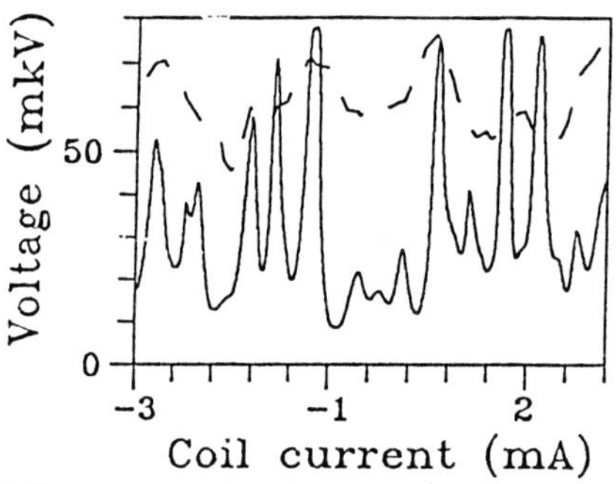

Figure 2. Voltage vs. flux curves for the interferometers 604-1 (ZrO_2, dashed line) and 710-1 ($SrTiO_3$, solid line).

5. Conclusion

The obtained results show that dc SQUID fabrication using $SrTiO_3$ bicrystal substrates is more preferable than on ZrO_2.

The achieved level of the magnetic field sensitivity could be improved by using multiturn high-T_c thin film input coils.

References

1. P. Chaudhari, J. Manhart, D. Dimos, C.C. Tsuei, J. Chi, M.M. Oprysko, M. Scheuermann : Phys.Rev.Lett. **60**, 1653 (1988).
2. A.S. Kovalev, S.I Krasnosvobotsev, M.Yu. Kuprijanov, A.G. Maresov, A.A. Ozerenko, E.V. Pechen, V.G. Pirogov, I.S. Pogosova, O.V. Snigirev, I.I. Vengrus : IEEE Trans. on Magn. **27**, *N*2 (1991).
3. A.S Kovalev, S.I. Krasnosvobotsev, A.G. Maresov, A.A. Ozerenko, E.V. Pechen, V.G. Pirogov, I.S. Pogosova, O.V. Snigirev : Extend. Abstr. of ISEC'89, 538 (1989).
4. K.K. Likharev : in Dynamic of Josephson Junctions and Circuits (Gordon and Breach, New York, 1986).
5. K.K. Likharev, M.Yu. Kuprijanov : [Sov. Phys. - Uspekhi] **160**, 49 (1990).

Sub-Micron $YBa_2Cu_3O_{7-\delta}$ Grain Boundary Junction DC-SQUIDs

M. Kawasaki, P. Chaudhari, T. Newman, and A. Gupta

IBM Research Division, T.J. Watson Research Center,
P.O. Box 218, Yorktown Heights, NY 10598, USA

Abstract. We have fabricated and measured the properties of micron size single grain boundary junctions and dc superconducting quantum interference devices (SQUIDs). The width of the junctions was varied between 0.25 and 2 microns. The noise properties of SQUIDs made of these junctions have been characterized. The white noise level, 3×10^{-31}J/Hz measured at 77 K, approaches the theoretical limit. The source of 1/f noise is identified and associated with critical current fluctuations.

We have shown that relatively low noise dc SQUIDs can be formed in a reproducible and reliable manner by using an epitaxial film containing a single grain boundary [1]. The two junctions comprising the SQUID are formed when a microbridge is patterned across the isolated grain boundary [2]. The microbridge can be formed not only by laser ablation or photo lithography but also by electron beam lithography [3], which enabled us to make narrow junctions down to 0.25 μm. Junction parameters such as critical current and normal resistance were predetermined from our master curve [4]. Using optimal values, we show that the white noise level approaches the theoretical limit. The 1/f noise generated in the SQUID is well characterized and identified to be associated with critical current fluctuations [5].

There are some recent reports where reasonable junctions and SQUIDs have been fabricated on epitaxial films containing a 45° grain boundary or on step-edge substrates [6]. However, the product of critical current I_C and normal resistance R_N in these junctions ranges between 1 μV and 40 μV at 77K. These values are low and limit the maximum flux to voltage transfer function, H, achievable in a SQUID. This in turn requires the use of a low noise preamplifier.

In this paper, we describe the use of lower angle bicrystals in order to obtain a larger $I_C R_N$ product. By using electron beam lithography, we have designed and fabricated submicron SQUIDs in which we have achieved the highest flux to voltage transfer function and lowest white noise at 77K reported to date. The amplitude of 1/f noise is characterized and we explain the origin of the peculiar temperature dependence of 1/f noise reported in our earlier papers.

The master curve which we used to design submicron SQUIDs is replotted in Fig. 1 by using critical current density, j_C , and junction resistivity, ρ_N, as

Springer Proceedings in Physics, Vol. 64 Superconducting Devices and Their Applications
Editors: H. Koch, H. Lübbig

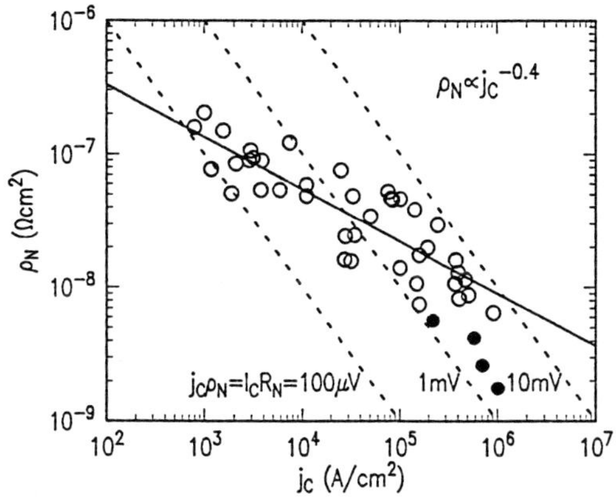

Fig. 1 The relationship between critical current density (j_C) and resistivity (ρ_N) of grain boundary junctions measured at 4.2K. The tilt angle of the bicrystals are varied from 15° to 45°. Open symbols represent the data taken for wide (4μm-40μm) junctions. The data obtained on small junctions is denoted by solid symbols. (From left to right, 1/4μm, 2μm, 1μm, and 1/2μm.)

the coordinate axes. The scaling behavior of grain boundary junctions constrains the choice of j_C and ρ_N (if an external shunt resistance is not used). Another constraint is the screening parameter, $\beta_L \equiv 2LI_C/\Phi_O$ where L is loop inductance and Φ_O is flux quanta. This has to be close to unity for optimal operation of a SQUID. Accordingly, we have to make narrow junctions in a small loop to achieve the appropriate I_C as well as small loop inductance.

The details of sample preparation and noise measurement have been described in our earlier papers [1 – 3]. Examples of our devices are shown in Fig. 2. The j_C and ρ_N of these submicron SQUIDs are plotted in Fig. 1. There is a small but systematic deviation from master curve, which we do not understand. However, the transfer functions as expected, are large and values as high as 300 $\mu V/\Phi_O$ were observed at 77K. In the intermediate temperature regime between 60 and 82 K, the variation of transfer function is found to be proportional to I_C.

The white noise measured at 71kHz for 2 μm junction SQUID is plotted as a function of L/R_N for this SQUID as well as for those reported earlier for the square-washer geometry SQUIDs [7] in Fig. 3. Over the wide range investigated so far, the bicrystal SQUIDs follow the theoretical limit.

The 1/f noise is analyzed on the basis of our results reported in an earlier paper [5]. At a fixed temperature, the amplitude of voltage noise, S_V is found to be proportional to dynamic resistance squared, $R_D{}^2$. Since the amplitude of critical current fluctuation, $a \equiv \delta I_C/I_C$ is constant in this temperature range, S_V can be written as $a^2R_d{}^2I_C{}^2$. From a numerical calculation of R_D we find that the ratio of R_D/R_N scales as $R_D/R_N \propto I_C{}^{0.3}$

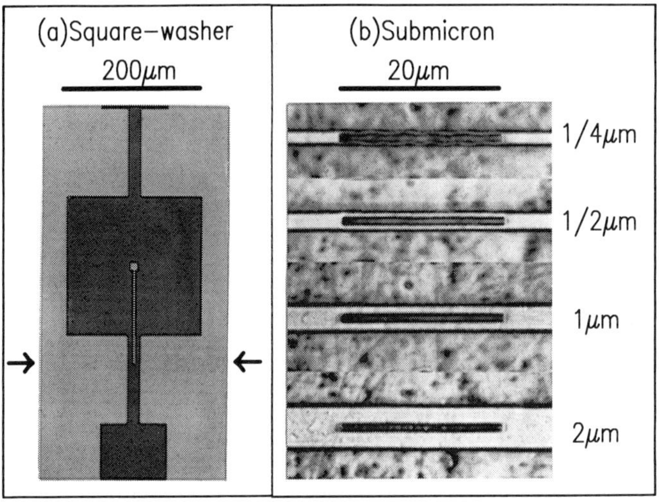

Fig. 2 Optical microscope photograph for square-washer SQUID and submicron SQUIDs made on bicrystal substrates.

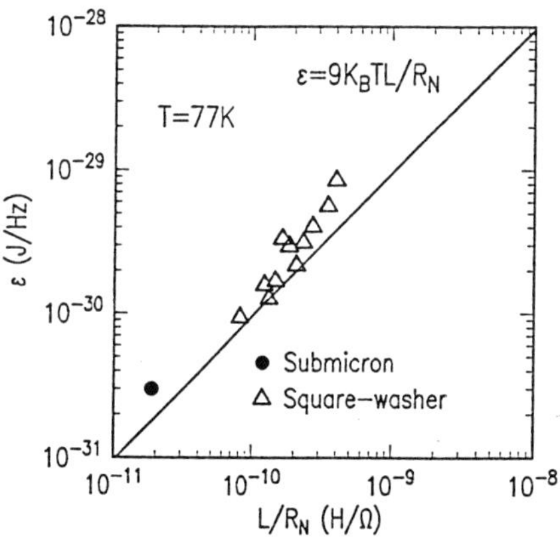

Fig. 3 The white noise measured at 77K is plotted as a function of L/R_N. The theoretical limit is also shown by the solid line.

Therefore, S_V should scale as $a^2 R_N{}^2 I_C{}^{2.6}$. The data from our previous papers [3,5] is replotted in Fig. 4 and indeed, this scaling behavior is seen both for the submicron SQUIDs as well as the square-washer SQUIDs. The rapid increase in S_V as the temperature is lowered noted in all grain boundary bicrystal junctions can be attributed to two effects: an increase of I_C and the resultant increase of maximum dynamic resistance.

The weak temperature dependence of flux noise spectrum, $S_\Phi \equiv S_V/H^2$, can now be readily explained. For a given SQUID with screening parameter close to 1, the flux to voltage transfer function is given by bI_CR_N where b is

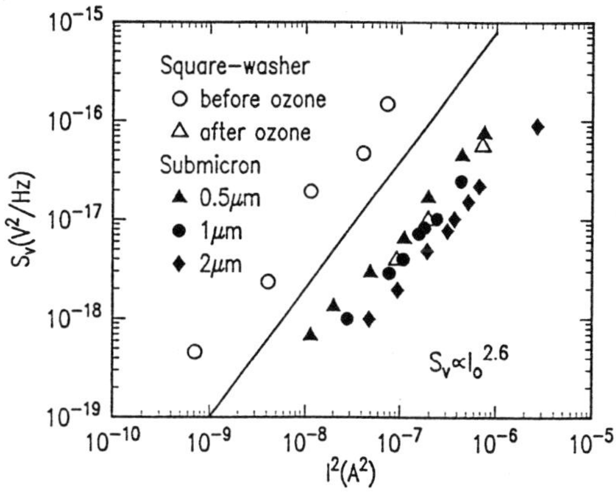

Fig. 4 The voltage noise amplitude, S_V, and the critical current was measured at various temperatures between 60K and 82K. The SQUID was biased at approximately its critical current, where the maximum transfer function and the minimum flux noise were obtained.

a numerical factor of the order unity. Dividing the voltage noise spectrum by H^2 suggests that S_Φ scales as $a^2 I_C{}^{0.6}/b^2$ and this has a much weaker dependence on temperature than S_V. An important conclusion from this analysis is that the only way to reduce low frequency 1/f noise in optimized grain boundary junction SQUIDs is to lower critical current fluctuations of the junctions, assuming that double modulation scheme are not to be used. We have shown that the low temperature annealing in ozone reduces critical current fluctuation [5] . This result indicates that the dominant source of fluctuations is related to oxygen deficiency at the grain boundary. The next step is to understand the microscopic origin of these fluctuations and to find suitable dopants which reside at grain boundaries.

In summary, we have designed, fabricated, and measured submicron SQUIDs. These show very high values of the flux to voltage transfer functions and very low white noise. The source and characteristics of 1/f noise is analyzed. We demonstrate a way to decrease 1/f noise by ozone annealing.

References

1. R. Gross, P. Chaudhari, M. Kawasaki, M. B. Ketchen, and A. Gupta, Appl. Phys. Lett., **57**, 727 (1990); see also Physica C, **170**, 315 (1990).

2. P. Chaudhari, J. Mannhart, D. Dimos, C. C. Tsuei, J. Chi, M. M. Oprysko, and M. Scheuermann, Phys. Rev. Lett., **60**, 1653 (1988).

3. M. Kawasaki, P. Chaudhari, T. H. Newman, and A. Gupta, Appl. Phys. Lett., **58**, 2555 (1991).

4. R. Gross, P. Chaudhari, M. Kawasaki, and A. Gupta, Phys. Rev. B **42**, 10735 (1990).

5. M. Kawasaki, P. Chaudhari, A. Gupta, submitted to Phys. Rev. Lett.

6. K. Char, M. S. Colclough, S. M. Garrison, N. Newman, and G. Zaharchuk, unpublished.; M. S. Dilorio, S. Yoshizumi, K-Y. Yang, J. Zhang, and M. Maung, Appl. Phys. Lett., **58**, 2552 (1991).; see also this proceedings SQUID-91.

7. R. Gross, P. Chaudhari, M. Kawasaki, M. B. Ketchen, and A. Gupta, IEEE Trans. Mag. **MAG-27**, 2565 (1991).

Thin Film BiPbSrCaCuO DC-SQUIDs Operating at 78 K

H.D. Xu[1], *H.Z. Wang*[1], *S.Q. Xue*[2], *X.Q. Shi*[3], *X.M. Huang*[3], *C.B. Yang*[3], and *X.N. Cao*[3]

[1]Institute of Semiconductors, Academia Sinica, 100083 Beijing, P.R. of China
[2]National Institute of Metrology, 100013 Beijing, P.R. of China
[3]Institute of Electronics, Academia Sinica, 100080 Beijing, P.R. of China

Abstract. dc SQUIDs (Superconducting Quantum Interference Devices) have been successfully fabricated by using Pb doped BiSrCaCuO superconducting thin film made by mixed evaporation from a single source with resistance heater. The dc SQUID comprises a square-washer with a small hole. These SQUIDs show perfectly periodic voltage-flux characteristics without magnetic shield, typically, the flux noise and the energy resolution at a frequency range from DC to 1 Hz and at 78K are 1.7×10^{-3} $\Phi_0/\sqrt{Hz}$ and 3.6×10^{-26} J/Hz, respectively. Meanwhile, we have found that one of our SQUIDs operated on flux-locked mode produces a voltage-flux second harmonic under no bias currents in the measuring circuit. This phenomenon is not well understood, but it is most probably related to the I-V (current-voltage) characteristic of the dc SQUID.

1. Introduction

Since high T_c superconductors were discovered, the applications of these oxide materials were also investigated. The first application is that of different types of superconducting quantum interference devices (SQUIDs) made from high T_c superconductors. Bulk $Y_1Ba_2Cu_3O_{7-\delta}$ rf SQUIDs were produced and operated at 77K, their performances were comparable to that of commercial rf SQUIDs operating at 4.2 K[1-4]. However, the largest progress was made in Y-based thin film dc SQUID studies[5,6]. Thin film[7-9] and bulk[10] dc SQUIDs of the TlBaCaCuO have already been produced and operated at 77K. The flux noise of dc SQUIDs of Tl-base material presented is better than that of YBaCuO. Recently a number of groups made large efforts on high T_c phase BiSrCaCuO thin films, because their performance is in the middle of those of both materials, e. g. anti-wet and less toxic properties. Dc SQUIDs of Bi-based thin film oxide material have only two brief reports. One relates to the dc

SQUID produced with BiSrCaCuO (2212, T_{co}= 85K) and operated at 30K[11] and the other (2223, $T_{co} \leqslant 100K$) up to 75K[12]. In this paper we report our results obtained on dc SQUIDs fabricated from BiPbSrCaCuO superconducting thin films and operated at 78K.

2. Device preparation

The Pb doped BiSrCaCuO oxide superconducting thin films used in this work were made by mixed evaporation from a single source with resistance heater. We changed the atom ratio of $Bi_2Sr_2Ca_2Cu_3$ oxide composiion into the weight ratio of Bi, SrF_2, CaF_2 and Cu. After weighing them, we put them into Mo boats and heated them to evaporate completely and deposited the material onto MgO (100) single crystal substrate. This process was repeated twelve times. When the pressure in the chamber was pumped down to 5×10^{-6} Torr, the oxygen gas was introduced into the chamber and directly blown to the surface of the MgO until the material in the Mo boats was completely evaporated . During the deposition, the oxygen partial pressure was kept to $2-3 \times 10^{-5}$ Torr. The as-deposited films were annealed in flowing oxygen at below 250℃ for 4 hours, in this case Bi and Pb could be fully oxidized, then the films were kept at 850℃ in wet oxygen for 1 hr. , and finally cooled down at about 10℃/min. This process produced high quality thin film which had a large fraction of (2223) high T_c(110K) phase with a T_{co}(R=0) of above 100K and a critical current density of 10^3 A/cm^2 at 77K. The film thickness was about 1-2μm.

The dc SQUID pattern comprises a square-washer with a 60-μm-diam hole and two slits yielding two narrow sections of different widths (50μm, 80μm or 100μm) on each side of the hole. The SQUIDs were patterned by standard photolithography using S-1400-27 resist, MF312 developer, and by wet etching in dilute nitric acid (1HNO_3 : 6H_2O by volume at 60℃). Two voltage pads and two current pads which were part of the same film were connected to the SQUID. External leads were connected through pressing Indium balls to the pads of the thin film. After the pattern processing, the zero resistance temperature of the weak link section of the film was usually dropped to 1-2K.

3. Device Measurement

A schematic of the electronic circuit used for dc SQUID measurements is shown in Fig. 1. The gain of the measuring system on the flux-locked mode is 120 dB. The measured voltage noise spectrum density is 0. 5 $nV/\sqrt{Hz}$ at 1 kHz with the source resistance R_g=10 ohms. To avoid the difficulty caused by the induction between signal loop and modulation-feedback coil, a low modulation frequency of 1 kHz is adopted. The magnetic modulation field is applied by a coil 1 mm in diameter, 3 mm in length which is fixed to the back of the dc SQUID. In order to alter the static magnetic field in the SQUID loop, a direct current bias magnetic field was applied. The bias current was provided by a constant current source with a current stability of 2×10^{-5}/hr. and a range from 1 μA to 100 mA.

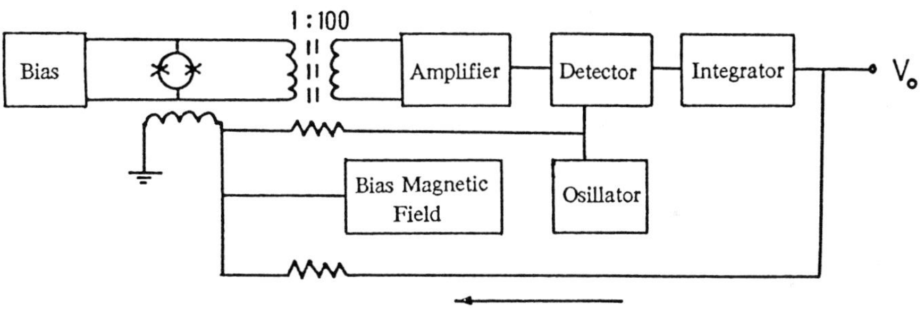

FIG. 1. Schematic diagram of the electronic system for thin film BiPb-SrCaCuO dc SQUIDs operating on flux-locked mode.

4. Experimental results and discussion

The typical I-V characteristic of the dc SQUID is shown in Fig. 2. Due to the different widths of the SQUIDs bridges, the critical current I_c varied in the range from a few μA to 4 mA. All devices exhibited SQUID behaviour. We have noted that even for similar widths of the bridge, the critical current I_c still had large differences. In some dc SQUIDs dynamic resistences were large, and their I-V curves presented ohmic performences at the larger voltage. This all means that the homogeneity of the film is not so good. However, regular triangular voltage patterns appeared in all the dc SQUIDs with differing I_c at 78K , it implies that the amount of critical current seems not to be the unique factor for producing quantum interference effects in high T_c SQUIDs.

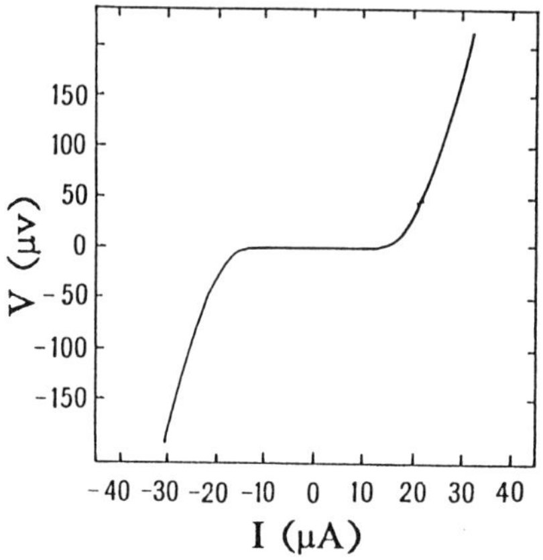

FIG. 2.
Voltage-current characteristic of thin film BiPbSrCaCuO dc SQUIDs at 78K.

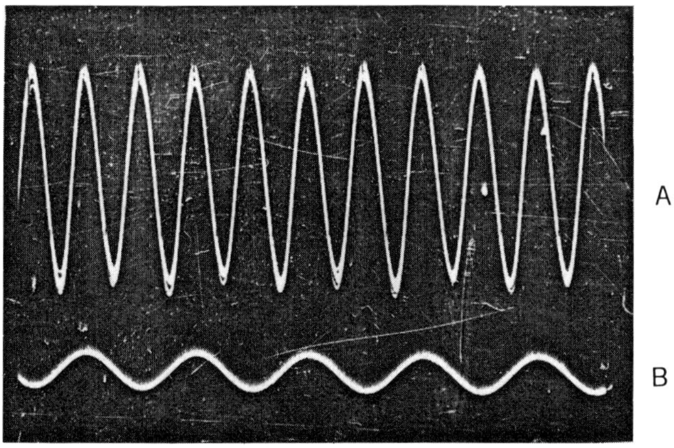

FIG. 3. Thin film BiPbSrCaCuO dc SQUID signal without magnetic shield; a; dependence of output voltage on applied flux at 78K, peak to peak value of the voltage $V_{p-p} \approx 11\mu V$.
b; modulation field.

Under fixed magnetic modulation field, it was usually observed that the voltage varied in response to an applied flux with various bias currents. Due to the missing magnetic shield, the dc SQUID made a quick response to earth field and an amplitude of the output voltage dramatically varied with the azimuth of earth magnetic field. Fig. 3 shows the typical voltage-flux characteristic of the SQUID at 78K. The peak to peak value V_{p-p} of the output voltage is $11\mu V$. According to the current modulation $\Delta I_c \sim \Phi_o/L$, in our case, $L = 170$ pH, $\Delta I_c \sim 12\mu A$. The corresponding

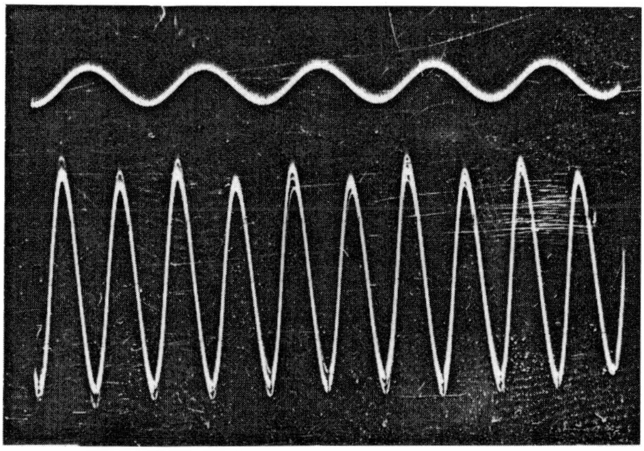

FIG. 4. Flux signal of a thin film BiPbSrCaCuO dc SQUID operating on flux-locked mode at 78K with no applied bias current.
a; modulation field. b; applied flux dependence of the output second harmonic signal. $V_{p-p} \approx 2\mu V$.

voltage modulation amplitude $\Delta V \sim \Delta I_c R_d$, where R_d represents the dynamic resistance, ΔV is ranging from 1 μV to 12μV for our dc SQUIDs. When the SQUID was optimally operated on the flux-locked mode, the output voltage was proportional to the flux applied to the SQUID loop and the output voltage per flux quantum was measured to be 1.2 V/Φ_o. With no magnetic shield the flux noise of $1.7 \times 10^{-3}\, \Phi_o/\sqrt{Hz}$ was measured in the frequency bandwidth of 0—1 Hz at 78K. Here, 1/f noise is predominant, but the nature of the 1/f noise is not well understood yet. It may be intrinsic to the superconductive properties of the bridge area. The energy resolution of the dc SQUID at 78K is 3.6×10^{-26} J/Hz.

Among the measured dc SQUIDs, we have found that one of them still showed the output voltage triangular pattern at applied magnetic modulation field providing no bias current to the SQUID. Fig. 4 shows the voltage harmonic in response to the varying modulation flux. Its peak to peak V_{p-p} value was 2 μV. We had strictly verified the measuring system and found no bias current leaked out of the current source ($I_{bias} \equiv 0$). This SQUID has survived three thermal cycles between 78K and 300K, and kept its properties except a slight decrease of V_{p-p}. The essence of this abnormal effect is not found out yet. A similar phenomenon has also been observed in a Tl-base film dc SQUID. Both of them had a similar I-V characteristic at 78K as shown in Fig. 5, the critical current is quite

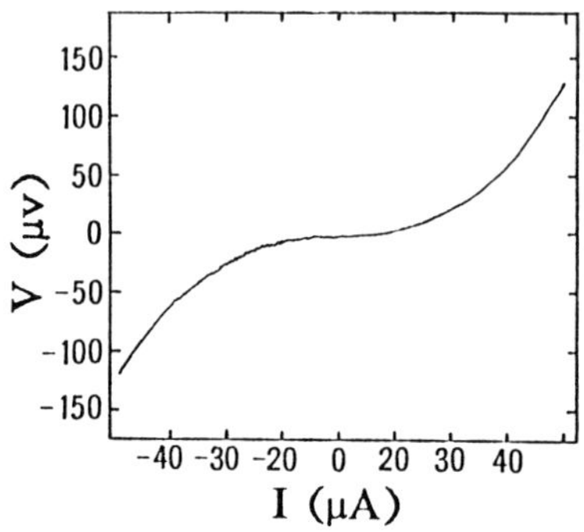

FIG. 5. Voltage-current characteristic of a thin film BiPbSrCaCuO dc SQUID having flux response signal without bias current.

small, about a few μA. It is difficult to distinguish normal current from super-current. The symmetry of the voltage-current curve has a little distinction. An explanation of this phenomenon is under further study.

References

1. Y. Zhang, M. Diegel, and C. Heiden, IEEE Trans. Magn. 25, 869 (1989).
2. V. N. Polushkin and B. V. Vasiliev, Mod. Phys. Lett. B3(17), 1327(1989).
3. S. Q. Xue, J. Zhang, J. Z. Li, P. Y. Yang, and C. J. Wang , Cryogenics 30, 925(1990).
4. K. P. Daly, W. D. Dozier, J. F. Burch, S. B. Coons, R. Hu, C. E. Platt, and R. W. Simon, Appl. Phys. Lett. 58, 543 (1991).
5. A. Z. Lin, H. Q. Li, F. W. Liu, and L. Tang, IEEE Trans. Magn. MAG25, 885(1989).
6. R. Gross, P. Chaudhari, M. Kawasaki, M. B. Ketchen, and A. Gupta, Appl. Phys. Lett. 57, 727(1990).
7. V. Foglietti, R. H. Koch, W. J. Gallagher, B. Oh, B. Bumble, and W. Y. Lee, Appl. Phys. Lett. 54, 2259(1989).
8. Y. H. Huo, S. L. Yan, H. L. Cao, F. Z. Hao, and S. Q. Xue, Solid State Comm. 75, 991(1990).

9. X. H. Zeng, S. G. Wang, J. Z. Li, R. J. Liu, Z. Z. Wang, Y. D. Dai, S. L. Yan, and L. Chen, Cryogenics **30**, 915(1990).
10. D. Robbes, M. Lam Chok Sing, Y. Monfort, D. Bloyet, J. Provost, and B. Raveau, Appl. Phys. Lett. **54**, 1172(1989).
11. I. S. Gergis, J. A. Titus, P. H. Kbrin, and A. B. Harker, Appl. Phys. Lett. **53**, 2226(1988).
12. D. W. Face, J. M. Graybeal, T. P. Orlando, and D. A. Rudman, Appl. Phys. Lett. **56**, 1493(1990).

Thin Film High-Temperature Superconducting Flux Transformers Coupled to SQUIDs

F.C. Wellstood[1;*], *A.H. Miklich*[1], *J.J. Kingston*[1], *M.J. Ferrari*[1], *J. Clarke*[1], *M.S. Colclough*[2], *K. Char*[2], *and G. Zaharchuk*[2]

[1]Department of Physics, University of California, and
Lawrence Berkeley Laboratory, Berkeley, CA 94720, USA
[2]Conductus Inc., 969 West Maude Ave., Sunnyvale, CA 94086, USA
*Present address: Center for Superconductivity Research,
Department of Physics, University of Maryland,
College Park, Maryland 20742, USA

Abstract. Using photolithographic patterning and laser deposition, we have constructed a sensitive thin-film SQUID magnetometer from $YBa_2Cu_3O_{7-x}$ (YBCO). The device consists of a flux transformer, deposited on one substrate, which is coupled to a thin-film YBCO SQUID, made with bi-epitaxial junctions, deposited on another substrate. At 77K, the magnetometer has attained a sensitivity of $1.8 pTHz^{-1/2}$ at 1Hz, and has been used to measure the magnetic signal from the human heart.

1. Introduction

The development of a thin-film superconducting microelectronics technology using high transition temperature (T_c) superconductors requires the ability to produce just five basic circuit components. These are: (1) thin-film wires, (2) superconducting contacts or connections between wires, (3) superconducting to normal metal contacts, (4) multilayer wiring or crossovers, and (5) Josephson weak links or tunnel junctions. It is somewhat remarkable that this is all that is required to produce almost any superconducting circuit of interest, provided that all the components can be reliably constructed on the same substrate.

For a variety of reasons, our investigation of circuit construction techniques utilizing YBCO has focussed on the development of magnetometers based on the Superconducting Quantum Interference Device (SQUID). Most importantly, the magnetometers utilize all five of the basic components, but require only a small total number of components. In addition, because the SQUID can be utilized to measure current, voltage or flux, the circuit itself can provide a sensitive measure of the integrity of the circuit. Finally, a sensitive 77K magnetometer may be useful in a number of applications and may well find early commercialization.

2. YBCO Flux Transformers

A flux transformer is a closed superconducting circuit consisting of a relatively large area pickup coil connected to a smaller area multiturn input coil. Application of a

Springer Proceedings in Physics, Vol. 64 Superconducting Devices and Their Applications
Editors: H. Koch, H. Lübbig

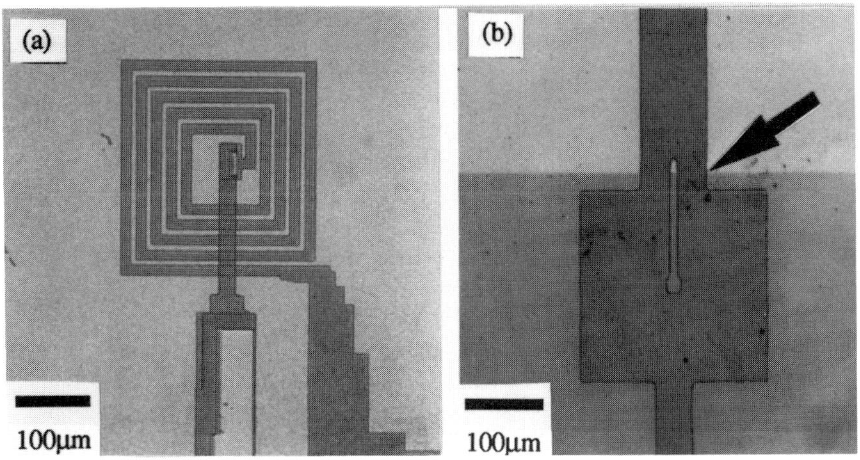

FIG. 1. (a) 5-turn input coil of a photolithographically patterned YBCO flux transformer; (b) YBCO dc SQUID with bi-epitaxial junctions. Arrow points to side of one junction formed at boundary between YBCO grown on MgO seed layer (light, upper portion of photograph) and YBCO grown without seed layer (darker, lower portion of photograph).

magnetic field to the pickup coil induces a screening current which flows through the input coil, thereby coupling flux into the SQUID. For maximum sensitivity to magnetic field, the pickup coil should be a single loop which is made as large as possible, the input coil should be the same size as the SQUID, and the number of turns on the input coil should give the input and pickup coil the same inductance. The main difficulty in making a flux transformer is that a thin-film superconducting connection must be made to the inner turn of the input coil, see Fig. 1(a). This connecting line must not electrically short circuit the turns, and thus crossovers are needed.

In our initial development of flux transformers we used *in situ* pulsed laser deposition of $SrTiO_3$ insulating layers and the YBCO films.[1] We patterned the transformers using shadow masks for the first two layers, followed by photolithography on the last layer. The advantages of the shadow masks are that they are simple to use, no chemicals are required to do the patterning, and the masks produce smoothly sloping edge profiles which are easily covered by layers deposited subsequently. This work was quite successful in that we constructed flux transformers which operated up to 77K.[1] However, it is difficult to extend the use of shadow masks to make small features, complex patterns, or patterns which would require unsupported masks.

The shortcomings of shadow masking have lead us to develop more general techniques which use photolithography and etching to define the layers.[2] The input coil of a photolithographically patterned flux transformer is shown in Fig. 1(a). The size of the input coil was chosen to fit a typical high T_c SQUID; it is about 250μm on a side, and has 5 turns of 10μm linewidth and an inductance of approximately 5nH. The pickup coil is approximately 10mm on a side, and has an effective pickup area of about $81mm^2$ and an inductance of about 20nH.

To build the flux transformer we first deposit 300nm of YBCO onto a (100) MgO substrate.[2] We use photolithography and a weak nitric acid etch to form the contact line which connects to the inner turn of the input coil. The resist is stripped with ethanol and the surface of the YBCO is etched clean using a bromine in methanol solution. We deposit a 400nm layer of $SrTiO_3$; this thickness is required to prevent shorts arising from the sharp edge remaining after etching the first YBCO layer. The substrate is covered with a layer of photoresist about 2μm thick, which is then developed to open two small windows for contacts to the connecting line. We use an Ar ion beam to etch through the $SrTiO_3$ to expose the surface of the YBCO in the two contact regions. The etching is done at an angle of 30° to the surface using beam parameters of approximately $0.5mAcm^{-2}$ and 600V. This produces a shallow, 8° sloped $SrTiO_3$ edge on one side of the windows. Finally, we deposit 400nm of YBCO and pattern it to form the pickup loop and the turns of the coil using photolithography and a weak acid etch.

The shallow slope of the $SrTiO_3$ edge produced by ion milling at 30° yields contact structures which can support critical current densities as high as 10^6Acm^{-2} at 77K (referred to the cross-sectional area of the thin-film wires making the contacts).[2] The completed flux transformer has a critical current of about 2mA at 77K, corresponding to a critical current density of 10^5Acm^{-2} in the turns of the coil. We believe that this reduced critical current density is due to portions of the last YBCO layer which cross $SrTiO_3$ deposited over a lower YBCO edge.

3. Bi-epitaxial Junction SQUIDs

At present, there are no techniques for producing Josephson tunnel junctions in YBCO with hysteretic current-voltage characteristics. Nonetheless, significant advances have been made in the construction of non-hysteretic Josephson junctions. In particular, Char *et al.* [3] have used "seed layers" of MgO to grow YBCO with different orientations on a sapphire substrate. The technique allows 45°, "bi-epitaxial" grain boundary junctions to be placed anywhere on a substrate using photolithography.

The SQUIDs are formed by first using standard *in situ* pulsed laser deposition to deposit a 3 to 10nm layer of MgO on an r-plane sapphire substrate. The MgO is patterned photolithographically and ion beam etched to leave MgO on a region of the substrate. A 10 to 100nm layer of $SrTiO_3$ is then laser deposited, followed by a 300nm layer of c-axis oriented YBCO. The $SrTiO_3$ grows with one orientation on the MgO and with its axes rotated by 45° about the c-axis on the sapphire. The YBCO replicates the orientation of the underlying $SrTiO_3$, giving rise to 45° grain boundary junctions where the two regions meet. The YBCO layer is patterned using photolithography and an acid etch to form the SQUID.

Figure 1b shows a completed SQUID which has outside and inside dimensions of 250μm and 25μm, respectively, and an inductance of approximately 100pH. The SQUID used in the following measurements operates up to about 83K. At 77K, the device showed a heavily noise rounded current-voltage characteristic, with a critical current of about 1μA and a dynamic resistance of 2Ω.

4. All High-Tc Magnetometer

To form a magnetometer,[4] the SQUID chip was covered with a 3μm thick sheet of mylar and the flux transformer chip was pressed against it, face-to-face. The SQUID was visually aligned with the input coil of the flux transformer to achieve good coupling. We find that the flux transformer typically increases the magnetic field response of the SQUID by a factor of from 40 to 80, the variation probably arising from small differences in the spacing or alignment between the SQUID and input coil.

We determined the magnetic field sensitivity by measuring the performance of the system when operated in a flux-locked loop and surrounded by a mu-metal shield. At frequencies, f, below 50Hz, we find low frequency noise with a 1/f power spectral density. At 77K the rms noise magnitude is typically about $10^{-3}\Phi_0 Hz^{-1/2}$ at 1Hz, where Φ_0 is the flux quantum. With a magnetometer gain of 83, this yields a magnetic field sensitivity of about $1.8pTHz^{-1/2}$ at 1 Hz. The low frequency noise depends only weakly on temperature; the sensitivity is approximately $1pTHz^{-1/2}$ at both 4.2K and 68K. [4]

We had earlier measured the noise in flux transformers (made with shadow masks) using a low noise, low T_c SQUID.[5] Noise arising from the motion of vortices in the input coil of the flux transformer was readily observed and greatly dominated any noise from the SQUID itself. In the present arrangement, however, the noise (expressed as flux units in the SQUID) was the same whether or not the flux transformer was attached, provided the system was magnetically well-shielded. Thus the sensitivity of this high T_c magnetometer is dominated by the noise arising in the SQUID. This difference occurs, first, because the noise power at 1Hz in the high T_c SQUID is 4 orders of magnitude higher than in the low T_c SQUID, and second, because the geometry of the present flux transformer results in significantly less noise coupled into the SQUID.

This magnetic field sensitivity is sufficient for some applications. As an example, we have used our magnetometer to obtain magnetocardiograms from human subjects. To shield against external noise sources, we surrounded the subject with two concentric cylinders of Hypernom which provided a low frequency magnetic field attenuation of about 300. The shields were placed in a Cu and steel plate shielded room to minimize rf interference, and the SQUID output was bandpass filtered to the range of 2Hz to 50Hz. In the absence of a subject, the noise level of the SQUID in this range was identical to that found in a well-shielded environment, with the exception of a few small peaks from external sources.

The SQUID and flux transformer were immersed in liquid nitrogen kept in a thin-walled glass dewar. The magnetometer was rigidly held by a support with the pickup coil at a distance of 10mm to 20mm from the subject's chest, and oriented with the plane of the pickup coil parallel to the surface of the chest. To prevent spurious magnetic signals, we removed all metallic objects from the subject, who was in a standing position leaning against a backrest to minimize motion.

Figure 2 shows a magnetocardiogram taken with the magnetometer. The signal is consistent with the expected magnetic signal from the human heart;[6] the large peaks were found to be synchronous with the subject's pulse and to correspond to the main stroke or QRS part of the heartbeat. While the magnetic signal is well-resolved, noise is still evident in the traces. Most of this arises from 60Hz powerline interference and the low frequency noise from the SQUID. For clinical use, another

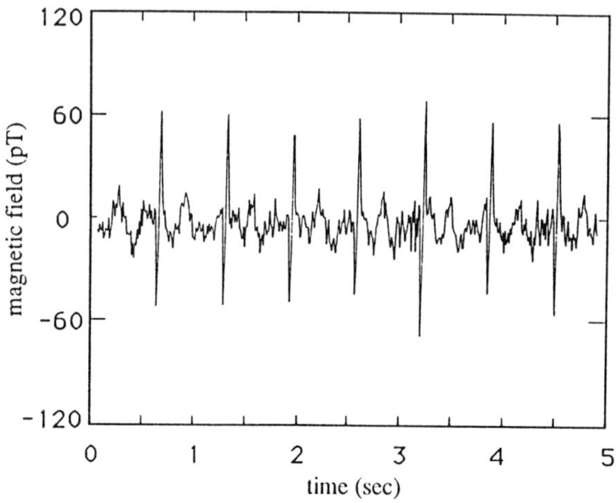

FIG. 2. Magnetocardiogram from a healthy adult male subject. The magnetometer was positioned 10 to 20 cm from the subject's chest, below the rib cage and slightly to the left of center.

order of magnitude increase in the sensitivity would be preferable. We believe that such an increased sensitivity will be readily achievable, for example, by increasing the size of the pickup coil to 25mm and using a matched and well-coupled input coil. In addition, an understanding of the low frequency noise and how it is affected by the fabrication technique may well lead to devices with better performance.

We note in concluding that very recently, using a 7-layer process, L. Lee *et al.* [7] have succeeded in constructing an integrated magnetometer, i.e. with the SQUID and flux transformer on the same chip. The device functions as a magnetometer, but at the time of writing the performance was still being evaluated.

5. Acknowledgements

We thank S. Ross and E. Commins for the loan of the Hypernom magnetic shielding cylinders. F. C. Wellstood acknowledges support from the Center for Superconductivity Research at the University of Maryland, and A. H. Miklich thanks the Fannie and John Hertz foundation for a fellowship. This work was supported by the California Competitive Technology Program, Conductus, Inc., and the Office of Basic Energy Sciences, USDOE, under contract number AC03-76SF00098.

References

[1] F.C. Wellstood, J. J. Kingston, M. J. Ferrari and J. Clarke, Appl. Phys. Lett. **57**, 1930 (1990).
[2] J.J. Kingston, F. C. Wellstood, D. Quan and J. Clarke, IEEE Trans. on Magn. **27**, 974 (1991).

[3] K. Char, M. S. Colclough, S. M. Garrison, N. Newman and G. Zaharchuk, (to be published in Appl. Phys. Lett.).
[4] A.H. Miklich, J.J. Kingston, F.C. Wellstood, J. Clarke, M.S. Colclough, K. Char and G. Zaharchuk, (to be published in Appl. Phys. Lett.).
[5] M.J. Ferrari, J.J. Kingston, F. C. Wellstood and J. Clarke, Appl. Phys. Lett. **58**, 1106 (1991).
[6] S.J. Williamson and L. Kaufmann, Journal of Magn. and Magn. Mat. **22**, 129 (1981).
[7] L. Lee *et al.*, (unpublished).

$YBa_2Cu_3O_{7-\delta}$ Thin Film Flux Transformers Made by Wet Etching

W. Eidelloth, B. Oh, R.K. Koch, W.J. Gallagher, and R.P. Robertazzi

IBM Research Division, Thomas J. Watson Research Center,
P.O. Box 218, Yorktown Heights, NY 10598, USA

Abstract. We have fabricated and tested multiturn input coils for SQUIDs. All levels were patterned by wet etching. Electrical measurements on these structures showed $T_c(R=0) > 77$ K for 5, 10, and 20-turn coils. Flux transformers in the form of gradiometers were coupled to $YBa_2Cu_3O_{7-\delta}$ dc SQUIDs and operated above 77 K.

1. Introduction

Remarkable progress in the area of thin film deposition and processing of high T_c superconductors during the last 5 years has enabled several researchers to produce dc SQUIDs which operate at or above 77 K [1,2]. To exploit the inherent sensitivity of such devices it is necessary to couple them to superconducting flux transformers which operate at similarly elevated temperatures. A couple of groups have reported on the successful fabrication of flux transformer [3-5]. In this work we report further progress on the fabrication of multilevel transformer structures. In contrast to earlier reports, our flux transformers were patterned exclusively by selective wet etching techniques. After a brief description of the fabrication both electrical and magnetic results of such coils and flux transformers are presented.

2. Fabrication

The devices reported here were fabricated either on (100) $SrTiO_3$ or on (110) $NdGaO_3$ substrates. These structures consist of three layers: a base layer of $YBa_2Cu_3O_{7-\delta}$ (YBCO), an insulating layer of $SrTiO_3$ (STO), and a top layer of YBCO. All three layers were patterned individually by wet etching.

Both the YBCO and the STO thin films were deposited by pulsed laser ablation using a frequency-tripled Nd:YAG laser. The YBCO (STO) films were grown in-situ at a block temperature of about 750 (700) °C at a rate of about 750 (500) Å/min. The base (top) layer of these devices was nominally 3000 (3800) Å thick, the insulating layer was nominally 3000 Å thick. The YBCO (STO) thin films had thickness variations of about 10 % (20 %) across a 10 mm × 10 mm wafer.

The patterning of the individual layers is described in detail in Ref. 6. Briefly, photoresist (1 μm thick) was spun onto the first YBCO layer and a

Springer Proceedings in Physics, Vol. 64 Superconducting Devices and Their Applications
Editors: H. Koch, H. Lübbig

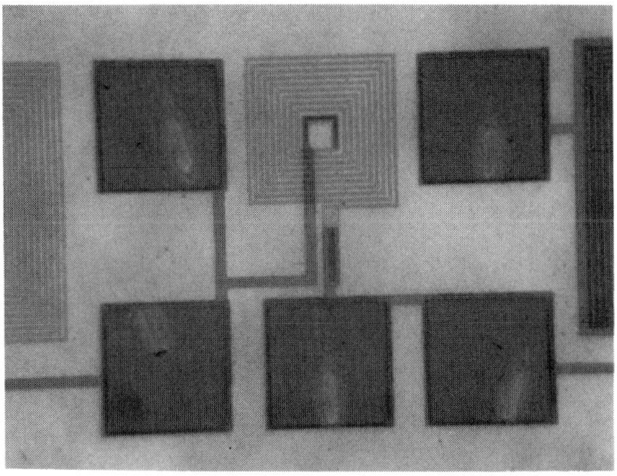

Fig. 1: Photograph of a 10 turn coil.

stencil was defined by standard lithography. The YBCO thin film was then etched in EDTA (ethylenediaminetetraacetic acid) [7] at an etch rate of about 1200 Å/min. After chemical removal of the photo resist (acetone, isopropanol, ethanol) the structures were exposed to an oxygen plasma (500 mTorr, 100 W, 5 min) to assure complete removal of resist residues. It was necessary to remove the resist-damaged surface layer of YBCO before high quality STO could be deposited on top. This was accomplished by either a quick etch in EDTA or a 3 min etch in a bromine solution (1 % in methanol). Both techniques gave satisfactory results. Next, a layer of STO was deposited. This insulating layer had to be patterned in order to facilitate electrical contacting of the top layer to the base layer. Standard lithography was used to define the stencil. The film was then etched in an acequous HF solution (3 % in deionized water) [6]. The etch rate was found to be about 4000 Å/min. This etchant is highly selective and stops at the underlying YBCO layer. In addition to the high selectivity the YBCO film darkens upon contact to the etchant thereby providing a convenient etch stop signal. As described in detail in Ref. 6 the high selectivity of the etchant seems to originate from a very effective reaction that passivates the YBCO surface. To remove the darkend YBCO passivation layer and to render it electrically active again a quick dip into EDTA (20 sec) was sufficient. Subsequently, the top layer of YBCO was deposited. It was then patterned analogously to the base layer as described above. In this way we fabricated 5,10, and 20 turn input coils with and without pick-up loops. The linewidths were 10 μm for the 5-turn coils and 5 μm for the other coils. Fig. 1 shows a 10-turn coil.

3. Electrical Performance

On each chip there were several flux transformers (i.e. input coils which are closed by a pick-up loop) and also seperate input coils where the pick-up loops

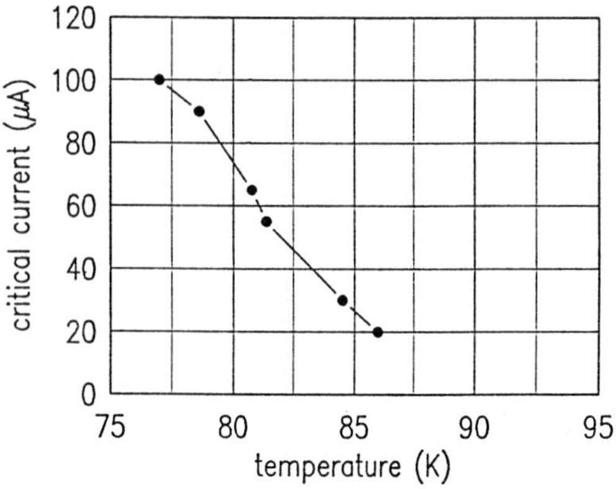

Fig. 2: Critical current vs. temperature for a 10 turn coil.

were replaced by contact pads. Using wire bonding to these pads the electical characteristics (i.e. resistance and critical current vs. temperature) of the input coils could be measured. We also calculated the nominal resistivities of the structures at room temperature. All measurements were done in a four-point configuration. The critical current criterion was 20 μV and the coil lines were 2.7 mm, 5.6 mm, 11.5 mm, and 19.5 mm long.

Separate test structures which were also incorporated on these chips allowed us to measure the electrical isolation of the base and top layers by the STO film. We found this insulator to have a resistivity larger than 200 kΩcm below 100 K. No electrical shorts through the STO layer were observed.

The 5-turn coils had R(300K) between 2.7 and 5.7 kΩ, corresponding to resistivities between 280 and 600 $\mu\Omega$cm. They had completed the superconducting transition at or above 87 K. The critical current at 77 K varied between 100 μA and 2.5 mA. The yield of 5-turn coils with $T_c(R=0) \geq$ 87 K was about 80 %.

Most of the coils with 10 or more turns had transition temperatures below 77 K. Fig. 2 shows the critical current vs. temperature of our best 10-turn coil which had T_c (R = 0) = 87 K. This particular device had a room temperature resistance of 15.8 kΩ which corresponds to a nominal resistivity of 416 $\mu\Omega$ cm. Our best 20-turn coil had T_c(R = 0) = 79 K and I_c(77 K) = 6 μA.

4. Magnetic Performance

Several of our input coils formed the center of first order gradiometers. The two pick-up loops consisted of 0.6 mm wide lines enclosing an area of about 6 mm^2. These gradiometers were mounted in a flip-chip configuration on top of YBCO bicrystal dc SQUIDs [2] to investigate their performance. The SQUIDs operated up to about 80 K whereas the best gradiometer worked at even higher temperatures. The coupling factor between the SQUID and the input coil was found to be about 0.45. We measured a responsivity of

$8.5 \times 10^{-10} m^3$. Detailed measurements on the magnetic performance of these gradiometers will be published elsewhere.

5. Summary

We have demonstrated that YBCO thin film multiturn coils and flux transformers can be fabricated exclusively by wet etching. The patterning of structures with line widths as small as 5 μm and a total length of 7.2 mm has been accomplished. The reported process technology shows a very high selectivity which is a strong advantage over dry etch processes such as ion-milling. This feature becomes more and more important in the upper layers of multilevel structures. Non-uniformities in film thickness can be accomodated much easier than with dry etching technology.

Acknowledgments

We are grateful to S. Andrek, C.D. Jessen, and R.L. Sandstrom for help with the fabrication of the bicrystal SQUIDs. This work was partially supported by the U.S. Office of Naval Research under contract number N00014-88-C-0439.

References

1. See for example R. B. Laibowitz, R. H. Koch, A. Gupta, G. Koren, W. J. Gallagher, V. Foglietti, B. Oh, J. M. Viggiano, Appl. Phys. Lett. **56**, 686 (1990); B. Häuser, M. Diegel, H. Rogalla, Appl. Phys. Lett. **52** , 844 (1988).
2. R. Gross, P. Chaudari, M. Kawasaki, M.B. Ketchen, A. Gupta, Appl. Phys. Lett. **57** , 727 (1990).
3. B. Oh, R. H. Koch, W.J. Gallagher, V. Foglietti, A. Gupta, G. Koren, W.Y. Lee, Appl. Phys. Lett. **56** , 2575 (1990).
4. B. Oh, R.H. Koch, W.J. Gallagher, R.P. Robertazzi, W. Eidelloth, to be published in Appl. Phys. Lett., July 1991.
5. F.C. Wellstood, J.J. Kingston, M.J. Ferrari, J. Clarke, Appl. Phys. Lett. **57** , 1930 (1990).
6. W. Eidelloth, W.J. Gallagher, R.P. Robertazzi, R.H. Koch, B. Oh, R.L. Sandstrom, submitted to Appl. Phys. Lett., March 1991.
7. F.K. Shokoohi, L.M. Schiavone, C.T. Rogers, A. Inam, X.D. Wu, L. Nazar, T. Venkatesan, Appl. Phys. Lett. **55** , 2661 (1990).

Progress Towards an Integrated HTS SQUID Magnetometer

K. Char, M.S. Colclough, L.P. Lee, and G. Zaharchuk
Conductus Inc., 969 West Maude Ave., Sunnyvale, CA 94086, USA

To date, thin film HTS SQUID magnetometers have been made by using two discrete components in a flip-chip arrangement. These components are the SQUID flux detector, with its two Josephson junctions, and the pick-up loop/input coil. It is very desirable to integrate these two components on a single substrate. This requires a multilayer structure that is more complex than any previously fabricated in high T_c superconductors. Our successful progress towards this goal is described briefly in this paper.

Introduction

Thin film dc SQUID magnetometers have become the device of choice for the measurement of small magnetic fields. They are incorporated into reliable products that are finding an increasing number of commercial applications, for example in research instrumentation and medical markets. Unfortunately, those devices operate only at or close to 4.2K. Although great strides have been taken towards making the use of liquid helium an almost trivial matter, for example by building dewars with hold times longer than one year, nevertheless for some applications a SQUID cooled with liquid nitrogen is very desirable. The discovery of the oxide superconductors in 1986 and 87 made this a possibility. To make it a reality required major progress in at least 3 areas, the growth of epitaxial multilayer structures, the invention of a suitable Josephson junction and the processing of the oxide films into the required patterns.

A SQUID magnetometer is comprised of two components. The first is a SQUID loop, say of the Jaycox-Ketchen washer type, which contains two Josephson junctions. Although the current through this washer is modulated by a single quantum of magnetic flux, its area is so small that it is not a particularly sensitive detector of magnetic field. To increase the sensitivity, a much larger pick-up loop is added and coupled to the SQUID with an input coil which is overlaid on the

Springer Proceedings in Physics, Vol. 64 **Superconducting Devices and Their Applications**
Editors: H. Koch, H. Lübbig

SQUID washer. The input coil and pick-up loop form a flux transformer with a gain of tens or hundreds.

The first approach to creating an HTS magnetometer was to fabricate the SQUID and the flux transformer on two substrates. These were then turned to face each other and brought together as closely as possible while still maintaining electrical isolation between them. This was accomplished by two groups, that at IBM [1] and a joint program (2) in which SQUIDs made at Conductus were used with flux transformers made at U.C. Berkeley. The performance of such "flip-chip magnetometers" is described elsewhere in this Proceedings, but can be summarised by noting that the sensitivity at 77K of the UCB/Conductus magnetometer was $1pTHz^{-1/2}$ at 1Hz and $0.2pTHz^{-1/2}$ at 100Hz, with 1/f noise as a dominant feature below 5kHz. The next logical step, given the most encouraging results obtained from these flip-chip devices, is to integrate both the SQUID and pick-up loop/input coil onto one substrate. This has now been achieved. The main challenge has been in multilayer growth and processing, as the basics of the two components have remained much as in the UCB/Conductus work described by Wellstood et al.[2] in this Proceedings.

Experimental

All the multilayer film structures were deposited by laser ablation in a system which allows any one of six different targets to be moved into position in front of the laser beam while the substrate remains at the appropriate temperature in the presence of oxygen. Patterning of the layers is accomplished with optical lithography and ion beam milling. In Fig. 1 we show a multi-turn input coil made in this way.

Fig. 1. SEM image of a multi-turn input coil with a crossover.

The substrates are 1cm x 1cm, which presently defines the limit of the size of the pick-up loop.

Structure 1

In the first integrated SQUID structure, the SQUID was fabricated first. This was necessary because of the constraints imposed by the initial bi-epitaxial junction process. As previously described [3] the junctions, which are 45° grain boundaries in an otherwise completely in-plane aligned film, were created by a seed and buffer layer deposited directly on to the substrate. The first choice of substrate was Al_2O_3, with MgO as the seed layer that creates a 45° rotation of the in-plane alignment of subsequently deposited $SrTiO_3$ (buffer) and YBCO films. This bi-epitaxial process is illustrated in Fig. 2.

In this design, the first YBCO film was patterned into both the SQUID washer and also the crossunder of the input coil, as shown in Fig. 3. This crossunder runs down the slit in the washer.

Following completion of the SQUID layer, an insulating layer of $SrTiO_3$ was deposited, and then patterned and milled to open a contact to the two ends of the crossunder. A second YBCO layer was then deposited, and patterned to form the pick-up loop and input coil.

The constraint of having to make the junctions at the beginning of the multilayer sequence was deemed undesirable, in that the grain boundary is then present in all films that cross the seed layer. Also, the choice of substrates is limited. Therefore we generalised the bi-epitaxial process so that the junctions can be placed at any level of a multilayer structure.

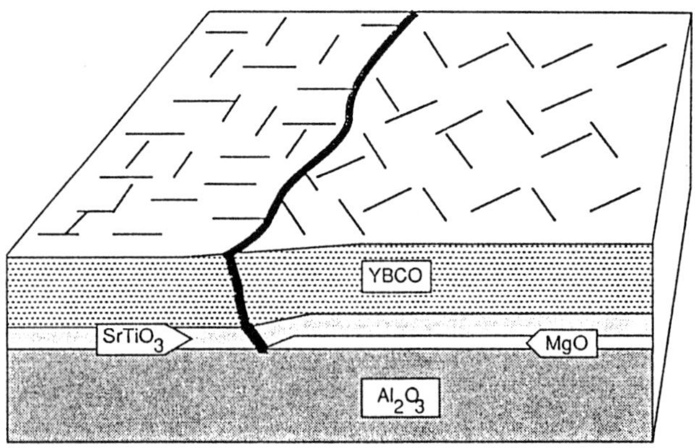

Fig. 2. An example of bi-epitaxial junction structure on r-plane sapphire substrates.

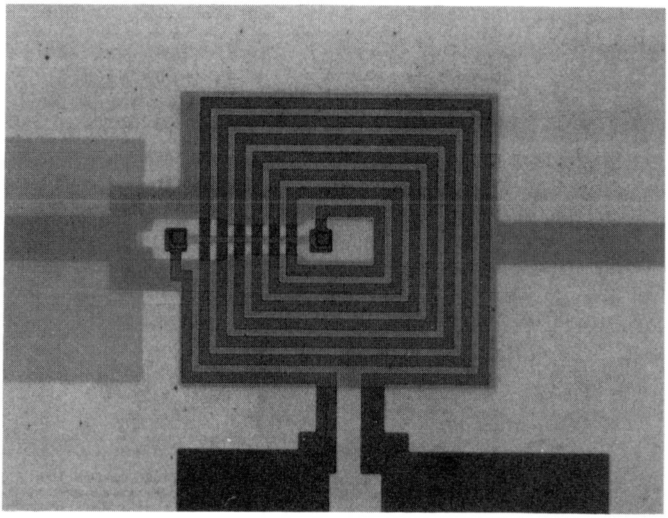

Fig. 3. Optical microscope image of a SQUID magnetometer made with two superconducting layer process (Structure 1).

Structure 2

As described recently [4], the bi-epitaxial junctions are now created on a base layer of $SrTiO_3$ using a seed layer of MgO and buffer layer of CeO_2. As the base layer of $SrTiO_3$ can be placed at any point in a sequence of epitaxial oxides (both superconducting and insulating), this greatly extends the applicability of the bi-epitaxial junction process, for example to digital circuits. In the case of the SQUID magnetometer, it allows us to make the pick-up loop and input coil first, followed by the insulating (base) layer, seed layer, buffer layer and SQUID washer. The junctions are formed in the SQUID washer during its deposition. A plan view of this alternative structure is shown in Fig. 4 and a cross sectional view in Fig. 5.

A major advantage of this generalised bi-epitaxial process is that a much wider range of substrate materials can be used. We have found that it is important to minimise stress in multilayer structures, particularly as the number of layers becomes large. Yttria stabilised Zirconia (YSZ) is a suitable substrate in this regard.

In its simplest form, this magnetometer design requires 7 epitaxial layers of oxides, plus a Ag contact layer. However, an alternative version of the same design we have completed recently uses 15 layers of oxides, plus Ag. It is shown in cross section in Fig. 6.

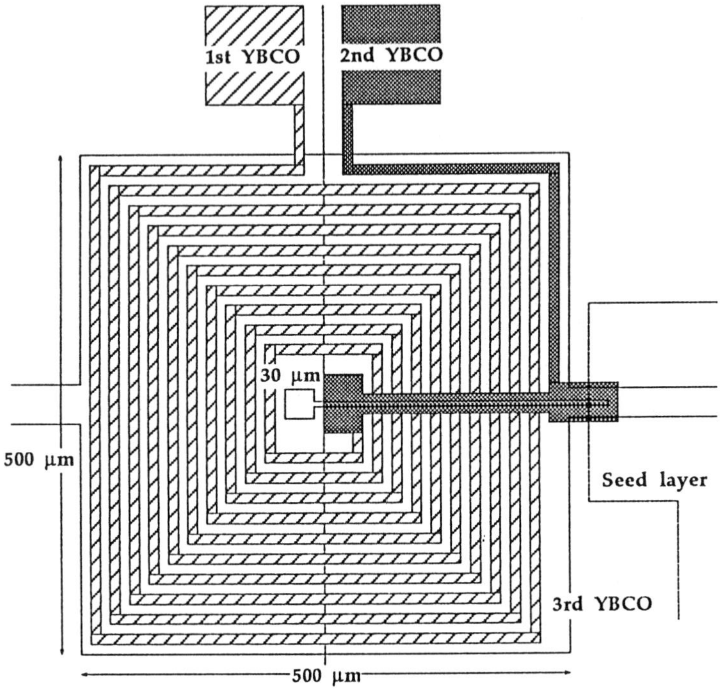

Fig. 4. Plan view of the three superconducting layer SQUID magnetometer (Structure 2).

A Useful Measurement Technique

It is clear that the processing of such multilayer device structures is not a trivial matter, and takes some days to accomplish. It is also undesirable to make contact to the chip to check the properties of the individual layers as the process progresses. Thus it is quite possible to spend days processing a chip in which a layer deposited early in the sequence is inadequate (e.g. T_c below 80K). To some extent, this difficulty has been overcome by the use of a mutual inductance probe as a screening measurement. Details will be described more fully elsewhere [5], but it is possible, for example, to measure the J_c of grain boundaries before they are patterned into junctions, and also to determine T_c and J_c of the flux transformer. Obviously, as the mutual inductance method requires no contacts to the chip, this results in a very significant saving of time and effort. X-ray analysis is also used routinely to confirm the presence of the 45° grain boundary as well as epitaxial qualities of all the layers. This work would have been impossible without the availability of an X-ray machine capable of taking such φ scans.

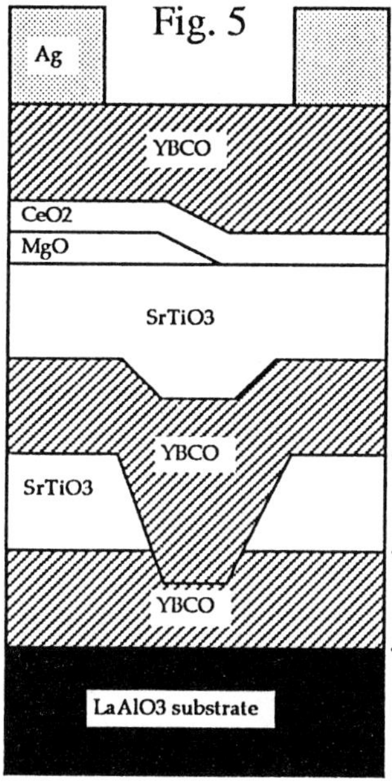

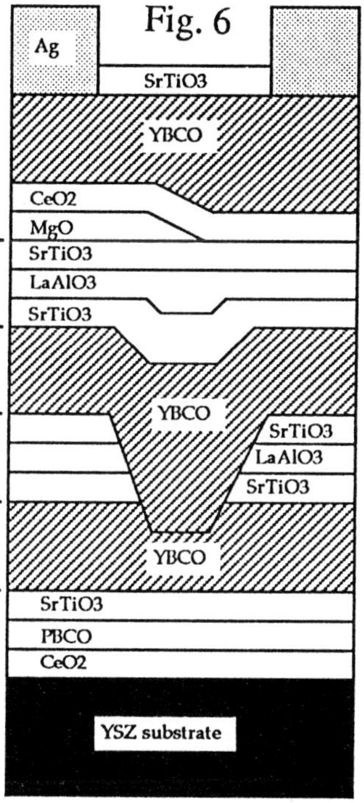

Fig. 5. Cross sectional view of the seven oxide layer SQUID magnetometer on $LaAlO_3$ substrates (Structure 2).

Fig. 6. Cross sectional view of the 15 oxide layer SQUID magnetometer on YSZ substrates (Structure 2).

Magnetometer Design and Performance

The first integrated SQUID magnetometer chip to operate close to 77K (actually 84K for the SQUID, 76K for the flux transformer) was designed to have a square washer 500μm x 500μm with a 30μm x 30μm hole. The junction width was 10μm. The input coil had 10 turns and the pick-up loop was $6x9mm^2$. The estimated flux transformer gain was 73, the measured gain was 170. The SQUID modulation at 77K was 1.2μV, as shown in Fig. 7 (a). The RSJ-like I-V characteristics of the SQUID are shown in Fig. 7 (b). A complete noise characterisation of such magnetometers will be performed when operation of the complete device at 77K is achieved.

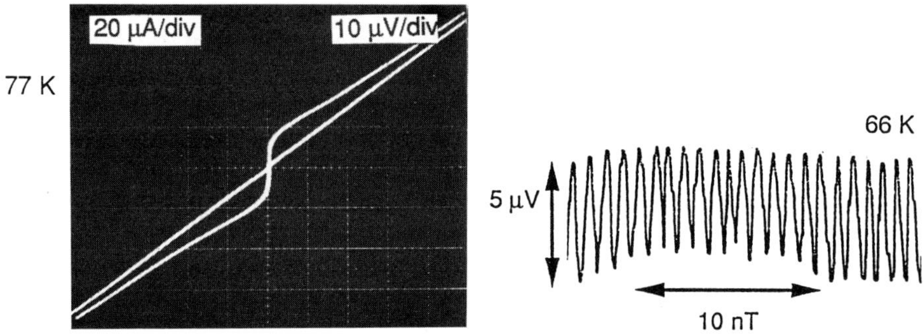

Fig. 7 (a) Fig. 7 (b)

Fig. 7. (a) Voltage versus flux curve of a SQUID magnetometer at 77 K. (b) Current-voltage characteristics of a SQUID magnetometer at 66 K.

Summary

The performance of HTS SQUID magnetometers at 77K is already adequate for many sensor applications that have so far been the domain only of SQUIDs working at 4.2K. Indeed, only the measurement of the signals from the brain is beyond the sensitivity that can be fairly confidently predicted for HTS magnetometers. For example, the improved coupling in the integrated SQUID magnetometer, plus the use of larger pick-up loops, should give a sensitivity of $0.1 pTHz^{-1/2}$ at 1 Hz, without any improvement in 1/f noise being required. The larger coils appear quite feasible, in that 5 cm diameter films of YBCO are now routinely produced at Conductus for microwave applications [6]. In addition, as we understand more about the origin of 1/f noise, it is likely that reduction in 1/f noise will be made in the future. During the last two years we learned a great deal about how these oxide thin films behave structurally, chemically and electrically. Therefore progress towards genuinely valuable SQUID performance at 77K appears to be largely a matter of hard work in the processing of these very unusual materials.

References

1. B. Oh, R. H. Koch, W. J. Gallagher, R. P. Robertazzi, and W. Eidelloth, Appl. Phys. Lett. **59**, 123 (1991).
2. A. H. Miklich, J. J. Kingston, F. C. Wellstood, John Clarke, M. S. Colclough, K. Char, and G. Zaharchuk, Appl. Phys. Lett. **59**, 988 (1991).

3. K. Char, M. S. Colclough, S. M. Garrison, N. Newman, and G. Zaharchuk, Appl. Phys. Lett. **59**, 733 (1991).
4. K. Char, M. S. Colclough, L. P. Lee, and G. Zaharchuk, to appear in Appl. Phys. Lett. Oct. (1991)
5. K. Char, unpublished.
6. N. Newman, B. F. Cole, S. M. Garrison, K. Char, and R. C. Taber, IEEE Trans. Magn. **27**, 1276 (1991).

Light Detection Using High-T_c Microstrip Lines

J.C. Culbertson, H.S. Newman, U. Strom, J.M. Pond, D.B. Chrisey, J.S. Horwitz, and S.A. Wolf

Naval Research Laboratory, Washington, DC 20375, USA

Abstract. We have measured the effect of light on the transmission of microwaves through $YBa_2Cu_3O_{7-\delta}$ microstrip lines. The microstrip line geometries discussed include meander paths and an asymmetric ring interferometer in which a metallic segment has been introduced in the shorter leg of the interferometer. The photoresponse measurements performed on the meander paths allow the probing of the spatial homogeneity of the microwave properties of the superconducting film. The destructive interference nulls for superconductor–metal asymmetric ring interferometer were observed to shift with microwave frequency due to the kinetic inductance effect; the depth of the nulls were observed change rapidly with changing temperature due to the different dependencies of the surface resistance for the superconductor and metal component parts.

1. Introduction: The Photoresponse of Superconducting Films

Many seemingly conflicting observations have been reported concerning the photoresponse of superconducting films to light [1 – 13]. Mechanisms have been suggested to explain these observations. The reason for these differing observations may lie in differences in the films themselves; in different physical systems, the photoresponse may be dominated by different mechanisms (e.g. bolometric, flux flow, flux creep, hot electron, non-equilibrium quasiparticle distributions, etc.).

There are several processes which can give rise to fast photoresponses in superconductors. The fields of a photon can directly modulate the currents in a Josephson Junction; to yield a significant effect, the energy of such a photon must be on the order of or smaller than the superconducting energy gap(s). For more energetic photons, the absorption process results in about 300 fs [10] of avalanche production of hot carriers through carrier thermalization, followed by about 5 ps [10] during which quasiparticles recombine to form Cooper pairs. The latter time is thought to be limited by the rate of optical-phonon decay. Thus, unless there are thermalization bottlenecks, any non-bolometric photoresponse mechanism must involve other excitations that relax more slowly, or collective excitations. Measurements in granular superconductors commonly exhibit the Josephson Junction detection at grain boundary weak links for photons in the millimeter wavelength region [7,8,11,13], and can exhibit a Cooper–pair–breaking fast response whose relaxation time is limited by the electron–phonon scattering time τ_{ep} [11]; in such measurements, τ_{ep} has been

Springer Proceedings in Physics, Vol. 64 **Superconducting Devices and Their Applications**
Editors: H. Koch, H. Lübbig

found to vary as 1/T below 40 K from 4 ps at 40 K to 80 ps at 2 K. Measurements of apparently nonbolometric photoresponses in granular systems have been reported by numerous authors [1,3,8,11,12,13]; mechanisms offered to explain the photoresponses not associated with the above mentioned direct coupling to grain-boundary-weak-links fall into several categories: those which involve the weakening of the intergrain coupling and thus photoinduced phase slip [1,12]; and those which involve magnetic flux motion (e.g. an association with the magnetic flux vortex–antivortex binding phase transition which is describable in terms of the Kosterlitz–Thouless formalism [3]). Nonbolometric photoresponse measurements performed on "high quality" films [4,6,9] have been interpreted in terms of photoenhanced flux creep [4] and, photoactivated flux creep and flux flow [9]. Others have reported seeing only a bolometric photoresponse [2,5]. Braginski et al [14] believe that some of the reported non-bolometric photoresponses are in fact bolometric.

2. The Photoresponse of Microstrip Lines

We have measured the photoresponse of microstrip lines made of high T_c superconducting films. This was done by measuring the change in microwave transmission caused by shining light on the microstrip lines. Given the conditions under which these initial measurements were performed, a bolometric response was expected. A bolometric response was expected because (i) the films tended to be thick compared to the absorption depths of the light and microwaves which were incident on the superconducting film from opposite sides. (ii) The time scale of the modulation of the light source was too slow to expect nonequilibrium superconductivity phenomena to be important unless they dealt with collective excitations such as magnetic flux vortices. (iii) However, the films were of good crystalline quality grown using the laser ablation technique, so such photoresponses that seem characteristic of granular or at least polycrystalline samples were not expected.

2.1 Meander Path Microstrip Line

Figure 1a and 2a show the photoresponse measured as a function of temperature for different positions along two adjacent straight legs of a meander path. Also plotted in these figures are the measured microwave transmission curves F (dashed lines); the widths of the dashed lines were chosen so that the displayed line covers all of the microwave transmission measurements that were made concomitantly with each photoresponse measurement - thus illustrating the precision of the temperature measurement (~0.01 K). A microwave frequency of 3 GHz was used. The detection system included a mixer and intermediate frequency amplifier. The microstrip was patterned from a 5000 Å thick film of $YBa_2Cu_3O_{7-\delta}$ grown on an MgO substrate with a silver ground plane evaporated on the opposite MgO surface. All photoresponse measurements shown here were measured using a 0.63 μm wavelength light source with a power of 1 mW that was mechanically chopped at

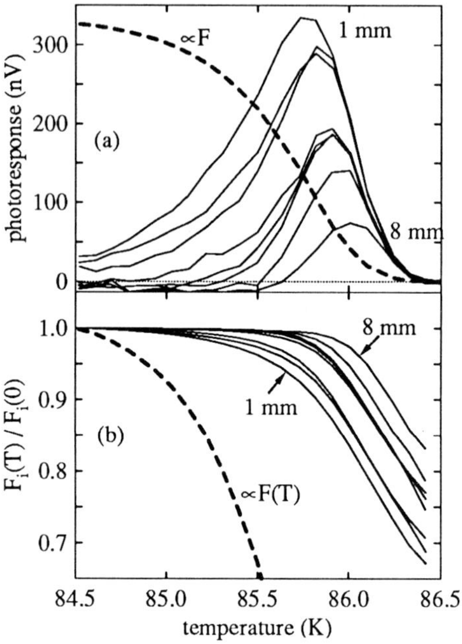

Figure 1. (a) Photoresponse measured vs temperature for 8 positions. Microwave transmission (dashed line). (b) Local microwave transmissions inferred from data in Fig. 1(a).

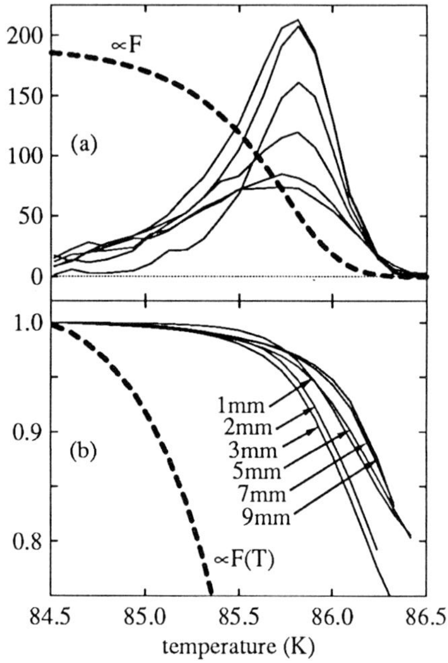

Figure 2. (a) Photoresponse measured vs temperature for 6 positions along a different leg of the meander path. Microwave transmission (dashed line). (b) Local microwave transmissions inferred from data in Fig. 2(a).

about 1 kHz. A bolometric photoresponse δF would be related to the microwave power transmission F by the relation $\delta F = dF/dT\ \delta T$, where δT is the modulation in temperature induced by the incident light at the frequency of the measurement. The photoresponses shown in Fig. 1a peak at distinctly different temperatures; in general they do not peak at the temperature where dF/dT changes most rapidly. This brings the assumption of a bolometric photoresponse into question. However, other measurements, such as the speed of the photoresponse and its variation with microwave frequency seem consistent with a bolometric mechanism. Fig. 1a shows that the measured photoresponse changes when a different part of the microstrip line is illuminated. This suggests that a model incorporating the possibility of spatially inhomogeneous film should be considered. Consider the microstrip line to be composed of n microstrip segments connected in series, where the size of the segment is determined by the length of the line that is effected by the illumination of light (this determines n, ~30 here). Thus $F = F_1 F_2 \ldots F_i \ldots F_n$ where F_i is the fraction of the microwave power incident on the microstrip segment i which is transmitted through. If light is incident only on the i'th segment, then only that segment's transmission will be affected: $\delta F = F_1 F_2 \ldots dF_i/dT\ \delta T \ldots F_n = F / F_i\ dF_i/dT\ \delta T =$

$F\, d(\ln(F_i))/dT\, \delta T$. This explains the observations in Fig. 1a; if the local microwave transmission varies with position, then the temperature where the photoresponse peaks will shift correspondingly. The correlation, seen in Fig. 1a, between the magnitude of the photoresponse and the temperature where the photoresponse peaks, is explainable if the local microwave transmissions differ from each other mostly by a rigid shift along the temperature axis. The local microwave transmissions F_i, inferred from the data in Fig. 1a using the above relation for the photoresponse δF are shown in Fig. 1b. The magnitude of the temperature modulation δT is estimated using the above expression for δF and the zero'th order approximation $F = F_i{}^n$ to be $\delta T \approx 0.4$ mK. The analogous calculation for the data in Fig. 2a is shown in Fig. 2b. It is seen that for the meander leg measured in Fig. 1, the inhomogeneity corresponds mostly to a shift in T_c, however the shape of the local transmission does change some. For the meander leg measured in Fig. 2, the shift in T_c is not sufficiently large compared to the shape changes of the F_i to result in the correlation observed in Fig. 1. These observations point out the danger of comparing a local property (such as photoresponse) with a global property (such as microwave transmission) and making conclusions concerning the bolometric character of the observed photoresponse. A straight forward analogous situation arises in low frequency resistive photoresponse measurements.

2.2 Superconductor–Metal Interferometer Microstrip Line

Figure 3 shows measurements, made as a function of microwave frequency, of the microwave transmission through the asymmetric ring interferometer shown schematically in the inset of this figure; measurement are shown for temperatures 0.1 K apart from 86.2 K to 87 K. The interferometer was patterned out of a 2000 Å thick Y-Ba-Cu-O film grown on a MgO substrate. The interferometer is all superconductor except for a short metallic segment in the shorter leg of the interferometer. When the temperature is carefully set to a value between 86.5 K and 86.6 K the destructive interference null at about 3.6 GHz drops down another 20 to 30 dB, becoming much sharper. The frequency shift on the null is due to the kinetic inductance effect. The rapid change in depth of the null with changing temperature is due to the difference between the behaviors of the metal and the superconductor. The loss in the normal metal depends only weakly on temperature, while the loss in the YBCO can depend strongly on temperature. The dependence of surface resistance R_s on microwave frequency is $R_s \propto f^{1/2}$ for metal, while it is $R_s \propto f^2$ for superconductor. This device yields a much faster null depth and frequency shift with changing temperature than does an all superconductor interferometer, and as such provides an interesting alternative geometry for a bolometric detector. The device shown has a shift of ~1 GHz/K near the null.

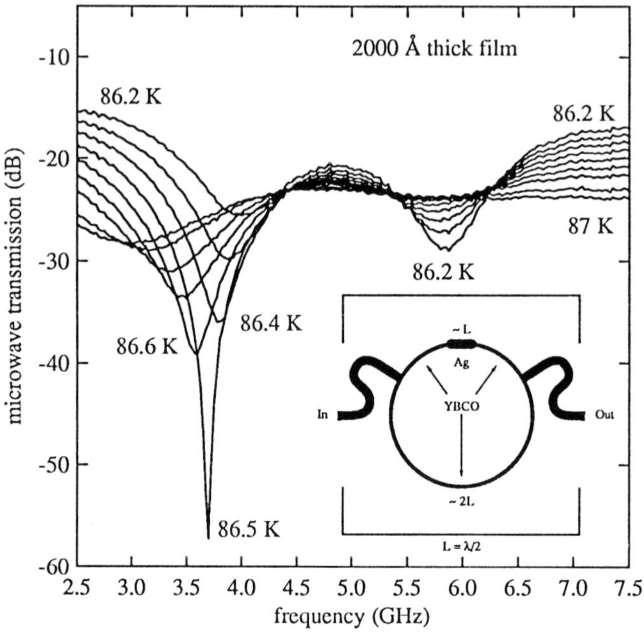

Figure 3. Microwave transmission through metal–superconductor asymmetric ring interferometer microstrip line measured as a function of microwave frequency for nine different temperatures 0.1 K apart. Inset shows schematic of interferometer.

3. References

1. M. Leung, P.R. Broussard, J.H. Claassen, M. Osofsky, S.A. Wolf, and U. Strom, Appl. Phys. Lett. **51**, 2046 (1987).
2. M. G. Forrester, M. Gottlieb, J.R. Gavaler, and A.I. Braginski, Appl. Phys. Lett. **53**, 1332 (1988).
3. J.C. Culbertson, U. Strom, S.A. Wolf, P. Skeath, E.J. West, and W.K. Burns, Phys. Rev B **39**, 12359 (1989).
4. E. Zeldov, N. M. Amer, G. Koren, and A. Gupta, Phys. Rev. B **39**, 9712 (1989).
5. W.S. Brocklesby, Don Monroe, A.F.J. Levi, M. Hong, S.H. Liou, J. Kwo, and C.E. Rice, Appl. Phys. Lett **54**, 1175 (1989).
6. H.S. Kwok, J.P. Zheng, and Q.Y. Ying, Appl. Phys. Lett. **54**, 2427 (1989).
7. U. Strom, J. C. Culbertson, and S. A. Wolf, SPIE **1240**, 516 (1989).
8. A.S. Afanasyev, A.F. Volkov, V.N. Gubankov, Yu. Ya. Divin, and P.M. Shadrin, IEEE Transactions on Magnetics **25**, 2571 (1989).
9. A. Frenkel, M.A. Saifi, T. Venkatesan, P. England, X.D. Wu, and A. Inam, Appl. Phys. Lett. **54**, 1594 (1989); J. Appl. Phys. **67** 3054 (1990).

10. S.G. Han, Z.V. Vardeny, K.S. Wong, O.G. Symko, and G. Koren, Phys. Rev. Lett. **65**, 2708 (1990).
11. E.M. Gershenzon, G.N. Gol'tsman, A.D. Semenov, and A.V. Sergeev, Solid State Commun. **76**, 493 (1990).
12. Keiichi Tanabe, Youichi Enomoto, Minoru suzuki, Tsunekazu Iwata, and Akihiko Yamaji, Jpn. J. Appl. Phys. **29**, L466 (1990); Keiichi Tanabe, Syugo Kubo, Youichi Enomoto, Hidefumi Asano and Akihiko Yamaji, Jpn. J. Appl. Phys. **30**, L110 (1991).
13. G. Schneider, H. Lengfellner, J. Betz, K.F. Renk, and W. Prettl, International Journal of Infrared and Millimeter Waves, **12**, 1 (1991).
14. A.I. Braginski, M.G. Forrester, and J. Talvacchio, Proceedings International Superconductivity Electronics Conference (ISEC '89), Tokyo, June 1989.

The Superconducting Flux Flow Transistor: Models and Applications

J.S. Martens, V.M. Hietala, T.E. Zipperian, D.S. Ginley, and C.P. Tigges

Sandia National Laboratories, Department 1140, Albuquerque, NM 87185, USA

The superconducting flux flow transistor (SFFT) is a single film, active device generally made from high temperature superconducting (HTS) materials. The device is based on the magnetic control of flux motion, has gain and can operate to over 40 GHz with 3 μm feature size. Existing models are reviewed along with the latest equivalent circuit values derived from S-parameter measurements. Noise performance is compared to that predicted by a relatively simple flux noise model and recent amplifier performance is discussed.

1. Introduction

The SFFT is being investigated (e.g., [1]-[3]) as an HTS device for high speed applications. Toward that goal, extensive simulations and microwave characterizations have been performed to refine the device and determine the device's ultimate performance limits. A more immediate result is the ability to better predict behavior in current applications.

2. Device Basics

The basic structure of the devices tested is shown in Fig. 1 and consists of an SFFT cell embedded in coplanar waveguide for microwave measurements. The body of the device consists of a parallel array of weak links through which the main bias current (I_{bdy}) passes. Each link is typically 10 μm by 3 μm and is 50-100 nm thick (etched from a film thickness of 300-400 nm). A separate control line provides a local magnetic field that modulates the flux density in the link system.

When the body of the device is biased below the critical current (on the order of 1 mA) no flux is admitted into the link system. Above the critical current both vortex/anti-vortex pairs (self-field) and single vortices (from external field) are admitted. The vortices are subject to 'Lorentz' forces from bias and external fields, viscous damping, pinning forces (which are undesirable in that the average vortex speed is reduced) and surface barriers at the edges of the links which hamper flux entry and exit. The balance of these forces determines the flux motion and hence the terminal voltage.

Springer Proceedings in Physics, Vol. 64 Superconducting Devices and Their Applications
Editors: H. Koch, H. Lübbig

Figure 1

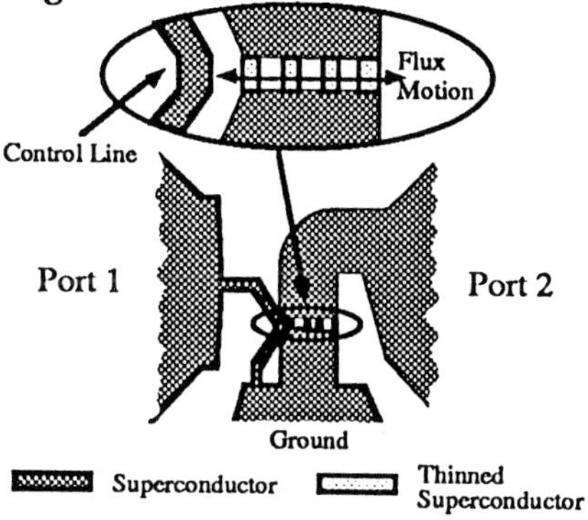

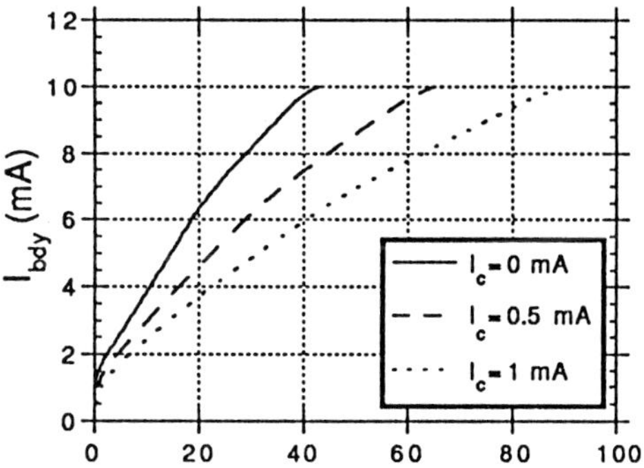

Figure 2

Figure 1. A layout of the superconducting flux flow transistor. Port 1 is the control line and port 2 is the device body. The transmission line ground planes are above and below the section shown.

Figure 2. IV curves of a TlCaBaCuO SFFT. I_{bdy} is the current through the device body (link region) and I_c is the control current. The voltage V is measured across the link system.

The device behavior is closely correlated to the quality of the film remaining in the link region. High pinning reduces sensitivity to external magnetic fields (and hence gain) and reduces the flux speed and frequency response of the SFFT. A low H_{c1} allows easier flux entry into the system, increasing sensitivity to the control fields. The penetration depth strongly influences the surface barriers and hence control-field sensitivity [4]. Maximum gain occurs for links thinner than a penetration depth but not so thin that the superconductivity is destroyed or the defect density rises enough to induce excessive pinning centers.

Sample IV curves for a $Tl_2Ca_2Ba_2Cu_3O_{10}$-based device at 77K are shown in Fig. 2. The zero-voltage and flux flow states are clearly illustrated. At high bias currents, the links are driven completely normal. Upon application of a control current, a given bias point suffers a nearly horizontal translation as is consistent with a flux density change. A transresistance element is therefore the logical choice for the active element in the equivalent circuit.

3. An Equivalent Circuit

The primary equivalent circuit used is shown in Fig. 3. The input or control line side is a superconducting short with some inherent inductance and resistance (the latter arising from microwave surface resistance and some contact resistance). The transresistance element discussed above is shown on the output side. There is a finite output resistance arising from the presence of the normal vortex cores. The output inductance has significant kinetic contributions because of the extreme

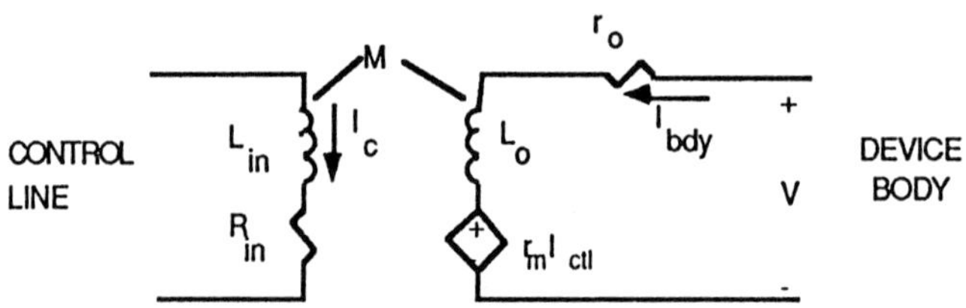

Figure 3. An equivalent circuit of the SFFT. The transresistance element is the active element. Other interesting features include the non-linearity of the output inductor and the low amount of cross-talk (M is typically 10 pH). Other typical component values are r_m=21 Ω (transresistance), r_0=4 Ω (output resistance), L_{in}=100 pH, L_{out} = 70 pH, R_{in}= 0.05-0.1 Ω (at moderate bias).

thinness of the links. The output resistance, output inductance and transresistance are bias dependent. The non-linearity of the link (or output) inductance allows the development of efficient, low-loss phase shifters [2]. The values of the parameters are derived through mechanistic simulation (of flux motion), moment method calculations [3], DC measurements and through fitting microwave S-parameter data [2].

4. Noise Mechanisms and Performance

We will concentrate on broadband flux noise and its effects on microwave performance. Based on simulations and some flux speed measurements [3], the vortices tend to move through the link system in bundles of variable size. The noise can then be thought of as a doubly stochastic Poisson process with the two random variables: the size of a bundle and the time of its occurrence. The key parts of this argument have been previously derived by Huebner [5]. In the frequency range far from DC (so that $1/f^{\alpha}$ effects are small) but well below the frequencies where transit time effects are important (≈ 30 GHz with the present devices), the spectral density of this flux noise can be approximated as [6]

$$G_{midband} \approx \left(\frac{\hbar\pi}{2e}\right)^2 \left[2B + \frac{B^2}{K}\right] \quad V^2 \text{ in a 1 Hz BW} \quad , \qquad (1)$$

where B = number of vortices moving past an observation point per unit time within a bundle (an average density of vortices within a bundle) and K = number of bundles moving past an observer per unit time. Estimates for these two parameters have been obtained using the mechanistic simulations discussed above and flux speed experiments [3]. For a typical Tl device at moderate bias levels (about half way between the zero voltage branch and the normal switch-over), B≈ 1-2 x 10^{11} 1/s while K ≈ 0.5-1 x 10^{11} 1/s. It should be noted that both of these parameters will be bias dependent (both increase with bias). These estimates should be valid

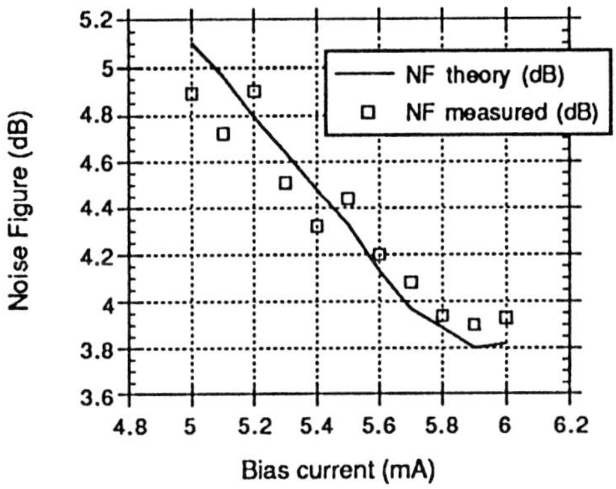

Figure 4. Theoretical and measured noise figure of an unmatched SFFT as a function of bias (2 GHz).

from a few MHz to at least 10 GHz based on the relevant time constants and estimates of the extent of the 1/f noise from other work (e.g., [7]).

The noise factor of an unmatched SFFT amplifier at low microwave frequencies (up to a few GHz perhaps) would then be

$$F \approx 1 + \frac{R_{source}}{4kTr_m^2}\left\{\left(\frac{\hbar\pi}{2e}\right)^2\left[2B + \frac{B^2}{K}\right] + 4kT'r_0\right\}, \tag{2}$$

where R_{source} is the source resistance, T=300K (source resistance temperature) and T'=77K (SFFT output resistance temperature). For the parameter estimates listed above , the noise figure ≈ 4.8 - 7 dB. This ignores the filtering effects of the parasitics in the circuit.

To test this hypothesis, the noise of several SFFTs was measured as a function of bias. The bias range was kept small to avoid large shifts in the model parameters (small changes were accounted for in the calculations). The device was probed with spring-loaded coplanar probes and the noise measured using a Cascade Microtech Noise Measurement system. Plots of the measured noise figure of a TlCaBaCuO SFFT (2 GHz, 77K) and that predicted from simulations are shown in Fig. 4. The simulations used the equivalent circuit of Fig. 3 (with de-embedded parameter values), the noise voltage of Eq. 1 and experimentally determined launch parasitics. At least over this bias range, the model fits well. Shot-like flux noise is probably the dominant contributor to device noise at these frequencies.

5. Applications and Model Predictions

One obvious application is a matched amplifier (a single SFFT with input and output passive impedance-matching networks). The gain of such a circuit along

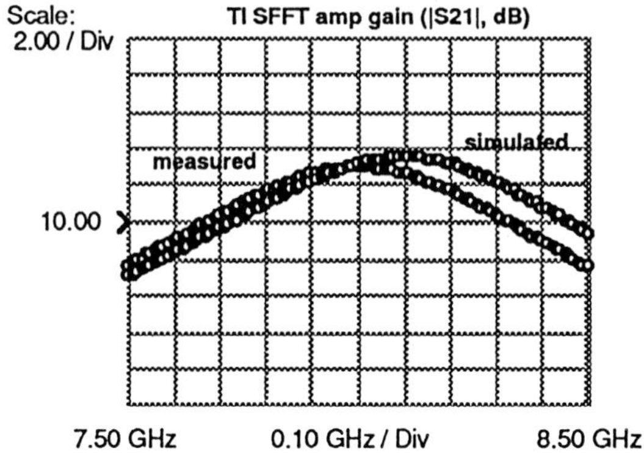

Figure 5. Performance of a matched TI SFFT amplifier (77K).

with that predicted by the model is shown in Fig. 5. Maximum available gain, computed from device S-parameter measurements, is on the order of 20 dB, so improvement is possible. The noise figure for this amplifier in a 50 Ω system was about 3.1 dB (estimated from minimum noise figure measurements using the Cascade system) while that predicted by the model was 3.45 dB.

6. Conclusions

We have presented some recent results of work on the SFFT. The latest designs have tried to minimize excess inductances while increasing transresistance. The results has been verified by S-parameter measurements which show input inductances ≈ 100 pH and transresistance > 20 Ω. Noise calculations and measurements have been performed suggesting that shot-like flux noise is dominant for linear microwave applications and that typical noise figures are in the several dB range. Matched amplifiers are being made with gains > 10 dB over a bandwidth > 1 GHz.

7. Acknowledgments

This work was supported by the U.S. Department of Energy under contract No. DE-AC04-76P00789.

8. References

[1] J. S. Martens, J. B. Beyer, J. E. Nordman, G. K. G. Hohenwarter and D. S. Ginley, IEEE Trans. on Mag, vol. MAG-27, 3284 (1991).

[2] J. S. Martens, V. M. Hietala, T. E. Zipperian, D. S. Ginley, C. P. Tigges, and J. M. Phillips, "S-parameter Measurements and Applications of Superconducting Flux Flow Transistors," presented at the 1991 IEEE MTT-S International Microwave Symposium, Boston, MA USA, 11-13 June 1991.
[3] J. S. Martens, D. S. Ginley, J. B. Beyer, J. E. Nordman and G. K. G. Hohenwarter, IEEE Trans. Appl. Super., vol. 1, (June 1991).
[4] T. Van Duzer and C. W. Turner, Principles of Superconductive Devices and Circuits (Elsevier, New York, 1981), chp. 3.
[5] R. P. Huebner, Magnetic Flux Structures in Superconductors, (Springer-Verlag, Berlin, 1979).
[6] J. W. Goodman, Statistical Optics, (Wiley, New York, 1985), chps. 2-3.
[7] F. C. Wellstood, J. J. Kingston, M. J. Ferrari and J. Clarke, IEEE Trans. on Mag., vol. 27, 2569 (1991).

Preliminary Studies of HTS Far-Infrared Micro-Bolometers

C. Barholm-Hansen[1] *and M.T. Levinsen*[2]

[1]Topsil Semiconductor Materials A/S, DK-3600 Frederikssund, Denmark

[2]Physics Laboratory, H.C. Ørsted Institute, Universitetsparken 5, DK-2100 Copenhagen, Denmark

Abstract. We are developing an antenna coupled micro-bolometer using high temperature superconductors (HTS) for detection in the far infrared. A prototype consisting of a thin-film strip of $Y_1Ba_2Cu_3O_x$ on MgO, coupled to a logarithmic periodic silver antenna has been fabricated. The measured responsivity was 10 V/W, giving an NEP of $8 \cdot 10^{-10}\, W/Hz^{1/2}$ at 500 Hz. Measurements of the DC responsivity indicate that the intrinsic responsivity is larger, pointing to a poor coupling efficiency of the antenna. Excess noise is found to be proportional to the bolometer resistance at a fixed bias current. Currently we investigate the possibility of using self-supporting bufferlayers for high sensitivity microbolometers on silicon. Some success with $Y_1Ba_2Cu_3O_x$ on silicon with an YSZ bufferlayer has been achieved.

1. Introduction

Interest in intermediate temperature thermal detectors was reactivated soon after the discovery of high T_c superconductors, mainly because of their possible application as space-born sensors. Microbolometers, based on bismuth thin-films for use at ambient temperatures, have been described by Neikirk *et al.*[1]. The theoretical design and performance of HTS based microbolometers was discussed by Hu and Richards[2]. An actual bolometer was constructed by Nahum *et al.*[3]. Here we present some results on an antenna coupled microbolometer and discuss some ways of improving the performance.

2. The Microbolometer

A microbolometer consists of two equally important parts, the bolometer strip, and the antenna. The bolometer strip connects the two halves of the antenna, so that radiation received by the antenna is thermalized in the strip, changing its temperature. A device is shown in Fig. 1. The antenna is needed because the bolometer strip is much smaller than the wavelengths we detect. The resistive bolometer monitors the resistance as a function of temperature, and to get a sensitive device a large change in resistance for small changes in temperature is important. Superconductors are known to have a very abrupt change in resistance at the transition from normal- to the superconducting phase. The dependence of resistance, R, on temperature, T, is characterized by the temperature resistance coefficient defined as $\alpha \equiv R^{-1} dR/dT$. For HTS materials, at the mid-point of their transition, α is about 1 K^{-1}. The incoming radiation is modulated by means of a

Springer Proceedings in Physics, Vol. 64 Superconducting Devices and Their Applications
Editors: H. Koch, H. Lübbig

Fig. 1. The picture shows a logarithmic periodic antenna prior metal deposition. The antenna is designed for the wavelength band from 90 µm to 600 µm. The thin-film is seen to be translucent, as it is only approx. 100 nm thick. Current is passed through the device via the wider lines, and the voltage over the bolometer is measured at the thinner lines.

mechanical chopper, introducing the angular frequency ω. The resistance is monitored by a four probe measurement using a constant bias current, I_b. In[1] the responsivity ***R*** of a bolometer, i.e. the signal voltage per unit incident power, is shown to be expressed as

$$\boldsymbol{R} = \alpha R \eta I_b / G[1 + \omega^2 (C_b/G)^2]^{1/2}, \qquad (1)$$

where η describes the coupling efficiency of the antenna, C_b is the heat capacity of the bolometer, G is the effective thermal conductivity expressed as $G = \kappa\sqrt{2\pi A}\,[1 + (\tau\omega)^{1/2}]$, and τ is the thermal time constant of the bolometer, given as $\tau = C_s A / 2\pi\kappa$. C_s is the specific heat capacity per unit volume of the substrate, A is the microbolometer area, and κ is the thermal conductivity of the substrate material. From Eq. (1) it can be seen that for small values of G, we will have large responsivities as long as ω is kept small relatively to the thermal mass of the bolometer. The thermal properties of the substrate is described by τ, which in practice sets the upper limit on ω. Note that decreasing the sensor area A, we get an increase in responsivity, as we at the same time push the limit on ω upwards. Small values of A are thus advantageous.

3. Experimental

Films of $Y_1Ba_2Cu_3O_x$ were prepared on MgO(100) substrates by *in situ* DC magnetron sputtering in an off-axis geometry using a resistive substrate heater. Film thickness was typically 0.1 µm. The films were patterned using standard photo lithography, then wet etched in a 2% phosphoric acid for 10 seconds. Silver was then evaporated onto the thin-film, leaving only a strip of 4 µm in width and

20 μm in length free, using a lift-off technique. The strip connects the two halves of a logarithmic periodic antenna[4], and its resistance can be monitored by a four point measurement. A photograph of the structure prior to metal deposition is shown in Fig. 1. $T_{c,0}$ of the device was 92 K. J_c was measured inductively to be larger than 10^6 A/cm^2 (70K, zero field). The films grew epitaxially with the c-axis perpendicular to the substrate, showing a FWHM of 1.35°.

The microbolometers were placed on a cold finger in an optical cryostat facing a 3 mm thick teflon window. Liquid nitrogen was used as the cooling agent. The temperature of the cold finger is controlled to within 0.1 K. The responsivity measurements were performed using an aluminium plate heated to 570 K as an approximate blackbody source. The incident radiation was modulated with a chopper having a duty cycle of 50%. The antenna is designed[4] for the wavelength band of 90 μm to 600 μm. The teflon window is chosen accordingly, as it transmits radiation in this interval and absorbs radiation of shorter wavelengths. From the geometry, the incident power can be estimated. Assuming total absorption in the antenna in this range, and transmission of the whole black body frequency spectrum onto the active bolometer area, a conservative estimate for the input power is $12 \cdot 10^{-9}$ *Watt*.

The bolometer signal was recorded as a function of temperature for several values of bias current, using a lock-in amplifier. The antenna impedance R, only depends on the dielectric constant of the substrate. For MgO R is about 80 Ω. This should be matched by the resistance of the bolometer strip.

The responsivity of the device was measured at 500 Hz to be a modest 10 V/W. However it is based on a calculation of the incident power, assuming an η of 1, and 100% transmission through the teflon window. A calculation of the DC responsivity from electrically heating the bolometer, typically yields values an order of magnitude larger. The noise was found to scale as $1/f^\alpha$ with α being close

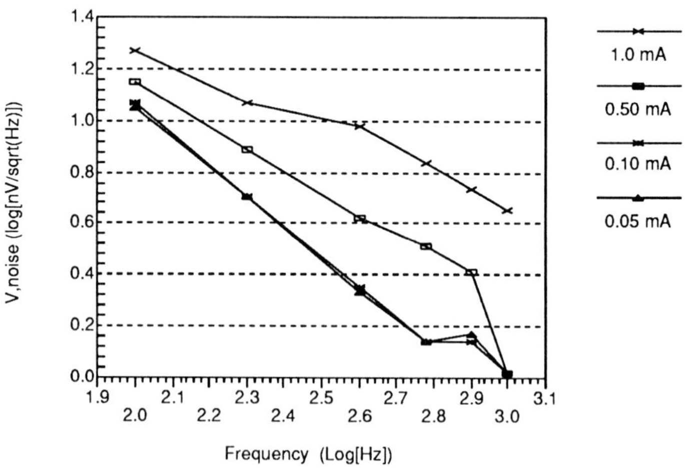

Fig. 2. Noise of a microbolometer versus chopping frequency is seen to scale as $1/f^\alpha$ with α close to one. The noise is found to increase with I_b, and is an order of magnitude larger at 500 Hz than the expected Nyquist noise.

to one, see Fig. 2. The noise increases with I_b. The noise level found was higher than expected. This is attributed to excess noise in the superconducting thin-film. At higher frequencies the noise reduces, and at 500 Hz a noise equivalent power (NEP) of $8 \cdot 10^{-10}\ W/Hz^{1/2}$ was achieved. This is an order of magnitude higher than estimated by Hu and Richards[2]. This is partly due to the larger size of our bolometer strip. Decreasing the size of the bolometer to 5*1 μm^2 would yield an improvement of a factor 4. However, the important improvements have to be made on the coupling into the antenna. Further, one could make the antenna of a material with a higher T_c than the bolometer, cutting down noise and ohmic losses.

4. Future Work and Acknowledgements

As microbolometers can be made more sensitive by depositing the bolometer strip on a thin membrane, it is of interest to investigate the possibility of growing HTS thin-films on membranes. Silicon is then an obvious choice of substrate, since it can be anisotropically back-etched to form the desired membranes. It is necessary to have a bufferlayer between the HTS thin-film and the silicon to prevent damaging interaction, and yttria stabilized zirconia (YSZ) is the preferred choice. The first attempts to grow $Y_1Ba_2Cu_3O_x$ thin-films on silicon with a YSZ bufferlayer have resulted in films with a T_c of 60 K. Obviously some optimization is still needed.

We like to thank Per Vase from NKT for performing the critical current density measurement. We would also like to thank Rasmus Kromann from Risø for investigating the epitaxial nature of our thin-films.

5. References

1. D.P. Neikirk, W.W. Lam, and D.B. Rutledge, Int. Journal of Infrared and Millimeter Waves **5** (3), 245 (1984).
2. Qing Hu and P.L.Richards, Appl. Phys. Lett. **55** (23), 2444 (1989).
3. M. Nahum, Qing Hu, P.L. Richards, S.A. Sachtjen, N. Newman, and B.F. Cole, IEEE Trans. on Magn. **27** (2), 3081 (1991).
4. R.H. DuHamel and D.E. Isbell, I.R.E. National Convention Record Part 1, Antenna and Prop., 119 march 18-21 (1957).

Current Injected Thick Film HTS Magnetic Sensors

P.J. Gielisse[1], *H. Niculescu*[1], *B. Roy*[1], *P. Pernambuco-Wise*[2], *J.E. Crow*[2], *G. Sykora*[3], *and R. Wahlers*[3]

[1]Florida A&M University, Florida State University, College of Engineering, Tallahassee, FL 32316-2175, USA
[2]MARTECH Center for Materials Research and Technology, Florida State University, Tallahassee, FL 32306, USA
[3]Electro-Science Laboratories Inc., P.O. Box 1059, King of Prussia, PA 19406, USA

Abstract. The performance of current injected HTS thick film sensors for low magnetic field measurements is reported. The influence of material properties, processing and design parameters as well as driving currents on sensitivity is discussed.

1. Introduction

The nonlinear magnetic response of high temperature superconductors to alternating magnetic fields, can be employed for measurements of weak *dc* fields in the nanotesla range, as previously shown [1-3].

In low density polycrystalline materials, critical currents cannot shield the bulk of the sample against the applied field. Fields below 1 Oe will easily penetrate deeply between grains along the grain boundaries. For alternating fields with amplitudes larger than the lower critical field of the intergranular region, the flux lines are assumed to sweep in and out of the material, lagging behind the driving field. Low intensity, nanotesla range *dc* fields, slowly swept (0.1 - 1 Hz) and superimposed on the alternating carrier, can be detected and measured.

In the thick film devices reported on here, we have applied the driving current directly to the sensor. The only alternating magnetic field is that generated by the transport current. We have investigated the performance of a thick film sensor as a function of frequency and amplitude of the *ac* and *dc* fields, and device parameters.

2. Experimental

A magnetometer configuration can be seen in Fig. 1. The superconducting component consists of a thin disk formed by screen printing commercially available $YBa_2Cu_3O_7$ powder onto a zirconia substrate shown schematically in Fig. 2. Electrical connections to the film were made via printed silver tabs. The resistance of the contacts was aproximately 1 $\mu\Omega$. A wire wound coil

Springer Proceedings in Physics, Vol. 64 Superconducting Devices and Their Applications
Editors: H. Koch, H. Lübbig

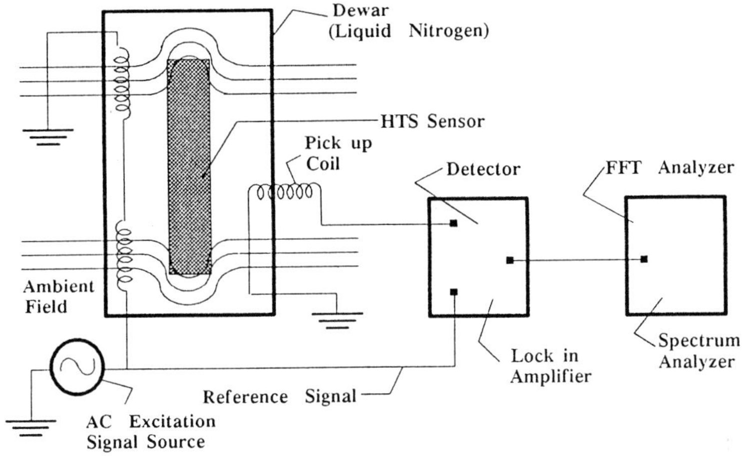

Fig. 1. Experimental test set-up.

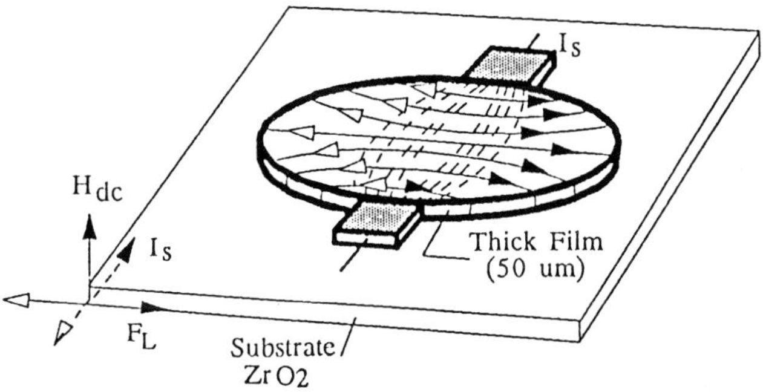

Fig. 2. Schematic of a thick film sensor.

placed against the surface of the sample allowed for signal detection via the induced voltage from the resultant magnetic field.

Sensors were suspended by their connection leads into a liquid nitrogen dewar and centrally positioned within the magnetic field space of a Helmholtz coil pair. A low frequency magnetic field (0.02 - 1 Hz) could thus be applied perpendicular to the sensor and to the *ac* driving field. This field simulated the *dc* background to be detected. Signal processing consisted of a lock-in amplifier, detecting at the second harmonic of the applied field, with the output being on-line Fourier transformed with a spectrum analyzer and plotted in spectral form.

3. Results and discussion

Assuming a nonlinear response of the superconductor to a sinusoidal field ($H_{ac}\sin\omega t$), the components of the voltage developed across the detection coil at harmonics of the applied frequency are:

$$V_n = \omega AN\{a_n \sin(n\omega t) + b_n \cos(n\omega t)\}.$$

where N indicates the number of turns in the detection coil and A the effective sensor area. The values of a_n and b_n depend on the relation between *ac* and *dc* field amplitudes and sensor parameters.

The signal shows a perfectly linear dependence on the *dc* field, as shown in Fig. 3 for field strength in the range 10^{-3} to 5×10^{-1} Oe at 1 Hz and a driving *ac* field at 50 kHz. The sensitivity of thick film devices, evaluated at 1 nT, is currently lower than that for devices based on bulk type materials.

The influence of the *ac* field is referenced in Fig. 4. At a *dc* field of 0.02 Oe, one observes linear dependence between signal and *ac* current (slope = 1). At higher fields, 0.2 Oe, the initial slope increases to 1.6 while saturation sets in at *ac* field levels above 5 mA. The penetration field, H^*, for our 50 μm thick films at $J_c = 5$ A/cm^2 (measured on identical thick film strips) is about 16 mOe. The field produced by an I_{ac} current as evaluated, considering that superconducting electrons travel the shortest path between the electrical contacts, is: $H_{ac} = \mu_o I_{ac}/2w$, where w is the width of contacts. At $I_{ac} = 5$ mA, $H_{ac} = H^*$. For $I_{ac} > 5$ mA, $H_{ac} > H^*$ and saturation occurs as predicted by the Bean model [4].

A linear dependence of signal amplitude on driving field frequency has been found between 10^3 and 10^5 Hz. Our results with bulk sensors [5] indicate that linearity holds at even higher frequencies. An increase in frequency can, therefore, have a significant impact on device performance.

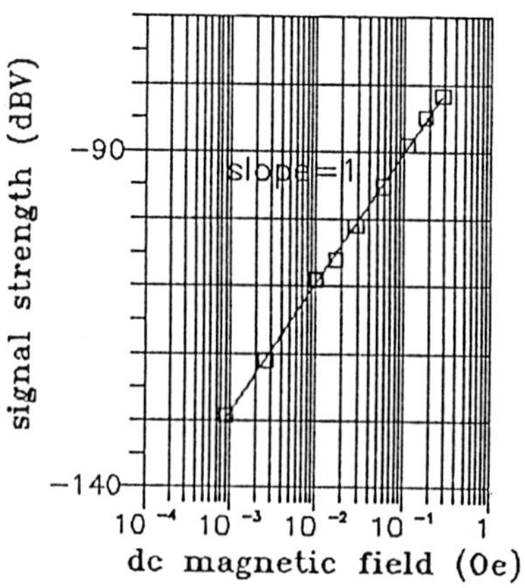

Fig. 3. Linear dependence of signal on *dc* field.

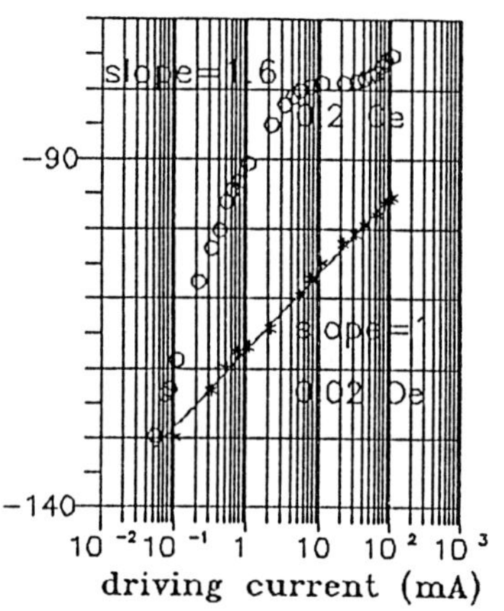

Fig. 4. Signal versus *ac* driving current.

References

1. J. C. Gallop, S. Lilleyman, C. D. Langham, W. J. Radcliffe and M. Stewart, IEEE Trans. Magn. **25**, 897 (1989).
2. M. Gershenson, ibid. **27**, 3055 (1991).
3. P. J. Gielisse, H. Niculescu, B. Roy and H. Shin, Cont. No. N 61331-89-D-0021, NCSC, Panama City, FL.(1990).
4. Charles P. Bean, Rev. Mod. Phys. **36**, 31 (1964).
5. P. J. Gielisse, H. Niculescu, B. Roy, K. W. Jones, G. Larkin and Z. Hu, ISEC'91, Glasgow, Scotland. To be published in Supercond. Sci.& Technol.

Acknowledgments. This work was supported in part by DARPA, Contract No.MDA 972-88-J-1006 and US Army-SDI, Contract No.DASG 60-90-C-D138.

Part II

Low-T_c Devices

Comparative Study on Two All-Refractory Josephson Tunnel Junctions: $Nb/AlO_x/NB$ and Nb/NbN/MgO/NbN/Nb

D.J. Adelerhof[1], *B. David*[2], *M.E. Bijlsma*[1], *J. Flokstra*[1], *and H. Rogalla*[1]

[1]University of Twente, Faculty of Applied Physics,
Low Temperature Group, P.O. Box 217,
NL-7500 AE Enschede, The Netherlands

[2]Philips Forschungslaboratorium, W-2000 Hamburg, Fed. Rep. of Germany

Abstract. Two types of all refractory Josephson tunnel junctions have been studied: $Nb/AlO_x/Nb$ and NbN/MgO/NbN. Conductance-voltage measurements and subgap current analyses have yielded information about the barrier shape and homogeneity of the oxide layers in both types of junctions. The AlO_x in SNINS $Nb/AlO_x/Nb$ junctions is intrinsically homogeneous, which results in a subgap current dominated by temperature dependent single-particle tunneling at 4.2 K. Barrier inhomogeneities which are probably introduced during the trilayer anodization cause excess tunneling currents.

The MgO barrier in NbN/MgO/NbN junctions is inhomogeneous. Due to this, the subgap current at 4.2 K is dominated by temperature independent tunneling currents possibly in combination with leakage currents. The barrier height of the artificial MgO is significantly lower than in thermally oxidized Mg. During fabrication, reactions of deposited MgO with residual gasses affect the barrier properties considerably. It is suggested that residual H_2O reacts with MgO during and after deposition.

1. Introduction

$Nb/AlO_x/Nb$ and NbN/MgO/NbN Josephson tunnel junctions are all-refractory junctions, which are stable against thermal cycling and offer long-term stability. In 1981 Gurvitch *et al.* initiated the research on $Nb/AlO_x/Nb$ junctions, which has led to the routine fabrication of high quality Josephson junctions [1,2]. In these junctions energy gaps and T_c values approaching the Nb bulk values of 1.56 meV and 9.3 K, have been obtained [3].

Because of its high energy gap (3.1 meV) and high T_c (15-17 K) NbN is also an interesting material for application in Josephson junctions [4,5]. The use of artificial MgO in NbN tunnel junctions, first reported by Shoji *et al.* [6], turned out to be an important step which has recently led to the fabrication of junctions with sumgap voltages up to 5.8 mV [7]. The junction quality, however, strongly depends on the fabrication of the NbN, probably due to its small coherence length (2-7 nm [5]), and on the deposition of the MgO.

In this study the shape and homogeneity of the potential barriers in $Nb/AlO_x/Nb$ and NbN/MgO/NbN junctions will be compared.

Springer Proceedings in Physics, Vol. 64 **Superconducting Devices and Their Applications**
Editors: H. Koch, H. Lübbig

2. Fabrication

The trilayers of the Nb/AlO_x/Nb junctions were made by DC magnetron sputtering in a turbo pumped vacuum system [2,8]. Three types of Nb/AlO_x/Nb junctions have been fabricated: asymmetric SNIS, symmetric SNINS, and double oxide SNIIS junctions, where 'S' is 150 nm Nb, 'N' is 3 nm of Al and 'I' is thermally oxidized Al. In SNIIS junctions the first AlO_x layer is an oxidized top layer of Al, the second AlO_x layer is a completely oxidized Al layer of about 10 Å. The junction areas are defined by SNAP [9].

Nb/NbN/MgO/NbN/Nb layers are deposited in a turbo pumped vacuum system with liquid N_2 trap [10]. NbN is reactively sputtered by DC magnetron sputtering. The T_c of the NbN films is typically about 15 K. MgO films are deposited by RF magnetron sputtering in pure Ar at a rate of 1.4 nm/min. The H_2O background pressure is 10^{-7} mbar. The time between MgO sputtering and the deposition of the NbN counter electrode was varied. In junctions PH 518 the waiting period was 4 minutes, in junctions PH 513 11 minutes. The junction areas were defined by RIE with CF_4. On one chip 6 junctions were made with areas of 400, 100, 25, 16, 9, and 4 μm^2.

3. Experimental results on Nb/AlO_x

In fig. 1a typical I-V characteristics of a SNINS junction at 4.2 and 1.5 K are shown, together with the theoretical BCS single-particle subgap current. To fit the BCS curve to the measured curve at 4.2 K the normal conductivity was chosen 1.4 times the measured conductivity at 4 mV. This adjusted conductivity was also used to calculate the BCS curve at 1.5 K. From fig. 1a it is clear that at 4.2 K the subgap current is dominated by single particle tunneling. At 1.5 K other tunneling mechanisms with onset voltages of $V_g/2$ and $V_g/3$ become very pronounced. Both multi-particle tunneling and self-

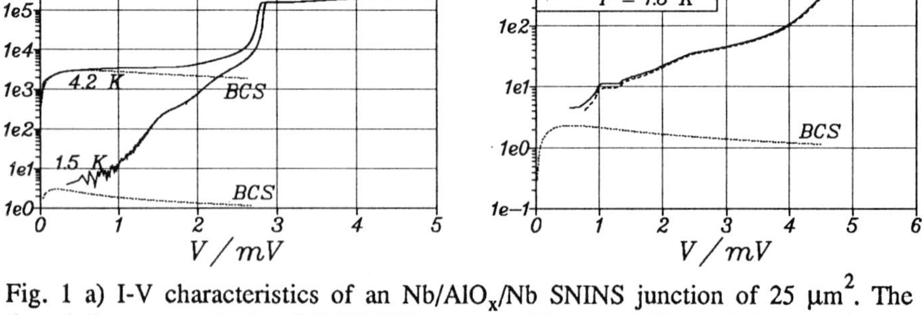

Fig. 1 a) I-V characteristics of an Nb/AlO_x/Nb SNINS junction of 25 μm^2. The dotted lines are calculated BCS I-V curves with Δ = 1.37 meV. b) I-V characteristics of a NbN/MgO/NbN junction of 9 μm^2 (PH 518). The dotted line is the BCS current, calculated for T = 4.2 K and Δ = 2.3 meV.

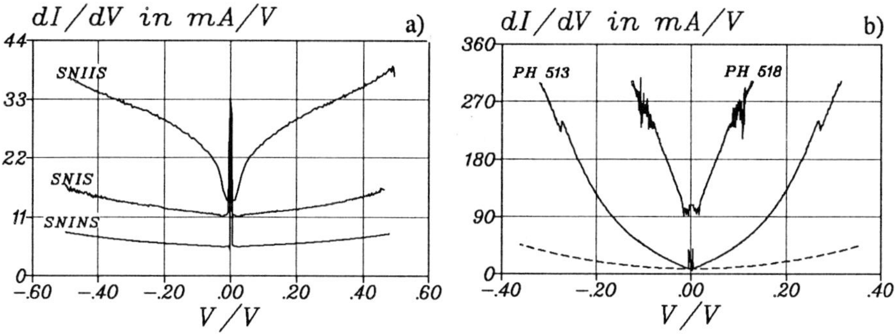

Fig. 2 a) Conductance-voltage curves of three different types of Nb/AlO_x/Nb junctions (area 25 μm^2). b) Conductance-voltage characteristics of two NbN/MgO/NbN junctions, measured at 4.2 K (area = 9 μm^2). The dashed line is a theoretical curve. The waiting periods were 11 and 4 minutes for PH 513 and PH 518, respectively.

coupling can account for these excess currents, provided that the barrier is not completely homogeneous [11,12]. These inhomogeneities are probably introduced during the anodization process.

Asymmetric SNIS and double oxide SNIIS junctions also show a strong temperature dependence of the subgap current between 4.2 and 1.5 K. However, sometimes microshorts are present, which form high resistance shunts and introduce small leakage currents. In none of the SNINS junctions significant leakage currents were observed. Apparently, microshorts in asymmetric junctions are introduced due to interactions at the AlO_x/Nb interface. H_2O is believed to play an important role in these interactions.

Figure 2a shows conductance-voltage curves of Nb/AlO_x/Nb junctions. In SNIS and SNINS junctions the AlO_x barrier can reasonably well be represented by a homogeneous, trapezoidal barrier with an average height of 1.3 - 1.6 eV, a small asymmetry and a thickness of 8 - 10 Å [13]. In SNIIS junctions the barrier can not be represented by a homogeneous potential barrier. It is very likely that the second Al layer does not form a closed layer, but nucleates in 'islands' during deposition.

In spite of the inhomogeneous barrier, high quality double oxide junctions with low subgap leakage currents can be made reproducibly. Apparently, the oxidation conditions to form a thin layer of AlO_x with good isolating properties are not too critical.

4. Experimental results on NbN/MgO

At 4.2 K the gap voltage V_g of the NbN/MgO/NbN junctions, measured at 50 % of the gap current is 4.4 - 4.5 mV. The gap width (80 % - 20 %) is about 0.6 mV. V_m at 2 mV is about equal in both junction series and ranges between 25 and

54 mV. The normal square resistances are 1.6k$\Omega\mu m^2$ for junction series PH 513 and 80 $\Omega\mu m^2$ for series PH 518.

In fig. 1b the I-V characteristics at 4.2 and 1.6 K of a NbN/MgO/NbN junction are depicted. The BCS curve at 4.2 K has been calculated, assuming an energy gap of the electrodes of 2.3 meV. The conductivity used in these calculations is 1.4 times the conductivity at 6 mV. Fig. 1b shows clearly that the subgap current hardly decreases with decreasing temperature. This behaviour is typical for all NbN/MgO/NbN junctions we measured. The current below V_g is dominated by temperature independent tunneling currents and, possibly, leakage currents due to microshorts. From this it is concluded that the barrier is inhomogeneous. It is interesting to mention that the maximum value of V_m at 2 mV predicted by the BCS curve is over 350 mV.

Typical dI/dV-V characteristics of both types of NbN/MgO/NbN junctions are shown in fig. 2b. For comparison a theoretical curve has been calculated for a potential barrier with height 0.3 eV and thickness 17.5 Å [13]. This theoretical curve indicates that the effective average potential barrier in both junctions is significantly lower than 0.3 eV. The measured characteristics have also been fitted to a barrier model which is valid for $V > \phi/e$ [14]. However, this model could not describe the measured characteristics either.

Fig. 2b clearly shows, that during the waiting period after the MgO deposition the barrier changes significantly. Because MgO is very hygroscopic, we think that the as deposited barrier reacts with residual H_2O. It is known that H_2O is important for oxidation of Mg, which might have been deposited during MgO sputtering [15].

The fabrication method of the, to our knowledge, best results on NbN/MgO/NbN junctions which have been reported, support this idea [7]. The effective deposition rate of the MgO used by Shoji is low enough to accomplish saturation with H_2O during deposition. This probably enables a reproducible control of the zero voltage current by the deposition time [16].

5. Conclusions

The AlO_x insulator in SNIS and SNINS Nb/AlO_x/Nb junctions can be represented by a homogeneous trapezoidal potential barrier, with barrier height 1.3 - 1.6 eV. Thermally oxidized Al layers are intrinsically closed, which results in subgap currents dominated by BCS single-particle tunneling at 4.2 K. In SNINS junctions no leakage currents due to microshorts have been observed.

The AlO_x in double oxide junctions is inhomogeneous. Despite of this, high quality SNIIS junctions, with very low leakage currents can be made.

The MgO insulator in NbN/MgO/NbN junctions can not be represented by a homogeneous trapezoidal potential barrier. The low effective barrier height suggests that the insulating layer does not consist of pure MgO [15]. The deposition of artificial MgO in NbN/MgO/NbN junctions leads to inhomogeneous barriers. The results reported in literature strongly suggest that this is an intrinsic property of the barrier in these type of junctions.

Reactions of deposited MgO with residual gases in the sputter chamber seem to affect the barrier properties considerably. It is suggested that saturation of MgO with H_2O, during or after sputtering, enables a reproducible NbN/MgO/NbN junction fabrication.

Acknowledgements: These investigations in the program of the Foundation for Fundamental Research on Matter (FOM) have been supported (in part) by the Netherlands Technology Foundation (STW).

References

[1] M. Gurvitch, M.A. Washington, H.A. Huggins, Appl. Phys. Lett. **42**, 472 (1983).
[2] D.J. Adelerhof, E.P. Houwman, P.B.M. Fransen, D. Veldhuis, J. Flokstra, H.Rogalla, IEEE Trans. Mag. **27**, 3153 (1991).
[3] A.I. Braginski, J.R. Gavaler, M.A. Janocko, J. Talvacchio, SQUID'85, editors H.D. Hahlbohm and H. Lübbig, W. de Gruyter, Berlin-New York.
[4] M. Gurvitch, J.P. Remeika, J.M. Rowell, J. Geerk, W.P. Lowe, IEEE Trans. Mag. **21**, 509 (1985).
[5] M. P. Mathur, D.W. Deis, J.R. Gavaler, J.Appl.Phys. **43**, 3158 (1972).
[6] A. Shoji, M. Aoyagi, S. Kosaka, F. Shinoki, H. Hayakawa, Appl. Phys. Lett. **46**, 1098 (1985).
[7] A. Shoji, IEEE Trans. Mag. **27**, 3184 (1991).
[8] E.P. Houwman, D. Veldhuis, J. Flokstra, H. Rogalla, J. Appl. Phys. **67**, 1991 (1990).
[9] H. Kroger, L.N. Smith, D.W. Jillie, Appl. Phys. Lett. **39**, 280 (1981).
[10] B. David, O. Dössel, G. Kürsten, G. Rabe, to be presented at the ISEC conference Glasgow, (1991).
[11] J.R. Schrieffer and J.W. Wilkins, Phys. Rev. Lett. **10**, 17 (1963).
[12] L.-E. Hasselberg, M.T. Levinsen, M.R.Samuelsen, Phys. Rev. B **9**, 3757 (1974).
[13] W.F. Brinkman, R.C. Dynes, J.M. Rowell, J. Appl. Phys. **41**, 1915 (1970).
[14] J.G. Simmons, J. Appl. Phys. **34**, 1793 (1963).
[15] W. Plesiewicz and J.G. Adler, Phys. Rev. B **34**, 4583 (1986).
[16] A.Shoji, M.Aoyagi, S.Kosaka, F.Shinoki, IEEE Trans. Mag.**23**, 1464 (1987).

Preparation and Characteristics of Full-Epitaxial $NbC_xN_{1-x}/MgO/NbC_xN_{1-x}$ Josephon Tunnel Junctions

A. Shoji, S. Kiryu, S. Kashiwaya, S. Kohjiro, S. Kosaka, and M. Koyanagi

Electrotechnical Laboratory, 1-1-4 Umezono, Tsukuba-shi, Ibaraki 305, Japan

Abstract. All refractory Josephson tunnel junctions composed of niobium carbonitride base electrodes, magnesium oxide barriers, and niobium carbonitride counterelectrodes have been prepared on single-crystal MgO substrates in a full-epitaxial growth manner. High-quality tunneling characteristics with large gap voltages(6.2mV-6.3mV), small subgap leakage currents(V_m>30mV), and small gap transition widths(~80μV, from 30% to 80%) have been obtained for junctions fabricated depositing their counterelectrodes at substrate temperatures in the range of 280°C-320°C.

1. Introduction

All refractory Josephson tunnel junctions with niobium nitride films or niobium carbonitride films as electrodes and very thin magnesium oxide films(~1nm) as tunnel barriers (NbN/MgO/NbN and $NbC_xN_{1-x}/MgO/NbC_xN_{1-x}$ Josephson tunnel junctions) have been attracting much attention in superconducting microelectronics, because of their extremely high stability for thermal cyclings, large gap voltages(>5mV), and high critical operating temperatures($T_{c(op.)} = 0.5T_c$>8K)[1-5].

However, fabrication of high-quality NbN/MgO/NbN or $NbC_xN_{1-x}/MgO/NbC_xN_{1-x}$ Josephson tunnel junctions with sharp gap structures is prevented by short coherence lengths(<10nm) and structure sensitive superconducting properties of niobium nitride and niobium carbonitride.

To overcome this problem, one of the present authors(A.Shoji) has recently fabricated $NbC_xN_{1-x}/MgO/NbC_xN_{1-x}$ Josephson junctions using single-crystal NbC_xN_{1-x} films, which were hetero-epitaxially grown on single-crystal MgO substrates, as base electrodes[6]. Fabricated junctions showed good tunneling characteristics with gap voltages over 5.6mV, small subgap leakage currents(V_m>20mV), and much smaller gap transition widths(<200μV) than those for NbN/MgO/NbN Josephson junctions with polycrystalline NbN films as base and counterelectrodes. The successful results strongly suggested that we can obtain high-quality tunneling characteristics for $NbC_xN_{1-x}/MgO/NbC_xN_{1-x}$ Josephson junctions, if we prepare them on the substrates in full-epitaxial growth manners.

In this paper, we report the results of a study which has been done to prove the suggestion experimentally. First, we report a preparation of $NbC_xN_{1-x}/MgO/NbC_xN_{1-x}$ Josephson junctions on single-crystal MgO substrates in a full-epitaxial growth manner. Then, we report the results of measurements of tunneling characteristics of prepared junctions.

Springer Proceedings in Physics, Vol. 64 Superconducting Devices and Their Applications
Editors: H. Koch, H. Lübbig

2. Preparation of junctions

We prepared NbC_xN_{1-x}/MgO/NbC_xN_{1-x} junctions in an rf-sputtering system which is equipped with 6inch-diam. magnetron-type targets of purity 99.9%Nb and purity 99.99%MgO, a stainless steel shutter, a stainless steel substrate holder, a substrate heater, and three evacuation pumps(a cryo pump, a turbo-molecular pump, and an oil-rotary pump). The base pressure of the system was 3×10^{-5}Pa. The substrates were polished (100) MgO single-crystals with a dimension of 2cm×2cm×0.05cm. The target to substrate distance was 4cm.

Niobium carbonitride films as base electrodes of junctions were deposited on the substrates by sputtering the Nb target in a mixture of 93.3vol.%argon, 5.6vol.%nitrogen, and 1.1vol.%acetylene at a total pressure of 0.67Pa. During the deposition of the base NbC_xN_{1-x} films, the substrates were heated up to 420°C to grow the films epitaxially on the substrates. The deposition rate for the base NbC_xN_{1-x} films was 50nm/min. The typical thickness of deposited base NbC_xN_{1-x} films was 150nm.

After the deposition of the base NbC_xN_{1-x} films, the substrates were cooled down to temperatures below 100°C in a high vacuum($<5\times10^{-5}$Pa), and 1.1nm-thick MgO barriers were deposited on the base NbC_xN_{1-x} films in 1.3Pa argon atmosphere. During the deposition of the barriers, we did not intentionally heat the substrates to grow the barriers uniformly on the base NbC_xN_{1-x} films. The deposition rate for the barriers was 4.2nm/min. The thickness of the barriers was estimated from the deposition rate and the deposition time.

The NbC_xN_{1-x} films as counterelectrodes of junctions were deposited in a mixture of 95.4vol.%argon, 3.9vol.%nitrogen, and 0.7vol.%acetylene at a total pressure of 2.0Pa at various substrate temperatures in the range of 280°C-400°C(280°C, 300°C, 320°C, 380°C and 400°C). This was done to find an optimum substrate temperature for epitaxially growing the counter NbC_xN_{1-x} films on the MgO barriers without damaging the barriers. The deposition rate for the counter NbC_xN_{1-x} films was 60nm/min. The typical thickness of the counter NbC_xN_{1-x} films was 120nm.

Crystalline properties of deposited base NbC_xN_{1-x} films, MgO barriers, and counter NbC_xN_{1-x} films were analyzed by reflection high-energy electron diffraction(RHEED). Figure 1 show a series of RHEED patterns from the surfaces of (a) a base NbC_xN_{1-x} film, (b) an MgO barrier, and (c) a counter NbC_xN_{1-x} film deposited at a substrate temperature of 280°C. We recorded these RHEED patterns with 20keV electrons at the azimuthal angle corresponding to a [010] direction in a crystal. In all of the RHEED patterns of Fig.1(a), (b), and (c), diffraction spots are seen clearly. This indicates that the base NbC_xN_{1-x} film, the MgO barrier, and the counter NbC_xN_{1-x} film in Fig.1 grew epitaxially. The results of RHEED measurements for other counter NbC_xN_{1-x} films deposited at higher substrate temperatures than 280°C also indicated that all of the counter NbC_xN_{1-x} films grew epitaxially.

Critical temperatures of deposited base NbC_xN_{1-x} films were measured by a conventional four-terminal method and found to be in the range of 17.4K-17.5K.

Critical temperatures of counter NbC_xN_{1-x} films were estimated to be in the range of 17.0K-17.4K from T_c values of NbC_xN_{1-x} films which had been deposited

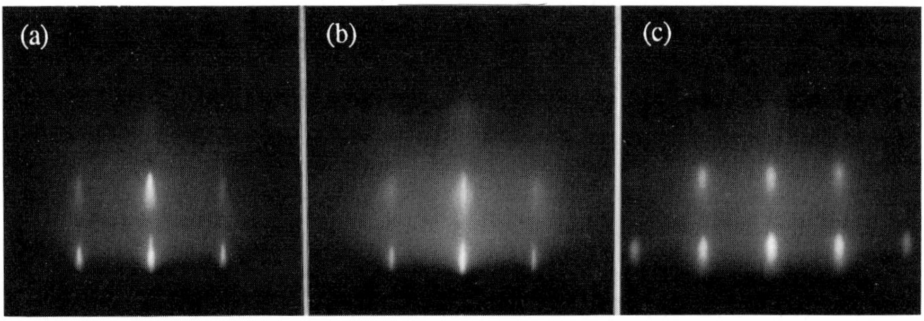

Figure 1. RHEED patterns from the surfaces of (a) a base NbC_xN_{1-x} film, (b) an MgO barrier, and (c) a counter NbC_xN_{1-x} film. All patterns were recorded with 20keV electrons at the azimuthal angle corresponding to a [010] direction in a crystal.

on single-crystal MgO substrates with the same sputtering conditions as those for the counter NbC_xN_{1-x} films.

From a junction which had been prepared on the whole surface of a substrate, we defined individual junctions with small areas(5μm×5μm-40μm×40μm) using a process named SNIP(Self-aligned Niobium Isolation Process)[7]. In the process, photoresist processes and reactive-ion-etching techniques were used for defining the areas and electrodes of junctions and silicon mono-oxide(SiO) films were used for isolating the electrodes of junctions from the wirings.

We call the fabricated junctions "full-epitaxial junctions" to distinguish them from junctions with only their base electrodes epitaxially grown. We call the latter junctions "semi-epitaxial junctions".

3. Tunneling characteristics of junctions

Figure 2 shows a current-voltage(I-V) characteristic for a 20μm×20μm full-epitaxial NbC_xN_{1-x}/MgO/NbC_xN_{1-x} junction fabricated depositing its counterelectrode at a substrate temperature of 280°C.

As shown in this figure, high-quality tunneling characteristics with sharp gap structures and small subgap leakage currents have been obtained for full-epitaxial NbC_xN_{1-x}/MgO/NbC_xN_{1-x} junctions which had been fabricated depositing their counterelectrodes at substrate temperatures of 280°C, 300°C, and 320°C.

In Table 1, we summarize the gap voltages V_g, the products of the maximum critical currents I_c and the normal-tunneling resistances R_n(static resistances at 10mV), the subgap-leakage parameters V_m(products of I_c and subgap resistances R_{sg} measured at 3mV), and the gap transition widths ΔV_g(voltage widths within which the currents of junctions increased from 30% to 80% of the values at 6.5mV) for those full-epitaxial NbC_xN_{1-x}/MgO/NbC_xN_{1-x} junctions.

The gap voltages for these full-epitaxial NbC_xN_{1-x}/MgO/NbC_xN_{1-x} junctions are about 0.5mV larger than those(5.6mV-5.8mV) for semi-epitaxial NbC_xN_{1-x}/

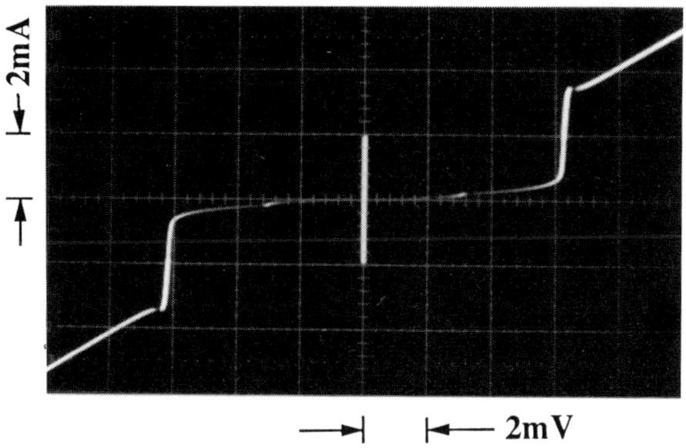

Figure 2. I-V characteristic for a full-epitaxial $NbC_xN_{1-x}/MgO/NbC_xN_{1-x}$ junction at 4.2K. The counterelectrode of this junction was deposited at a substrate temperature of 280°C. The critical current density of this junction is 500A/cm^2. The V_g, I_cR_n, and V_m values for this junction are 6.3, 3.9, and 38mV, respectively.

Table 1. Junction parameters for full-epitaxial $NbC_xN_{1-x}/MgO/NbC_xN_{1-x}$ junctions fabricated depositing their counterelectrodes at substrate temperatures(T_s) of 280°C, 300°C, and 320°C.

Junction	T_s (°C)	Area (μm^2)	J_c (A/cm^2)	V_g (mV)	I_cR_n (mV)	V_m (mV)	ΔV_g (μV)
$NbC_xN_{1-x}/MgO/NbC_xN_{1-x}$	280	20×20	500	6.3	3.9	38	~ 80
$NbC_xN_{1-x}/MgO/NbC_xN_{1-x}$	300	20×20	350	6.2	3.7	38	~ 80
$NbC_xN_{1-x}/MgO/NbC_xN_{1-x}$	320	20×20	370	6.3	3.7	35	~ 80

MgO/NbC_xN_{1-x} junctions[6]. This indicates that a high-quality superconducting property is obtained not only for the base electrodes but also for the counterelectrodes of these full-epitaxial $NbC_xN_{1-x}/MgO/NbC_xN_{1-x}$ junctions.

The gap transition widths for these full-epitaxial $NbC_xN_{1-x}/MgO/NbC_xN_{1-x}$ junctions are as large as a half of those(100μV-200μV) for semi-epitaxial NbC_xN_{1-x} $/MgO/NbC_xN_{1-x}$ junctions[6]. This indicates that a high degree of energy gap uniformity is achieved in the both of base and counterelectrodes of these full-epitaxial $NbC_xN_{1-x}/MgO/NbC_xN_{1-x}$ junctions.

The V_m values obtained for these full-epitaxial $NbC_xN_{1-x}/MgO/NbC_xN_{1-x}$ junctions are larger than those(20mV-30mV) obtained for semi-epitaxial $NbC_xN_{1-x}/$

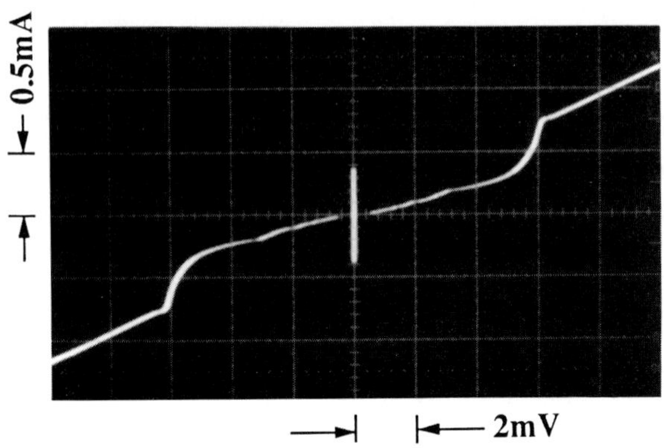

Figure 3. I-V characteristic for a full-epitaxial $NbC_xN_{1-x}/MgO/NbC_xN_{1-x}$ junction at 4.2K. The counterelectrode of this junction was deposited at a substrate temperature of 380°C. The critical current density of this junction is 1600A/cm^2. The V_g, I_cR_n, and V_m values for this junction are 6.0, 3.4, and 7mV, respectively.

MgO/NbC_xN_{1-x} junctions, but much smaller than those(>2×10^{4}mV) expected from a simple tunnel theory[8].

Figure 3 shows the I-V characteristic for a 5μm×5μm full-epitaxial NbC_xN_{1-x} /MgO/NbC_xN_{1-x} junction fabricated depositing its counterelectrode at a substrate temperature of 380°C.

As shown in this figure, low-quality tunneling characteristics with large subgap leakage currents and large gap transition widths were obtained for full-epitaxial $NbC_xN_{1-x}/MgO/NbC_xN_{1-x}$ junctions which had been fabricated depositing their counterelectrodes at substrate temperatures of 380°C and 400°C. For these junctions, were also obtained much larger critical current densities(>1500A/cm^2) than those(350A/cm^2-500A/cm^2) for full-epitaxial $NbC_xN_{1-x}/MgO/NbC_xN_{1-x}$ junctions fabricated depositing their counterelectrodes at substrate temperatures in the range of 280°C-320°C. These suggested that interdiffusion of atoms occurred between the tunnel barriers and the counterelectrodes of these full-epitaxial $NbC_xN_{1-x}/MgO/NbC_xN_{1-x}$ junctions during the deposition of the counterelectrodes.

4. Summary

We have prepared all-niobium carbonitride Josephson tunnel junctions with thin magnesium oxide barriers(~1nm) on single-crystal MgO substrates in a full-epitaxial growth manner. High-quality tunneling characteristics with large gap voltages (6.2mV-6.3mV), small subgap leakage currents(V_m>30mV), and small gap transition widths(~80μV) were obtained for junctions fabricated depositing their counterelectrodes at substrate temperatures in the range of 280°C-320°C. However, junctions fabricated depositing their counterelectrodes at substrate temperatures higher than 380°C showed low-quality tunneling characteristics with large subgap leakage currents and large gap transition widths, suggesting that interdiffusion of atoms occurred between the MgO barriers and the counterelectrodes of these junctions during the deposition of the counterelectrodes.

Acknowledgment

The authors would like to thank Dr. K. Kajimura of Electrotechnical Laboratory for his encouragement to this study.

References

[1] S.R. Whiteley, F. Kuo, M. Radparvar, and S.M. Faris, IEEE Trans. Magn. **25**, 770 (1989).
[2] J.A. Stern, B.D. Hunt, H.G. LeDuc, A. Judas, W.R. McGrath, S.R. Cypher, and S.K. Khanna, IEEE Trans. Magn. **25**, 1054 (1989).
[3] R.P. Robertazzi and R.A. Buhrman, IEEE Trans. Magn. **25**, 1384 (1989).
[4] B.D. Hunt, H.G. LeDuc, S.R. Cypher, and J.A. Stern, Appl. Phys. Lett. **55**, 81 (1989).
[5] M. Aoyagi, H. Nakagawa, I. Kurosawa, and S. Takada, IEEE Trans. Magn. **27**, 3180 (1991).
[6] A. Shoji, IEEE Trans. Magn. **27**, 3184 (1991).
[7] A. Shoji, M. Aoyagi, S. Kosaka, and F. Shinoki, IEEE Trans. Magn. **MAG-23**, 1464 (1987).
[8] M. Gurvitch and J. Kwo, Advances in Cryogenic Engineering Materials **30**, Edited by A.F. Clark and R.P. Reed, 509 (1984).

Lead/Bismuth- and Niobium-Based SIN Tunnel Junctions

T. Lehnert, K.H. Gundlach, and H. Kohlstedt

IRAM, Institut de Radioastronomie Millimétrique, 300, rue de la Piscine, Domaine Universitaire, F-38406 Sant-Martin d'Hères, France

Abstract. This paper describes the preparation and properties of various types of SIN devices : Pb/Bi/In–oxide–Ag junctions, evaporated through metal masks and three slightly different types of Nb/Al-based junctions, applying "trilayer-technique", similar as for Nb–Aloxide–Nb SIS junctions. These three types are: Al(thick)–Alox/Al–Nb1, Nb–Alox–Al1(thick), Nb–Alox–Al(thick)–Nb1. We show the anodization profiles of these trilayers and compare experimental I-V characteristics with theory. Photon steps were observed at 370 GHz.

1. Introduction

With increasing operation frequency, Josephson noise becomes increasingly a concern for low-noise quasiparticle mixing in SIS junctions. SIN (Superconductor-Insulator-Normal metal) junctions have no Josephson current. Using experimental I-V curves of SIN junctions, theoretical analysis yields good performance.[2,3] Such calculations even predict conversion gain for Pb/Bi/In–oxide–Ag junctions.[4] An experimental SIN mixer receiver at 220-230 GHz gave promising results.[5] These observations have motivated us, to study different types of SIN junctions.

2. Pb/Bi/In–oxide–Ag junctions

These devices were fabricated using metal masks. First, a pre-alloyed Pb/Bi (29wt % Bi) layer, about 300nm thick, was evaporated onto a quartz substrate at room temperature. This was followed by the evaporation of a 15 to 25 nm thick In film without breaking the vacuum. After growing the oxide tunnel barrier, the vacuum system was opened to interchange the masks for the evaporation of the about 400nm thick Ag counter electrode. Evaporations were made from thermally heated tantalum boats in a vacuum of about $5{\cdot}10^{-6}$ mbar. Pairs of metal masks were used; one mask for the base, the other for the counter electrode.[6]

Each mask has an opening for one side of the low-pass filter for the IF output and slots, about 8μm long and 1–2 μm wide, which define the junction

Springer Proceedings in Physics, Vol. 64 Superconducting Devices and Their Applications
Editors: H. Koch, H. Lübbig

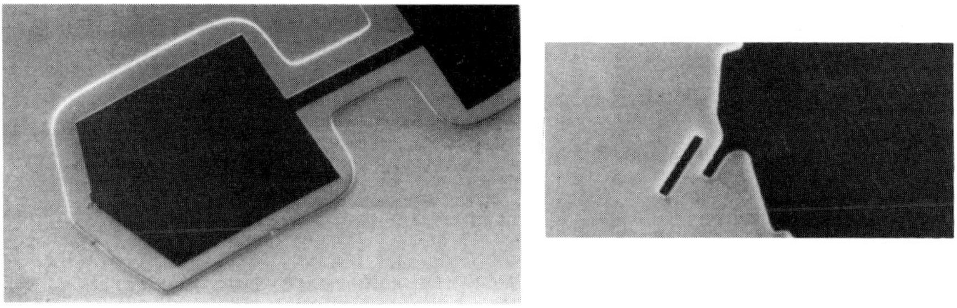

Fig.1 Metal mask for the evaporation of Pb/Bi/In–oxide–Ag junctions. The slots in the right part define the junction base (or counter) electrodes for a three junction series array. The reduced magnification in the left part reveals the opening for the low-pass filter.[6]

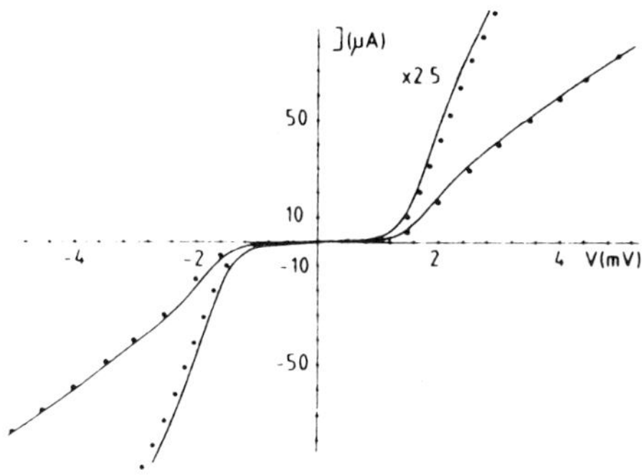

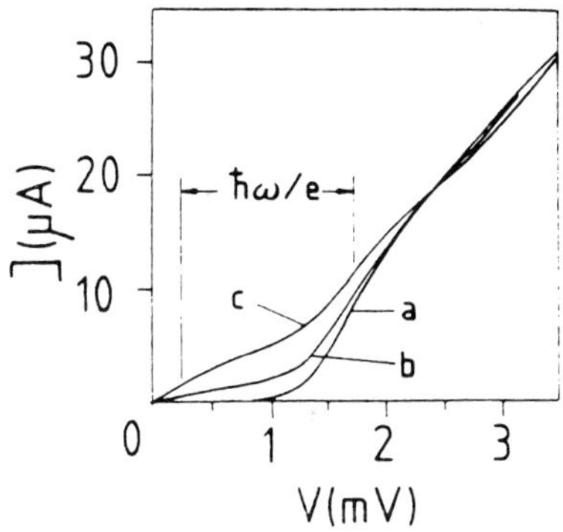

Fig.2 *(left)* I-V curve of a Pb/Bi/In–oxide–Ag junction at 2.5K. Junction area is $2\mu m^2$ and R_N=60Ω. The points represent theoretical values for Δ=1.7meV.
Fig.3 *(right)* The effect of 370GHz irradiation on a Pb/Bi/In–oxide–Ag junction with an area of $2.5\mu m^2$, R_N=115Ω. (a) without, (b) and (c) with irradiation.

electrodes. Single junctions and series arrays of three and five junctions have been made. Fig.1 shows a photo of such a mask.

In fig.2 an experimental J–V curve is compared with points evaluated numerically with the BCS density of states for an energy gap Δ=1.7 meV and a normal state resistance R_N=60Ω. Fig.3 shows a photon step obtained at 370 GHz.[7]

3. Nb-based SIN junctions

The fabrication of all Nb-based SIN junctions is based on in-situ sputtered trilayers, similar as for Nb–Aloxide–Nb SIS junctions. The trilayer is prepared using photoresist stencils and lift-off. The trilayers remaining after lift-off form the rf low-pass filter. Anodization spectroscopy was used to diagnose the trilayers.

3.1 Al(thick)–Aloxide/Al–Nb1

First a 100nm Nb layer (henceforth Nb1) was deposited, followed by a 8nm thick Al film. After exposing the Al film to oxygen, to form the tunnel barrier, the trilayer was completed by a 30nm thick Al counter electrode. The junction

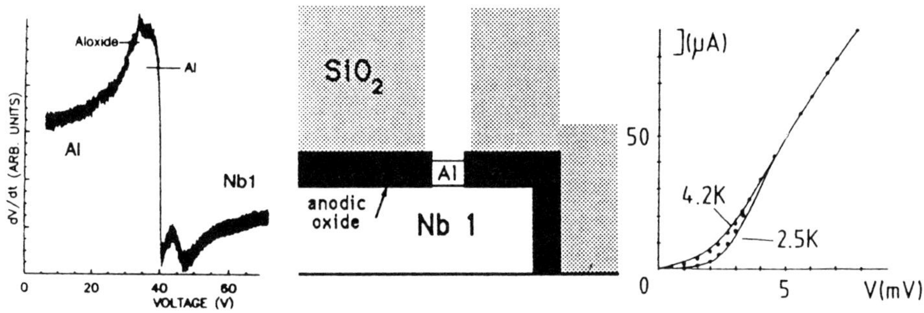

Fig.4 *(left)*: Anodization profile of a Al(thick)–Aloxide/Al–Nb1 trilayer. *(middle)*: Cross section of the junction. The Nb wiring layer is not shown. *(right)*: I-V curves of a two-junction array. The single junction has an area of 2.5μm^2. The dots represent BCS values for Δ=1.37meV at 2.5K and 1.40meV at 2.5K.

3.2 Nb–Aloxide–Al1(thick)

First a 100nm Al layer (henceforth Al1) was deposited and exposed to oxygen. The subsequent Nb counter electrode is 30nm thick. Reactive ion etching was used to obtain the junction mesa which was then sealed by light liquid anodization and evaporation of SiO_2 (fig.5).

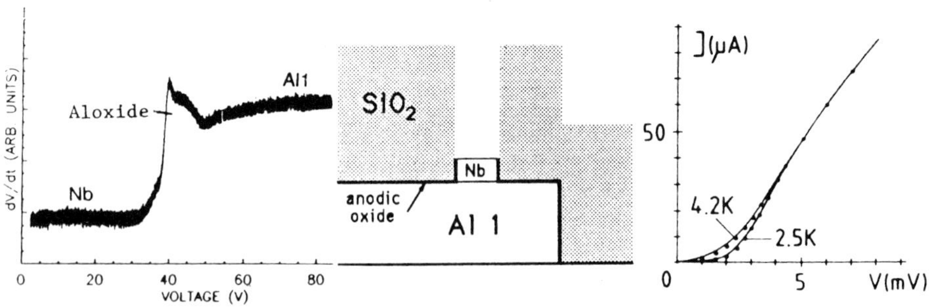

Fig.5 *(left)*: Anodization profile of a Nb–Aloxide–Al1 trilayer. *(middle)*: Cross-section of the junction. *(right)*: I-V curves of a two-junction array. The area of a single junction is about 1.6μm^2. The dots represent BCS values for Δ=1.37meV at 4.2K and 1.40meV at 2.5K.

3.3 Nb–Aloxide–Al(thick)–Nb1

The 100nm thick Nb1 layer is followed by an Al layer whose thickness can be varied from about 5nm to 100nm. In this way, we intend to study the transition from SIS to SIN junction performance. The mesa preparation and sealing is very similar as under (3.2). Fig.6 clearly shows that the current onset shifts to smaller voltages (V_{on}) and that the transition to the ohmic region becomes broader with increasing Al thickness.

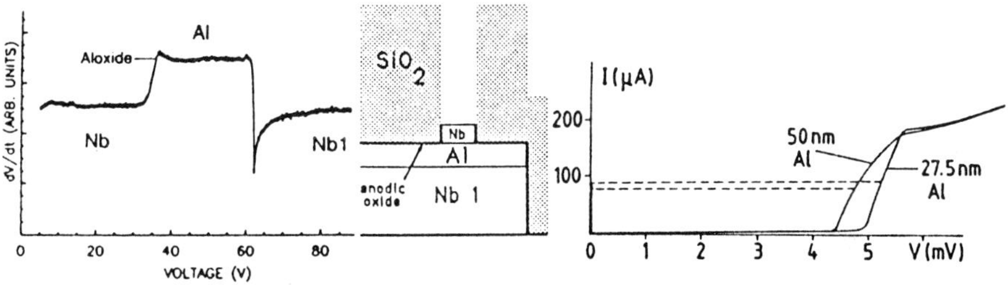

Fig.6 *(left)*: Anodization profile of a Nb1-Al(27.5nm)-Aloxide-Nb trilayer. *(middle)*: Cross section of the junction. *(right)*: I-V curves of a two-junction array at 2.5K, with $V_{on} \simeq 4.92$mV for $d_{Al} \simeq 27.5$nm and $V_{on} \simeq 4.40$mV for $d_{Al} \simeq 50$nm. The area of each junction is $2\mu m^2$.

References

[1] present address: Advantest Laboratories Ltd, Otaru, 047-02 Hokkaido, Japan.

[2] C.E. Tong, L.M. Chernin and R. Blundell, in J. Appl. Phys. **68**, 4192 (1990).

[3] L.M. Chernin and R. Blundell, in J. Appl. Phys. **69**, 2682 (1991).

[4] R. Blundell, to be published.

[5] R. Blundell and K.H. Gundlach, Int. J. of Infrared and Millimeter Waves, **8**, 1573 (1987).

[6] Metal masks are fabricated by Dr. Johannes Heidenhain, G.m.b.H., 8225 Traunreut, Germany.

[7] R. Blundell and K.H. Gundlach, unpublished results.

Relation Between Tunnel Junction Quality, Anodization Profiles, and Low-Frequency Noise

H. Kohlstedt[1,*], *S. Kuriki*[2], *T. Lehnert*[3], *D. Billon-Pierron*[3], *and K.-H. Gundlach*[3]

[1]Advantest Laboratories Ltd., 250-1, Hoshino-cho, Otaru, 047-02 Hokkaido, Japan

[2]Research Institute of Applied Electricity, Hokkaido-University, 060 Sapporo, Japan

[3]Institut de Radioastronomie Millimètrique, Domaine Universitaire, F-38406 St.-Martin-d'Hères, France

*Present address: 1. Physikalisches Institut der Universität Köln, Zülpicher Straße 77, W-5000 Köln, Fed. Rep. of Germany

Abstract. Anodization profiles and low-frequency noise was investigated for low and high quality Nb-AlO_x/Al-Nb (SIS) and, to our knowledge for the first time, for Al-AlO_x/Al-Nb (SIN) tunnel junctions.

Our comparison shows a relation among the junction quality (determined by the V_m parameter [1], low frequency noise, and anodization profile.

1. Introduction

Telegraph noise and anodization spectroscopy are two relatively new techniques to characterize interface layers in tunnel junctions [2,3].

Biasing small area (typically less than 2 μm^2) superconducting tunnel junctions, often two or more level random switching events in the voltage-time traces are observed. This so-called telegraph noise is usually explained by the capture and emission of electrons in barriers [4]. Rogers and Buhrman [5] were able to link these switching events in small area diodes with 1/f noise in larger area junctions.

Anodization spectroscopy was first used by Gates et. al. [6] to investigate adjacent Nb-Al layers. This technique was evaluated by the Fujitsu group [3,7] and became a very useful tool to monitor the properties of Nb-AlO_x/Al-Nb trilayers and related sandwich structures as Nb/Ta multilayers [8].

2. Experimental Methods

Nb-AlO_x/Al-Nb (SIS) [9] or Al-AlO_x/Al-Nb (SIN) [10] diodes were fabricated on one inch rectangular wafers. Instead of the more common selective niobium etching process (SNEP) we used a lift-off technique [11].

The layers were deposited by dc magnetron sputtering in a cryopumped system. Using optical lithography and reactive ion etching, individual unshunted

Springer Proceedings in Physics, Vol. 64 Superconducting Devices and Their Applications
Editors: H. Koch, H. Lübbig

two-junction series arrays were fabricated. Each tunnel contact has an area of approximately 2 μm^2.

All noise measurements were performed at a temperature of 4.2 K in a common four point arrangement and with a double μ-metal magnetic shield to reduce flux noise. The junctions were biased with a current source and voltage fluctuations were measured with a low-noise preamplifier. The voltage-time traces and also the noise power spectra were registered with an Iwatsu 2100 signal analyzer.

A small substrate, which was placed beside the wafer during in-situ layer deposition, served as probe for anodization spectroscopy [12]. If an Nb2-AlO_x/Al-Nb1 trilayer is anodized, the anodization rate will change in successive stages from C_{Nb2} to C_{AlOx} to C_{Al} and C_{Nb1}. The anodization profile dV/dt vs V contains information about the interfaces of the layers. The derivative dV/dt was measured by an RC circuit.

3. Results

The two columns in Figure 1 show a comparison of a high and a low quality Nb-AlO_x/Al-Nb diode at the left and right side respectively. The rows from the top are the I-V curve, the anodization profile, the power spectral density and a voltage-time trace.

The high quality two-junction series array (Fig. 1a) with V_m = 30 mV has a sum gap voltage of 5.5 mV. The normal resistance is R_N = 118 Ω. The anodization profile (Fig. 1b) is "sharp" and contains two characteristic peaks. These maxima at 37 V and 44 V represent the Nb2-AlO_x and Al-Nb1 interface respectively. From our experience such an anodization profile is a necessary precondition of a trilayer to get high quality junctions from the lift-off process [13].

The voltage noise power spectral density (Fig. 1c) shows low 1/f noise. Above 100 Hz the noise was dominated by the preamp. This spectrum was measured at a bias voltage of V_B = + 50 mV (ground electrode positive). In between a bias voltage range of -100mV and +100 mV we could not find any indication of two or more level switching events in the voltage-time traces. An example of this "silent" behaviour is shown in Figure 1d.

In contrast, the low quality two-junction series array with a figure of merit V_m = 4 mV showed a different behaviour. The sum gap voltage of 4.5 mV is much lower than the theoretical value. The normal resistance was R_N = 175 Ω. The proximity structure at +/-5 mV originates from a normal layer near the tunnel barrier [14]. The anodization profile (Fig. 1f) has a "shoulder" instead of the first peak, which indicates a different AlO_x-Nb interface compared with that in the high quality array.

Figure 1g shows the voltage noise power spectral density of the low quality array. The Lorentzian like spectrum is a characteristic feature for a two-level system. The voltage-time trace of this array is shown in Figure 1h. This telgraph signal is probably a result of random parameter fluctuations of the junction resistance. The solid line in Figure 1g represents the voltage noise power spectral density S_V calculated from the Machlup equations [15].

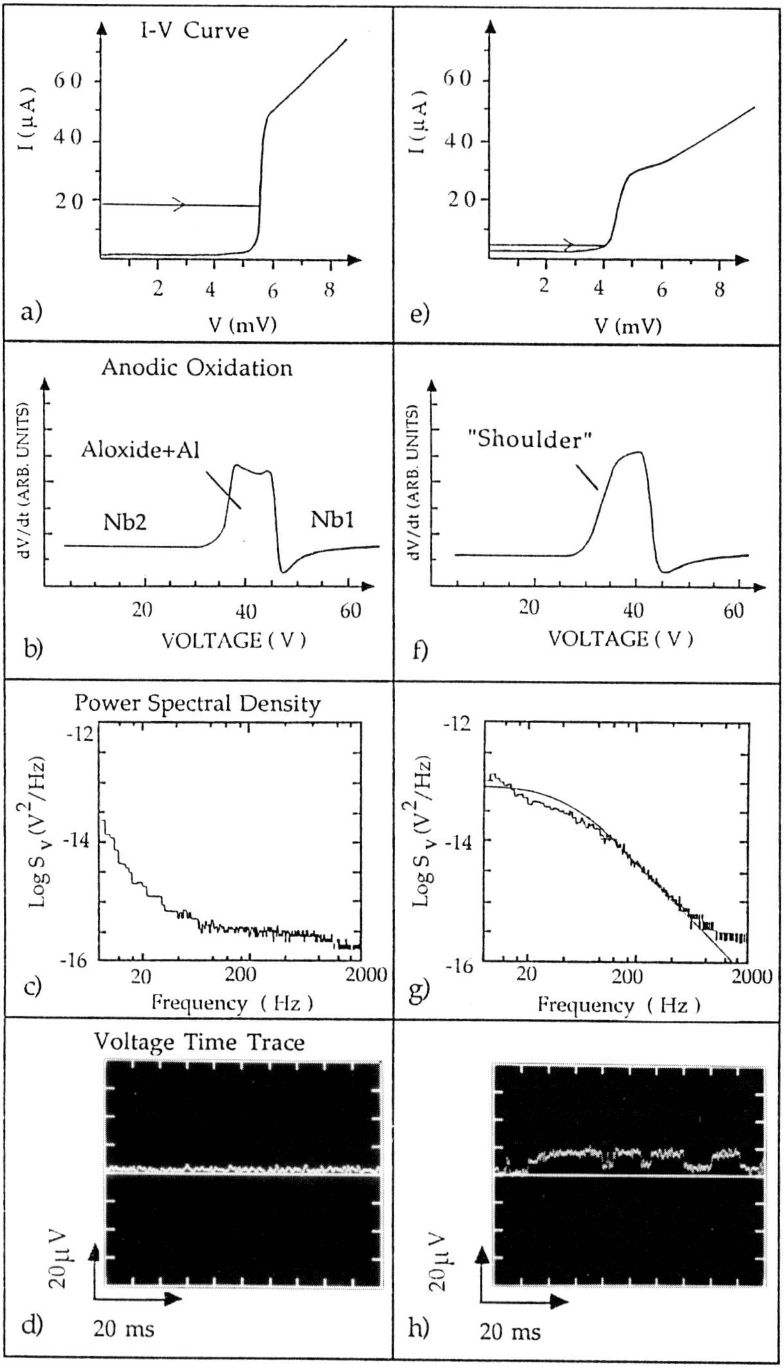

Fig. 1a-h. Comparison a high and a low quality Nb-AlO_X/Al-Nb two-junction array.

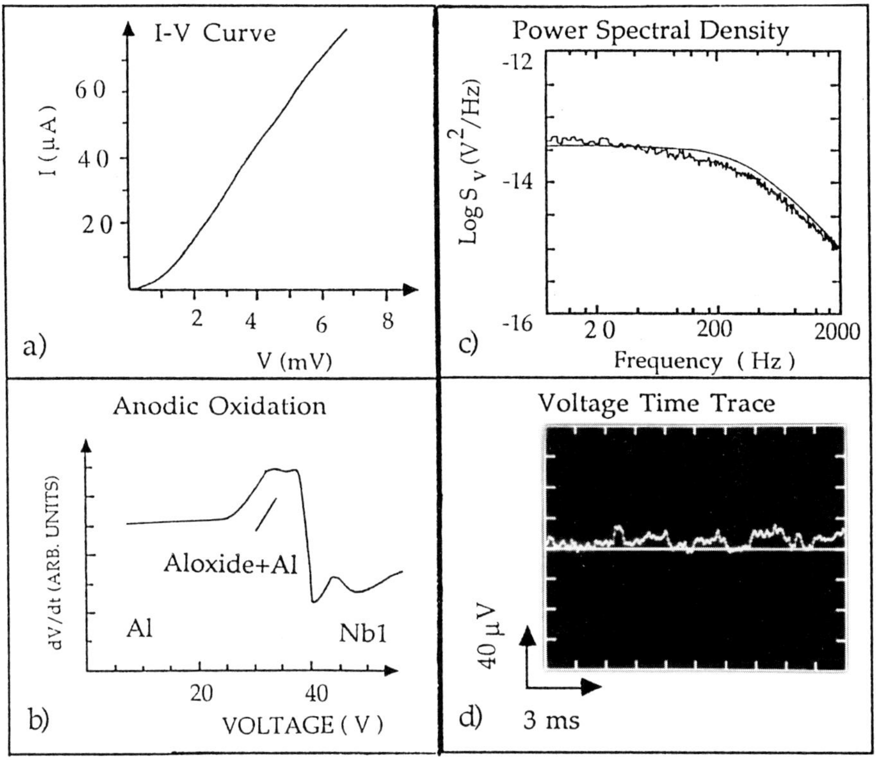

Fig. 2a-d. I-V curve, anodization profile, voltage noise power spectral density and voltage noise power spectral density and voltage-time trace respectively of an Nb-AlO_x/Al-Al (SIN) two-junction array.

$$S_V = 4\,(\delta V)^2\,(t_{eff}/(t_{up} + t_{down}))\,(t_{eff}/(1 + \omega^2\,t^2_{eff})), \tag{1}$$

$$1/t_{eff} = 1/t_{up} + 1/t_{down}, \tag{2}$$

where t_{up} and t_{down} are respectively the average times in the high and low resistance state and δV is the voltage pulse height. For this junction array biased at $V_B = +50$mV we observed $t_{up} = 10.44$ ms, $t_{down} = 4.58$ ms and $\delta V = 5.6$ μV. We found a good agreement between the calculated S_V from the time traces and the measured power spectral density.

In Figure 2 the same measurements were done for an Nb-AlO_x/Al-Al two-junction array. The I-V characteristic is of lower quality compared with that, for example, of a high quality Pb/Bi-InO_x-Ag (SIN) diode [16]. The gap voltage in Figure 2a is difficult to determine, but estimated to 2.4 mV +/-0.3 mV. The transition in the anodization profile, shown in Figure 2c, from the Al counter electrode to the Nb-Al ground electrode has a "shoulder" comparable to that profile of the low quality SIS diode shown in Figure 1g. By analyzing the low-frequency noise of

the SIN array we found telegraph signals in the voltage-time trace above a bias voltage of V_B = 30 mV. For the comparison with the Nb-AlO_x/Al-Nb diodes we measured the telegraph noise and the noise voltage power spectral density (see Figs. 2c and 2d respectively) at V_B = 50 mV.
The bump frequency is at approximately 320 Hz. The solid line in Figure 2c is the noise power spectrum calculated from equations (1) and (2) with the parameters t_{up} = 0.86 ms, t_{down} = 0.97 ms and δV = 8.9 μV. Again we obtained a good agreement with the measured spectrum.

4. Discussion

Next we try to explain the different behaviour of the low quality SIS diode array in respect to the I-V curve, the anodization profile and the noise spectrum S_V in comparison with the high quality SIS junction array.

Let us assume an Al layer that is not pinhole free as in the case of the low quality junction array. During oxidation the formation of Nb_2O_5 and suboxides will take place in the pinhole region. The result is an I-V curve with a leakage current and a reduced gap voltage of this tunnel contact as shown in Figure le. When the anodization front reaches such pinholes at the Nb-AlO_x/Al interface, the anodization rate is reduced and results in a "shoulder". In contrast to this, an Nb-Al interface with Nb completely covered by Al has a sharp peak in the anodization profile (Fig. 1b).

Rogers and Buhrman [4] investigated Nb-Nb_2O_5Pb/Bi junctions and found, that an oxygen vacancy in amorphous Nb_2O_5 barriers could possibly be a trap and ionic configurational tunneling is an obvious explication for random resistance fluctuations under 15 K.

From this point of view poor quality Nb-AlO_x/Al-Nb junctions likely have a higher probability to create low-frequency excess noise, due to the higher density of traps, than diodes with excellent I-V characteristics.

The reasons for the suppression of the first peak in the anodization profile (Fig. 1f) are yet unclear. A possible explication is the degradation of a not sufficiently postbaked photoresist during trilayer formation, the most serious draw-back of the lift-off process. Even a small amount of outgasing hydrocarbon molecules from the resist can easily affect the wetting of Al on Nb or the termal oxidation of the Al layer and consequently the tunnel barrier properties.

The Nb-AlO_x/Al-Al (SIN) junction showed telegraph noise and also a slope in the anodization profile. Due to these similarities to the Nb-AlO_x/Al-Nb junction of poor quality in Figure 1, we suggest that the Al of the ground electrode does not completely cover the Nb surface. Consequently, pinholes in the Nb-Al interface are responsible for the poor I-V characteristics of this SIN array.

5. Conclusions

We compared the anodization profile and low-frequency noise of Nb-AlO_x/Al-Nb and Nb-AlO_x/Al-Al two junction arrays. We observed a relation among the junction quality, the anodization and the low-frequency noise behaviour. Our low quality junctions show clear telegraph noise signals above certain bias voltages, which results in larger excess noise in comparison with an electrically high quality junction. We suggest that the anodization profiles could be a useful guide to predict the quality as well as the low-frequency noise properties of diodes based on the Nb-Al technique.

Acknowledgements. We would like to thank Prof. G. Matsumoto for his strong collaboration to this work. On of us (H.K.) is indebted to N. Hajimoto and T. Ogawa for the introduction to the telegraph noise measurement system.

References

[1] Calculated from $V_m = 1/2(0.7\ \Delta J_{qp}\ R_{(4\ mV)})$, due to the examined two-junction series array. ΔJ_{qp} is the quasi-particle current at the gap voltage.
[2] C.T. Rogers, R.A. Buhrman, H. Kroger and L.N. Smith, Appl. Phys. Lett. **49**, 1107 (1986).
[3] T. Imamura and S. Hasuo, IEEE Trans Mag. **MAG-25**, 1131 (1989).
[4] C.T. Rogers, R.A. Buhrman, Phys. Rev. Lett. **55**, 859 (1985).
[5] C.T. Rogers, R.A. Buhrman, Phys. Rev. Lett. **53**, 1272 (1984).
[6] J.-V. Gates, M.-A. Washington, M. Gurvitch, J. Appl. Phys. **55**, 1146 (1984).
[7] T. Imamura and S. Hasuo, Appl. Phys. Lett. **55**, 2550 (1989).
[8] M.G. Blamire, K.H. Huang, R.E. Somekh, E. Kirk, G.W. Mori and J.E. Evetts, Appl. Phys. Lett. **55**, 732 (1988).
[9] K.-H. Gundlach, IRAM Working Report, No. 203 (1990).
[10] T. Lehnert and K.-H. Gundlach, unpublished.
[11] M. Yuda, K. Kuroda and J. Nakano, Jpn. J. Appl. Phys. **26**, L166 (1987).
[12] H. Kroger, L.-N. Smith, D.-W. Jillie, Appl. Phys. Lett. **39**, 280 (1981).
[13] H. Kohlstedt, K.-H. Gundlach and A. Schneider, IEEE Trans. Mag. **MAG-27**, 3149 (1991).
[14] E.L. Wolf, Principles of Electron Tunneling Spectroscopy, Plenum Press (1985).
[15] S. Machlup, J. Appl. Phys. **25**, 341 (1954).
[16] R. Blundell and K.-H. Gundlach, Int. J. Infrared and Millimeter Waves, **8**, 1573 (1987).

Submicron Sized, High Current Density Nb/PbBi Window Tunnel Junctions

J. Edstam and H.K. Olsson

Physics Department, Chalmers University of Technology,
S-412 96 Göteborg, Sweden

Abstract. We have fabricated high quality, submicron sized Nb/PbBi window tunnel junctions suitable for sub-mm wave applications. The junctions had critical current densities in the range 20 - $4*10^4$ A/cm^2 and reproducible areas from 0.03 to 5 μm^2. The junction quality (V_m~20 mV) did not depend on the junction size down to 0.2 μm^2.

1. Introduction

The increasingly higher operating frequencies of superconducting detectors and generators require increasingly smaller capacitances, C (or junction areas, A) in order for the signals not to be short circuited ($1/\omega C$) or the device not to be narrow band ($Q=\omega RC$). For signal frequencies above ~ 100 GHz we are already stretching the photo-lithography techniques to its limits at ~ 1 µm. Sub-µm shadow mask junctions using electron-beam resists have been made using soft metals [1] whereas sub-µm Nb based junctions are made using the film edge [2,3] or are limited to about 1 μm^2 for planar structures. Neither of these methods are well suited in an application where integration with microstrips, shunting resistors, Nb, and good control over the junction area are required. We have developed a window junction process for Nb/PbBi junctions using PMMA electron-beam resists and achieved good quality tunnel junctions (V_m~20 mV) with areas down to 0.03 μm^2 and current densities up to $4*10^4$ A/cm^2.

2. Junction fabrication

Window tunnel junctions were fabricated using Nb groundplanes for base-electrodes [4], windows in SiO defining the junction area [5], and counterelectrodes of PbBi. Whereas the groundplane and counterelectrode were fabricated with standard evaporation, lift-off, and wet-etching techniques using photoresists, the window required electron-beam techniques.

Two different procedures in defining lift-off stencils for the window in SiO were adopted. The first method used a double resist layer consisting of 340 nm of P(MMA-MMA) and 70 nm of PMMA on top. Relatively small e-beam exposure doses (80 µC/cm^2 at 50kV) were used in order to avoid too large under-cuts due to back-scattered electrons. P(MMA-MMA) is more sensitive to electrons facilitating the undercut. The double-layer resist was subsequently developed for 40 sec in Toluene/2-propanol (1:3), 20 sec in 2-ethoxyethyl acetate/Ethanol (1:10), and

Springer Proceedings in Physics, Vol. 64 Superconducting Devices and Their Applications
Editors: H. Koch, H. Lübbig

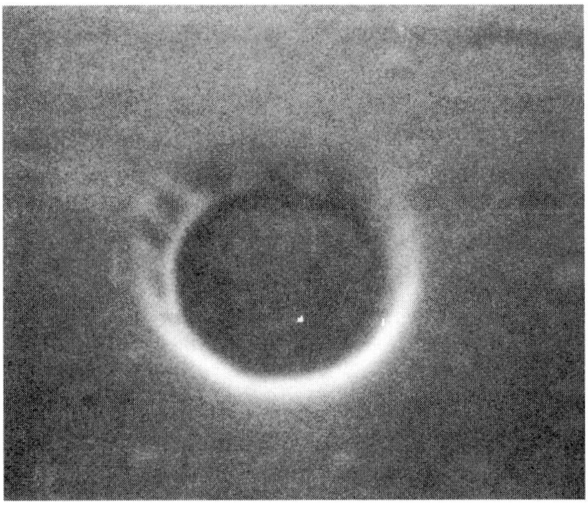

Fig 1. SEM photo of a window of diameter 1120 nm (1 μm^2) in a 300 nm thick SiO film. The sample is tilted ~ 30°.

finally rinsed in 2-propanol for 30 sec. This procedure gave well defined and reproducible undercuts. The second method used a sequence of 900 nm polyimide, 20 nm of evaporated NiCr and 300 nm of PMMA on top. The PMMA was exposed and developed to define the window and then by sequentially ion-beam etching with Ar and Ar+O, a well defined lift-off mask was produced. The substrate was rotating and tilted 5-10° during the etching to acheive an undercut in the polyimide. The use of a thicker resist in this method gave better edge-coverage and a more reliable lift-off.

300 nm of SiO was evaporated at high rates (2.5 nm/sec) at a pressure of $8*10^{-7}$ torr [5] and lifted off in acetone at 50°C. Figure 1 shows an example of a window, with a diameter of 1120 nm, fabricated using the first method. The windows are well defined and have a slight slope due to a lateral growth of SiO on the lift-off mask during deposition.

A standard recipe for ion-beam oxidation of the junctions was used [2] including an oxygen RIE (Reactive Ion Etching) cleaning step prior to the junction formation: cleaning in Ar (220 V) and low voltage (60 V) oxidation, immediately followed by thermal evaporation of the PbBi counterelectrode. The substrate was kept at - 60°C during these steps.

3. Results

Tunnel junctions were made using both window definition techniques described above though smaller junctions could be made with the "NiCr-technique". Windows of diameters down to 0.5 µm (0.2 μm^2) were made with the double-layer resist and diameters down to 0.2 µm (0.03 μm^2) with the NiCr as etch-stop. Both techniques produced good quality junctions with V_m~20 mV ($V_m = I_0$ R(at 2 mV)). Only slightly more leakage current was seen for the smallest junctions. However, no or poor quality junctions were produced if the windows were not RIE-cleaned from

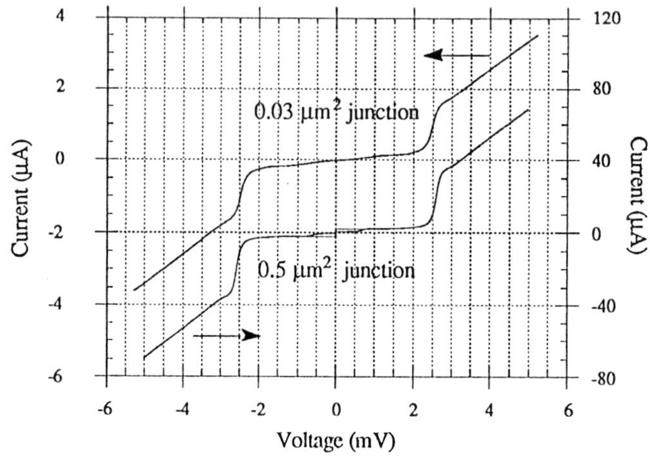

Fig 2. IV curves for a large and a small Nb/PbBi tunnel junction taken at 4 K. Junction areas are 0.03 and 0.5 μm^2 and $j_0 = 4000\ A/cm^2$.

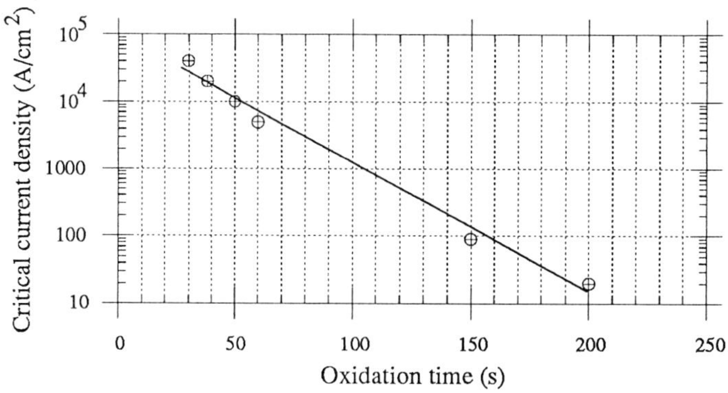

Fig 3. Critical current density vs oxidation time for Nb/PbBi window junctions at 4 K.

resist residues before barrier preparation. Figure 2 shows IV-curves for two junctions of the same critical current density ($j_0 = 4000\ A/cm^2$) but one large (0.5 μm^2) and the other small (0.03 μm^2). The absence of clear critical currents in the IV-curves is due to thermal smearing or accidental trapping of flux in the junction barrier. Junctions made with different oxidation-times gave critical current densities in the range 20 - $4*10^4$ A/cm², see Figure 3. The current density increases approximately exponentially with the inverse of oxidation time indicating that the barrier thickness grows roughly linearly with time. We find that the for sub-mm wave region important current range around 10^4 A/cm² can be obtained by choosing a proper oxidation time in the 30 - 60 sec interval. The same current density was obtained for both small and large junctions.

4. Conclusions

We conlude that it is possible to fabricate Nb based window tunnel junctions of good quality and reproducible dimensions down to 0.03 μm^2. Junctions with current densities of around 10^4 A/cm^2 can directly be implemented in integrated sub-mm wave detectors and generators. Smaller junctions would probably require using thinner dielectricas and resists with less undercuts.

Acknowledgements

Experimental assistance from B. Nilsson and E-K. Andersson and the use of the Swedish Nanometer Laboratory is gratefully acknowledged. This work was supported by the Swedish Board for Technical Development.

References

[1] T.A. Fulton and G.J. Dolan, Phys. Rev. Lett. **59**, 109, (1987).

[2] A.W. Kleinsasser and R.A. Buhrman, Appl. Phys. Lett., **37**, 841, (1980).

[3] J.M. Martinis and R.H. Ono, Appl. Phys. Lett., **57**, 629, (1990).

[4] S.Z. Wang, Cryogenics, **27**, 564, (1987).

[5] H.K. Olsson, IEEE Trans. Magn. **MAG-25**, 1115, (1989).

Barrier-Properties of All-Niobium Tunnel Junctions

G.M. Daalmans[1], *O. Eibl*[2], *and L.Bär*

[1]Siemens Central Research Laboratories,
W-8520 Erlangen, Fed. Rep. of Germany
[2]Siemens Central Research Laboratories,
W-8000 München, Fed. Rep. of Germany

Abstract. The barrier-region of all-niobium tunnel junctions has been imaged by means of high resolution transmission electron microscopy (atomic scale). Amorphous silicon barriers as well as amorphous Al_2O_3-barriers have been studied. Different behaviour concerning wetting and roughness have been found. Finally the TEM-data are correlated with the transport properties of the barriers.

1. Introduction

The main problem in making Josephson junction circuitry is the reproducible preparation of reliable tunnel junctions. In the beginning of Josephson microelectronics Josephson junctions of soft materials, like lead, have been used. Excellent quality Josephson junctions could be obtained even with simple vacuum systems. Unfortunately, these junctions were not recyclable due to thermal stress relaxation by hillock growth. Due to this problem junctions of the more technical superconductor niobium came into focus. It was obvious to make the barrier from native oxide by thermal or glow discharge oxidation. However this technical route was not very successful, because the thermally stable niobium tunneljunctions showed very leaky characteristics. As argued very early by Halbritter [1] the niobium oxide is not a very good choice as a barrier material, because it is not stochiometric and it can be reduced by the top electrode. As soon as the community was convinced about this problem artificial barriers became of interest. The introduction of artificial barriers was accompanied with new problems. Due to the exponential dependence of the transport properties of tunneling barriers from the barrier thickness very thin layers had to be deposited homogeneously on top of the ground electrode. Pinholes should be prevented by perfectly homogeneous deposition and by perfect wetting of the ground electrode. To meet these requirements the barrier material choice is strongly limited. A first successful approach was to prepare the barrier from a semiconductor like amorphous silicon [2]. The idea was to take advantage of the low barrier height of semiconductors for tunneling, allowing thicker tunneling barriers with relaxed requirements concerning homogeneity.

Another even more decisive issue was to cover the niobium with a normal metal layer obeying different requirements. First of all the normal metal layer should be superconducting by the proximity effect, secondly it should not diffuse

Springer Proceedings in Physics, Vol. 64 **Superconducting Devices and Their Applications**
Editors: H. Koch, H. Lübbig

or solve into the niobium and thirdly it should have a stochiometric oxide with a high formation energy. A metal that meets all the requirements is aluminium. It turned out to be a very good choice [3] concerning leakage currents and homogeneity. The large progress in Josephson junction circuitry during the last five years [4] is due to the introduction of Nb/Al/Al_2O_3/Nb junctions together with the SNAP-patterning process of Kroger [5].

We studied Josephson junctions with amorphous silicon barriers and aluminium oxide barriers by high resolution transmission microscopy and correlated these data with transport measurements.

2. Preparation of the samples

Three different samples were prepared. One sample concerned a Nb/Al/Al_2O_3/Nb sandwitch fabricated in a high vacuum system with AC- and DC-magnetron sputtering at rates and pressures to be stressfree. Another sample concerned a Nb/α-Si/Nb sandwich prepared in a ultra high vacuum system with e-beam evaporation in a one vacuum step process. The third sample was again a Nb/α-Si/Nb sandwich made in the same ultra high vacuum system with the same e-beam gun but with a two vacuum steps process. In the first step the ground electrode was made and in the second the barrier and the top electrode were prepared. In between there was a reactive ion etching process and an ion milling process. Details about layer thicknesses as determined from the deposition processes are shown in the table below.

3. X-ray diffraction and TEM-results

With x-ray diffraction it was observed that the e-beam evaporated films were randomly oriented, whereas the sputtered films were (110) textured. As a consequence the e-beam films had a greater surface roughness. The roughness of the sputtered films, as determined from REM-pictures, was about 2 nm. The e-beam films on the other hand showed a roughness of about 4 nm without ion milling and about 7 nm with the ion milling step included. The crystal size was for all the samples about 50 nm.

The sputtered Nb/Al/Al_2O_3/Nb sandwich shows a reduced roughness at the Al_2O_3-surface if compared with the roughness of the ground electrode leading to excellent coverage with very thin aluminium films. This phenomenon is to our opinion the main reason for the successful preparation of Nb/Al/Al_2O_3/Nb junctions. In the samples 1 and 2 however the roughness of the barrier surface is the same as the roughness of the ground electrode. This means that for good Nb/α-Si/Nb Josephson junctions the roughness of the ground electrode should be just a few percent of the barrier thickness.

The thickness as determined from TEM-pictures is in good agreement with the thickness obtained from the deposition technology. From the TEM-pictures we obtain for sample 1 about 5.0 nm α-Si, for sample 2 about 4.5 nm α-Si and

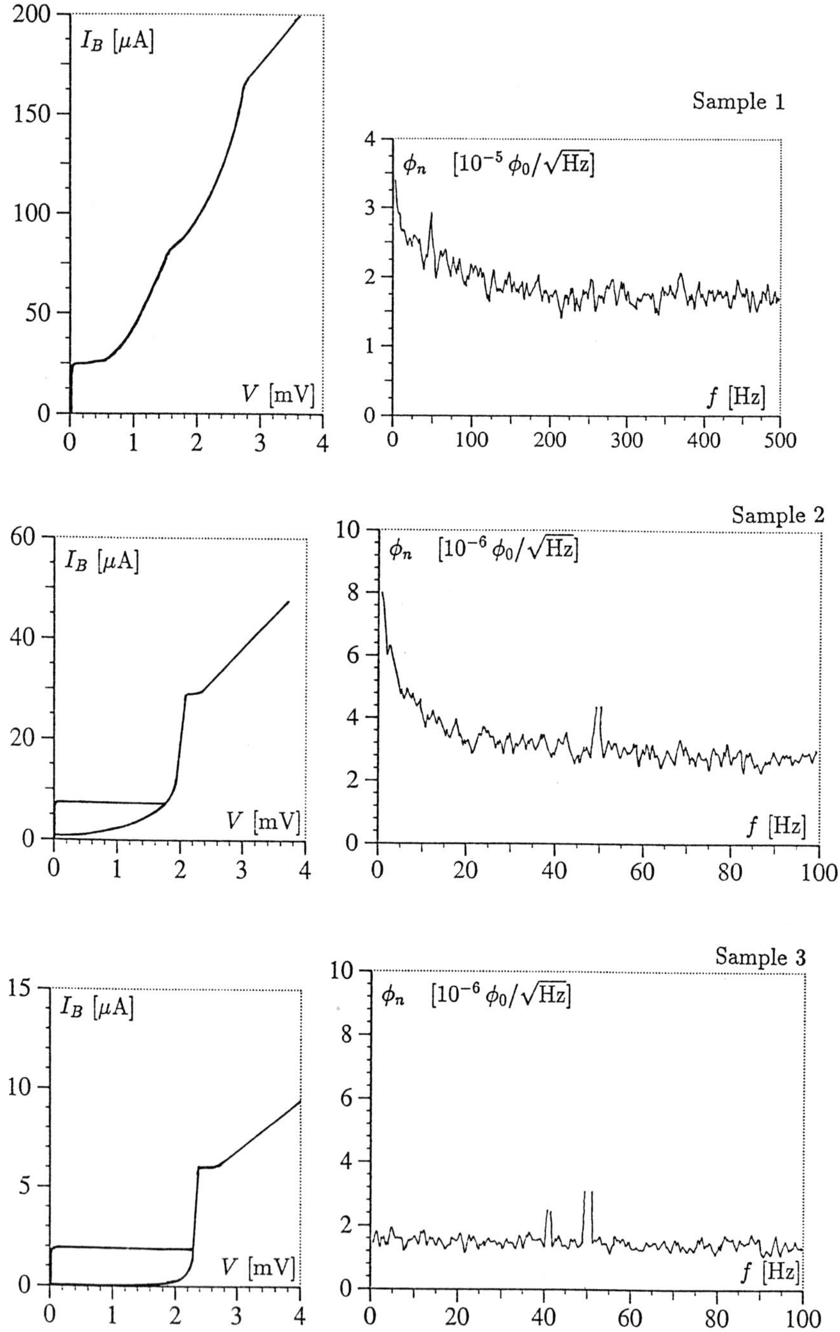

Figure 1 I-V-curves and equivalent flux noise for the samples 1, 2 and 3. The I-V-curves belong to unshunted Josephson junctions, whereas the noise data belong to shunted Josephson junctions.

Sample 1	Sample 2	Sample 3
Nb : 135 nm Si : 5.0 nm Nb : 50 nm Substrate (111) Si one step process	Nb : 135 nm Si : 4.5 nm Nb : 50 nm Substrate (111) Si two step process	Nb : 230 nm Al : 10 nm Nb : 50 nm Substrate (111) Si coated with SiO_2

for sample 3 about 12.5 nm aluminium. So we conclude that we don't loose material by diffusion in grain boundaries as discussed in previous papers [6]. The Al_2O_3-barrier is clearly seen as an amorphous layer of about 15 Å.

4. Transport properties

The transport properties of the different samples, referring to the leakage currents, are rather different (Figure 1). Sample three, with the Nb/Al/Nb stack, has the smallest leakage current corresponding with a $V_m = I_C x R(2mV)$ value up to 80 mV for 2 µm x 2 µm junctions. For sample 1, the single step Nb/α-Si/Nb stack, the leakage currents are very high resulting in V_m-values below 1 mV. The two step Nb/α-Si/Nb stack, shows intermediate leakage currents with V_m-values up to 10 mV. It should be stressed here that the current-density versus α–silicon layer thickness is shifted over about 2 nm in sample two towards thinner α–silicon films. So in order to obtain a current density of 100 A/cm^2 we need about 5 nm of α-silicon instead of the usual 7 nm in the single step process, despite of the greater roughness for the two step process. The noise of the samples in strongly correlated with the leakage currents. Leakage currents cause an increased noise level.

5. Discussion

From the TEM-results it is very likely that the high surface mobility of aluminium at room temperatur results in perfect coverage of the underlaying niobium film.It must be relatively easy therefore to obtain a pinhole free Al_2O_3-barrier. Due to the high barrier-hight of Al_2O_3 and the existence of just one stochiometric oxide low leakage currents have to be expected for Nb/Al/Al_2O_3/Nb stacks. The low surface mobility of silicon at room temperature leads to the transformation of niobium surface roughness in barrier thickness variations. As soon as the surface roughness of the niobium film is comparable to the nominal silicon-thickness, pinholes can not be avoided. The pinholes, filled with niobium oxide, cause leakage currents. Of course leakage currents can also be due to resonant tunneling in the barrier layer itself. If the niobium oxide is reduced, for instance by evaporating the top electrode, it can behave as a normal metal and contribute to excess currents above the gapvoltage by Andreev reflections [7].

The reduced conductance of the two step barrier on a niobium ground electrode with increased roughness, can be explained by the existence of a composite barrier. Unfortunately the TEM-method can not discriminate between the elements Si, O and C due to the relatively low mass differences.

References

[1] J. Halbritter, Proc. of SQUID '85, Walter de Gruyter, Berlin, New York, 1985, p. 31.

[2] H. Kroger, IEEE Trans. on Electron. Devices **ED-27**, 1980, p. 2016
B.J. Bending, M.R. Bearly, Phys. Rev. Lett. **55**, 1985, p. 324.
G.M. Daalmans, Proc. SQUIDs '80, Walter de Gruyter, Berlin, New York, 1980, p. 399.

[3] M. Gurvitch, M.A. Washington, H.A. Huggins, Appl. Phys. Lett. **42**, 1983, p. 472.

[4] S. Hasuo, S. Kotani, A. Inoue, N. Fujimaki, IEEE Trans. on Magn. **MAG 27**, (2), 1991, p. 2602.

[5] H. Kroger, L. Smith, D. Jillie, Appl. Phys. Lett. **39** (1981), p. 280

[6] J. Kwo, G.K. Wertheim, M.Gurvitch, D.N.E. Buchanan, Appl. Phys. Lett. **40** (8), 1982, p. 675.

[7] T.M. Klapwijk, Proc. SQUID '85, Walter de Gruyter, Berlin, New York, 1980, p. 1.

Fabrication and Performances of NbZr/Oxide/NbZr Thin Film Josephson Junctions

U. Gambardella[1], *D. Di Gioacchino*[1], *S. Frigerio*[1], *G. Paternò*[2], and *M. Cirillo*[3]

[1]INFN-Laboratori Nazionali di Frascati, P.O. Box 13, I-00044RM Frascati, Italy
[2]ENEA-CRE Frascati, P.O. Box 65, I-00044RM Frascati, Italy
[3]II Università di Roma, via E. Carnevale, I-00173RM Roma, Italy

Abstract. We report on the fabrication of symmetrical NbZr-Ox-NbZr Josephson junctions by means of thin film RF magnetron sputtering deposition. The insulating barrier consisted of natural oxide thermally grown in air at room temperature. The variation of the superconducting gap Δ as a function of the temperature was measured by recording the I-V characteristics. The experimental data are in good agreement with the dependence numerically computed from the BCS theory. The magnetic behavior of the maximum Josephson current was measured and both the magnetic penetration depth and the Josephson penetration depth were computed. By means of the Kulik theory on the junction self-induced resonances the microwave quality factor Q of the junctions, and the surface impedance of the films have been estimated.

1. Introduction

The literature reports tunneling measurements on non symmetrical junctions ($NbZr/Al/Al_2O_3/In$ or NbZr/Ox/Pb), both to determine the electron-phonon coupling factor [1], and the superconducting energy gap [1,2] Δ. However the reported data are from tunnel junctions realized on thin NbZr foils. Very few data deals with NbZr thin films [3] which, in any case, have been fabricated by using different methods, i.e. co-sputtering from two cathodes or electron-beam evaporation, and the reported critical temperatures T_c are below 9 K. The use of artificial barriers is accounted for the poor insulating properties of the NbZr natural oxide.

To allow for a direct measurement of the superconducting properties of the NbZr films we realized symmetrical tunnel junctions. In fact the counterelectrode of the same material makes easier the measurements in a wider temperature range.

2. Sample fabrications

The samples were realized by means of RF magnetron sputtering. The $Nb_{75}Zr_{25}$ cathode, fabricated by melting, was shaped in a 3 inches diameter disc. Before the deposition the vacuum was ranging in 2÷4 x 10^{-7} mbar with a cold LN_2 trap. The junction geometry was of the cross type, defined by metallic masks. The film thickness was ranging between 3000 Å and 5000 Å, controlled by a quartz microbalance. The lower electrodes were usually thinner than the upper to get good overlapping films. The sputtering is operated in a 5 x 10^{-3} mbar atmosphere of pure Ar gas. The deposition rates to realize tunnel junctions were 13.1 Å/s and 10.1 Å/s respectively for the lower and upper electrodes. The critical temperatures of the NbZr films were not strongly affected by the deposition rate in this range. Moreover, before each deposition, a 20 minutes pre-sputtering on a movable shield has been performed. The substrates were either Corning glass 7059 or sapphire, previously cleaned in an ultrasonic bath of acetone kept at 40 °C and dryed under nitrogen flow. During the sputtering the substrates were not heated. The tunnel oxide barriers were grown at room temperature by natural oxidization in air for 17 hours approximatively.

3. Experiments

The the current voltage I-V curves of the tunnel junctions were recorded biasing the junctions with a programmable current source and measuring the corresponding voltage by means of a high impedance differential nanovoltmeter. The measurements were managed by a computer which also stored data. The temperature was measured by means of calibrated carbon glass thermometer fed at 10 μA constant current.
Temperatures above the LHe temperature were attained keeping the sample into the He cold vapour, and checking the temperature was stable within ± 20 mK. Then the temperature is recorded at the beginning and at the end of the data storage to make further control. Temperatures lower than 4.2 K were reached lowering the bath pressure. In Fig. 1 a typical I-V characteristic at 4.2 K is plotted. The tunnel resistance above the gap voltage is $R_{NN} \approx 0.06\ \Omega$, but the low voltage dynamical resistance exihibits a rather low value indicating that there was a significant current leakage. However the gap voltages are well defined. In the same figure the derivative, numerically computed, of the I-V curve is shown.

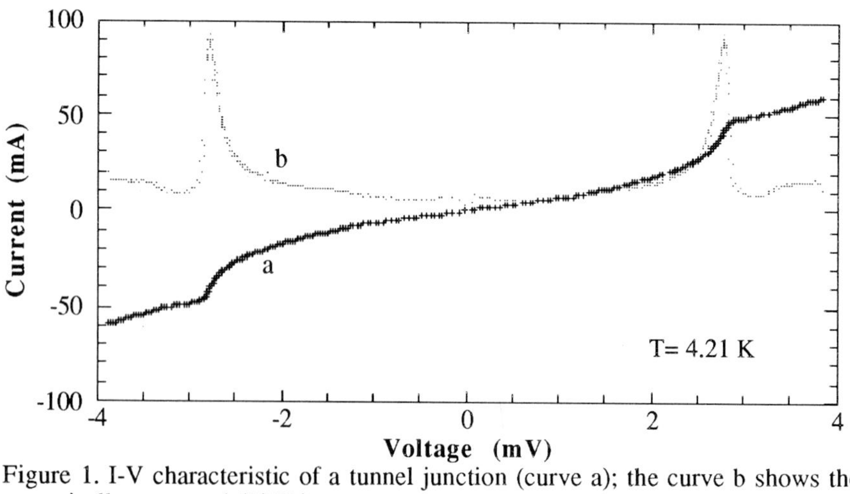

Figure 1. I-V characteristic of a tunnel junction (curve a); the curve b shows the numerically computed dI/dV in a.u..

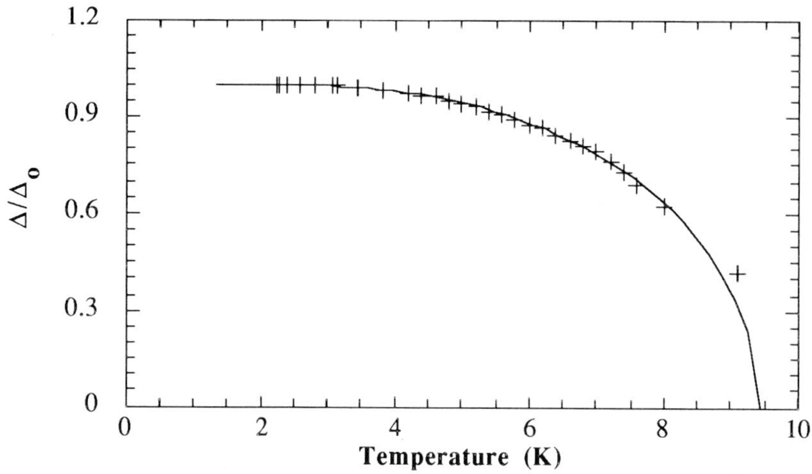

Figure 2. Temperature behavior of Δ: crosses are experiments; solid line represents the theoretical BCS curve.

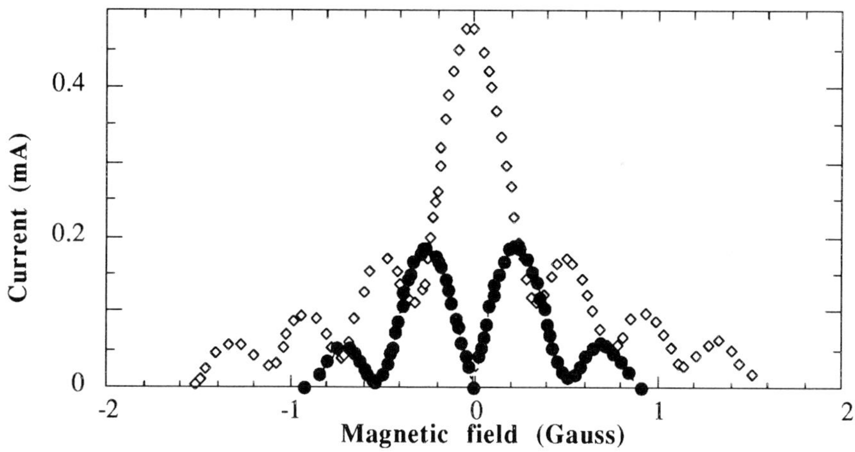

Figure 3. Magnetic field dependence of the Josephson current (open romb); magnetic field dependence of the 1st Fiske step (full circle).

In Fig. 2 it is reported the reduced gap as a function of the temperature for the sample of Fig. 1. The normalizing coefficient Δ_0 for the $Nb_{75}Zr_{25}$ is the value measured at 2.28 K, i. e. $\Delta(2.28)=1.63$ mV. In Fig. 2 it is also reported (solid line) the reduced BCS data theoretically computed [4]. The agreement between the theory and experiment is satisfactory. On different samples we observed the same BCS behavior but we measured lower Δ_0 and the T_c to be used to get good agreement with theory was lower. This may be accounted for surface effects on the NbZr film.

Finally we measured in a magnetic shielded room the Josephson current behavior of our samples. In Fig. 3 it is reported the magnetic behavior of both the maximum

Josephson amplitude and of the first Fiske step in a junction having $L \approx 200\ \mu m$ and $\lambda_J \approx 320\ \mu m$. From the magnetic pattern of the Josephson current we estimate a magnetic penetration depth $\lambda_{eff} \approx 1100$ Å. However in view of the non zero modulation of the current we recognize a non uniform current distribution through the barrier.

Finally from the Fiske step, occuring at a voltage $V_1 \approx 18\ \mu V$, we evaluate the junction quality factor Q as in ref. [5]. The computation gives $Q_1 \approx 70$, and considering as the main losses term the surface impedance, we get a surface resistance R_s of NbZr of $R_s \approx 230\ \mu\Omega$ at a frequency of 9 GHz and a T=4.2 K.

4. Conclusion

We have realized symmetrical NbZr Josephson tunnel junctions. From the I-V we measured the dependence of the energy gap on the temperature. Moreover from the Josephson features we get preliminary results both for the magnetic penetration depth and the microwave performance of the material. Further improvements on the fabrication process are in progress in order to match the requirements of device fabrications by using photolitography in patterning the first electrode and lift off technique for the second one.

References

1. E. L. Wolf, R. J. Noer, *J. Physique* **8** 39, C6-454 (1978)
2. I. Dietrich, *Phys. Lett.* **9** 3, 221 (1964)
3. R. Delesclefs, Ø. Fischer, *J. Low Temp. Phys.* **53** 3/4, 339 (1983)
4. B. Muhlschlegel, *Z. Physik* **155**, 313 (1959)
5. Y. S. Gou, R. I. Gayley, *Phys. Rev. B* **10** 11, 4584 (1974)

Some Aspects in the Tunneling Behavior of Photosensitive Junctions

C. Camerlingo, B. Ruggiero, M. Russo, and E. Sarnelli

Istituto di Cibernetica, Consiglio Nazionale delle Ricerche, I-80072 Arco Felice (Naples), Italy

Abstract Results on the tunneling behavior of light-sensitive superconductive junctions are discussed. The evolution of the barrier parameters by illumination have been investigated.

Superconducting tunnel junctions employing a photosensitive semiconductor barrier constitutes a feasible way to get a light-controllable device for superconductive electronics. In such a kind of structure, the tunneling parameters can in fact be modified by a suitable optical input. Both tunneling conductance and supercurrent can be increased, inducing a Josephson current even though it is absent in dark conditions. Moreover the occurrence of memory effects ("conductivity storage" at low temperatures) after the light input removal offers a further potential for applications.

Aspects concerning both basic and possible applications of light-sensitive Josephson tunnel junctions have been considered in literature so far [1]. Attention has been mostly focused on tunneling structures with a cadmium sulphide thin film barrier. Never the less exhaustive information on the barrier morphology is not yet available.

The fundamental role of the choice of the electrodes in determining the tunnel behavior of the junction has been discussed. For instance the dependence of the light-induced critical current on temperature and semiconductor thickness has been successfully described in terms of a simple model considering the role of the built-in potential barrier at the superconductor/semiconductor interfaces. Symmetrical (In-CdS-In) and asymmetrical (Pb-CdS-In) junctions have been considered taking into account the occurrence of an ohmic contact at the In/CdS boundary and a barrier at the Pb/CdS interface. Estimations of the height, E_B of this barrier are available in literature. They are obtained by measurements on Pb-CdS-Pb symmetrical structures. Since 1968 I. Giaver obtained $E_B = 20$ mV from experimental current-voltage (I-V) characteristics [2]. Assuming the occurrence of a Schottky barrier at the interfaces, Mehbod et al.[3] computed the same value from I-V characteristics of junctions with a highly doped, thick (t=200 nm) cadmium sulphide film. These authors also considered the behavior of the junctions for increasing illumination levels. Later on Andreozzi et al.[4] and Bobbio et al.[5] studied the case of thin (t= 10-50 nm) barrier film in good quality junctions. Barrier parameters (height and thickness) where computed from experimental (dI/dV) vs V curves by an inversion of the tunneling current expression due to Simmons [6] in the framework of the WKB approximation. His approach based on an equivalent rectangular average barrier leads to the approximate expression $dI/dV=\alpha(1+3\gamma V^2)$ for bias voltages relatively low with respect to the barrier height. Such an expression allows to compute the rectangular average barrier height $\langle E_B \rangle$ and the barrier thickness s (measured

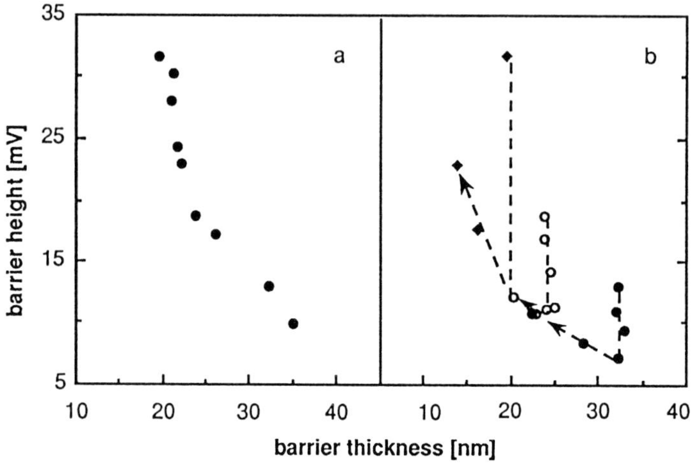

Fig. 1 Barrier height vs. thickness for a set of Pb-CdS-Pb junction in dark conditions (a). Evolution of the barrier parameters with the illumination for three different samples (b). The arrows indicate increasing illumination levels.

between its two classical turning points) from the experimental data. Details can be found in Refs[4] and [5]. For junctions in dark conditions values of $\langle E_B \rangle$ ranging from 10 to 32 mV have been obtained; they are in a reasonable agreement with previously reported estimations. If the value of the effective electron mass in the semiconductor is assumed to be very close to the tabulated one (m_i=0.17 m), a rather good agreement is obtained between computed values of s and the thickness of the CdS film measured by a profilometer. Moreover the dependence of $\langle E_B \rangle$ on s suggests a non-rectangular shape of the barrier though the equivalent rectangular average barrier well account for the tunneling data at low voltages. Such a situation seems to hold in a rather large range of conductance obtained on each sample by a suitable illumination. In fact the decrease in conductance results in the determination of a substantially constant value of s and a lowering of $\langle E_B \rangle$, see Fig. 1.For further illumination levels, the model used is no more adequate to get informations on the barrier parameter evolution. Hypothetical on the barrier shape have to be considered. The occurrence of a parabolic potential barrier (ideal Schottky barrier) at the metal semiconductor interfaces has been assumed in the framework of the theoretical approach due to Stratton [7]. Calculations still lead to an approximated parabolic expression for dI/dV allowing an estimation both of the barrier height E_B and of the parameter $N/\varepsilon m_i$ for highly illuminated samples. As expected E_B values compatible with previously reported values are obtained together with rather high values of the donor density in the semiconductor. This circumstance lead to a determination of the width of the two Schottky barriers of the order of a few nanometers. The effect of the light is a further increase of the parameter $N/\varepsilon m_i$ with a constant E_B which result in a thinning of the barriers at the interfaces. However the model seems to hold in a rather small range of illuminations. Its main interest is related to estimations of the depletion width which are in agreement with independent evaluations based on the occurrence of light-induced self-resonant modes in the junctions. In fact, the junction capacitance, as obtained from measurements of

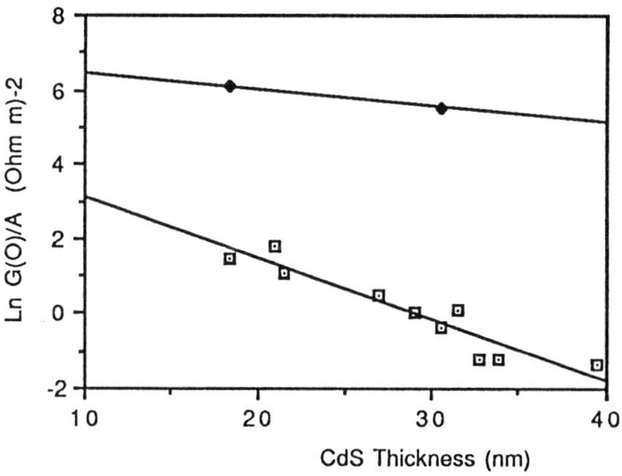

Fig. 2 Junction conductance vs. measured thickness of the CdS semiconductor layer for samples in dark conditions (lower curve). Quite preliminary measurements on samples at high illumination level are also reported. For such a level a relevant change of the rate occurs.

the voltage position of the Fiske-steps and magnetic penetration depth, indicates a barrier morphology similar to that reported by Lum and Van Duzer [8].

For highly illumination levels a further change in the junction behavior occurs. Preliminary measurements give a dependence of the junction conductance on the semiconductor film thickness exhibiting a slower rate. In Fig. 2 quite preliminary dependences are reported.

Further measurements are in progress to clarify the tunneling barrier behavior in this range of illumination levels.

This work is partially supported by the Consiglio Nazionale delle Ricerche under the Progetto Finalizzato "Tecnologie Superconduttive e Criogeniche".

References

[1] M. Russo, in " Josephson Effect: Achievements and Trends" A. Barone ed. (World Publishing, Singapore, 1986) p. 216.

[2] I. Giaever, Phys. Rev. Lett. **20**, 1286 (1968).

[3] M. Mehbod, W. Thijs, Y. Bruynseraede, Phys. Stat. Sol. **32**, 203 (1975).

[4] F. Andreozzi, A. Barone, R. Cristiano, M. Russo, and G. Peluso, IEEE Trans. Magn. **MAG-19**, 983 (1983).

[5] L. Bobbio, C. Camerlingo, R. Cristiano, M. Russo, and G. Peluso, IEEE Trans. Magn. **MAG-21**, 622 (1985).

[6] J. Simmons, J. Appl. Phys. **34**, 1793 (1963).

[7] R. Stratton, J. Phys. Chem. Solids **23**, 1177 (1962).

[8] W. Y. Lum and T. Van Duzer, J. Appl. Phys. **49**, 4560 (1978).

Construction and Performance of Wire-Junction RF-SQUIDs

A.C. Bruno and J.E. Zimmerman

Departamento de Física, Pontifícia Universidade Católica,
Rua Marquês de São Vicente 225, 22453 Rio de Janeiro, Brazil

Abstract. Two-hole symmetric rf-SQUID's were made by using a pair of wire segments to form the point contact Josephson junction. The SQUID's are reliable, easy to make and to adjust, and have a sensitivity of $2.5 \times 10^{-29}\ J/Hz$.

1. Introduction

Because of the high cost and limited reliability of commercially available rf-SQUID's, and the fact that when they fail they cannot be saved, we have constructed a number of rf-biased niobium point-contact SQUID's for use in a multichannel biomedical gradiometer. They are essentially the same to those introduced in 1970 and produced commercially for several years, except for a simplified design of the point contact. In place of the pair of tiny screws and lock-nuts, we used a pair of niobium wire segments held by friction in a closely fitting hole. The aim of this work is to develop a SQUID technology affordable for any laboratory having a simple machine shop.

2. Methods

The SQUID's were of the same geometry as those shown in reference [1]. They were 9.5 mm diameter by 25 mm long, with two parallel holes 2.5 mm diameter by 10 mm long separated by 2 mm. A slot was hand cut between the two holes using a 0.2 mm jeweler's saw. A 0.4 mm hole was drilled through the SQUID perpendicular to the slot to house a pair 0.4 mm wire segments, one flat on the end and one pointed, which formed a point contact Josephson junction bridging the slot. The wire segments, were 4 mm long and were slightly bent so that they were effectively held in place by friction. The segment with the flat end was oxidized by heating it in air at temperatures of $200°C$ to $300°C$. Contact adjustment at room temperature was made by using the high-frequency impedance of the SQUID as a measure of the contact resistance. Contact adjustment was performed by placing the flat-ended segment in the hole and by using a differential micrometer screw to push the pointed segment against it. The SQUID and the micrometer screw, along with a coil of 100 turns wound on a 2.2 mm plexiglass support, in parallel with a capacitor and an RF connector, were mounted in a suitably machined aluminum block, which was set up to allow motion of the pointed wire to be viewed with a low-power microscope.

Springer Proceedings in Physics, Vol. 64 Superconducting Devices and Their Applications
Editors: H. Koch, H. Lübbig

3. Results

Data presented here are typical of eight SQUIDs built and adjusted as described above. Figure 1 shows the impedance change as a function of the displacement of one segment of the contact pair relative to the other, as the two segments were forced together. The fact that the curve for 300°C oxidation lies mostly above the one for 250°C probably reflects a large uncertainty, owing to minute irregularities in the contact, in the zero of the displacement axis. The important parameter here is the slope of the curves, which decreases monotonically and rapidly with oxidation temperature. In order to obtain a proper critical current with an unoxidized pair, it is necessary to press one segment against the other by about $0.25\,\mu m$. This corresponds to an impedance change of about 0.7% if the pointed wire is sharpened to about 150°. Oxidizing the wire for 15 min. at 250°C, one wire segment must be pressed against the other by about $10\,\mu m$ in order to obtain about the same value of critical current. This corresponds to an impedance change of about 6%.

Figure 2 shows the SQUID impedance change, while closing the contact pair, as a function of the oxidizing temperature, that must be obtained in order to get a critical current of about $10\,\mu A$. Heating the flat-ended wire segment in air at 200°C, an impedance change of about 1% is required, at 250°C 6% and at 300°C, 20%. A good tradeoff between oxidation temperature and a fairly high impedance change, when the contact is closed, was 250°C for 15 min. and 6% of impedance change. These settings depend on the angle of the pointed wire segment. For instance, if it has an angle of about 100°, instead of 150°, the same critical current is obtained with an impedance change of 3% for the 250°C oxidation. When the oxidizing temperature is 300°C and thus requiring a larger impedance change, the flat-ended segment tends to be pushed out of the hole as the pointed segment is forced against it.

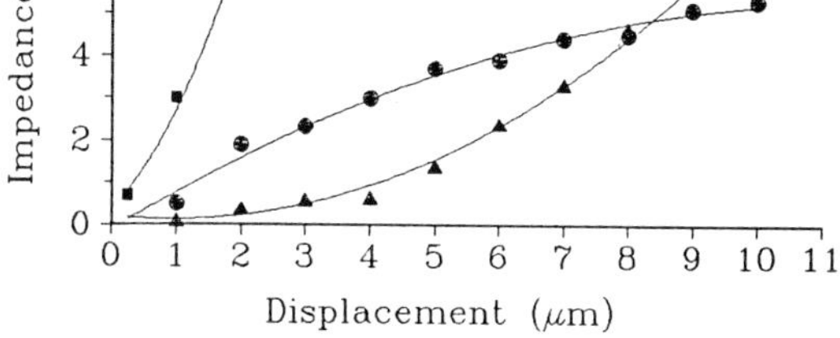

Fig. 1 - Impedance change *vs* displacement of one wire segment relative to the other, for various temperatures.

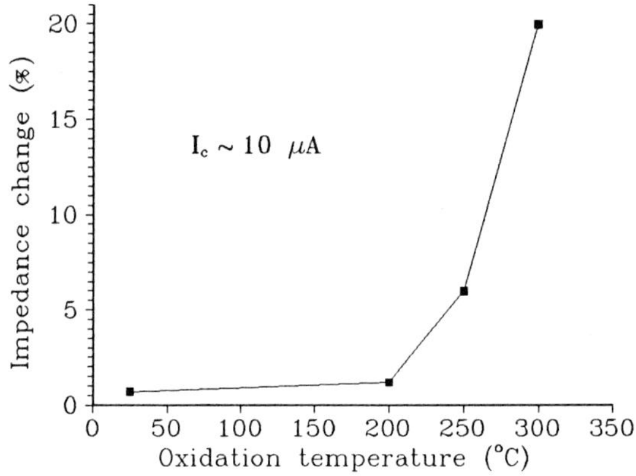

Fig. 2 - Impedance change *vs* oxidation temperature of the flat-ended wire.

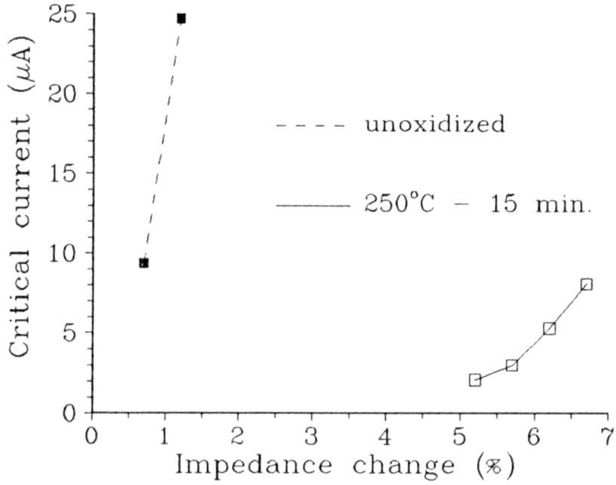

Fig. 3 - Critical current *vs* impedance change for oxidized and unoxidized wires.

Figure 3 shows critical current as a function of impedance change for oxidized and unoxidized wire segments. A half percent change for an unoxidized wire, roughly doubles the SQUID critical current, from $10\,\mu A$ to $25\,\mu A$. For an oxidized wire, as a lower initial setting is possible, the contact adjustment can be done more smoothly for the same half percent change, i.e., from $2.3\,\mu A$ to $3.2\,\mu A$, to $6.1\,\mu A$ and to $9.4\,\mu A$.

SQUID's made in this way have been stored in room-temperature air and have been recycled for more than 70 times during a one-year period, without a measurable change of performance. They have been kept in a helium atmosphere while recycling to avoid moisture condensation.

Using S.H.E. electronics [2], these SQUIDs have flat noise spectra, above $0.1\,Hz$, tipically of $5 \times 10^{-5}\Phi_0/\sqrt{Hz}$. The SQUID inductance was about $4 \times 10^{-10}\,H$. For a coupling coefficient of $25\,nH$ between the SQUID and the

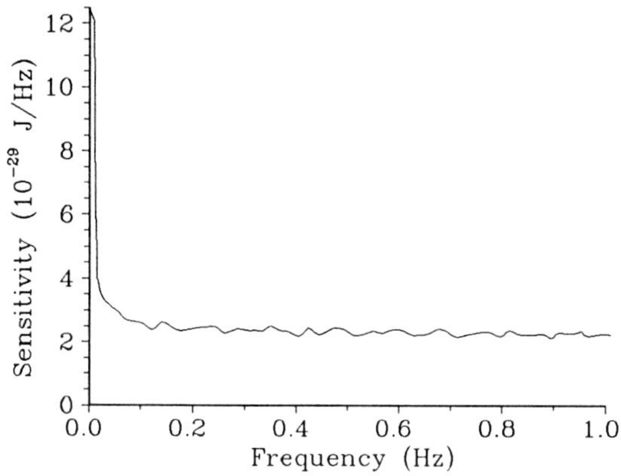

Fig. 4 - Typical noise spectra of the wire-junction SQUID's.

secondary coil (100 turns of 80 μm niobium wire wound on a 2.2 mm plexiglass support) a sensitivity of $2.5 \times 10^{-29}\, J/Hz$ referred to the secondary coil is obtained, as shown in Fig. 4.

4. Acknowledgements

This work was partially supported by FAPERJ and CNPq. We thank Fundação de Tecnologia Industrial, Lorena, SP, Brasil, which kindly provided the niobium rods used in the experiments.

References

[1] J.E. Zimmerman, P. Thiene and J.T. Harding, *J. Appl. Phys.* **41,** 1572 (1970).

[2] S.H.E. Inc., now Biomagnetic Technologies Inc., San Diego, CA 92121, U.S.A.

Self-Consistent Theory of the Voltage-Current Characteristic and Intrinsic Noise of the Hysteretic RF-SQUID

Ya.S. Greenberg

Novosibirsk Electrotechnical Institute, 20 K. Marx Ave.,
Novosibirsk, Russia

Abstract. We suggest a selfconsistent theory of the rf-SQUID voltage-current characteristic (VCC) fine structure with account of the intrinsic noise of the interferometer which allows to investigate analytically the VCC plateau structure with the aid of relatively simple equations for slow variables. The outline of the noiseless version of the theory has been published in [1].

As well known the tank circuit voltage V_t of the rf-SQUID depends both on bias current amplitude I_b and on external magnetic flux Φ_x. At the working regime the voltage-current curve (VCC) of the rf-SQUID, i.e., the dependence $V_t(I_b)$ at Φ_x = const, is a gently sloping line (plateau).

The plateau slope dV_t/dI_b is a very important parameter since it is directly connected through the current amplitude fluctuations, thermal noise of tank circuit and current noise of preamplifier to the device sensitivity [2]. Experiments have shown [3, 4, 5,] that the structure of the plateau is rather complicated: the slope is not constant but depends both on I_b and on Φ_x.

The exact dynamics equations which govern the interaction of the Josephson interferometer with the tank circuit are [6]:

$$\ddot{\nu} + (1/Q)\dot{\nu} + \lambda\nu = k^2\lambda\dot{\varphi} + \epsilon(t)\,, \tag{1}$$

$$q\dot{\varphi} + l\sin\varphi = \varphi_x - \varphi - \dot{v} + \epsilon(t)\,. \tag{2}$$

The dimensionless variables are: tank circuit voltage ν, in units of $\Phi_0/2\pi MC_t\omega_t$, where M = mutual inductance between the tank circuit inductance L_t and the interferometer loop inductance L, C_t = tank circuit capacitance, ω_t = its resonance frequency, the internal φ and external φ_x magnetic flux are in units of $\Phi_0/2\pi$. The excitation amplitude $\epsilon(t) = \epsilon\cos\omega t$, where $\epsilon = 2\pi MI_b/\Phi_0$, $l > 1$, $k^2 = M^2/L_tL$, $\lambda^{-1} = 1 - k^2$, $q = \omega L/R$ (R = ohmic resistance of weak link), $Q = Q_t(1-k^2)$ (Q_t = tank circuit quality factor). The dots in (1), (2) are the derivatives relative to the dimensionless time $\tau = \omega_t t$.

The main difficulty in solving (1), (2) is connected with the correct choice of $\varphi(t)$-solution of the nonlinear equation (2). It is known that at the plateau $\varphi(t)$ is a sequence of short pulses which describe the quick ($\tau \simeq 1/q$) transitions between neighbouring stable flux states of the interferometer. Computer simulations of equations (1) and (2) show that the behaviour of the internal flux could be described by two parameters: the pulse period T_0 and the pulse width T_1. T_0 depends

Springer Proceedings in Physics, Vol. 64 **Superconducting Devices and Their Applications**
Editors: H. Koch, H. Lübbig

on excitation amplitude I_b and decreases when I_b increases, reaching at the end of the plateau the value $T_o \approx 2\pi/\omega$. The quantity T_1 is approximately a constant along the plateau. Hence we take $\varphi(t)$ in the form of a sequence of square pulses:

$$\varphi(t) = \alpha\varphi_e(t) + 2\pi(1 - \alpha)\, s(t), \qquad (\varphi_x < \pi) \tag{3}$$

where

$$s(t) = \sum_{n=-\infty}^{n=+\infty} [\theta(t - nT_o + \eta_n^0) - \theta(t - nT_o - T_1 + \eta_n^1)] \tag{4}$$

and $\alpha = \arccos(-1/l)/\varphi_c$, $\theta(x)$: a step function, η_n^0, η_n^1 : stochastic time variables.

As seen from (4) the flux transitions occur at the moments nT_o and $nT_o + T_1$, i.e., every time the external flux $\varphi_e(t) = \varphi_x - \dot{v} + \varepsilon$ reaches one of its critical values φ_c and $2\pi - \varphi_c$, where $\varphi_c = (l^2-1)^{1/2} + \arccos(-1/l)$.

The quantities η_n^i $(i = 0,1)$ are uncorrelated for different i or n, since the transitions between adjacent flux states occur independently of each other. So the probability density of the random process (4) could be written as follows:

$$\Psi(\{\eta^0\}, \{\eta^1\}) = \prod_{n=-\infty}^{n=+\infty} \psi_o(\eta_n^0)\, \psi_1(\eta_n^1) \,, \tag{5}$$

where ψ_o and ψ_1 are normalized density functions of the time distribution of the moments of the flux transitions.

The functions $\psi_o(\tau)$, $\psi_1(\tau)$ are directly connected to the transition probability $p(t)$ between flux states:

$$\psi_o(\tau) = c_o p_\uparrow(\tau) \exp\{-\int_0^\tau p_\uparrow(\tau')\, d\tau'\} \,, \tag{6}$$

$$\psi_1(\tau) = c_1 p_\downarrow(\tau) \exp\{-\int_0^\tau p_\downarrow(\tau')\, d\tau'\} \,, \tag{7}$$

where $p_\uparrow(\tau)$ = transition probability from flux state $\varphi = 0$ to flux state $\varphi = 2\pi$ which occurs near point $\varphi - \varphi_c$, $p_\downarrow(\tau)$ = transition probability from flux state $\varphi = 2\pi$ to flux state $\varphi = 0$ which occurs near the point $\varphi \simeq 2\pi - \varphi_c$; c_o, c_1 = normalizing factors.

As shown in (7):

$$p_{\uparrow;\downarrow}(\tau) = (\omega/2\pi q)\, (\Delta\varphi_{\uparrow;\downarrow})^{1/2} \exp\{-G\, (\Delta\varphi_{\uparrow;\downarrow})^{3/2}\} \,, \tag{8}$$

where $G = \dfrac{4\sqrt{2}}{3\, l(l^2-1)^{1/4}} \dfrac{\epsilon_c}{k_B T}$, $\Delta\varphi_\uparrow = \varphi_c - \varphi_e(\tau)$, $\Delta\varphi_\downarrow = 2\pi - \varphi_c - \varphi_e(\tau)$.

Here $\varphi_e(\tau)$ is taken near φ_c for transition $\uparrow$ and near $2\pi - \varphi_c$ for transition $\downarrow$. The expressions for $\varphi_e(t)$ near the transition points are:

$$\Delta\varphi_\uparrow = \tau^2/2\ (\varphi_c - \varphi_x); \quad \Delta\varphi_\downarrow = 2\sqrt{(\varphi_c - \pi)(\pi - \varphi_x)}\, \tau - (2\pi - \varphi_c - \varphi_x)\, \tau^2/2.$$

The unknown parameters of the pulse sequence T_1 and T_0 have to be found in a self-consistent way.

With the aid of (4) we solved equation (1) by the harmonic balance technique that proved to be successful in analyzing rf-SQUID dynamics (6). We obtained for a VCC with $\alpha(\epsilon,\varphi_x)$ and $v(t) = \alpha(\epsilon)\cos(\omega t + \Theta)$ two coupled equations:

$$\alpha^2 = F^2 + 2\bar{b}F\Delta + \Delta^2(\bar{b}^2 + \bar{c}^2)\ , \tag{9}$$

$$\Delta^2(\bar{b}^2 + \bar{c}^2) + \Delta(2/Q)F(\bar{c} - \chi Q\bar{b}) + \epsilon_1^2 - \epsilon^2 = 0\ , \tag{10}$$

where $\bar{c} = \gamma_\uparrow - \gamma_\downarrow(1-c) - \delta_\downarrow b$; $\bar{b} = \delta_\uparrow + \gamma_\downarrow b - \delta_\downarrow(1-c)$;

$$\begin{Bmatrix}\gamma_\uparrow\\ \gamma_\downarrow\end{Bmatrix} = \int_0^\infty \begin{Bmatrix}\psi_0(\tau)\\ \psi_1(\tau)\end{Bmatrix}\cos(\tau)d\tau\ ; \qquad \begin{Bmatrix}\delta_\uparrow\\ \delta_\downarrow\end{Bmatrix} = \int_0^\infty \begin{Bmatrix}\psi_0(\tau)\\ \psi_1(\tau)\end{Bmatrix}\sin(\tau)d\tau\ ;$$

ϵ_1 is the high frequency amplitude at the beginning of the plateau,
$\epsilon_1 = (1 + \chi^2Q^2)^{1/2}F/Q$, $\chi = (\omega^2 - \omega_t^2)/\omega^2$, $F = \varphi_c - \varphi_x$, $\Delta = 2\Lambda/\tau_0$,
$c = 2(\varphi_c - \pi)/F$, $b = 2[(\varphi_c - \pi)(\pi - \varphi_x)]^{1/2}/F$, $\Lambda = 2\pi k^2(1-\alpha)/[1 - k^2(1-\alpha)]$.

In (10) we must choose the solution which vanishes at the beginning of the plateau ($\Delta(\epsilon_1)=0$). This means physically the absence of flux transitions at the beginning of the plateau ($2\pi/T_0 = 0$). At the end of the plateau ($\epsilon = \epsilon_2$) the pulse sequence frequency is comparable with the bias frequency ($\tau_0 = 2\pi$) so at the end of the plateau ϵ_2 can be found from (10) at $\Delta = \Lambda/\pi$.

We have shown in (1) that for large l>>1 and in the case of no noise equation (10) has an exact solution

$$\Delta = 1/2\,(\epsilon - \epsilon_1) \tag{11}$$

that gives a noiseless plateau slope:

$$\alpha^2 = F^2 + 2\sqrt{(\varphi_c - \pi)(\pi - \varphi_x)}\,(\epsilon - \epsilon_1) + (\epsilon - \epsilon_1)^2\ . \tag{12}$$

If $\varphi_x > \pi$ then simple arguments show that in above expressions one should make a change $\varphi_c \to 2\pi - \varphi_c$.

Our theory allows to find the spectral density of the intrinsic noise in the band width of the tank circuit which is defined by a stationary part of the stochastic process (4). In flux quantum units:

$$S_{\Phi_0}(\omega) = \pi\Delta\Lambda Q^2\, 1/\omega\, (\Phi_0/2\pi)^2 f(T,\varphi_x)\ , \tag{13}$$

where

$$f(T,\varphi_x) = [2 - \gamma_\uparrow^2 - \gamma_\downarrow^2 - \delta_\downarrow^2 - \delta_\uparrow^2]\ , \tag{14}$$

From (13) we see that the resulting flux noise is proportional to the frequency of the pulse sequence $\Delta \simeq 2\pi/\tau_0$ as it should be from the physical point of view. As we have seen before at the beginning of the plateau $\Delta = 0$ and in this point there is no flux noise due to intrinsic fluctuations since there are no transitions between flux states. Along the plateu Δ increases with the rf-amplitude ϵ and the flux noise increases as well, reaching its maximum at the end of the plateau where the frequency of the flux transitions is maximum.

Our result (13), (14) differs from the known formulae [2], [8] that do not account for the plateau structure, since our expressions for the flux noise depend both on φ_x and ϵ.

As an example, we can estimate from expression (13) the typical value of flux noise of the rf-SQUID due to the intrinsic fluctuations of the interferometer. Let the working point be at the middle of the plateau, i.e., $2\pi/\tau_0 \simeq 0.5$ and take the following typical values: $k^2Q \simeq 3$, $\omega/2\pi \simeq 20$ MHz, $l \simeq 3$, $\epsilon_c/k_BT \simeq 100$. With these values the quantity $f(T,\varphi_x) \simeq 0.05$ and has very weak dependence on φ_x. Substituting the above values in expression (13) we obtain for the flux noise the result $S_{\Phi_0}{}^{1/2} \simeq 2.5\times10^{-5}\ \Phi_0/\mathrm{Hz}^{1/2}$, a value that very well corresponds with the ultimate sensitivity of practical rf-SQUIDs.

References

[1] Ya. S. Greenberg, J. Techn. Phys. **57**, 2392 (1987) (in Russian).

[2] L. D. Jackel, R. A. Buhrmann, J. Low Temp. Phys. **19**, 201 (1975).

[3] I. M. Dmitrenko, V. A. Khlus, G. M. Tsoi, V. I. Snyrkov, IEEE Trans. Magn. **19**. 576 (1983)

[4] I. M. Dmitrenko, D. A. Konotop, G. M. Tsoi, V. I. Snyrkov, Proc. 10th Int. Conf. Helsinki, Finland, 31 July - 3 Aug., 746 (1984).

[5] N. V. Golyshev, S. V. Motorin, in: "Measuring and Computing Systems" Novosibirsk Electrotechnical Institute, 40 (1989) (in Russian).

[6] V. V. Danilov, K. K. Likharev, O. V. Snigirev, in: SQUID'80, Eds.: H.-D. Hahlbohm, H. Lübbig, Berlin, New York: W. de Gruyter, 473 (1980).

[7] J. Kurkijärvi, Phys. Rev. **6B**, 832 (1972).

[8] J. Kurkijärvi, J. Appl. Phys. **44**, 3729 (1973).

An Analysis of Non-Linear Behaviour in the Radio Frequency SQUID Magnetometer

J.F. Ralph, T.P. Spiller, T.D. Clark, R.J. Prance, H. Prance, A.J. Clippingdale, D.J. Rathbone, and M.E. Brooks

School of Mathematical and Physical Sciences, Physics Division, University of Sussex, Falmer, Brighton, Sussex, BN1 9QH, UK

Abstract. Using a model rf-SQUID magnetometer; taken to be a linear classical circuit coupled reactively to a macroscopic quantum circuit, in the adiabatic limit, we investigate the effect of quantum mechanical transitions on the non-linear behaviour found in such a system.

The system under study consists of a thick (compared to magnetic penetration depth) superconducting ring, containing a weak link constriction, which we couple reactively to a driven LC oscillator circuit via a mutual inductance, M.

The superconducting (SQUID) ring is considered to be a macroscopic quantum circuit, with geometric inductance Λ, weak link capacitance C_s and pair tunnelling matrix element $\frac{\hbar\nu}{2}$ for the weak link [1]. We assume that any fluctuations in applied flux occur at frequencies much lower than any energy level separation of the quantum circuit to ensure that it remains adiabatically in an energy eigenstate. This constraint is frequently satisfied in rf-SQUID experiments.

When we reactively couple such a SQUID ring to a classical linear circuit, via a mutual inductance, the flux from the circuit is coupled to the ring which induces a screening current. This screening current is then coupled back to the circuit, which introduces a non-linearity into the dynamical equation for the otherwise linear circuit. Because of the quantum mechanical nature of the SQUID ring, it is necessary to equate the expectation value of the screening current with an equivalent classical current, via the Correspondence principle.

In the limit, $\nu \ll \omega_0$, $\omega_0 = (\Lambda C_s)^{-\frac{1}{2}}$, it is found [2] that the expectation value of the screening current for a particular energy state, $|\kappa\rangle$, is given by the simple analytic expression,

$$\langle\kappa|I_s|\kappa\rangle = -q\nu \sin(\frac{2\pi\Phi_X}{\Phi_0})L_\kappa(2\pi^2\beta)e^{-\pi^2\beta} \,, \qquad (1)$$

where $L_\kappa(x)$ is the Laguerre polynomial of order κ, and $\beta = \frac{\hbar\omega_0\Lambda}{\Phi_0^2}$. Notice that each energy eigenstate has the same sinusoidal form for the screening current, differing only in amplitude.

Springer Proceedings in Physics, Vol. 64 **Superconducting Devices and Their Applications**
Editors: H. Koch, H. Lübbig

The model tank circuit has a resistive element R_t, an inductive element L_t and a capacitive element C_t, which are connected in parallel and driven by a single harmonic current source at the resonant frequency of the bare tank circuit, $\omega_R = (L_t C_t)^{-\frac{1}{2}}$.

Taking the limit where Eqn.1 is valid, the dynamical equation for the flux, Φ, threading the oscillator becomes,

$$C_t\ddot{\Phi} + \frac{\dot{\Phi}}{R_t} + \frac{\Phi}{L_t} = I_{IN}\cos(\omega_R t + \delta) - \mu I_\kappa \sin(\frac{2\pi(\Phi_x + \mu\Phi)}{\Phi_0}) \text{ , } \quad (2)$$

where $\mu = \frac{M}{L_t}$ and

$$I_\kappa = q\nu L_\kappa(2\pi^2\beta)e^{-\pi^2\beta} \text{ , } \quad (3)$$

where δ is an initial phase factor and Φ_x is an external D.C. flux.

Putting Eqn.2 into dimensionless form, with; $Q = \omega_R R_t C_t$ $i_{IN} = \frac{I_{IN}}{C_t\omega_R^2\Phi_0}$, $i_\kappa = \frac{I_\kappa}{C_t\omega_R^2\Phi_0}$, $\phi = \frac{\Phi}{\Phi_0}$, $\phi_x = \frac{\Phi_x}{\Phi_0}$, $\tau = \frac{\omega_R t}{2\pi}$, we obtain the equation,

$$\frac{d^2\phi}{d\tau^2} + \frac{2\pi}{Q}\frac{d\phi}{d\tau} + (2\pi)^2\phi = (2\pi)^2 i_{IN}\cos(2\pi\tau + \delta) - (2\pi)^2\mu i_\kappa \sin(2\pi\phi_x + 2\pi\mu\phi) \quad (4)$$

This equation may be solved for particular parameter values, and given initial conditions, using standard numerical techniques.

In experiments, measurements of the in- and out-of-phase components of the output voltage across the tank circuit are taken whilst the amplitude of the current source is varied quasi-statically. Here we are interested in the effect that the non-linear term has on these voltage characteristics. We therefore solve Eqn.4 for different values of the coupling parameter, μ, and vary the forcing amplitude, i_{IN}, quasi-statically. All other parameters are kept constant. The values $i_0 = 1$ and $Q = 50$ are used, which are consistent with real systems, and we set $\delta = 0$ and $\phi_x = 0$ for convenience.

In the uncoupled limit, $\mu = 0$, Eqn.4 becomes the usual linear damped oscilator equation and has an unique analytic solution. This leads to Ohmic response for the in-phase component of voltage whilst the out-of-phase component is zero (ignoring transience).

Once $\mu \neq 0$, however, Eqn.4 does not necessarily have a simple analytic solution, and any solution which is present is not necessarily unique. Of course, when μ is very small, we may attempt to expand perturbatively about the linear solution and hope this gives a good approximation to the behaviour of the non-linear system in the very weakly coupled region. Indeed, numerical calculations show that, upto $\mu \simeq 0.02$, first order

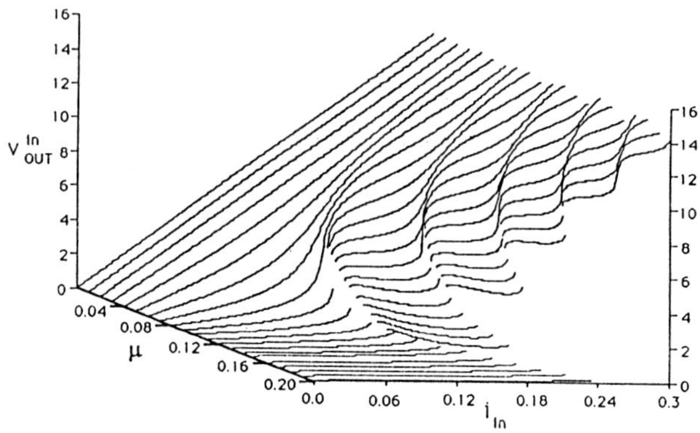

Figure 1: In-phase voltage surface for coupling $0.0 < \mu < 0.2$ and forcing current amplitude $0.0 < i_{IN} < 0.3$

perturbation theory does give an accurate indication of the behaviour of the system.

Past $\mu \simeq 0.02$ the differences between the perturbed solution and the numerical results become significant [3]. Fig.1 shows the in-phase voltage surface for $0.0 < \mu < 0.2$ and $0.0 < i_{IN} < 0.3$. To first order, the in-phase voltage should remain linear, whilst the out-of-phase component becomes non-zero. Fig.1 shows that as the coupling is increased the in-phase voltage is seen to be suppressed for low values of i_{IN}, and that as this suppression becomes more marked the surface 'folds' over itself. This fold bifurcation occurs at $\mu = 0.086$, $i_{IN} = 0.124$ and leads to two coexisting stable solutions. Since we vary the forcing amplitude quasi-statically, these coexisting solutions lead to hysteretic behaviour as the system moves from a region with two stable solutions to one of only one stable solution.

One of these stable solutions is very much smaller (typically by one or more orders of magnitude) than the other and gives rise to the 'dead' region seen in Fig.1, where a quasi-static increase in i_{IN} gives no observable increase in voltage (either in- or out-of-phase).

To take account of the quantum mechanical nature of the SQUID ring we must also include the possiblity of transitions between the energy eigenstates. It was noted above that, in the limit chosen here ($\nu \ll \omega_0$), the expectation value of the screening current in each energy eigenstate has the same sinusoidal dependence on external flux, and varies only in amplitude. This leads, in the perturbative regime, to the same Ohmic (in-phase) voltage response for each energy eigenstate. Beyond the perturbative regime, however, variations in the ampiltude of the non-linear

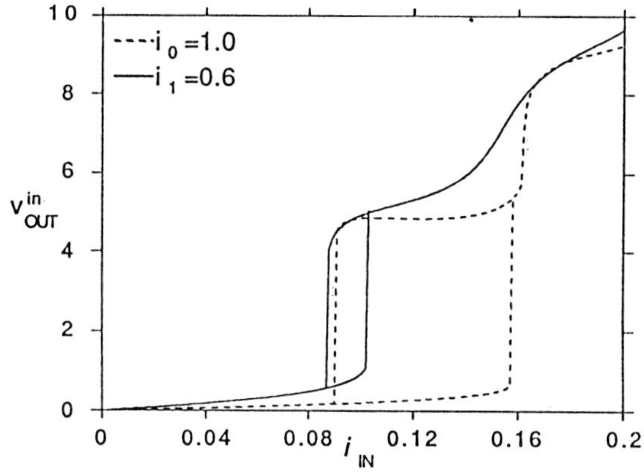

Figure 2: In-phase voltage versus forcing current amplitude for the ground state ($i_0 = 1.0$) and an excited state ($i_1 = 0.6$)

term can have a marked effect on the dynamics, both in- and out-of-phase. Fig.2 shows the in-phase voltage charcteristics for two values of the screening current amplitude. For $i_1 = 0.6$ we can see that the hysteretic region is far smaller and the suppression far more evident than that found for $i_0 = 1.0$. The two voltage characteristics are easily distinguished given the behavior for a sufficiently large range of i_{IN}.

REFERENCES

1. B.D.Josephson, *Phys. Lett.* **1** (1962) 251.
2. T.P.Spiller, D.A.Poulton, T.D.Clark, R.J.Prance, H.Prance 'The dynamics of a quantum circuit connected to a classical lossy oscillator' *to be published in Int.Jour.Mod.Phys.B*
3. J.F.Ralph, T.P.Spiller, T.D.Clark, H.Prance, R.J.Prance, A.J. Clippingdale 'Non-linear behaviour in the rf-SQUID magnetometer' *submitted for publication*

Low Noise UHF Thin Film SQUID with Cryogenic HEMT Preamplifier

A. Cavalleri[2], *M. Cerdonio*[1,2], *G. Fontana*[1], *G. Jung*[1,3], *R. Macchietto*[1], *R. Mezzena*[1], *S. Vitale*[1], *and J.P. Zendri*[1]

[1]Dipartimento di Fisica Università degli Studi di Trento, I-38050 Povo Trento, Italy
[2]Centro di Fisica degli Stati Aggregati e Impianto Ionico del CNR, I-38050 Povo, Trento, Italy
[3]Instytut Fizyki PAN, PL-02-668 Warszawa, Poland

Abstract. The design and performance of UHF thin film RF SQUID system employing a cryogenic RF preamplifier is described. The flux noise of the magnetometer is $7 \times 10^{-6}\Phi_0/\sqrt{Hz}$ at 4.2 K and $4.8 \times 10^{-6}\Phi_0/\sqrt{Hz}$ at 1.4 K. The temperature dependance of the fractional step rise α was measured showing a good agreement with the Kurkijärvi-Webb theory.

1. Introduction

The noise performance of RF SQUIDs, associated with intrinsic thermodynamic fluctuations, improves with increasing pump frequency [1]. Further advantages are obtained reducing the noise due to the amplifier chain, the tank circuit and flux trapping effects, which usually prevails the intrinsic noise and determines therefore the ultimate sensitivity.
In this paper we describe an improved version of our first UHF SQUID system [2] that now employs a thin film sensor and a revised mechanical and electrical layout. The system noise is limited by the intrinsic noise of the SQUID. Only at temperatures as low as 1.5 K the preamplifier noise becomes significant.

2. Experimental equipment: mechanical and electronic set-up

Our system consists of a probe at whose extremities are located respectively the input/output electrical connections and a double room container immersed in the liquid helium reservoir. At the top is also mounted a three way vacuum valve which allows the probe volume evacuation and its refilling up to a controlled pressure. The lead-coated double room container is the housing for the cryogenic preamplifier and the SQUID that is surrounded by an additional niobium screen. The sensor is a commercially available RF thin film SQUID [3] designed to operate between 20 and 200 MHz. We are able to increase the pump frequency up to 316

Springer Proceedings in Physics, Vol. 64 Superconducting Devices and Their Applications
Editors: H. Koch, H. Lübbig

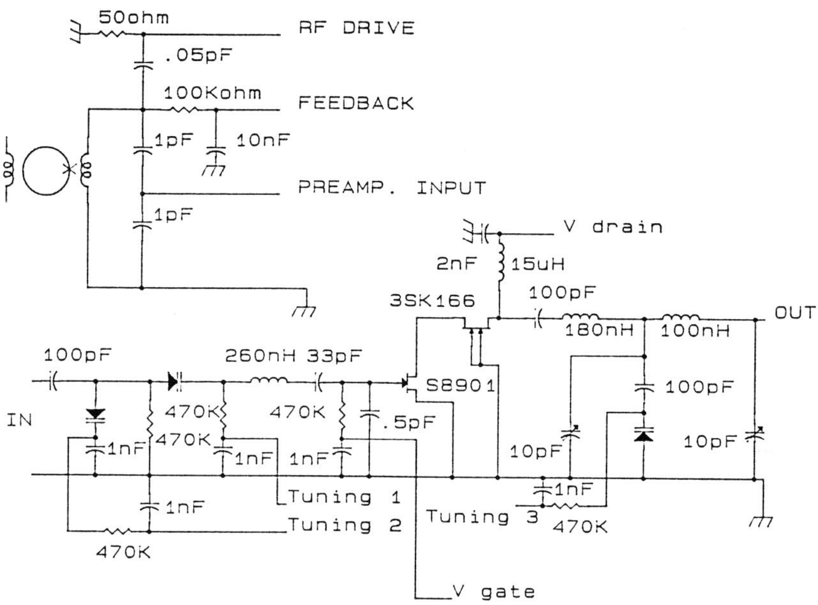

Figure 1: Circuit diagram of the cryogenic amplifier and the tank circuit.

MHz by minimizing the stray capacitance and the length between the tank circuit and the preamplifier. The active elements of the preamplifier are arranged in a cascode mode composed of a microwave HEMT and a dual-gate MESFET [2]. This configuration offers the possibility to obtain the best noise matching of the preamplifier to the SQUID tank circuit and eliminates severe stability problems that plague cryogenic UHF amplifiers. Another advantage of this configuration is its higher gain obtained at lower dc power dissipation in the input device (lower drain source voltage with respect to a single FET amplifier). The circuit diagram of our amplifier and tank circuit is shown in fig. 1. The input and output impedance matching of the amplifier to the tank circuit and transmission lines is obtained via standard matching networks [4] using superconducting inductances and GaAs varactor diodes. The output signal of the preamplifier is delivered to a two stage room temperature amplifier (HP 8447F). Between these two stages there is a resonant 316 MHz band-pass filter. The amplified SQUID signal is finally demodulated by a diode. As the tuning of the amplifier should be adjusted to keep the optimum performances over periodes of weeks for uninterrupted operation, we have developed a computerized optimization technique to continuosly control the bias of the varactor diodes. The SQUID is operated in flux-locked mode with one ϕ_0 flux modulation in order to estimate the p-p amplitude

V_p of the triangular pattern. It is easy to show that the feedback loop can be closed with a lock-in amplifier in the 2f mode. The noise level, estimated by Fourier analysis of the diode demodulated signal, is divided by V_p^2 and minimized.

3. Experimental results and discussion

The preamplifier has been tested separately at 4.2 K. The measurements have confirmed the tuning abilities of the GaAs varactor diodes and the stability of the chosen configuration. The preamplifier noise temperature has been measured with the hot-cold method employing a resistor with adjustable temperature as noise source. The measurement of the noise temperature gives $6 \pm .5$ K as upper limit at 300 MHz. The maximum stable voltage gain as high as 140 is also due to an accurate amplifier layout. The total equivalent flux noise in the white regime is $7 \times 10^{-6}\Phi_0/\sqrt{Hz}$ at 4.2 K in the open loop configuration. The reduction of the intrinsic noise of the SQUID as a function of the temperature lowers the total noise to $4.8 \times 10^{-6}\Phi_0/\sqrt{Hz}$ at 1.5 K corresponding to an energy resolution of 1×10^{-31}J/Hz. Since theoretical previsions show that the intrinsic noise is proportional to the fractional step rise α [5] this latter has been measured over the $4.3 \div 1.5$ K temperature range. In fig. 2 are shown the experimental results and the one parameter fit to the data of the Kurkijärvi-Webb prediction. The comparison of the intrinsic

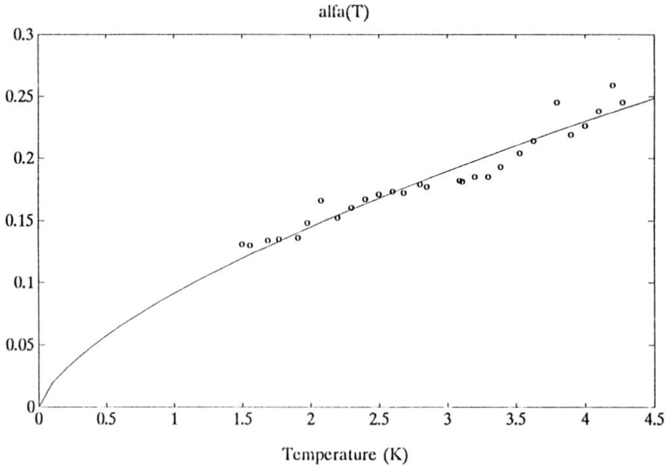

Figure 2: Fractional step rise α as a function of the temperature; the solid line is a one-parameter fit to the data of the Kurkijärvi-Webb prevision.

noise values obtained from α and the total noise values shows that the system sensitivity is dominated by the intrinsic noise of the SQUID. The flux noise measurement at 4.2K in the closed loop configuration gives the value $1 \times 10^{-5}\Phi_0/\sqrt{Hz}$ both in standard and in 2f lock-in mode in the band $1 \div 200$ Hz.
The RF SQUID offers interesting performance both in flux noise and slew rate if operated in combination with advanced electronic and computer control systems. As our system is very near the intrinsic noise of the SQUID, the energy resolution can be enhanced by raising the pump frequency towards a right compromise among the parameters affecting the SQUID response and the HEMT behaviours [6].

Acknowlegment

This work has been supported by Progetto Finalizzato CNR "Tecnologie Superconduttive e Criogeniche". The authors wish to thank also Mr. Fabrizio Gottardi for his precious technical support and assistance.

References

[1] J.Kurkijarvi and W.Webb, in Proc.ASC.'72 ,IEEE New York 1973 p.581
[2] A.Cavalleri, M.Cerdonio, G.Fontana, G.Jung, R.Mezzena and S.Vitale, in Proc. of LTEC'90, Inst. of Cryogenics, Southampton 1990 p.09.3
[3] Cryogenic Cons. Lim., model SQS6
[4] A.P. Long, T.D. Clark and R.J. Prance,Rev. Scien. Instrum.,**51**,8(1980)
[5] L.D. Jackel and R.A. Buhrman, J. Low Temp. Phys.,**19**,201(1974)
[6] K.Duh M.W.Pospieszalski, W.F.Kopp, P.HO A.A. Jabra, P.C.Chao, P.M.Smith, L.F.Lester, J.M.Ballingal and S.Weinreb ,IEEE Trans.on El. Dev., **35**,249(1988)

Design and Fabrication Considerations for Extending Integrated DC-SQUIDs to the Deep Sub-Micron Regime

M.B. Ketchen

IBM Research Division, T.J. Watson Research Center,
P.O. Box 218, Yorktown Heights, NY 10598, USA

Abstract - Over the past decade many groups around the world have successfully produced planar low-Tc dc SQUIDs with integrated input coils. While a number of dc SQUIDs with sub-μm junctions have been demonstrated, practical integrated dc SQUIDs have generally been fabricated with minimum lithographic features of 2-5μm. In this paper some of the important design and fabrication issues associated with scaling practical integrated dc SQUIDs to deep sub-μm dimensions are discussed.

1. Introduction

The first dc SQUID demonstrated in 1964 was a planar thin-film device [1]. Designs incorporating useful input coils soon appeared in the form of 3-dimensional structures such as cylinders and toroids [2,3]. Planar designs with integrated input coils were introduced in the early 1980s [4,5] and are now fabricated by many groups. The most prevalent design [4] consists of a superconducting washer with a slit from its central hole to its outer edge where two Josephson junctions are incorporated. The washer is tightly coupled to a spiral input coil fabricated in close proximity. Although dc SQUIDs with sub-μm junctions have been built, the minimum lithographic feature size in practical integrated dc SQUIDs has typically been 2-5μm. At IBM and within the Consortium for Superconducting Electronics (CSE) we have been addressing the issues of extending the design and fabrication of such dc SQUIDs to deep sub-μm dimensions. Recent calculations indicate that the basic coupling scheme continues to work well for spiral linewidths down to 0.25μm and below [6]. Combining this with 0.1μm junctions could lead to a practical dc SQUID less than 0.05mm^2 in area with a tightly coupled 2μH input coil, minimal flux creep related hysteresis, and coupled energy sensitivity approaching the quantum limit at 4.2K. The technology needed to fabricate such a dc SQUID is indeed very complex. To meet this challenge, we have developed a planarized all-refractory Nb-AlOx-Nb trilayer-based process for 125mm wafers [7] that, with the exception of the preparation of the CSE trilayers [8], is practiced in an advanced Si Technology fabrication line. This approach has allowed us to leverage highly off of existing state-of-the-art lithography, metal reactive ion etching, oxide deposition, and planarization capabilities. We have successfully processed wafers using 1X lithography (at the 2μm level),

Springer Proceedings in Physics, Vol. 64 Superconducting Devices and Their Applications
Editors: H. Koch, H. Lübbig

where the entire wafer is personalized and 5X lithography (with 0.7μm minimum junction size), where a 1cm^2 area chip is stepped across the wafer. This planarized technology has yielded fully functional dc SQUIDs, gradiometers, soliton oscillators, and other devices. Junction quality is excellent with the figure of merit V_m typically 60 - 90mV for a Josephson critical current density j_c of 10^3A/cm^2. The vertical dimensions in our existing process are consistent with lateral dimensions of 0.25μm or less, and we are actively working to scale our technology to the deep sub-μm regime. In the following paper our approach to deep sub-μm SQUID technology is summarized.

2. Design

The theory for white noise in tunnel junction dc SQUIDs was developed in the 1970s[9]. Noise scaling predictions based on this theory have been subsequently verified down to the vicinity of the quantum limit [10,11]. Throughout the 1980s Nb-AlOx-Nb trilayer technology assumed a dominant role in the low-Tc fabrication area. Nevertheless in late 1989, at the time of the inception of the CSE, there was no convincing demonstration of a Nb-AlOx-Nb SQUID with white and 1/f noise on a par with what could be obtained with Nb-NbOx-Pb/alloy edge junction technology [12,13]. As a joint IBM-MIT Lincoln Laboratory project within the CSE we designed and fabricated a matrix experiment of dc SQUIDs at 2μm minimum feature size [14] using 50mm Nb-AlOx-Nb trilayers [8] and a Selective Nb Anodization Process (SNAP) [15]. Figure 1 shows a section of one of our 6mm chips. The SQUIDs were all of the basic square washer design [4] with no input coil and a fractional turn modulation coil that did not overlay the washer. We analyzed in considerable detail a nearly optimized SQUID having a critical current per junction I_o of 16μA, a resistance per shunt R of 14Ω, and a SQUID inductance L of 50pH. Our noise measurements of this SQUID along with

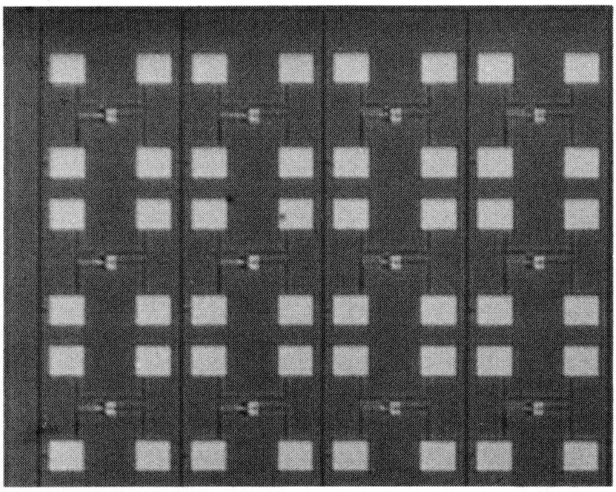

Fig. 1: Portion of a 6mm SQUID chip fabricated with SNAP.

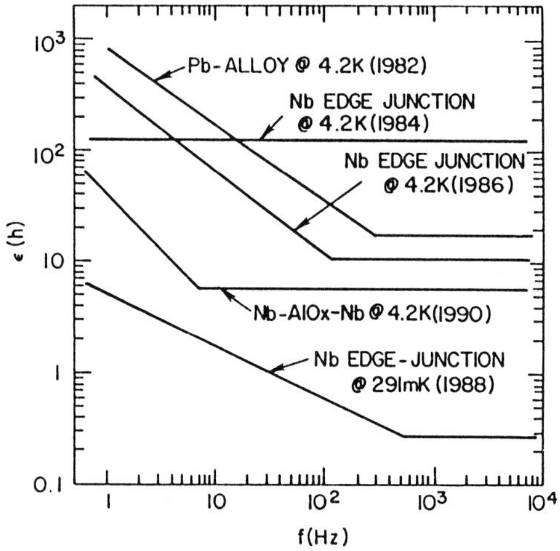

Fig. 2: Energy sensitivity ε as a function of frequency f for the Nb-AlOx-Nb trilayer SQUID and several other SQUIDs fabricated in different technologies.

results from a number of other SQUIDs previously demonstrated at IBM [4,11,12,13] are shown in Fig. 2. The 1/f noise is very low and the white noise is in good agreement with theory which predicts an intrinsic energy sensitivity in the thermal noise limit of

$$\varepsilon = \gamma h\left(\frac{k_B T}{e I_0 R}\right), \tag{1}$$

with γ of order 3-5. Results from this SQUID can be used as a basis for predicting Nb-AlOx-Nb trilayer SQUID performance as we scale to smaller dimensions. For SQUIDs with integrated input coils 100pH turns out to be an appropriate value for L. With an optimized design this leads to $I_o = 10\mu A$, corresponding to a critical current density j_c of $10^3 A/cm^2$ for a junction of area $\mathscr{A} = 1\mu m^2$. As the junction size is reduced j_c must increase as

$$j_c = 10^3 A/cm^2\left(\frac{1\mu m^2}{\mathscr{A}}\right), \tag{2}$$

and the energy sensitivity should improve as

$$\varepsilon(4.2K) = 5h\left(\frac{\mathscr{A}}{1\mu m^2}\right)^{1/2} \tag{3}$$

For $\mathscr{A} = 0.01\ \mu m^2$ we will require $j_c = 10^5 A/cm^2$ and ε should be well below h. A high sheet rho (10's of ohms per □) resistor technology must be developed in parallel to provide low parasitic shunt resistors and any damping resistors required to damp resonances.

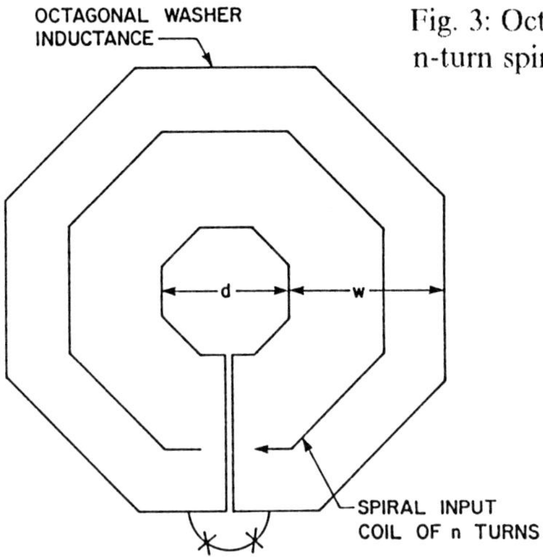

Fig. 3: Octagonal washer SQUID design with n-turn spiral input coil.

The introduction of thin film coupling schemes has enabled a new generation of planar SQUIDs with integrated input coils. Figure 3 shows a prototypical design consisting of an octagonal washer with an n-turn spiral input coil [4,6]. The washer is slit, and the electrical path around the hole is completed through two tunnel junctions remotely located at the outside edge. This design is not unique, but its principals of operation are fairly general in nature and relevant to a broad class of SQUID layouts.

If the washer width w is greater than the hole size d, the inductance of the hole L_h is about $1.05\mu_o$d essentially independent of w. For an open (ungroundplaned) slit the slit inductance L_t is approximately w x 0.3pH/μm. The parasitic inductance L_j associated with getting in and out of the junctions is a few pH. A set of equations that describes the inductive properties of this SQUID is [16]

$$L = L_h + L_t + L_j\,, \tag{4}$$

$$M_i = n(L_h + L_t/2)\,, \tag{5}$$

$$L_i = n^2(L_h + L_t/3) + L_s\,, \tag{6}$$

$$k^2 \approx 1 - \frac{L_t}{3L_h} - \frac{L_j}{L_h} - \frac{L_s}{n^2 L_h}\,, \tag{7}$$

where L_i is the inductance of the n-turn spiral input coil, M_i is the mutual inductance between L_i and L, L_s is the stripline inductance of the spiral with respect to the washer, and k^2 is the coupling constant. For a given washer size it is a reasonable approximation to treat L_s/n^2 as a constant down to dimensions where the cross-section of the spiral line is on the order of the square of the London penetration depth [6]. As a specific example consider an octagonal SQUID as shown in Fig. 3 with d = 60μm and w = 80μm. For this SQUID $L_h = 80pH$, $L_t = 24pH$, $L_j = 3pH$, and $k^2 = 0.85$. M_i,L_i, and L as

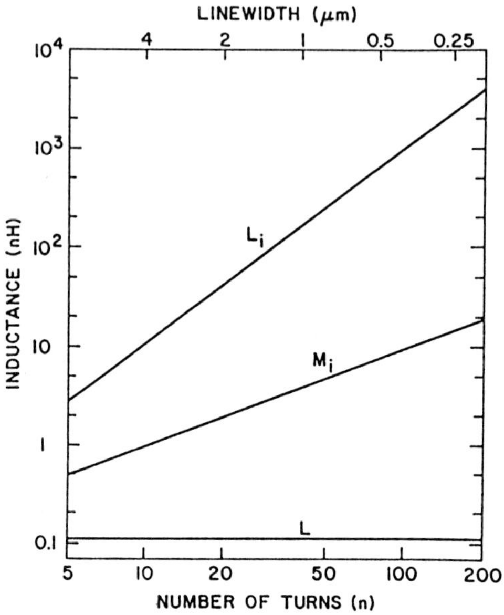

Fig. 4: M_i, L_i, and L as functions of the number of spiral input coil turns n and spiral linewidth for a washer with d = 60μm and w = 80μm.

functions of n and spiral linewidth (equal lines and spaces) are shown in Fig. 4. With a 160-turn, 0.25μm linewidth input coil we have L = 107pH, $M_i = 14.7nH$, and $L_i = 2.4\mu H$.M_i and L_i are not far from the industry standard values for commercial SQUIDs, but the size of this SQUID is less than 0.05 mm^2 and its energy sensitivity for 0.25μm junctions should be of order 2h at 4.2K. The small washer size helps to provide a favorable aspect ratio for operation in the earth's field with minimal hysteresis.

3. Fabrication

The fabrication of Josephson devices at sub-μm dimensions presents some unique challenges. It is not by accident that most integrated Josephson process work to date has been at minimum feature size above 1μm. The equipment, facilities, and other resources required to practice sub-μm technology are complex, expensive, and, in the absence of any return revenue, difficult to justify. As an alternative approach IBM and the CSE have undertaken a Josephson process development program in collaboration with Si technology. High quality 125mm Nb-AlOx-Nb trilayers are prepared at MIT Lincoln Laboratory [8] and the rest of the fabrication is carried out in an advanced Si technology fabrication facility at IBM Research. Using this approach we have demonstrated a novel planarized process that is scalable to deep sub-μm dimensions [7].

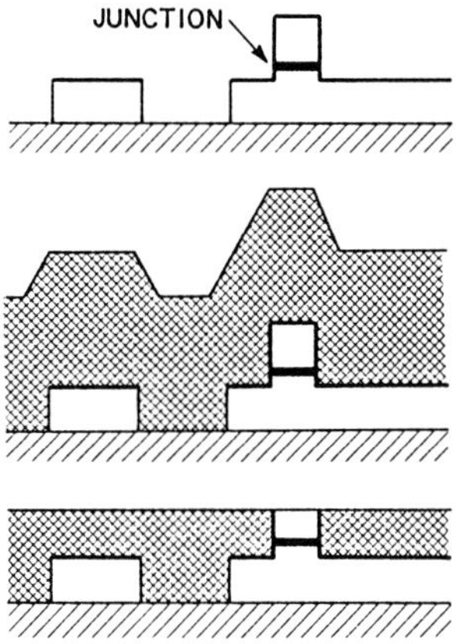

Fig. 5: Cross-sections in the vicinity of a junction at key points during the planar process: (a) after two-step RIE, (b) after sputtered quartz deposition, and (c) after planarization.

At the heart of our process is the ability to define junction areas with sub-μm lithography and Reactive Ion Etching (RIE) and later to provide a planar low-dielectric constant insulator self aligned to these areas. Three key points in our process flow are depicted in Figs. 5. Figure 5(a) shows a cross-section in the vicinity of a junction following a two-step RIE process that requires two separate resist stencils. In the first resist-RIE cycle we use a CI RIE process to etch vertically down to and through the trilayer tunnel barrier to form individual small junctions as well as larger junctions that will later serve as contacts [16]. In the second resist-RIE cycle we pattern the remaining base electrode that also serves as a wiring level. Figure 5(b) shows a cross section of the same area following a sputtered quartz deposition that completely buries the junction structures. Figure 5(c) shows this area after planarization (by chemical-mechanical polish) which results in the desired self-aligned low-dielectric constant insulator. Evaporated or sputtered resistors (patterned by liftoff) and the final sputtered Nb wiring level (patterned by liftoff or RIE) complete the process. A representative cross-section of a completed structure depicting junctions, resistors, and two levels of wiring (A and B) is shown in Fig. 6 [7].

With our precision CI RIE capability the size and control of lateral dimensions is largely a function of lithography. We have fabricated wafers using 1X contact lithography at 2μm minimum feature size and 5X i-line lithography at 0.7μm minimum feature size. We are presently extending the 5X work to 0.5μm and also pursuing e-beam lithography for features of 0.25μm and smaller. A photograph of a 125mm wafer processed with 5X lithography is shown in Fig. 7. This design consists of a 1 x 2 array of 7mm chips stepped across the wafer. Figure 8 shows the central area of an octagonal spiral coil patterned with i-line lithography following the completion of the two-step

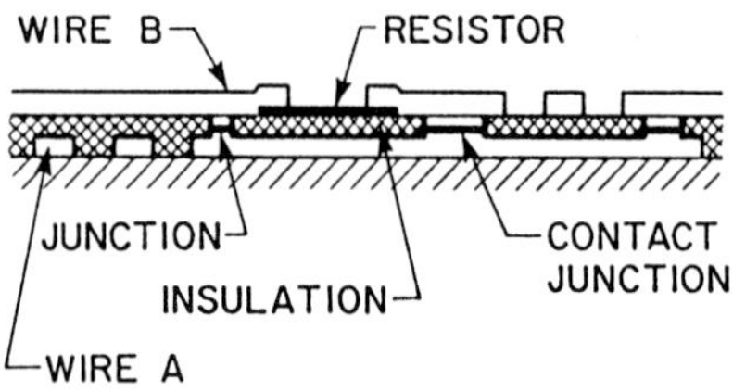

Fig. 6: Representative cross-section at the completion of the planar process.

Fig. 7: Photograph of a 125mm wafer processed using 5X lithography and the planar process.

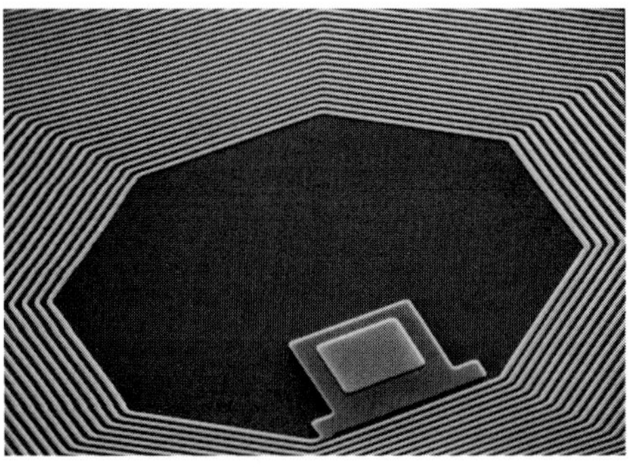

Fig. 8: Portion of a spiral input coil of 0.7μm linewidth.

etch. The lines and spaces in this spiral are 0.7μm wide. In the case of 1X lithography the entire wafer can be personalized providing a vast amount of real estate for trying out new designs. This technology has yielded fully functional SQUIDs, gradiometers, soliton oscillators and other devices. Junction characteristics are excellent with V_m typically in the range of 60 - 90mV [7]. Fig. 9 shows the IV characteristic of two series-connected 5μm junctions.

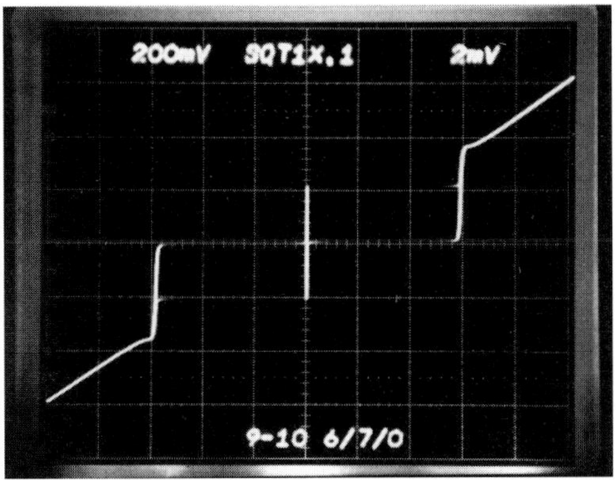

Fig. 9: Current-voltage characteristic of two series connected $5\mu m$ junctions. The vertical scale is 200 μA/division and the horizontal scale is 2mV/division.

4. Conclusions

The IBM/CSE approach to deep sub-μm SQUID technology has been discussed. There are clearly significant SQUID performance improvements inherent in scaling to very small dimensions. Effective coupling to state-of-the-art semiconductor fabrication technology is providing a fast and economical route to the realization of deep sub-μm SQUIDs.

References

[1.] R.C. Jaklevic, J. Lambe, A.H. Silver, and J.E. Mercerau, Phys. Rev. Lett. 12, 159 (1964).

[2] R.P. Giffard, R.A. Webb, and J.C. Wheatley, J. Low Temp. Phys. 6, 533 (1972).

[3] J. Clarke, W.J. Goubau, and M.B. Ketchen, J. Low Temp. Phys. 25, 99 (1976).

[4] J.M. Jaycox and M.B. Ketchen, IEEE Trans. Magn. 17, 400 (1981); M.B. Ketchen and J.M. Jaycox, Appl. Phys Lett. 40, 736 (1982).

[5] M.W. Cromar and P. Carelli, Appl. Phys. Lett. 38, 723 (1981); P. Carelli and V. Foglietti, J. Appl. Phys. 53, 7592 (1982).

[6] M.B. Ketchen, IEEE Trans. Magn. 27, 2916 (1991).

[7] M.B. Ketchen, D. Pearson, M. Bhushan, A. Kleinsasser, C.-K. Hu, M. Smyth, J. Logan, K. Stawiasz, E. Baran, M. Jaso, T. Ross, K. Petrillo, M. Manny, S. Basavaiah, S. Brodsky, S.B. Kaplan, and W.J. Gallagher, Appl. Phys. Lett., to be published (1991).

[8] M. Bhushan and E.M. Macedo, Appl. Phys Lett. 28, 1323 (1991).

[9] C.D. Tesche and J. Clarke, J. Low Temp. Phys. 29, 301 (1977).
[10] D.J. Van Harlingen, R.H. Koch, and J. Clarke, Appl. Phys. Lett. 41, 197 (1982).
[11] D.D. Awschalom, J.R. Rozen, M.B. Ketchen, W.J. Gallagher, A.W. Kleinsasser, R.L. Sandstrom, B. Bumble, Appl. Phys. Lett. 53, 2108 (1988).
[12] C.D. Tesche, K.H. Brown, A.C. Callegari, M.-M. Chen, J.H. Greiner, H.C. Jones, M.B. Ketchen, K.K.Kim, A.W. Kleinsasser, H.A. Notarys, G. Proto, R.H. Wang, and T. Yogi, IEEE Trans. Magn. 21, 1032 (1985).
[13] V. Foglietti, W.J. Gallagher, M.B. Ketchen, A.W. Kleinsasser, R.H. Koch, S.I. Raider and R.L. Sandstrom, Appl. Phys. Lett. 49, 1393, (1986).
[14] M.B. Ketchen, M. Bhushan, S.B. Kaplan, and W.J. Gallagher, IEEE Trans. Magn. 27, 3005 (1991).
[15] H. Kroger, L.N. Smith and D.W. Jillie, Appl. Phys. Lett. 39, 280 (1981).
[16] C.-K. Hu, N. Mazzeo, S.J. Wind, D.J.Pearson, and M.B. Ketchen to be published in the proceedings of the International Conference on Metallurgical Coatings and Thin Films, San Diego, CA (1991).

Nb-AlO$_x$-Nb SNAP Technology for 125 mm Wafers Developed in Partnership with Silicon Technology

M. Bhushan[1], *M.B. Ketchen*[2], *C.-K. Hu*[2], *K. Stawiasz*[2], *C. Cabral Jr.*[2], *M. Smyth*[2], *E. Baran*[2], *and D. Pearson*[2]

[1]Lincoln Laboratory, Massachusetts Institute of Technology, Lexington, MA 02173, USA
[2]IBM Research Division, T.J. Watson Research Center, P.O. Box 218, Yorktown Heights, NY 10598, USA

Abstract - We have demonstrated a 4-mask-level selective niobium anodization process (SNAP) technology for 125 mm wafers. With the exception of the initial Nb-AlO$_x$-Nb trilayer preparation, this process is carried out almost exclusively within a VLSI fabrication area and utilizes existing 1× lithography, Cl-based reactive ion etching, and materials deposition capability developed for Si technology. Our design consists of a personalized wafer layout with a number of 6 × 6 mm diagnostic chips and a wide variety of SQUIDs, magnetometers (as large as 12 × 12 mm), gradiometers (as large as 12 × 24 mm), soliton oscillators, and other Josephson devices at minimum design rules of 2 µm. We have used this process to produce high quality Josephson junctions with the figure of merit V_m > 60 mV for a Josephson critical current density J_c of 500-1000 A/cm^2. Resistors were fabricated with 35-nm-thick e-beam evaporated Pt having a sheet resistance of about 4 Ω/□. This approach to Josephson fabrication provides a low overhead, fast turn-around process particularly well suited for trilayer development and for production of many physically large, application specific integrated Josephson devices.

1. Introduction

The overhead associated with fabrication of thin film superconductive circuits can be reduced considerably by coupling to existing semiconductor VLSI fabrication capability. This requires that the thin film materials (metals and insulators) and the deposition and patterning techniques be compatible with semiconductor processing. The large wafer size characteristic of Si technology can provide the advantage of allowing multiple copies of a wide variety of designs as well as physically large devices. A planarized all- refractory process for superconductive devices, based on Nb-AlO$_x$-Nb trilayers [1] and developed in a Si technology environment, has recently been reported [2]. In this paper, we describe an extension of the selective Nb anodization process [3] (SNAP) to 125-mm-diameter wafers that was developed in partnership with Si technology. The four-mask-level SNAP described here is less complex than the planarized process and can be used for a rapid evaluation of the trilayers and for making devices and circuits in which the large parasitic capacitance associated with the Nb_2O_5 dielectric can be tolerated.

2. Process Description

The Nb-AlO_x-Nb trilayers were deposited in a MRC dc magnetron sputtering system. This system has been used to deposit trilayers on 50-mm-diameter substrates and has produced high quality Josephson junctions (V_m > 50 mV) and low-noise dc SQUIDs [4,5]. The Nb and Al targets were 200 mm in diameter, and the film thickness was uniform within ± 15% over a 125-mm-diameter wafer. Although this particular sputtering system was not located within the Si fabrication area, depositions of refractory metals and Al are routinely performed in VLSI fabrication processes.

The parameters for Nb and Al depositions are described elsewhere [4]. They were selected to produce low stress Nb films and very uniform, pin-hole-free thin Al films. The Nb base-electrode and counterelectrode thicknesses were 200 and 35 nm, respectively. The Al film was 5 nm in thickness and was partially oxidized by backfilling the sputtering chamber with pure O_2 to produce the thin AlO_x tunnel barrier. Si wafers, 125 mm in diameter with a 300-nm-thick thermal oxide, were used as substrates. No substrate cooling was provided during the trilayer deposition; the wafer temperature remained below 150 °C.

The process flow is shown in Figure 1. The trilayer was patterned using a 0.5-μm-thick positive photoresist on a 1× contact optical exposure system. The photoresist was developed to give a tapered edge profile which is ultimately transferred to the Nb trilayer during etch. The patterned trilayer was etched using Cl_2 gas in a single-wafer reactive ion etch system routinely used for defining Al(Cu) VLSI interconnects. The process is carried out at a power of 100 W, a pressure of 13 mTorr, and a

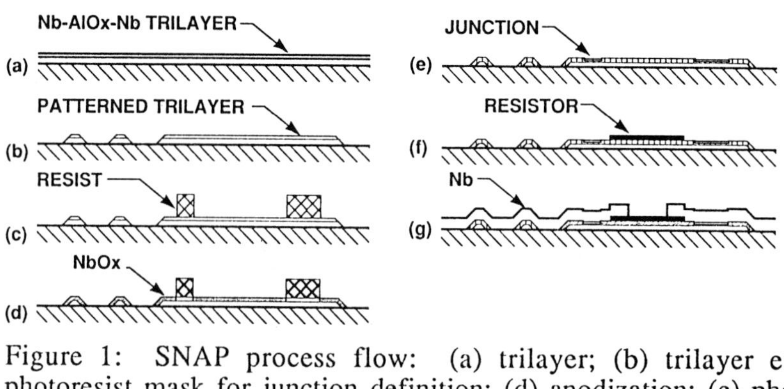

Figure 1: SNAP process flow: (a) trilayer; (b) trilayer etch; (c) photoresist mask for junction definition; (d) anodization; (e) photoresist strip; (f) resistor deposition; (g) Nb wiring.

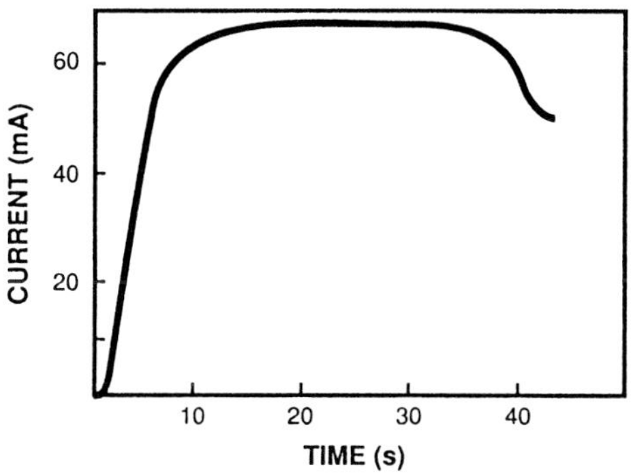

Figure 2: Anodization current as a function of time.

dc self-bias of -150 V; it removes both Nb and AlO_x, allowing a single-step etching of the trilayer [6]. While these conditions result in an anisotropic etch, the thin tapered photoresist results in a Nb sidewall angle of 50° due to photoresist erosion from ion bombardment. This etch process exhibits a selectivity for Nb over SiO_2 of 13:1. To maintain the wafer temperature below 100 °C, the etching process is interrupted every 20 s and the wafers are allowed to cool down for 2 min. After completing the trilayer etch, the photoresist is stripped in N-methyl pyrillodine.

The Josephson junction areas were delineated by SNAP using 1.2-µm-thick positive photoresist, after which the wafers were anodized in a solution of ethylene glycol and ammonium pentaborate. All the Nb areas to be anodized were connected to grid lines which were in turn connected to a ring on the periphery of the wafer for making electrical contact to the anode. Anodization was carried out at a constant voltage ramp rate, and the anodization current was monitored. The current remained constant while the Nb counterelectrode was being anodized and dropped at completion because of the lower anodization rate of AlO_x. A typical plot of the anodization current as a function of time is shown in Figure 2. The sharpness of the decrease in the anodization current is an indication of a clean Nb/AlO_x interface and has been used to monitor the quality of the trilayer [7]. The anodization was carried out at a rate of 1 V/s. At this rate, the anodization current density was 4 mA/cm^2. For defining junction sizes less than 2 µm, it is necessary to increase the voltage ramp rate as high as 8 V/s to maintain a high junction quality [4], corresponding to an anodization current density of 32 mA/cm^2. It is essential to lay out the grid line pattern so that the voltage drop between the edge and the center of the wafer is < 0.2 V to ensure a uniform Nb_2O_5 thickness.

Following the anodization, the photoresist was stripped in acetone. The wafers were then patterned for Pt resistor deposition using an image reversal process to produce an undercut photoresist profile suitable for lift-off processing. To improve the adhesion of the Pt film to the substrate, the wafers were sputter cleaned in a vacuum chamber to remove ~ 50 Å of the surface and then transferred in air into an electron-beam evaporator for Pt deposition. The Pt film was deposited at a rate of 0.15 nm/s to a thickness

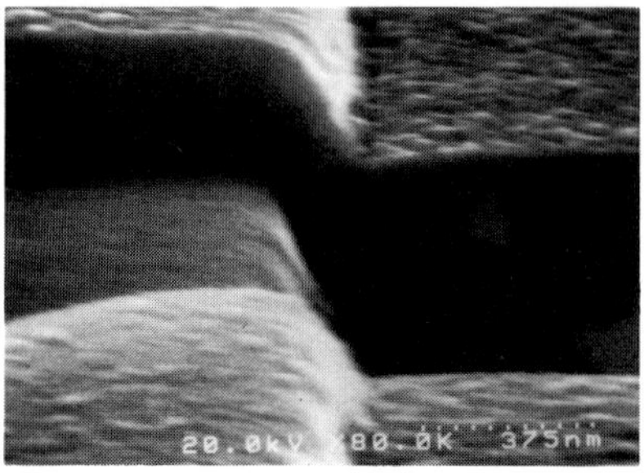

Figure 3. Scanning electron micrograph showing the trilayer step coverage by the Nb wiring layer.

of about 35 nm to obtain a sheet resistance of about 4 Ω/□ at 4.2 K. Lift-off was performed in a solvent bath using ultrasonic agitation.

The wafers were next coated with 250 nm of Nb after sputter cleaning the surface to provide intimate Nb-Nb and Nb-Pt contacts. This Nb layer was patterned and etched in the Cl_2-based reactive ion etch system described earlier. The initial etch was done at a low pressure (13 mTorr) to endpoint and was followed by a 30-s high-pressure overetch to remove any remaining Nb from the tapered edges of the underlying topography without damage to the thin Pt film. At higher pressures, the dc self-bias is reduced in order to lower the ion bombardment and resputtering of material from the wafer so that damage to the Pt resistors is minimized. A scanning electron micrograph of the step coverage is shown in Figure 3.

3. Wafer Layout and Experimental Results

A photograph of the wafer is shown in Figure 4. There are seventeen 6 × 6 mm chips distributed across the wafer for testing circuit components such as Josephson junctions and resistors. They also contain diagnostic structures for process evaluation such as contact resistances and dielectric isolation. The other chips contain a variety of devices including soliton oscillators, a number of different SQUID designs, and large-area devices such as magnetometers (as large as 12 × 12 mm) and gradiometers (as large as 12 × 24 mm).

The I-V characteristic of a Josephson junction with 2 × 2 μm^2 area at 4.2 K is shown in Figure 5. The critical current density is 550 A/cm^2 and the junction quality factor V_m (the product of critical current I_c and the subgap resistance, assuming an I_cR_n product of 2 mV, where R_n is the normal state resistance) is 60 mV. The junction quality was the same across the wafer. The sheet resistance R_s of the Pt resistor material was 3.8 Ω/□. The contact resistance between the Nb and Pt was less than 0.5 Ω for a contact area of 10 μm^2. The spread in the J_c and R_s was about

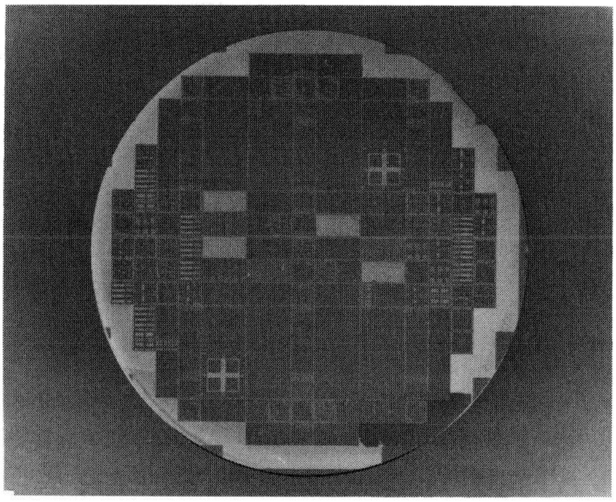

Figure 4. Photograph of a completed 125-mm-diameter wafer.

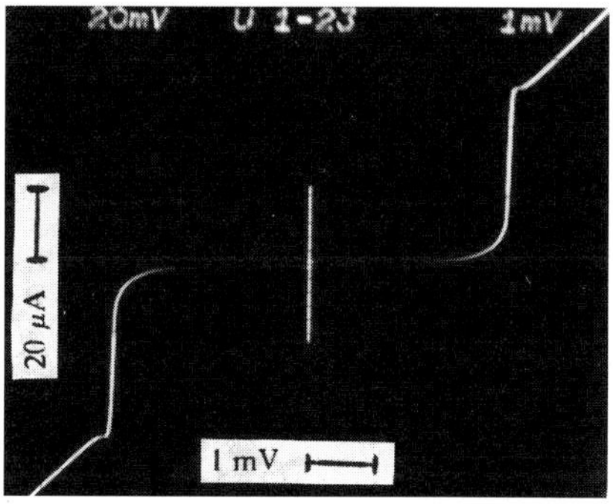

Figure 5. I-V characteristic of a Josephson junction with 2 × 2 μm^2 area at 4.2 K.

± 15% across the wafer. Functional SQUIDs have been identified and evaluation is in progress.

4. Acknowledgments

The authors wish to thank E.M. Macedo and R. Brown for depositing the Nb-AlO$_x$-Nb trilayers, T. Ross for SEM analysis, and D. Gambrel and T.H. Field for fabrication of the anodization power supply and for initial SQUID measurements. This work was conducted under the auspices of the Consortium for Superconducting Electronics. The Lincoln Laboratory work was supported by the Defense Advanced Research Projects Agency.

References

[1] M. Gurvitch, M.A. Washington, and H.A. Huggins, Appl. Phys. Lett. **42**, 472 (1983).
[2] M.B. Ketchen, D. Pearson, A. Kleinsasser, C.-K. Hu, M. Smyth, J. Logan, K. Stawiasz, E. Baran, M. Jaso, T. Ross, K. Petrillo, M. Manny, S. Basavaiah, S. Brodsky, S.B. Kaplan, W.J. Gallagher, and M. Bhushan, accepted for publication in Appl. Phys. Lett., 1991.
[3] H. Kroger, L.N. Smith, and D.W. Jillie, Appl. Phys. Lett. **39**, 280 (1981).
[4] M. Bhushan and E.M. Macedo, Appl. Phys. Lett. **58**, 1323 (1991).
[5] M.B. Ketchen, M. Bhushan, S.B. Kaplan, and W.J. Gallagher, IEEE Trans. Magn. **27**, 3005 (1991).
[6] C.-K. Hu, N. Mazzeo, S.J. Wind, D.J. Pearson, and M.B. Ketchen, to be published in the Proceedings of the International Conference on Metallurgical Coatings and Thin Films held in San Diego, California, 1991.
[7] T. Imamura and S. Hasuo, IEEE Trans. Magn. **25**, 1131 (1989).

DC-SQUIDs Fabricated with Nb/Al-oxide/Nb Tunnel Junctions

H.C. Kwon, Y.H. Lee, J.M. Kim, and J.C. Park

Quantum Physics Laboratory, Korea Standards Research Institute,
P.O. Box 3, Taedok Science Town, Taejon 305-606, Republic of Korea

Abstract. We have fabricated a planar type dc SQUID which incorporates high quality Nb/Al-oxide/Nb Josephson junctions prepared by rf sputtering and SNAP method. DC SQUIDs were designed to have double transformer structure for inductance matching. Preliminary tests showed that our non-optimized SQUIDs had equivalent flux noise ϕ_n of $6\mu\phi_o/\sqrt{Hz}$, corresponding to intrinsic energy sensivity $\varepsilon = 5.9 \times 10^{-31}$ J/Hz in the flux-locked loop mode.

1. Introduction

Recently, Nb/Al-oxide/Nb Josephson junctions became the most attractive candidates for high performance SQUIDs[1]. Particularly, such junctions have low levels of excess noise at low frequency and thus may be very desirable for SQUIDs applicable to biomagnetism[2].

In this report,we describe a fabrication process of high quality Nb/Al-oxide/Nb junctions, the construction of dc SQUIDs, and preliminary tests of device performance.

2. Experimental

The fabrication process was based on a Nb/Al-oxide/Nb trilayer junctions using chlorobenzene lift-off and SNAP method[3]. All metal layers were deposited by rf magnetron sputtering with a base pressure of, when throttled, typically 1×10^{-5} Pa. We have found the optimal sputtering condition by investigating the stress and surface morphology dependence of junction quality[4].

The fabrication steps of the process are illustrated in Fig.1. First, the W and Nb film were deposited on the thermally oxidized Si wafer covered with the lift-off stencil mask for the shunt resistor. This Nb film was used to protect the W film so that the resistance of the W film should not be changed during the anodization process. Next, the photoresist mask frames the base electrode and the Nb/Al-oxide/Nb trilayer was fabricated without breaking vacuum. The deposition rates of the Nb and Al films were 100 and 40 nm/min.,respectively. The Ar pressure was 1.7 Pa which was slightly below 2.2 Pa, stress-free condition for Nb film in our system. The tunnel barrier formation was done in a pure oxygen at a pressure of 500 Pa at room temperature for 30 min. After the trilayer deposition, anodization using photoresist mask was carried out.

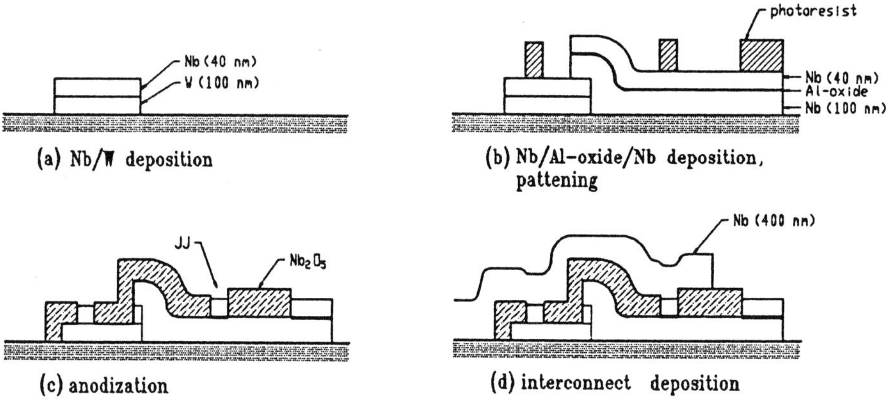

Fig. 1. Junction fabrication process.

Because of the underanodization,the junction size was reduced by 1 $\sim$ 2μm from the photomask. Since we used no additional insulation layer, the single anodization process didn't give a good insulation layer often yielding electrical shorts between the SQUID and input coil. To solve this problem, we used double anodization process using two separate photoresist masks. In the second anodization process, the photoresist covered a much larger area around the junction to prevent the junction from becoming smaller and anodization time was short. Last, the Nb metallization layer was deposited.

The structure of a dc SQUID is similar to that of Muhlfelder[5], but without groundplane beneath the SQUID loop. Its schematic electrical diagram and micrograph are shown in Fig.2. It consists of a washer type dc SQUID loop with 50μm hole size and 5 turn input coil in connection with an intermediate transformer for inductance matching. Intermediate transformer has the same shape, i.e., single turn square loop coupled with 19 turn coil. A single turn modulation coil for the modulation and feedback signal was also coupled to the SQUID loop. Our SQUID was designed to allow enough flexibility to meet the various criteria, but it was not optimized.

For testing, the chip was surrounded by Pb shielding tube and the probe was immersed directly in liquid helium. The four terminal method was used to measure the periodic modulation of the SQUID voltage with applied flux. The white noise voltage of the SQUID V_n was amplified by a cooled L_tC_t resonant circuit with quality factor Q_t as shown in Fig. 2(a) and measured by an external bandpass filter and ac voltmeter at 81 kHz. The equivalent flux noise ϕ_n was estimated as $V_n / \mid \partial V/\partial\phi \mid$ and used to calculate the intrinsic energy sensivity $\varepsilon = \phi_n^2/2L$. A small ac flux, whose amplitude and frequency were $10^{-3}\phi_o$ and 81 kHz, respectively, was also applied to the modulation coil in measuring a $Q_t(\partial V/\partial\phi - \phi)$ characteristic.

For flux-locked loop operation, we have made electronics using two parity biasing and square wave modulation [6] where the SQUID is a part of a bridge circuit. Its performance test is being carried out and will be reported later.

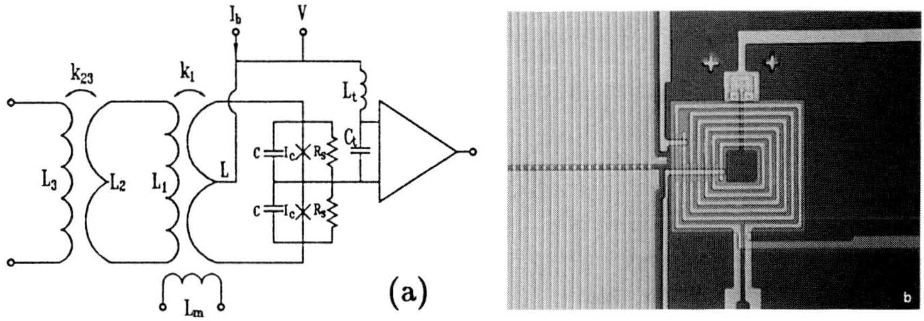

Fig. 2. (a) The electrical schematic diagram of the dc SQUID with intermediate transformer and (b) A micrograph of the SQUID loop with spiral coils.

3. Results and discussion

For optimized sputtering condition, junctions with high quality were reliably reproduced. Typical parameters of W shunted Nb/Al-oxide/Nb junctions from the I-V characteristics at 4.2 K are summarized in table 1. The junction quality was characterized using the ratio of I(4mV)/I(2mV) for unshunted junctions and it ranges up to 40 indicating the low subgap leakage current.

Table 1. Typical parameters of W shunted Nb/Al-oxide/Nb junction.

Energe gap	2Δ	$2.5 \sim 2.6$ mV
Junction area	A	$6 \sim 15\mu m^2$
W shunt resistance	R_s	$3 \sim 10\Omega$
Critical current	I_c	$7 \sim 60\ \mu A$
Current density	J_c	$100 - 500\ A/cm^2$
$I_c R_n$ product		$50 - 200\mu V$

Most of the measurements have been done on a SQUID with input coil opened, i.e., the intermediate transformer L_2 and L_3 was not connected to avoid possible external noise source(see Fig.2). The paramaters of the SQUID fabricated as mentioned above are : $2I_c \sim 15.5\mu A$, $I_c R_s \sim 70\mu V$.

Fig.3 shows the I-V curve at two flux biases . In the curve the bump around $59\mu V$ is due to the LC resonance of the SQUID and the bump around $115\mu V$ is due to the $\lambda/2$ resonance of the input coil. The modulation parameter β was estimated to be about 1.0 from the measured modulation depth of $2I_c$ with external flux giving the loop inductance L of 129 pH, while the designed value from the loop geometry was 80 pH. This difference could be attributed to the slit inductance. The junction capacitance C might be estimated from the LC resonance voltage. Then, the hysteresis parameter β_c was estimated to be 0.5. The measured mutual inductances were all within 10 % of the designed values

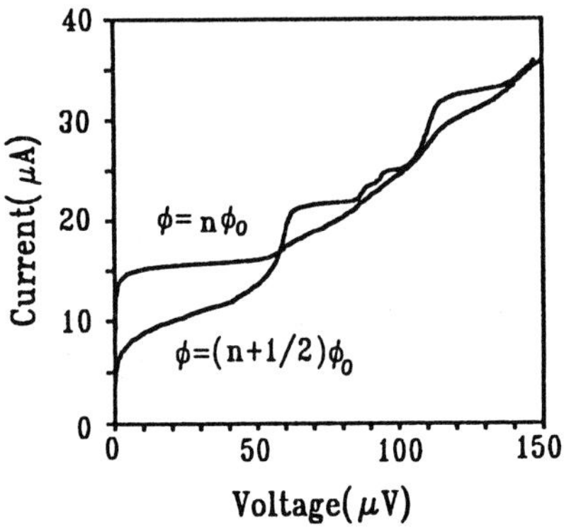

Fig. 3. I-V curve of a dc SQUID for $\phi = n\phi_o$ and $\phi = (n + 1/2)\phi_o$.

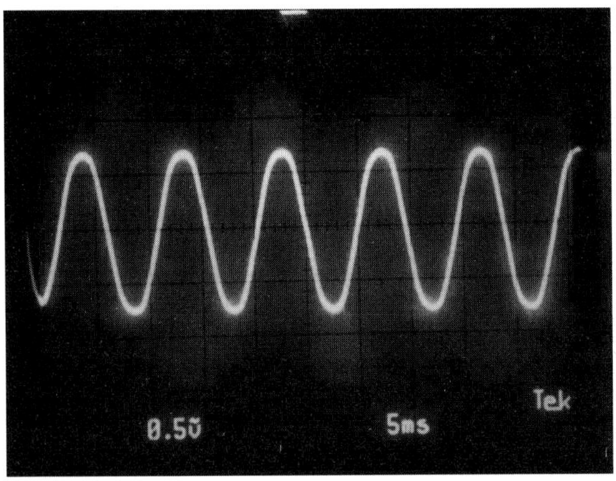

Fig. 4. $V-\phi$ curve of the dc SQUID with biasing current of 14μA. $\Delta V \sim 50\mu$V.

and the effective input inductance of our device was estimated to be about 0.5μH.

The $V - \phi$ curve is smooth enough to use a conventional readout scheme as shown in Fig.4, but ΔV and $\partial V/\partial\phi$ of 50 μV and 220μV/ϕ_o, respectively, were higher than expected. Since we used no additional insulation layer between the SQUID loop and input coil, the parasitic capacitance might cause these results[7]. At maximum sensitivity, a noise voltage V_n of 860 pV/$\sqrt{Hz}$ was obtained and the equivalent flux noise ϕ_n was estimated to be $3.9\mu\phi_o/\sqrt{Hz}$, corresponding to $\varepsilon = 2.4 \times 10^{-31}$ J/Hz. In flux-looked loop mode, ϕ_n was $6.1\mu\phi_o/\sqrt{Hz}$, corresponding to $\varepsilon = 5.9 \times 10^{-31}$ J/Hz.

References

1. M.W. Cromar, J.A.Beall, Diano Go, K.A.Masarie, R.H.Ono, and R.W.Simon, IEEE Trans.Magn. **MAG-25**, 1005 (1989).
2. B.Savo, F.C.Wellstood, and J.Clarke, Appl.Phys.Lett. **50**, 1757 (1987)
3. H.Kroger, L.N.Smith, and D.W.Jillie, Appl.Phys.Lett. **39**, 280 (1981).
4. Y.H.Lee, H.C.Kwon, J.M.Kim, and J.C.Park, New Physics, **31**, 182 (1991).
5. B.Muhlfelder, W.Johnson, and M.W.Cromar, IEEE Tran. Magn.**MAG-19**, 303 (1983).
6. V.Foglietti, W.J.Gallagher, and R.H.Koch, IEEE Trans. Magn. **MAG-23**, 1150 (1987).
7. D.Drung and W.Jutzi, SQUID'85, Superconducting Quantum Interference Devices and Their Applications, eds.H.D.Hahlbohm and H.Lubbig(Walter de Gruyter,Berlin,1985), 807

A Compact Very Low Noise DC-SQUID Magnetometer

R. Cantor[1], *T. Ryhänen*[2], *and H. Seppä*[2]

[1]Physikalisch-Technische Bundesanstalt, Institut Berlin, Abbestr. 2–12, W-1000 Berlin 10, Fed. Rep. of Germany

[2]Metrology Research Institute, Helsinki University of Technology and Technical Research Centre of Finland, Otakaari 7 B, SF-02150 Espoo, Finland

Abstract. We have designed and fabricated a very low noise integrated dc Superconducting QUantum Interference Device (SQUID) magnetometer on a 4×4 mm^2 chip. The white flux noise at 1 kHz is 6×10^{-7} $\Phi_0/\sqrt{Hz}$. This corresponds to a flux density noise of 3.4 $fT/\sqrt{Hz}$ and a coupled energy resolution of 32 h. The 1/f noise corner frequency is 10 Hz and the 1/f flux noise at 1 Hz is $2 \times 10^{-6}\Phi_0/\sqrt{Hz}$.

1. Introduction

It has been demonstrated[1,2] that dc Superconducting QUantum Interference Devices (SQUIDs) can have noise energies approaching the quantum limit. Such devices are therefore very attractive for practical applications, but coupling to the very low SQUID inductance can be difficult. This is normally accomplished by forming the SQUID loop into a square-shaped washer with a closely coupled, integrated input coil[3,4] on top. The presence of the input coil, however, introduces numerous parasitic effects which, if not taken into account in designing the SQUID, can lead to strong irregularities in the current-voltage $(I - V)$ and voltage-flux $(V - \Phi)$ characteristics and to a significant deterioration of the ultimate noise performance.[5,6,7] We have developed a design approach for the optimization of low noise SQUIDs with integrated coupling circuits which takes these parasitic effects into account.[5,6,8] This approach has been used to design an integrated dc SQUID magnetometer on a 4×4 mm^2 chip.[5] In this paper, we report recent results for an improved version of this magnetometer.

2. Magnetometer Design and Optimization

The input coil of a dc SQUID with integrated flux coupling circuit introduces a parasitic capacitance C_p which reduces the damping of the SQUID dynamics and, according to simulations,[7,9] leads to hysteretic characteristics. This has been shown[10] to cause a deterioration of the energy resolution. Thus, the SQUID should be designed so as to minimize C_p. In addition, the presence of the input coil introduces resonances:[6,11,12,13] the LC resonance involving the inductance of the input coil and its lumped capacitance to the SQUID

Springer Proceedings in Physics, Vol. 64 Superconducting Devices and Their Applications
Editors: H. Koch, H. Lübbig

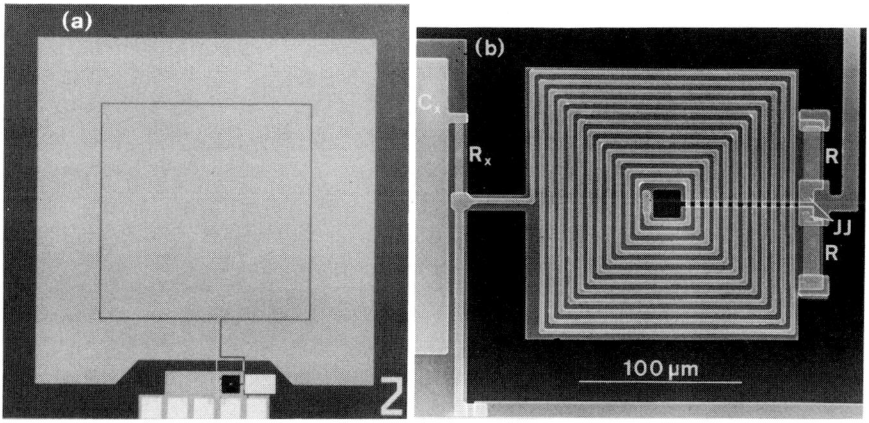

Figure 1. (a) Photograph of the 4×4 mm^2 magnetometer chip. The inner loop in the center is for operation using APF. (b) SEM micrograph of the SQUID area, showing the Josephson junctions JJ, junction shunt resistors R, and R_xC_x shunt resistor R_x and capacitor C_x. For this device, $R_x = 6.8\ \Omega$ and $C_x = 25$ pF.

washer, and the microwave transmission line resonance of the microstripline formed by the input coil and the washer. The $\lambda/2$ resonance of the washer is enhanced as well. By properly choosing the SQUID layout, these resonances can be moved to frequencies well away from the intended operating frequency of the SQUID, f_{op}. The low frequency resonances can be damped by inserting a series R_xC_x shunt in parallel with the input coil;[5,6] the shunt smoothes the SQUID characterisitics with only a minimal contribution to the total noise.

For magnetometer applications which require that the flux density noise B_n rather than the energy resolution be minimum, it is possible to accept a slight mismatch between the input coil and pick-up loop inductances, which diminishes the energy resolution, in order to increase the effective area of the pick-up loop. This enhances the field sensitivity and results in an improved flux density noise. We have used a computer optimization model to generate the electrical parameters of the improved magnetometer. For a fixed chip size and junction area, the minimization of B_n is carried out for successive numbers of input coil turns by varying the SQUID washer hole size, the critical current density, the junction shunt resistance, and the pick-up loop linewidth. The minimization is constrained by feasable limits of the film widths and thicknesses, and by the condition that the resonances be located well away from f_{op}, as discussed above. The R_xC_x shunt[6] integrated into the magnetometer is designed by matching the resistor R_x to the nominal transmission line impedance of the input coil microstripline, and by choosing the capacitor C_x so that the resonant frequency of the input circuit, consisting of the input coil in parallel with the pick-up loop and the R_xC_x shunt, is around 500 MHz.

The magnetometers are fabricated using a simple, four-level Nb/Si_xN_y/Nb Josephson junction technology.[5,14] The magnetometer and the SQUID area are shown in Fig. 1. A feedback loop for operation using additional positive feedback[15,16,17] has also been integrated into the design. The SQUID washer has a hole side length of 18 μm and a slit width of 3 μm. The area of each Josephson junction is about 14 μm^2. The junction shunts are connected to the base electrode via two high critical current junctions, and the current return path for the 12-turn input coil with 3 μm linewidth is routed through the SQUID washer. The capacitor C_x is integrated into the pick-up loop.

3. Performance

The magnetometer characteristics and noise performance are measured using a computer controlled, automatic measurement system.[18] The $I - V$ and $V - \Phi$ characteristics shown in Fig. 2 are smooth. For this device, $\beta_c = (2\pi/\Phi_0)I_cR^2C = 0.5$, where the junction critical current $I_c = 32$ μA, the junction shunt resistance $R = 2.4$ Ω, the junction capacitance $C = 0.9$ pF, and Φ_0 is the flux quantum, and $\beta = 2I_cL/\Phi_0 = 1.6$, where the SQUID inductance $L = 51$ pH. The input coil inductance $L_i = 6.2$ nH and the pick-up loop inductance $L_p = 8.5$ nH.

The flux noise is measured in an open loop configuration using a second magnetometer[5] as a low noise preamplifier. The voltage signal from the magnetometer to be measured is coupled to the preamplifier via two 4.7 Ω metal film resistors and a Cu wire-wound coil. The preamplifier voltage noise spectra

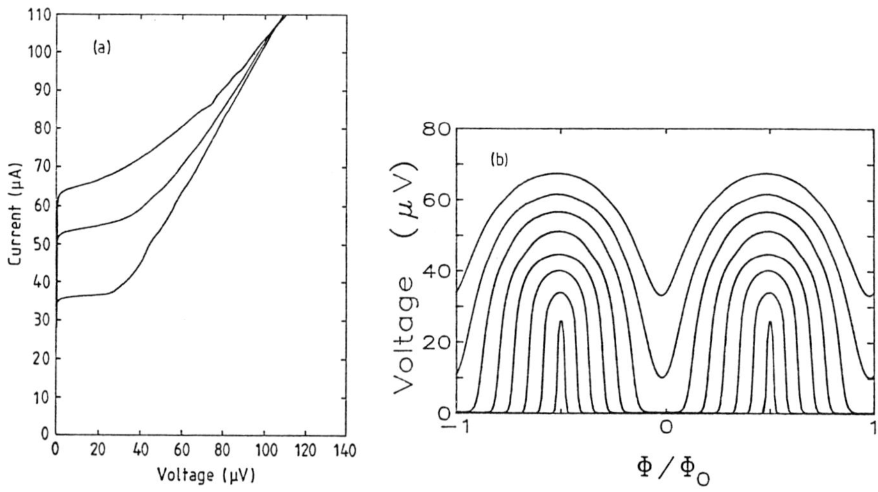

Figure 2. (a) Current-voltage characteristics for an applied flux Φ_a of 0 (upper curve), 0.25 Φ_0 (middle curve), and 0.5 Φ_0 (lower curve). (b) Voltage-flux characteristics for I_b ranging from 35 to 70 μA in 5 μA steps.

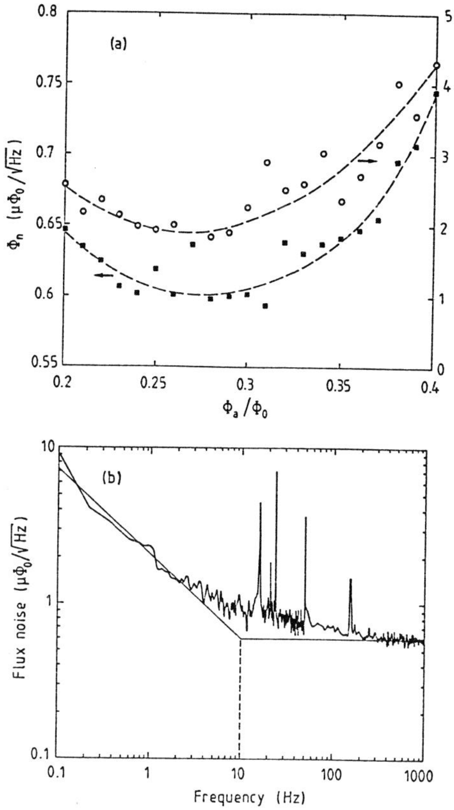

Figure 3. (a) Dependence of the white flux noise at 1 KHz (lower curve) and 1/f noise at 1 Hz (upper curve) on the applied flux, for $I_b = 54.8$ μA. The curves serve as guides to the eyes only. (b) Flux noise spectrum for $I_b = 54.8$ μA and $\Phi_a = 0.3\Phi_0$.

are recorded and used to reduce the total noise spectra in order to determine the intrinsic noise of the magnetometer. These measurements can be carried out over a large number of different working points by automatically varying the bias current and applied flux. In this way, a noise map for the magnetometer can be generated. For most working points of interest (i.e., the points where the flux noise Φ_n is lowest), the total voltage noise is at least 25 % higher than the preamplifier noise.

The noise mapping data for a bias current I_b of 54.8 μA through the SQUID and the flux noise spectrum for an applied flux of $0.3\Phi_0$ and the same bias current are shown in Fig. 3. The white flux noise (at 1 kHz) is 6×10^{-7} $\Phi_0/\sqrt{Hz}$ over a broad range of working points. Using the calibrated sensitivity of the magnetometer, 5.6 nT/Φ_0, this corresponds to a flux density noise of $3.4fT/\sqrt{Hz}$. The coupled energy resolution is 32 h. The 1/f noise corner frequency is 10 Hz and the 1/f noise at 1 Hz is $2 \times 10^{-6}\ \Phi_0/\sqrt{Hz}$. These results represent an improvement over the previous design.[19]

4. Conclusions

We have designed and fabricated a very low noise integrated dc SQUID magnetometer on a $4 \times 4\ \text{mm}^2$ chip. The compact size of these high performance magnetometers makes it possible to fabricate 105 chips per 2 inch diameter Si wafer, thus increasing yield.

References

[1] M.B. Ketchen and R.F. Voss, Appl. Phys. Lett. **35**, 812 (1979).
[2] M.W. Cromar and P. Carelli, Appl. Phys. Lett. **38**, 723 (1981).
[3] J.M. Jaycox and M.B. Ketchen, IEEE Trans. Magn. **MAG-17**, 400 (1981).
[4] M.B. Ketchen and J.M. Jaycox, Appl. Phys. Lett. **40**, 736 (1982).
[5] R. Cantor, T. Ryhänen, D. Drung, H. Koch, and H. Seppä, IEEE Trans. Magn. **MAG-27**, 2927 (1991).
[6] H. Seppä and T. Ryhänen, IEEE Trans. Magn. **MAG-23**, 1083 (1987).
[7] D. Drung and W. Jutzi, in: SQUID'85, Superconducting Quantum Interference Devices and Their Applications, eds. H.D. Hahlbohm and H. Lübbig (Walter de Gruyter, Berlin, 1985), p. 807.
[8] J. Knuutila, M. Kajola, H. Seppä, R. Mutikainen, and J. Salmi, J. Low Temp. Phys. **81**, 369 (1988).
[9] J.A. Ketoja, J. Kurkijärvi, and R.K. Ritala, Phys. Rev. **B30**, 3757 (1984).
[10] T. Ryhänen, R. Cantor, D. Drung, H. Koch, and H. Seppä, IEEE Trans. Magn. **MAG-27**, (1991).
[11] C. Hilbert and J. Clarke, J. Low Temp. Phys. **61**, 237 (1985).
[12] B. Muhlfelder, J.A. Beall, M.W. Cromar, R.H. Ono, and W.W. Johnson, IEEE Trans. Magn. **MAG-21**, 427 (1985).
[13] T. Ryhänen, H. Seppä, R. Ilmoniemi, and J. Knuutila, J. Low Temp. Phys.**76**, 287 (1989).
[14] R. Cantor, D. Drung, M. Peters, and H. Koch, J. Appl. Phys. **67**, 3038 (1990).
[15] T. Ryhänen, R. Cantor, D. Drung, and H. Koch, Appl. Phys. Lett. 59, 228 (1991).
[16] D. Drung, R. Cantor, M. Peters, H.J. Scheer, and H. Koch, Appl. Phys. Lett. **57**, 406 (1990).
[17] H. Seppä, to appear in Rev. Sci. Instrum.
[18] T. Ryhänen, H. Seppä, R. Cantor, D. Drung, and H. Koch, these proceedings.
[19] The calculation of the intrinsic flux noise in ref. 5 did not take the small dynamic resistance of the SQUID into account, which leads to an underdetermination of the noise. The corrected white flux noise at 1 kHz for the magnetometer in ref. 5 is $6.7 \times 10^{-7}\ \Phi_0/\sqrt{Hz}$.

A Low Noise DC-SQUID Based on Nb/Al-AlO_x/Nb Josephson Junctions

L. Grönberg[1], *H. Seppä*[2], *R. Cantor*[3], *M. Kiviranta*[1], *T. Ryhänen*[2], *J. Salmi*[1], and *I. Suni*[1]

[1]Technical Research Centre of Finland, Semiconductor Laboratory, Otakaari 7 B, SF-02150 Espoo, Finland

[2]Technical Research Centre of Finland and Helsinki University of Technology, Metrology Research Institute, Otakaari 7 B, SF-02150 Espoo, Finland

[3]Physikalisch-Technische Bundesanstalt, Institut Berlin, Abbestr. 2–12, W-1000 Berlin 10, Fed. Rep. of Germany

Abstract. Low noise dc SQUIDs based on a Nb/Al-AlO_x/Nb Josephson junction technology have been developed. The design has been optimized for improved control of the resonances. An eight-level process has been used for device fabrication. Flux noise levels of $2.8 \times 10^{-7}\Phi_0/\sqrt{Hz}$ at 1 kHz and $1.5 \times 10^{-6}\Phi_0/\sqrt{Hz}$ at 1 Hz have been measured for an integrated SQUID magnetometer having 3 μm^2 junctions, a 12-turn input coil, and a dc SQUID inductance $L_{dc} = 11$ pH.

1. Introduction

It has been shown that a careful optimization of the SQUID design yields very low noise devices [1,2,3]. The design of the autonomous SQUID is straightforward, but the presence of the input coil complicates the SQUID dynamics and makes the optimization of the device more difficult. We have designed several optimized SQUID structures and report here preliminary results for a dc SQUID magnetometer with 3 μm^2 Josephson junctions.

2. Design

The design is based on existing analytic models to minimize the energy resolution.[2,3] Most of the SQUID parameters are optimized: the SQUID inductance, the number of turns in the input coil, and the critical current. The effect of the input circuit is also taken into account. The fabrication technology limits are used as constraints for the optimization procedure. A major point in the design is the control of the resonances introduced by the input coil. They have to be damped and kept from interfering with the Josephson oscillations. Several devices with various geometries are included on the same wafer; each SQUID design has been optimized separately. Several points have been considered. The layout of the SQUID is chosen so that: (1) the parasitic capacitance across both junctions is smaller than the actual junction capacitance, and (2)

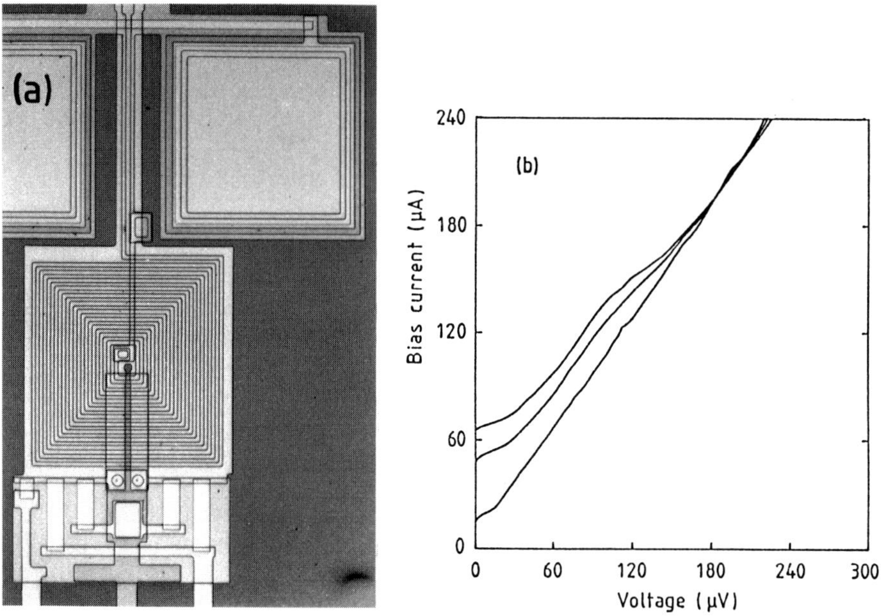

Figure 1. a) Photograph of a dc SQUID. The junction diameter is 2 μm. In the top right corner, one of the shunting capacitors can be seen. The 12-turn input coil is in the center and below these are the two Josephson junctions (the small dots). b) Current-to-voltage characteristics for three different values of the applied flux.

the resonant frequencies determined by the SQUID dimensions are such that $f_{strip} < f_{op}/4$ and $f_{washer} > 4f_{op}$, where f_{strip} is the resonant frequency of the input coil microstripline, f_{washer} is the SQUID washer resonant frequency, and f_{op} is the operating frequency of the SQUID. The latter ensures that the resonant frequencies are located well away from f_{op}. An R_xC_x shunt in parallel with the input coil damps the input coil resonances; C_x prevents noise currents at low frequencies, and R_x provides adequate damping of the input coil LC resonance. Consequently, the oscillations induced by the Josephson junctions decay rapidly. Simultaneously, R_x properly terminates the microstripline.

The layout includes SQUIDs with nominal junction diameters of 3 μm and 5 μm arranged in magnetometer and gradiometer configurations, and SQUIDs for which each Josephson junction is replaced by four junctions in series. A SQUID magnetometer with two 3 μm junctions and a 12-turn input coil is shown in Fig. 1a.

3. Fabrication

The fabrication process consists of 8 mask levels. The 100 mm oxidized Si wafers are mounted on water-cooled Cu blocks for the dc sputter deposition of the Nb and Al layers. First, a 150 nm Nb film is sputtered and patterned in a SF_6 plasma to define the input coil. In order to get a sloped etch profile, the photoresist is baked at 150 °C before plasma etching. This is essential for the input coil current return path (made from trilayer), which runs across the windings of the coil. A 200 nm PECVD SiO_2 film is then deposited to provide isolation between the input coil and the trilayer which follows. The SiO_2 layer is deposited in two steps; a 50 nm layer is deposited at 120°C, and then a 150 nm layer is deposited at 220°C. Contact openings are made by wet chemical etching in buffered HF. The trilayer is deposited next. A 200 nm Nb base electrode is sputtered first, followed by a 7 nm thick Al layer. The tunnel barrier is grown on the Al by oxidation for 1 hour in pure oxygen at 40 Pa. After evacuating the chamber, a 32 nm top Nb electrode is deposited. A 50 nm layer of PECVD SiO_2 is then deposited for subsequent use as an anodization mask. The Josephson junctions are defined by anodizing the top Nb layer. The dielectric for the capacitors is formed at this time as well. The trilayer is patterned by etching the top Nb layer using RIE, the Al-AlO_x layer by wet etching in phosphoric acid, and the bottom Nb layer again using RIE. A second 250 nm SiO_2 layer is then deposited in two steps as before. After defining the contact openings, a 250 nm Nb layer is sputtered and patterned using RIE. The resistors are 93 nm dc magnetron sputtered Mb, patterned by wet etching in phosphoric acid. Finally, a 300 nm layer of SiO_2 is deposited for passivation. Openings for the bonding pads are etched using buffered HF.

4. Results

We have fabricated a set of devices for which the critical current was the target value for junctions having a nominal diameter of 5 μm. The devices reported here have a junction diameter of 2 μm, giving an area of 3 μm^2. The critical current is thus too small for these junctions.

Figure 1b shows the $I - V$ characteristics of the SQUID magnetometer, and Fig. 2 shows the corresponding voltage vs. flux response. The flux noise spectral density function, measured using a second dc SQUID as a preamplifier [4], is shown in Fig. 3. Our experiments indicate that the flux noise, Φ_n, and the flux-to-voltage transfer function, $\partial V/\partial\Phi_a$, depend very weakly on the point of operation. Due to the low critical current, the SQUID is overdamped ($\beta_c = 0.2$, $\beta_L = 1.3$), which explains the smooth characteristics.

According to Fig. 3, the flux noise approaches $2.2 \times 10^{-7}\Phi_0/\sqrt{Hz}$ at high frequencies, and at 1 Hz the flux noise is $1.5 \times 10^{-6}\Phi_0/\sqrt{Hz}$. The high 1/f noise is probably due to the small Josephson junctions. The low loop inductance ($L_{dc} = 11$ pH), however, leads to very low flux noise, both at low and high frequencies. These SQUIDs were also tested using the the directly coupled feedback elctronics, and very similar results were observed [5].

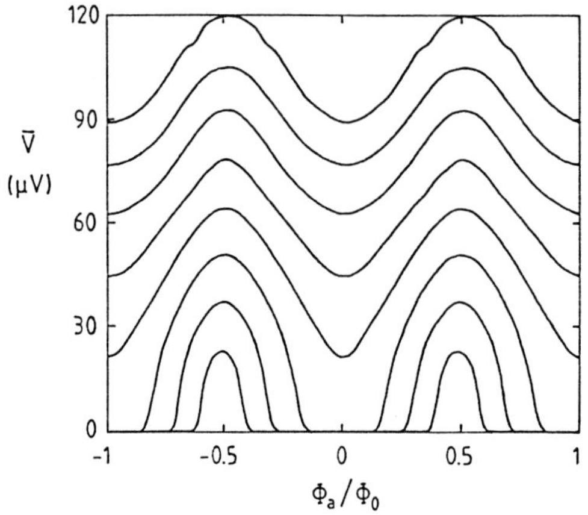

Figure 2. Voltage vs. flux characteristics for different bias currents.

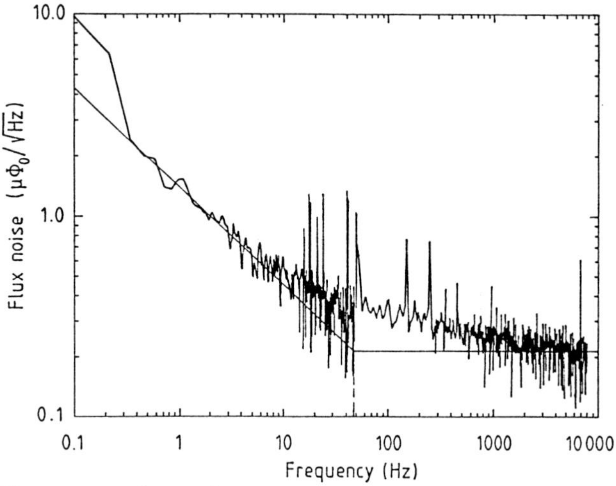

Figure 3. Flux noise spectral density function. The onset frequency of the 1/f noise is 50 Hz.

5. Conclusions

Very low noise dc SQUID magnetometes have been fabricated. Flux noise values of $2.8 \times 10^{-7}\Phi_0/\sqrt{Hz}$ at 1 kHz and $1.5 \times 10^{-7}\Phi_0/\sqrt{Hz}$ at 1 Hz have been observed. The small junction area is probably the main reason for the high 1/f noise.

Acknowledgements. The assistance of M. Kajola and J. Knuutila from the Helsinki University of Technology is gratefully acknowledged. This work has been financially supported by the Finnish National Fundation for Research (SITRA).

References

[1] M.B. Ketchen and R.F. Voss, Appl. Phys. Lett. **35**, 812 (1979).

[2] J. Knuutila, M. Kajola, H. Seppä, R. Mutikainen and J. Salmi, J. of Low Temp. Phys. **71**, 369 (1988).

[3] R. Cantor, T. Ryhänen, D. Drung, H. Koch and H. Seppä, IEEE Trans. Magn. **MAG-27**, 2927 (1991).

[4] T. Ryhänen, H. Seppä, R. Cantor, D. Drung, H. Koch, and D. Veldhuis, these proceedings.

[5] H. Seppä, these proceedings.

DC-SQUID with Aluminium Microbridges far from T_c

C. Chapelier[1], *M. El Khatib*[2], *P. Perrier*[2], *A. Benoit*[2], *and D. Mailly*[3]

[1]DRFMC/SPSMS/LCP, CENG, 85 X, F-38041 Grenoble Cedex, France
[2]CRTBT-CNRS, BP 166 X, F-38042 Grenoble Cedex, France
[3]L.M.M.-CNRS, 196, avenue H. Raverra, F-92220 Bagneux, France

Abstract. We have realized reliable DC-SQUIDs in aluminium with two Dayem bridges in only one lithographic operation. These bridges appear to behave like classic Josephson junctions even at low temperature far from T_c. They are very durable and reproducible upon temperature cycling. They exhibit a considerable hysteresis and critical currents as low as 10 µA. An electronic setup has been developed in order to measure the critical current of these hysteretic SQUIDs with a typical noise of $8\ 10^{-6}\ \phi_0/\mathrm{Hz}^{1/2}$. This non-destructive and damage free technique allows one to perform magnetization measurements on any material.

1. Introduction

The Josephson effect in microbridges has first been suggested in 1964 by Anderson and Dayem [1]. These superconductive weak links seemed to be very promising in order to design planar DC-SQUIDs with a one step thin film technology. But Dayem bridges exhibit an ideal Josephson current-phase relation only when their dimensions are small compared to the coherence length ξ [2]. Up to now, this condition was only fulfilled either near T_c where $\xi(T)$ diverges or after a metallurgical treatment of the bridge with the help of current pulses whose purpose is to reduce its zero temperature critical current [3].

Nowadays, electronic lithography allows to fabricate directly reliable planar DC-SQUIDs with two identical and repeatable microbridges which work far below their critical temperature. We have designed and constructed such DC-SQUIDs in aluminium and measured their magnetic flux response at 100 mK. The technique is based on a standard lift off operation performed on a GaAs substrate which is spin coated with a PMMA resist and exposed to the electron beam of a JEOL 5DIIE. The best resolution that we have reached with this technique is about 30 nm. Figure 1 shows one of such SQUIDs with an inner surface of a few microns square. We can easily recognize its two microbridges.

Springer Proceedings in Physics, Vol. 64 **Superconducting Devices and Their Applications**
Editors: H. Koch, H. Lübbig

Figure 1 - DC-SQUID

2. Measurement method

By measuring the critical current I_c of our DC-SQUID we are in principle able to detect very weak magnetic flux variation $\Delta\phi$ as I_c is a periodic function of ϕ with a period $\phi_0 = h/2e$. The main difficulty lies in the fact that such a SQUID has a hysteretic I-V curve and it is impossible to polarize it with a constant current near its transition and use it in a feedback loop. To overcome this difficulty one usually shunts the two weak links. Owing to simplicity considerations we have chosen the contrary to carry out the hysteretic cycle ABCDEFA (Fig. 2) for each measurement of I_c. The SQUID is supplied by a current source I(t) with a periodic time dependence shown in figure 3. The current is quickly brought near I_c(ABC), and then slowly increased (CD) until a sudden variation of the voltage from zero to a finite value is detected between the SQUID and the ground (DE). At that point the current is immediately reset to zero (EFA) in order to minimize Joule heating and a new cycle begins.

All these cycles are monitored by a micro-computer which measures the time t_c(CD) and converts it in I_c. Each cycle takes less than 100 µs. Every measurement is stored in an array of 256 values corresponding to 256 different times or values of I_0. The upper and lower limits of this array is abritrary chosen in order to filter unwanted results. After a time of accumulation τ, an histogram such as in figure 4 can be plotted. The mean value of t_c and the noise can be evaluated from it.

Another important advantage of this method lies in the possibility of measuring in a synchronous manner the response of the SQUID to a periodic signal such as an external magnetic field for instance. The computer modulates periodically this signal in order to perform n x 256 measurements of I_c during each period (n is an arbitrary integer). For more details see reference 4.

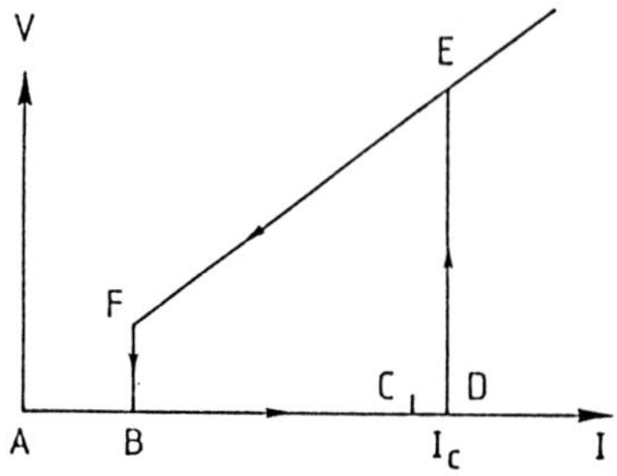

Figure 2 - Hysteretic cycle

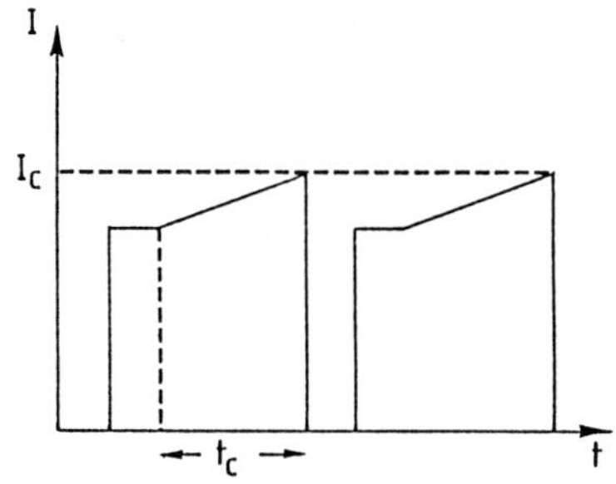

Figure 3 - Current cycle

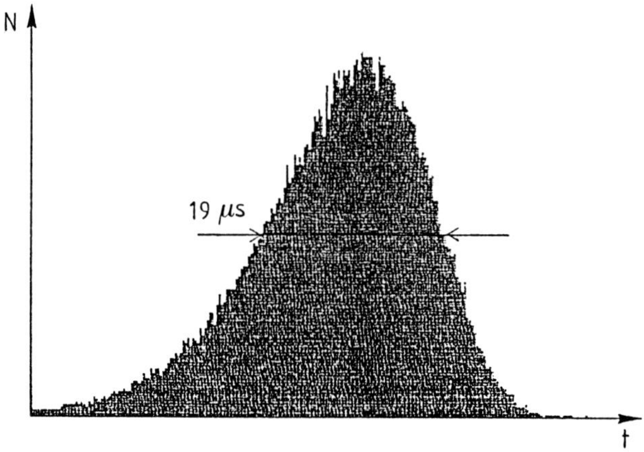

Figure 4 - Histogram

3. Results

3.1. Microbridges

As we have shown in figure 1, electronic lithography allows to realise microbridges with very small dimensions. These dimensions must be compared with the characteristic lengths of the superconductor.

First, we can define an effective geometrical length L_{eff} = Max(W,d), where W and d are respectively the width and the thickness of the film. As 30 nm $\leq$ W $\leq$ 50 nm and d $\approx$ 50 nm, we have L_{eff} = 50 nm. This length is the actually pertinent geometrical scale if we suppose that the currents are uniformly distributed in the bridge, that is to say that L_{eff} is smaller than the penetration length. At low temperature ($T \ll T_c$) and in the dirty limit ($\ell \ll \xi_0$), the bulk penetration length is [5] :

$$\lambda \approx \lambda_L(0) \sqrt{\xi_0 / \ell}$$

with $\lambda_L(0)$ the London penetration length at absolute zero temperature, ξ_0 the B.C.S. coherence length and ℓ the elastic mean free path of the electrons. For aluminium,

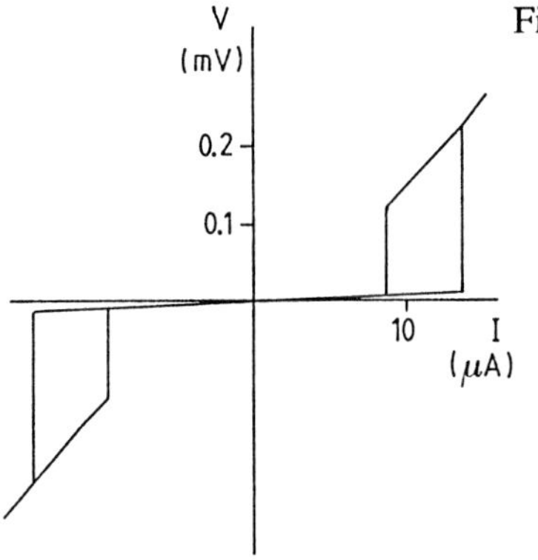

Figure 5 - V(I) of one bridge

typical values are [6] : $\lambda_L(0) = 16$ nm, $\xi_0 = 1.6$ μm, $\ell = 0.5$ μm. One finds $\lambda = 90$ nm. As $L_{eff} < \lambda$, the relevant penetration length for the microbridge is $\lambda_\perp = \lambda^2/L_{eff}$. In our case $\lambda_\perp = 160$ nm which is greater than L_{eff}. So the uniformly distributed current assumption is fulfilled.

Secondly, the size of the microbridge must be smaller than the coherence length $\xi(T)$ to exhibit the classic Josephson effect. In a dirty superconductor at low temperature we have the relation [5] :

$$\xi \approx \sqrt{\xi_0 \ell}.$$

One finds for aluminium $\xi = 300$ nm, which satisfies the required condition $L_{eff} < \xi$.

The typical values of the critical currents I_0 of our microbridges fluctuate between 10 μA and about 60 μA, depending on their width. In figure 5, we have recorded the characteristic of one of the narrowest one. The remaining ohmic slope at $I < I_0$ is due to the low impedance coaxial wire which connects the device to the current source through the dilution refrigerator. We measure in this case a normal resistance $R_N \approx 22\ \Omega$.

The hysteresis parameter

$$\alpha = \frac{I_{\min}}{I_0}$$

is about 0.6 which cannot be explained with the help of one of the Mc Cumber parameters [7] :

$$\beta_c = \frac{2\pi I_0 R_N^2 C_M}{\phi_0} \quad \text{and} \quad \beta_L = \frac{\phi_0}{2\pi L_M I_0},$$

where C_M and L_M are respectively the capacitance and the self-inductance of the microbridge. In our case we can evaluate C_M and L_M using [2] :

$$C_M = \frac{2\pi \varepsilon_r \varepsilon_0 a}{\ln(2\delta/\lambda_\perp)} \quad \text{and} \quad L_M = \frac{\mu_0 a}{2}\left(\ln\frac{4a}{\lambda_\perp} - 0.75\right),$$

where δ is the length of the bridge and a is the transverse distance between the bridge and the edges of the film. In figure 1, $\delta = 0.4$ µm and a = 0.2 µm, which gives $C_M \simeq 7\ 10^{-18}$F, $\beta_c \simeq 10^{-4}$ and $L_M \simeq 0.5$ pH, $\beta_L \simeq 67$ for $I_0 = 10$ µA.

So, the hysteresis is assumed to be dominated either by Joule self-heating [8] or by relaxation time effects [9]. We can also notice an excess current of about 0.3 I_0.

3.2. SQUID

The periodic critical current modulation of the DC-SQUID with the magnetic flux $\Delta I_c(\phi)$ is maximum when the modulation parameter β is unity, where :

$$\beta = \frac{2\,L\,I_0}{\phi_0}.$$

Peterson and Hamilton [10] have studied in details the relation between ΔI_c and β, but a rather good approximation is given by :

$$\frac{\Delta I_c}{2\,I_0} = \frac{1}{1 + \beta}.$$

In figure 6 we have recorded $I_c(\phi)$ for one of our SQUIDs whose inner surface is about 10 µm². The maximum of I_c gives 2 $I_0 \simeq 112$ µA and the modulation depth is fitted with $\beta = 0.8$, which is not very far from the optimum value. The self-inductance of the SQUID is found to be L $\simeq$ 15 pH. Another SQUID with a larger self-inductance L $\simeq$ 90 pH (for coupling considerations) is plotted on figure 7. As the critical currents are much lower the modulation parameter equals only two. The scale of the figure shows that, despite the very low value of I_0, the microbridges of the SQUID are only slightly asymmetric.

We have measured the flux equivalent noise amplitude with the help of the histogram of t_c (see the measurement method). If one supposes a Gaussian noise, its

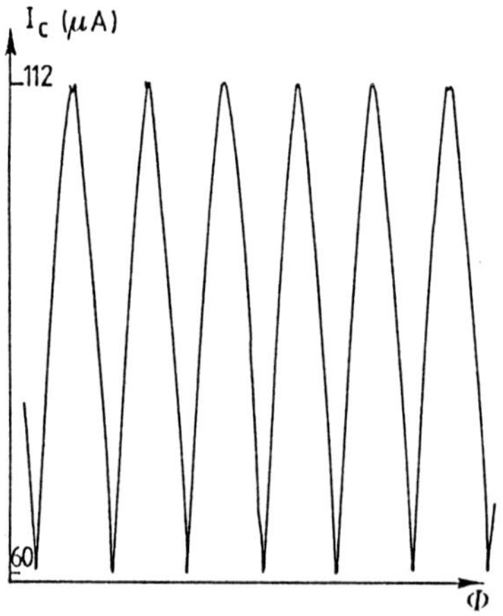

Figure 6 - Cricital current $\beta = 0.8$

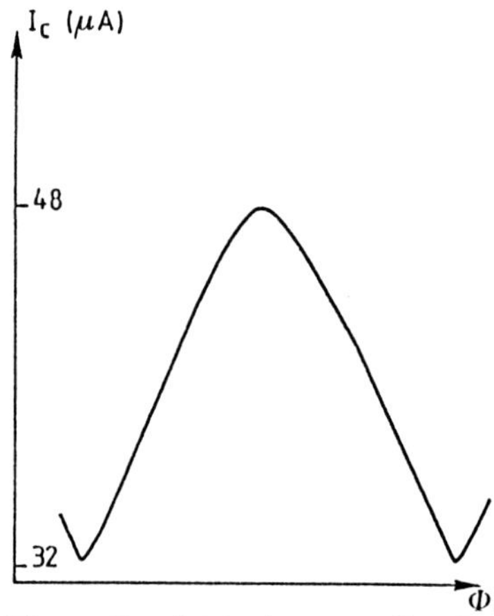

Figure 7 - Critical current $\beta = 2$

rms value is given by the width of the histogram at the half of its height. On figure 4 this noise is $\delta t_c = 19\ \mu s$. During this measurement the SQUID was polarized with an external field at a point of the $I_c(\phi)$ curve where the slope is maximum. Thus its sensitivity was of $3.1\ 10^{-5}\phi_0/\mu s$, which gives an equivalent flux noise of $6\ 10^{-4}\ \phi_0$. As the frequency measurement was about 5 kHz, we get finally a noise of $8\ 10^{-6}\ \phi_0/Hz^{1/2}$. This noise has actually been measured experimentaly and decreases like $\tau^{1/2}$ where τ is the accumulation time (see the measurement method). Theoretically, if one shunts such a SQUID, this would allow to use a DC method measurement with a weaker noise since the frequency limit is now given by the Josephson frequency.

4. Conclusion

We have demonstrated the capability of designing DC-SQUIDs with a one step electron beam lithographic operation. These SQUIDs allow us to perform magnetization measurements on samples whose typical size is one micron.

References

[1] Anderson P.W., and Dayem A.H., 1964, Phys. Rev. Lett. 13, 195.

[2] Lindelof P.E., 1981, Rep. Prog. Phys. 44, 949.

[3] Duret D., Bernard P. and Zenatti D., 1975, Rev. Sci. Inst. 46, 474.

[4] El-Khatib M., 1990, Ph.D Thesis, University of Grenoble (F).

[5] Burger J.P., 1975, in : "La supraconductivité des métaux, des alliages et des films minces", Masson et Cie Ed. (Paris).

[6] Romijn J., Klapwijk T.M., Renne MJ. and Mooij J.E., 1982, Phys. Rev. B26, 3648.

[7] Mc Cumber D.E., 1968, J. App. Phys. 39, 3113.

[8] Sckopol W.J., Beasley M.R., Tinkham M., 1974, J. Appl. Phys. 45, 4054.

[9] Song Y., 1976, J. Appl. Phys. 47, 2651.

[10] Peterson R.L. and Hamilton C.A., 1979, J. Appl. Phys. 50, 8135.

Accurate Determination of the Electrical Parameters of Thin Film SQUID Structures and Simulation of DC-SQUIDs with Coupling Coils

H. Uhlmann, H. Töpfer, F. Verwiebe, and J. Uhlig

Institute of Technology Ilmenau, PSF 327,
O-6300 Ilmenau, Fed. Rep. of Germany

Abstract. The modelling and simulation of superconducting circuits (e.g., SQUID structures) is a fundamental prerequisite for the elaboration of a design and dimensioning documentation. The suitable choice of the simulation model and the accuracy of the model parameters are of crucial importance for successfully carrying out design tasks by means of simulation methods.

After some general remarks on representing SQUID structures for simulation purposes and deriving electrical model parameters, computer programs for calculating the inductances and parasitic capacitances of three-dimensional superconducting structures are introduced in sections 2 and 3. As an example, in section 4 the simulation results for a dc-SQUID with coupling coil are described. Conclusions are drawn, providing hints for avoiding undesired resonances.

1. Introduction

For the purpose of simulation the structure under investigation is represented by an equivalent network consisting of lumped elements. In general, such elements are inductances or parasitic capacitances. The model network is described by a system of differential equations whose solution provides information about the dynamic behaviour of the system [5].

2. Inductance calculation

The investigation of real devices often requires a three-dimensional analysis of the electromagnetic field because of the small size of the structures. To calculate the inductance matrix L of three-dimensional superconducting structures the following method is used:

The inductance matrix L, consisting of L_{mag}, caused by the magnetic field, and the kinetic inductance L_{kin}, is calculated using a discretization algorithm. The current-carrying parts of the superconductor are discretized into prismatic elements with a line current flowing in the center of each element. The current distribution at the conductor surfaces is determined iteratively. The values of the internal currents result from appropriate relationships of the current decay into the superconductor volume (e.g., exponential decay in the London case). Each line current causes a magnetic induction B which can be determined using the Biot-Savart law.

Springer Proceedings in Physics, Vol. 64 **Superconducting Devices and Their Applications**
Editors: H. Koch, H. Lübbig

According to the required inductance matrix element, suitable planes are introduced on which the magnetic flux penetrating them has to be collected. After discretization of a plane, the magnetic flux density in the center of a discrete area element is derived by superposition of all line current contributions.The magnetic inductance and the kinetic inductance follow from:

$$L_{mag_{ij}} = \frac{\phi_i}{I_j} \Big|_{I_k=0,k\neq j} \quad ; \qquad L_{kin_j} = \frac{\mu_0}{I_j^2} \sum_{k=1}^{N} \lambda_j^2 J_{jk}^2 \cdot V_{jk} \ . \tag{1}$$

Thus, the required inductance L_{ij} of the inductance matrix L is

$$L_{ij} = L_{mag\,ij} + L_{kin_j} \cdot \delta_{ij} \ . \tag{2}$$

Based upon the method described above a computer program was developed to provide information about the inductance and the structure of the magnetic field in the circuit [1]. Furthermore, information about the current distribution can be derived. The substitution of the characteristic behaviour of current-decay inside the superconductor by an appropriate relationship enables the program to be applied to high-T_c superconducting circuits and devices as well.

3. Capacitance calculation

The parasitic capacitances are often inaccurately estimated by the analytical approximations available. Therefore a numerical solution is advantageous. The matrix of the capacitance coefficients C (size n × n) is frequently used to describe a network of capacitances with n nodes:

$$C \underline{\phi} = \underline{Q} \ ; \qquad C_{ij} = \frac{Q_i}{\phi_j} \Big|_{\phi_k=0,\, k\neq j} \ . \tag{4}$$

The Boundary Integral Equation Method is especially suitable for the numerical calculation of the capacitance coefficients, since the considered arrangements are fields with open boundaries and the permittivity is constant in finite areas. A system of integral equations for the charge density σ is needed [2]:

$$\phi(r) = \int_{A'} \sigma(r')\, G(r, r')\, dA' \ ; \quad r \in A_1 , \tag{5}$$

$$0 = \int_{A'} \frac{\partial G(r,r')}{\partial n} \sigma(r')\, dA' + \frac{\phi(r)}{2\varepsilon_0} \frac{1+k}{1-k} \ ; \qquad r \in A_2 ; \quad k = \frac{\varepsilon_a}{\varepsilon_b} \Big|_r , \tag{6}$$

($A = A_1 \cup A_2$; A_1: metallic surfaces; A_2: dielelectric boundaries). For the calculation of the capacitance coefficients it is necessary to define the nodes of the electrical network. Each boundary element p which is located on a metallic surface is connected with one of the nodes of the electrical network. Applying the boundary integration the elements of the matrix (4) will be given by:

$$C_{ij} = \frac{1}{\phi_j} \sum_{p \in i} \int_{A_p} \sigma(r)\, dA \; \Big|_{\phi_k = 0; k \neq j} \quad . \tag{7}$$

The computer program BIEM3D is based upon a collocation discretization of the system of integral equations (5), (6) and can be used for the calculation of capacitances. If there are dielectric 3d-halfspace and sandwich structures, the boundaries of permittivity are considered by modifications of the Green's function. Concerning sandwich structures this leads to a series statement, resulting from continuously applying the method of images [3]. The surfaces of the boundaries are discretized into rectangles. Constant and bilinear statements are available for the charge density. The elements of the matrix of coefficients are calculated by analytical integration. This integration can be connected with a distance dependent accuracy control. The factorization of the linear equations permits an effective solution of multiconductor problems, since (n-1) solutions of the system of equations are necessary because of Eq.(4).

4. Simulation of dc SQUID characteristics $\overline{U_{SQ}} = f(\phi/\phi_o)$

Multi-loop SQUIDs are often used. In this case, parasitic capacitances result from the overlap of striplines. In addition, the planar coupling coil arranged over the structure may detoriate the SQUID performance. These parasitic effects may suppress the SQUID signal completely. For investigating these effects an effective computer program for calculating the U-ϕ characteristics of SQUIDs has been developed.

For each point of the characteristic it is necessary to simulate the time behaviour of the voltage U_{SQ} of the autonomous network representing the SQUID and to determine the steady state. Subsequently, the average value of the voltage U_{SQ} is calculated over a period of the steady state. For determining the steady-state the "next-amplitude" method will be applied [4].

An equivalent network including the RSCJ-model taking into account the coupling coil with an inductance L_S and the parasitic capacitance C_S served as a basis for simulation [5]. Furthermore, an external damping resistance R_S has also been provided in the model. The capacitance C_p takes into consideration the parasitic capacitance of the SQUID structure. This model leads to a system of nonlinear differential equations of 6th order [5].

If the parasitic capacitance in the SQUID ring is taken into account, there will be regions without voltage modulations. For very large parasitic capacitances, however, this undesired region will be beyond the region of the operating current. This behaviour has to be attributed to resonance phenomena occuring in the SQUID ring. These phenomena are caused by the Josephson alternating current. The junction capacitances may have a similar influence, too. By inserting a damping resistance into the model under consideration, it is possible to damp these re-

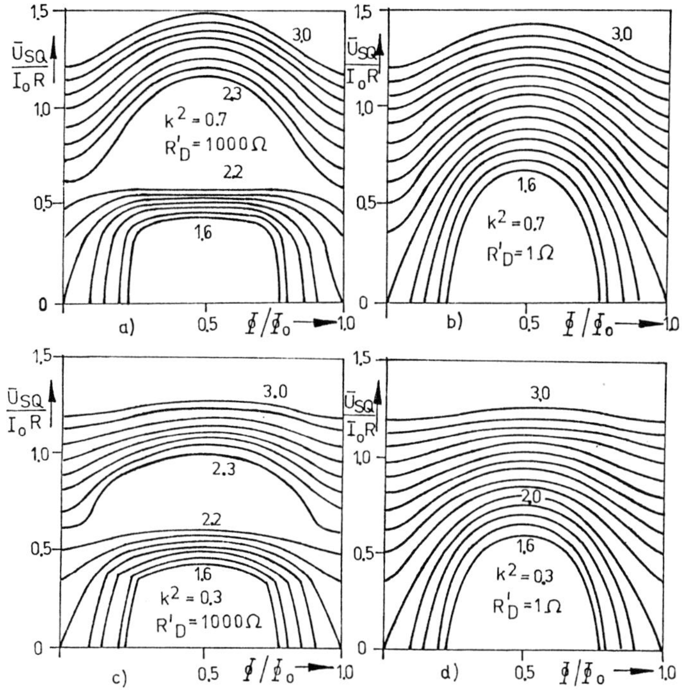

Fig. 1 Influence of the parasitic capacitance and of the coupling factor on the U-ϕ characteristic (parameter: $C_S/ü^2$ = 3 pF)

sonances. Including the coupling circuit, we can investigate the influence of resonances in the coupling circuit with respect to their dependence of the coupling factor k (or coupling ratio ü), (Fig. 1).

The simulation studies for investigating the influence of the parasitic elements showed that there are two fundamental possibilities for preventing undesired resonances:

1. inserting a resistance R_S in the coupling circuit for damping resonances, or
2. inserting a capacity C_S in the coupling circuit in order to remove the resonances from the desired operating range.

Both possibilities are applicable without necessitating any changes to the layout of the SQUID structure. Extensive simulations performed with the above-mentioned model of the 6th order confirm our results also for technically realizable coupling factors ranging from 0.7 to 0.9.

Acknowledgement

Part of this work is supported by the Bundesministerium für Forschung und Technologie (contract-no. 13N5920).

References

[1] Töpfer, H., Uhlmann, H.: Zur Bestimmung der Induktivität supraleitender Dünnschichtstrukturen. Supraleitung und Tieftemperaturtechnik. (VDI-Verlag GmbH Düsseldorf 1991) pp. 498-499.

[2] Rucker, M.: Anwendung der Boundary-Elemente-Methode zur numerischen Berechnung elektromagnetischer Felder. Thesis, TU Graz (1987).

[3] Ning, Z.,Q., Dewilde, P. M.: SPIDER: capacitance modelling for VLSI interconnections. IEEE-Trans. CAD-7 (1988) 12, pp. 1221-1228.

[4] Uhlmann, H., Uhlig, J., Tanzke, S.: Ein Verfahren zur rechnergestützten Bestimmung periodischer Lösungen von nichtlinearen autonomen Systemen. Wiss. Z. TH Ilmenau 34 (1988) 4, pp. 159 -166.

[5] Uhlig, J.: Ein Beitrag zur dynamischen Analyse von SQUIDs mit zwei Josephson-Tunnelkontakten. Thesis, TH Ilmenau (1989).

RF Properties of a DC-SQUID Coupled to a Multiturn Input Coil

K. Enpuku, T. Tanaka, and K. Yoshida

Department of Electronics, Kyushu University 36, Fukuoka 812, Japan

Abstract. In order to study the characteristics of a dc SQUID coupled to a multiturn input coil, rf properties of the coupling circuit between the SQUID and the input coil have been investigated by using an expanded model of the circuit. We observe resonant structures in the rf properties of the coupling circuit, which means that the SQUID coil in the presence the input coil can not be expressed by a simple inductance, in contrast to the isolated SQUID. The resonant structures can be suppressed with damping resistances. A model for the coupling circuit is also obtained.

1. Introduction

It has been pointed out that the characteristics of a SQUID with a multiturn input coil are degraded considerably, compared with the isolated SQUID [1-4]. Experimental studies have shown that the damping resistances are useful to prevent the degradation of the SQUID characteristics [3,4]. However, there have been few studies on the mechanism causing the degradation of the SQUID characteristics [3,5]. It is important to investigate the mechanism for the optimum operation of the SQUID coupled to an input coil.

In order to study the mechanism, we measure rf properties of the coupling circuit between the SQUID coil and the input coil. The rf properties of the circuit are important, since we can expect that rf currents generated from the Josephson junctions affect the SQUID characteristics through rf properties of the coupling circuit. In the experiment, we use an expanded model of the circuit with Cu electrodes [5,6]. Using the obtained results, we also propose a circuit model of the coupling circuit.

2. Experimental Results

In Fig. 1 (a), the SQUID coupled to the multiturn input coil is shown schematically. In order to study rf properties of the coupling circuit, we measure the impedance Z_{AB} as shown in Fig. 1 (a). The impedance Z_{AB} is important, since it dominates the characteristics of the SQUID. In the measurement, we adopt the so-called expanded model of the circuit [5,6]. The geometric size of each element of the circuit is enlarged by a factor of F = 500, compared with the original circuit. Rf

Springer Proceedings in Physics, Vol. 64 Superconducting Devices and Their Applications
Editors: H. Koch, H. Lübbig

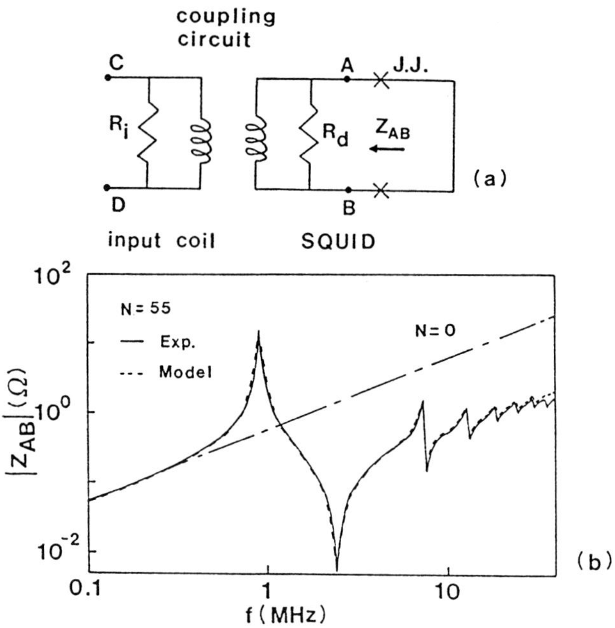

Fig. 1. (a) Schematic figure of the SQUID coupled to the multiturn input coil. (b) The solid line shows the impedance Z_{AB} measured from the expanded model with an expansion factor F = 500. The frequency range is reduced to 1/F = 1/500 in the expanded model, compared with the original circuit. The broken line is calculated from the circuit model of the coupling circuit shown in Fig. 3.

properties of the expanded model are measured up to f = 40 MHz, corresponding to f = 20 GHz in the practical SQUID. In Fig. 1 (b), the experimental result of Z_{AB} is shown when the SQUID is coupled to the input coil. The number of turns N of the input coil is N = 55, the line and space of the input coil are 1.0 mm, and the hole size of the SQUID coil is 5 cm; resistances R_d and R_i in Fig. 1 (a) are not connected in this case. For reference, we also show the impedance of the isolated SQUID, i.e., for N = 0.

As shown in Fig. 1 (b), the impedance can be expressed as $Z_{AB} = j2\pi fL_S$ for the case of the isolated SQUID, where L_S is the inductance of the SQUID coil. On the other hand, the impedance Z_{AB} changes significantly when the input coil is coupled to the SQUID. Resonant structures appear in Z_{AB} as shown in Fig. 1 (b), and the SQUID coil can not be expressed by a simple inductance. We can expect that the resonant structures in the impedance Z_{AB} cause the significant change in the SQUID characteristics. It must be mentioned that the resonant structures arise from the standing wave occurring between the input coil and the SQUID coil [5,6].

Next, we study the effect of the damping resistances R_d and R_i on the rf properties of the coupling circuit. In Fig. 2 (a), the impedance Z_{AB} is shown when the resistance R_d = 3 Ω is connected to the SQUID coil. As shown, the resonant

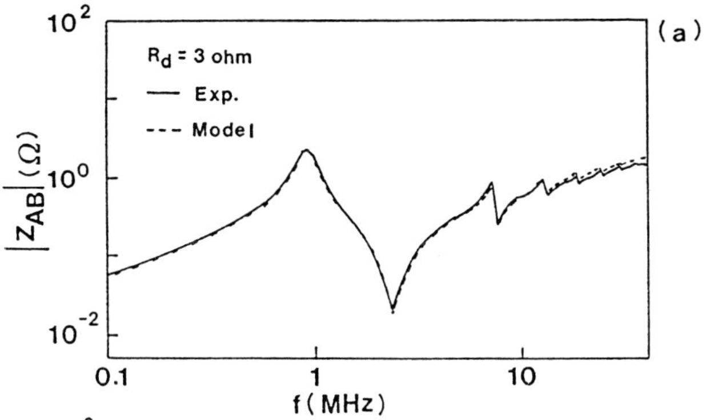

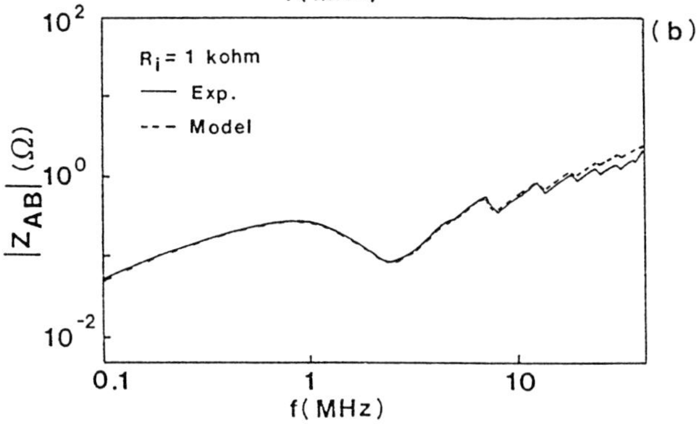

Fig. 2. Effect of the damping resistance on the impedance Z_{AB}. (a) $R_d = 3\ \Omega$, and (b) $R_i = 1\ k\Omega$. The resonant structures in Z_{AB} are suppressed with the damping resistances.

structures in Z_{AB} are suppressed by the damping resistor. The same result is also obtained when the resistance $R_i = 1\ k\Omega$ is connected to the input coil. As shown above, the damping resistances R_d and R_i suppress the resonant structures in Z_{AB}. This improvement of Z_{AB} can be expected to result in the improvement of the SQUID characteristics, since the characteristics of the SQUID are dominated by Z_{AB}. Therefore, the above results may explain the previous experimental results [3,4] that the characteristics of the SQUID coupled to the input coil are improved by the damping resistor.

3. Model of the Coupling Circuit

Using the experimental results, we propose a circuit model of the coupling circuit as shown in Fig. 3. The input coil is represented by a set of parallel resonant circuits, which is obtained by taking account the occurrence of the standing wave in the coupling circuit [5]. The resonant frequency of each resonant circuit is given by the frequency where the standing wave occurs, and corresponds to the frequency

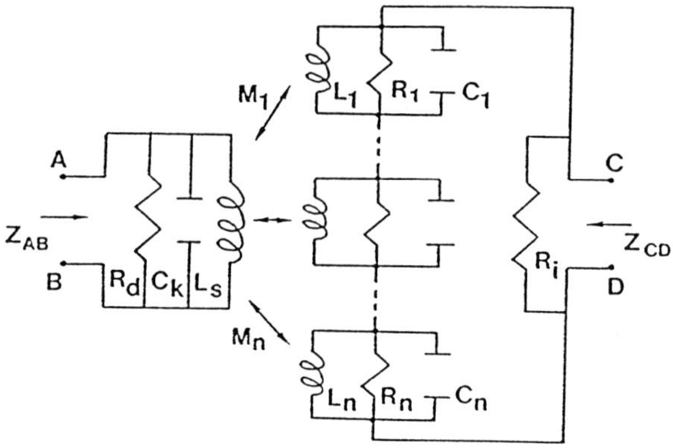

Fig. 3. A model of the coupling circuit. The multiturn input coil is expressed by a set of parallel resonant circuits. The SQUID coil couples to each resonant circuit with a mutual inductance.

where the impedance Z_{AB} has peak values. Each resonant circuit couples to the SQUID coil through a mutual inductance M_n. Details of the derivation of the circuit parameters will be mentioned elsewhere [5,6].

In Figs. 1 and 2, the impedances Z_{AB} calculated from the circuit model are shown by broken lines. As shown, the present circuit model explains well the experimental results of Z_{AB}. These agreements show the validity of the present model for the coupling circuit between the SQUID coil and the multiturn input coil.

4. Conclusion

Rf properties of the coupling circuit are studied by using the expanded model of the circuit. It is shown that the SQUID coil can not be expressed by a simple inductance due to the resonant structures in Z_{AB} caused by the input coil. The damping resistances are useful to suppress the resonant structures. The circuit model for the coupling circuit explains well the experimental results.

References

[1] V. J. de Waal, T. M. Klapwijk and P. van den Hamer, J. Low Temp. Phys. **53**, 287 (1983).
[2] C. Hilbert and J. Clarke, J. Low Temp. Phys. **61**, 237 (1985).
[3] J. Knuutila, A. Ahonen and C. Tesche, J. Low Temp. Phys. **68**, 269 (1987).
[4] V. Foglietti, W. J. Gallagher, M. B. Ketchen, A. W. Kleinsasser, R. H. Koch and R. L. Standstrom, Appl. Phys. Lett. **55**, 1451 (1989).
[5] K. Enpuku and K. Yoshida, to be pulished in J. Appl. Phys. (1991).
[6] K. Enpuku, T. Tanaka and K. Yoshida, to be published in IEICE of Japan, Trans. on Electronics (1991).

Resonances in All-Niobium DC-SQUIDs

G.M. Daalmans[1], *L. Bär*[1], *D. Uhl*[1], *F. Bömmel*[1], *R. Kress*[2], *and L. Warzemann*[3]

[1]Siemens Central Research Laboratories, W-8520 Erlangen, Fed. Rep. of Germany
[2]Siemens Medical Engineering Group, W-8520 Erlangen, Fed. Rep. of Germany
[3]Friedrich-Schiller-University Jena, O-6900 Jena, Fed. Rep. of Germany

Abstract. The influence of LC-Resonances on the I-V-curves and the noise of DC-SQUIDs has been studied on special purpose SQUID-chips. Every chip (5x10 mm) is provided with 10 Ketchen-SQUIDs of different type. A common modulation coil is integrated in order to operate the SQUIDs in a flux locked loop. The SQUIDs are all-niobium devices with either Al_2O_3-barriers or amorphous silicon barriers. The resonances are damped with gold resistors.

The resonances on the I-V-curves have been identified and the noise has been measured as a function of the bias current. The white noise component as well as the coloured noise component depend on the bias current and are images of the resonances. The absolute values of the coloured noise component depend strongly on the technology and are the best for Al_2O_3-barriers.

1. Introduction

The successful operation of DC-SQUIDs started with the introduction of a matching network between the SQUID and the amplifier and the flux locked loop operation of the SQUIDs by J.Clarke [1]. At that time the SQUIDs had flux to voltage transfer functions $\partial V/\partial \varnothing$ of just a few μV per flux quantum $\varnothing_0$. The SQUIDs were operated at very small voltages and the capacitances of the Josephson junctions did not play an important role. The SQUIDs were very well characterized by the so-called RSJ-model [1]. With the improvement of the preparation techniques [2] of Josephson junctions, SQUIDs with much higher I_CR-products became available [3], showing non RSJ-behaviour. The noise of those devices, especially at the lowest frequencies, could not be explained in terms of the RSJ-model and Niquist noise of the shunt resistors. Different groups [4] have simulated DC-SQUIDs taking into account the self-capacitance, C_{JJ}, of the Josephson Junctions. For the white noise level excellent agreement between the simulations and the measurements could be obtained [5]. For the lower frequencies the computer simulations are very time consuming and unable to resolve in the practical frequency range of SQUID-sensors.

A good solution to the problem is to study primitive SQUID devices [5] with well defined variations in one of the SQUID-parameters. In this study we varied the parasitic capacitance due to the coverplate of Ketchen-type DC-SQUIDs and studied the influence on the I-V-curves and the noise properties.

2. Layout and preparation techniques

Two types of testchips were designed and prepared, labeled as Type A and Type B (Figure 1). Both chips are 10 mm by 5 mm and contain 10 SQUIDs. The SQUIDs are of Ketchen-type with coupling holes of 100 µm by 100 µm. The slitted SQUID-washer has dimensions of 600 µm by 600 µm. The slit width was 10 µm. The Josephson junctions are squares of 4 µm by 4 µm and are positioned at the end of the slit. The slit is covered by a superconducting coverplate isolated from the washer by means of 200 nm of isolating material. The width of the coverplate is varied over a chip by a factor of 16. Every coverplate width is present twice on a chip, resulting in 5 different capacitors per chip. Every SQUID is provided with a superconducting screening ring [6] and a one turn modulation coil. A lumped circuit model for both designs is shown in Figure 2.

The testchip of Type B differs from the testchip of Type A in the Josephson junction region only as shown in Figure 1. In chip B the slit in the SQUID-washer is nearly completely covered in order to minimize the inductance L_S. Moreover the resistors are located on top of the superconductor in order to minimize the inductance L_R. For control measurements chip B was also realized without coverplate.

The chips have been prepared with a four level technology. The two superconducting levels are made of niobium and typically 100 nm to 200 nm thick. The insulation level is from anodised niobium backed with SiO. The Josephson junctions were prepared with a SNAP-process [2]. The resistive level was made from e-beam evaporated gold. The barrier material was either amorphous silicon or thermally oxidized aluminium. The deposition processes in the amorphous silicon devices were done with e-beam evaporation in ultra high vacuum. The Al_2O_3-devices were prepared with e-beam evaporation in ultra high vacuum as well as with sputtering in a high vacuum system.

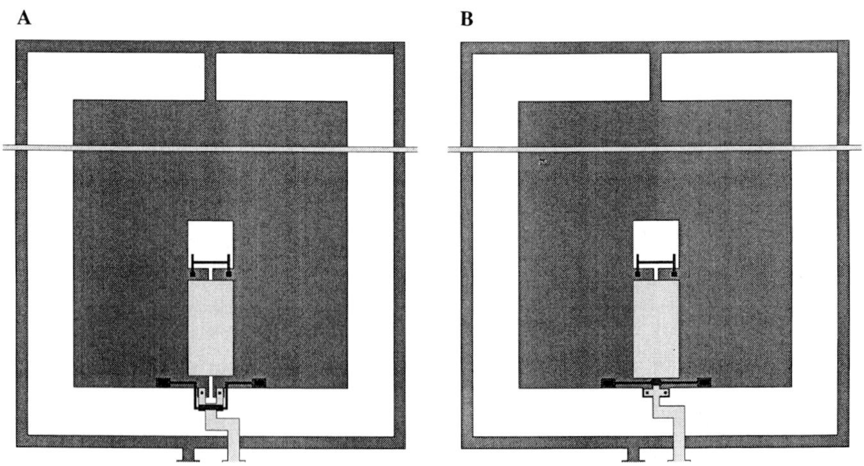

Figure 1 Layout of the testchips.

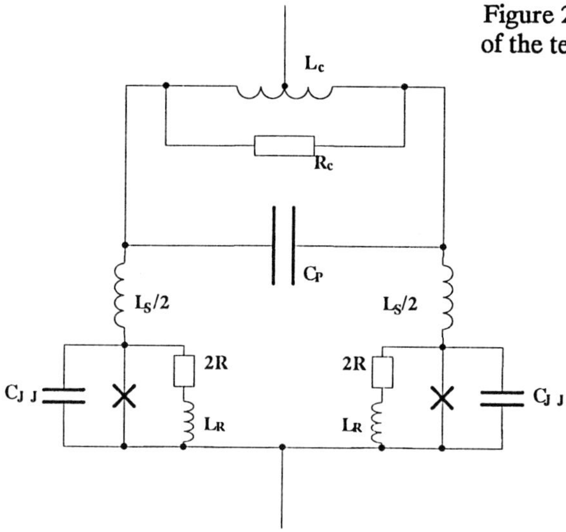

Figure 2 Lumped circuit model of the testchips.

The structuring was performed either with lift-off or with reactive ion etching in a CF_4-plasma. The photoresist applied was AZ 5218 used in the negative mode (lift-off) as well as in the positive mode (RIE).

3. Measurements

The electrical measurements were performed in a Faraday cage. The noise measurements were performed on SQUIDs driven in the flux locked loop mode with flux modulation at 170 kHz [1]. The SQUID-output was analysed with a fast fourier transformer.

The general character of the I-V-curves was nearly independent of the dimensions of the coverplate. The influence of the coverplate became visible at voltages below 20 μV as broad bumps in the I-V-curves.

We measured the equivalent white flux noise as a function of the bias current in order to study the influence of the resonances. A typical example, SQUID 5 of a Type A chip, is shown in Figure 3. For bias currents below the first cross-over we found a noise minimum where the transfer function is not maximum. In the region between the first and the second cross-over we obtained about the same noise level as below the first crossover at about the same value of the transfer function. The minimum, however, was much broader. Beyond the second cross-over the noise level was slightly higher as in the first two minima.

In order to study the influence of the coverplate capacitance on the SQUID performance we focused on the low voltage region, where the L_cC_p-resonance is present. We looked for the optimized equivalent flux noise for bias currents near the critical current. The measured equivalent flux noise was multiplied with the measured transfer function $\partial V/\partial \varnothing$ in order to obtain the

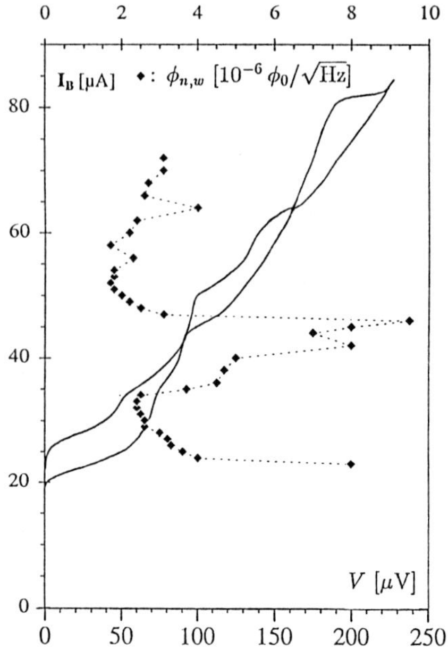

Figure 3 White equivalent flux noise of SQUID 5 with an amorphous Silicon-barrier.

noise voltage $(V_{n,th}^2)^{1/2}$. The noise voltage was scaled with the theoretically predicted value $(V_{n,th}^2)^{1/2}$ from Likharev's formula [7] valid for the RSJ-model. Four chips of Type A, three with an α–silicon barrier and one with an Al_2O_3-barrier, with in total 40 SQUIDs with different C_p were involved. The $ß_L$-parameter varied between 1.7 and 3.0 and the $ß_c$-parameter was between 0.1 and 0.6. We found a clear dependence of the scaled voltage noise from the transfer function ∂V/∂∅ as shown in Figure 4. It should be stressed here that devices with a certain coverplate capacitance are randomly distributed in this plot. So there is not a better fit to the RSJ-model for SQUIDs with small capacitances. From Figure 4 we conclude that for transfer function values below 100 $\mu V/\varnothing_0$ there is a good fit to the RSJ-model where as for values above 100 $\mu V/\varnothing_0$ application of the RSJ-model seems not to be very useful.

In order to study the influence of the coverplate on the frequency dependent noise we separated the frequency dependent noise from the total noise at the optimum biaspoint near I_C. Four chips of Type A with α–silicon barriers were involved together with one chip of Type B also with an α–silicon barrier. As in the case of the white noise we could not make a significant correlation between the noise level and the parasitic capacitance C_P. On the other hand a strong correlation was found between the coloured noise level and ∂V/∂∅ . As is shown in Figure 5 the coloured flux noise increases with the transfer function in a similar way as the white noise does.

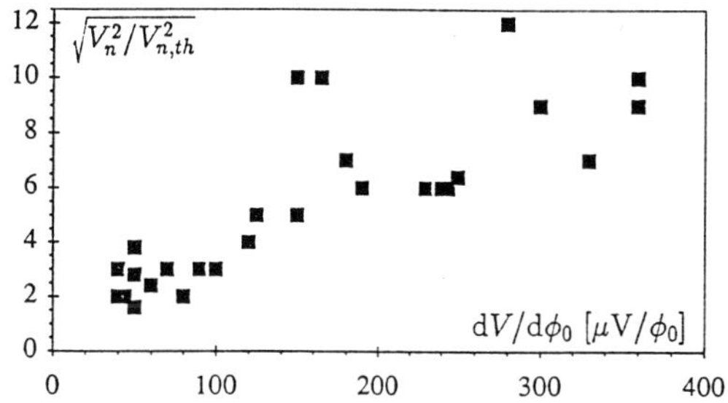

Figure 4 Normalized voltage noise of Type A SQUIDs with different coverplates as a function of ∂V/∂∅.

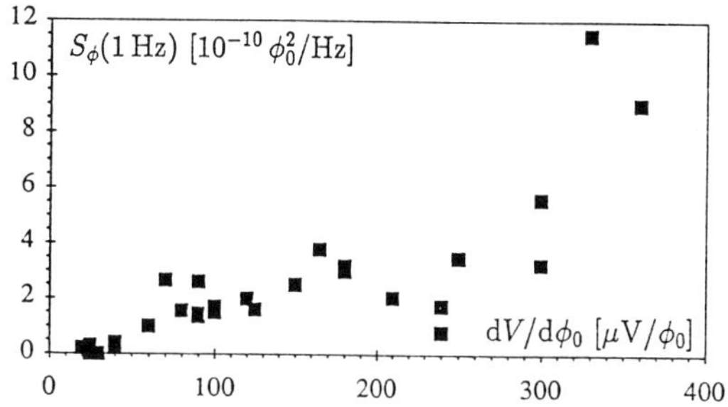

Figure 5 Coloured flux noise at 1 Hz of Type A and B SQUIDs with different coverplates as a function of ∂V/∂∅.

9. Conclusions

We could not find any correlation between the optimized white noise level at a bias current of about I_C and the value of the coverplate capacitance. This is in contrast to the results obtained by Ryhänen et al. [5]. An explanation could be that we were unable to resolve this dependence due to the limited bandwidth of the system. Normally we took the noise value at 500 Hz as the white noise level. May be this is not correct. Another explanation could be the large inductance of the devices and as a consequence the higher sensitivity for external noise. Also the different measurement techniques could be of influence.

References

[1] J. Clarke, W.M. Goubau, M.B. Ketchen, J. Low. Temp. Physics **25** (1/2) 99-144, 1976.

[2] H. Kroger, L. Smith, D. Jillie, Applied Physics Let. **39**, 280, 1981.

[3] R.F. Voss, R.B. Laibowitz, M.B. Ketchen, A.N. Broers, Proceedings SQUID '80, p. 365, Walter de Gruyter, Berlin, New York, 1980.

[4] V.J. de Waal, P. Schrijner, R. Llurba, J. Low. Temp. Physics. **54** (3/4), 215, 1984.
D. Drung, W. Jutzi, IEEE Trans. on Magn. **MAG-21** (2), p. 430, 1985.
[5] T. Ryhänen, R. Cantor, D. Drung, H. Koch, H. Seppä, IEEE Trans. on Magn. **MAG-27**, (2), p. 3013, 1991.
[6] G.M. Daalmans, H. Seifert, F. Bömmel, H.E. Hoenig, Extended Abstracts, 6th Int. Workshop on Future Electron Devices Conference, Hokkaido, June 7-9 1989, p. 219.
[7] A.N. Vystavkin, V.N. Gubankov, L.S. Kuzmin, K.K. Likharev, V.V. Migulin, V.K. Semenov, Rev. Phys. Appl. **9**, 79, 1974.

Noise in All-Niobium DC-SQUIDs

G.M. Daalmans[1], *L. Bär*[1], *F. Bömmel*[1], *D. Uhl*[1], *and R. Kress*[2]

[1]Siemens Central Research Laboratories, W-8520 Erlangen, Fed. Rep. of Germany

[2]Siemens Medical Engineering Group, W-8520 Erlangen, Fed. Rep. of Germany

Abstract. The equivalent flux noise of DC-SQUIDs fabricated with different technologies has been measured. The white noise is fitted very well with a thermal noise model. The coloured noise depends strongly on the preparation technique and the presence of a coupling coil. For Al_2O_3-barriers, for instance, we find an extremely low coloured noise level of typically $1\text{x}10^{-6}Ø_o/(\text{Hz})^{1/2}$ at 1 Hz for autonomous 100 pH-SQUIDs and about $3\text{x}10^{-6}Ø_o/(\text{Hz})^{1/2}$ at 1 Hz for SQUIDs with a coupling coil. For amorphous silicon barriers on the other hand these values are about one order of magnitude higher. The different behaviour of SQUIDs with amorphous silicon barriers compared to SQUIDs with Al_2O_3-barriers will be discussed.

1. Introduction

The noise voltage at the SQUID output in the white noise regime is determined by the shunt resistors of the Josephson junctions [1] and the internal dynamics of the SQUIDs governed by the hysteresis parameter $ß_C$ [2]. Unfortunately in most of the devices the noise level is not thermally limited, especially in the low frequency range. This frequency dependent excess noise normally has different origins. For instance LC- or microwave resonances [3] can influence the low frequency noise of the SQUID. Another source of low frequency noise is leakage current due to localized states in or at the barrier [4]. Also pinned flux in the SQUID-loop can contribute to frequency dependent noise if the pinning forces are small or the external electromagnetic noise is high. In this paper we have studied the noise properties of niobium DC-SQUIDs and tried to separate the influence of the different noise sources.

2. Layout and preparation

The SQUIDs were of Ketchen-type, integrated with 10 on a chip [5]. The SQUID inductance was 100 pH and every SQUID had one 50 turn input coil integrated on top of the SQUID-washer together with a one turn modulation coil. The Josephson junctions were made with a SNAP-Process [6] and they are 2 µm squares. The critical current of the devices was adapted to the SQUID-inductance in order to get an inductance parameter $ß_L = I_cL/Ø_o \leq 1$. The

Springer Proceedings in Physics, Vol. 64 Superconducting Devices and Their Applications
Editors: H. Koch, H. Lübbig

shunt-resistors for the Josephon junctions are chosen that way that hysteresis is avoided in the I-V-curves. Mostly we operated with McCumber-parameter values $\beta_c = 2\pi I_c R^2 C / \varnothing_o$ below 0.5.

The SQUIDs were made on silicon wafers in an eight level process [7]. The devices with amorphous silicon barriers were prepared with e-beam evaporation in ultra high vacuum and structured via lift-off. The devices with Al_2O_3-barriers where prepared with sputtering in high vacuum as well as with e-beam evaporation in ultra high vacuum.

3. Noise

We tried to separate the influence of the different noise sources on the noise properties of the devices. By studying the noise as a function of the bias current the influence of the resonances, localized at certain voltages, becomes evident.

In order to see the influence of pinned flux in the superconducting structures we trapped a field of about 10 Gauß in the SQUID-structure and studied the influence on the noise.

By applying an ac-current bias [8] we got information about the noise in the Josephson junctions themselves.

In the following we will present the results of these different experiments.

3.1 Noise as a function of the bias current

First of all we studied the influence of the input coil on the noise characteristics of the devices by comparing the results of SQUIDs with loaded input coils with those without loading. The loading was applied in different ways. For instance we loaded with resistances, capacitances and linear combinations of them. All types of loading circuitry gave more or less the same results. A typical result is shown in Figure 1a for a device with an amorphous silicon barrier. In the figure the white noise is separated from the coloured noise and behaves as one would expect from the RSJ-model. If we remove the load from the input coil we get the result shown in Figure 1b. As is evident from both curves there is a dramatic increase of the noise level due to the microwave resonance of the input coil.

Going to higher bias current we pass the LC-resonance of the inductance of the Josephson junction region and the junction capacitance at about 50 µV. At the resonance we found a reduction of the noise level, not only in these decives but also in different designed SQUIDs. For SQUIDs with an Al_2O_3-barrier these reduction is small in the white noise level as well as in the coloured noise level (Figure 2). For SQUIDs with an amorphous silicon barrier the reduction in the coloured noise level is much more important and can be one order of magnitude.

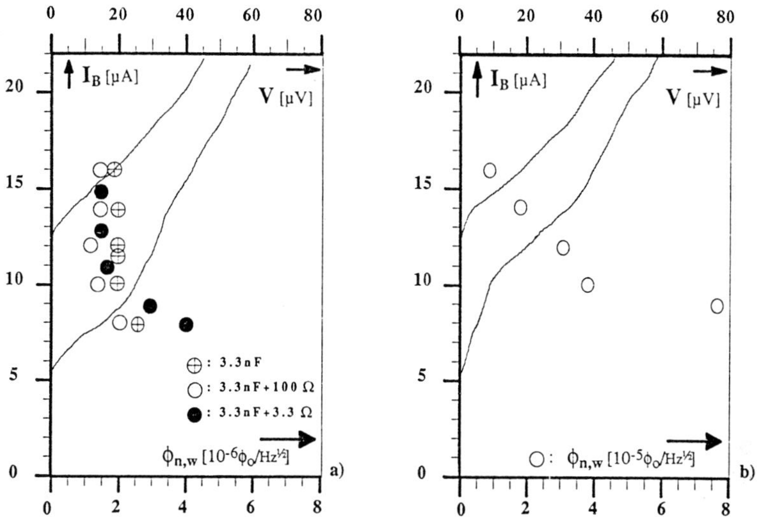

Figure 1 White equivalent flux noise as a function of the bias current a) with different loads at the input coil, b) without loading.

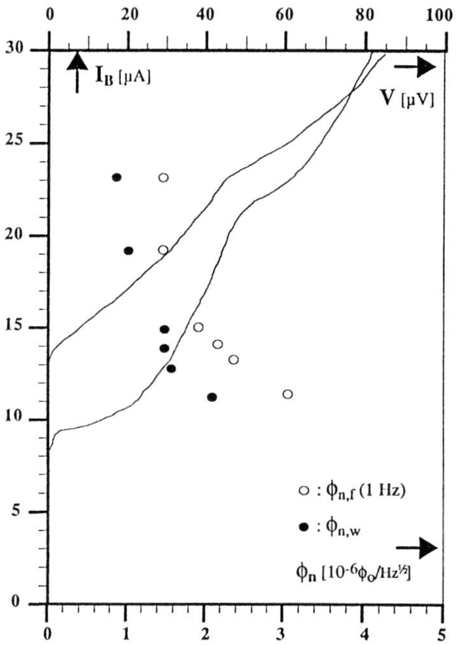

Figure 2 Equivalent flux noise of an Al_2O_3-SQUID as a function of the bias current.

3.2 Noise as function of trapped flux

We applied a field of about 10 Gauß to the SQUID-sensors above the critical temperature. With the field present we cooled down at 4 K. Then we studied the noise properties of the devices and compared with the results with no flux applied. For SQUIDs with Al_2O_3-barriers we found an increase in the white noise level as well as in the low frequency noise level of about one to two times 10^{-6} $Ø_0/(Hz)^{1/2}$. For SQUIDs with amorphous silicon barriers we could not detect any influence as long as the low frequency noise level was below $1x10^{-5}$ $Ø_0/(Hz)^{1/2}$ at 10 Hz.

3.3 Noise with ac-current bias

We applied an ac-bias modulation scheme [8] to the SQUIDs in order to reduce the influence of low frequency conductance fluctuations in the Josephson junctions. For the Al_2O_3-barrier SQUIDs we found the usual increase in the white noise level due to the reduced duty cycle of the modulation method. At the lower frequencies we could not observe a reduction of the coloured noise indicating an excellent barrier quality.

For devices with an amorphous silicon barrier on the other hand we observed a great reduction of the low frequency noise level as is shown in Figure 3.

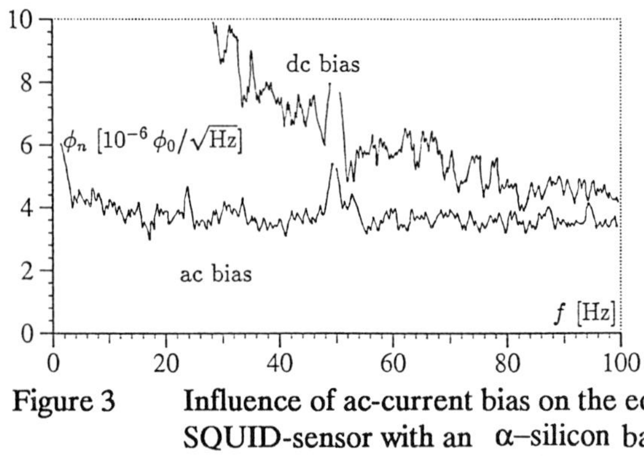

Figure 3 Influence of ac-current bias on the equivalent flux noise of a SQUID-sensor with an α–silicon barrier.

4. Conclusions

The input coil of niobium SQUIDs reduces the SQUID-signal and increases the noise level. The influence of the input coil can be reduced by proper loading. Depinning of trapped flux does not play an important role in our devices. Devices with an amorphous silicon barrier do have excess noise compared to devices with an Al_2O_3-barrier. This excess noise is related to the transport properties of the amorphous silicon barriers.

References

[1] J. Clarke, W.M. Goubau, M.B. Ketchen: Journal Low. Temp. Phys. **25** (1/2), p. 99, 1976.

[2] V.J.de Waal, P. Schrijner, R. Llurba: Journal Low. Temp. Phys. **54** (3/4), p. 215, 1984
D. Drung, W. Jutzi: IEEE Trans. on Magn. **MAG-21**, p. 430, 1985.

[3] J. Knuutila, A. Ahonen, C. Tesche: Journal Low. Temp. Phys. **68**, p. 269, 1987.

[4] C.T. Rogers, R.A. Buhrman: Phys. Rev. Lett. **53**, p. 1272, 1984.

[5] G.M. Daalmans, H. Seifert, F. Bömmel, H.E. Hoenig: Extended abstractsd 6th Int. Workshop on FED., p. 219, Hokkaido, Japan, June 7-9, 1989.

[6] H. Kroger, L. Smith, D. Jillie: Appl. Phys. Lett. **39**, p. 280, 1981

[7] G.M. Daalmans, L. Bär, F.R. Bömmel, R. Kreß, D. Uhl: IEEE Trans. on Magn. **MAG-27** (2) 1991, p. 2997

[8] V. Foglietti, W.J. Gallagher, R.H. Koch: IEEE Trans. on Magn. **MAG-23**, p. 1150, 1987.

Noise Measurements on DC-SQUIDs with Varied Design*,†

M.R. Condron II[1], *G.M. Gutt*[1], *B. Muhlfelder*[1], *J.M. Lockhart*[1,#], *J.P. Turneaure*[1], *M.E. Huber*[2], *M.W. Cromar*[2], *and E.K. Houseman*[2]

[1]Hansen Experimental Physics Laboratory, Stanford University, Stanford, CA 94305, USA

[2]National Institute of Standards and Technology, 325 Broadway, Boulder, CO 80303, USA

[#]Also with Physics and Astronomy Dept., San Francisco State University, San Francisco, CA, USA

Abstract. We have fabricated and tested two types of SQUIDs. The first, a low inductance stripline device, was characterized using a commercial dc SQUID as a following amplifier. The second type was well coupled to a 0.16μH input coil and was characterized and flux-locked using an ultralow noise room temperature preamplifier. We present device characterization and noise data for both types of devices.

1. Introduction

An important application for SQUID magnetometers is the detection of very small signals from inductive receivers. In our application, the relativity gyroscope experiment [1], a dc SQUID must efficiently couple to a 1μH source inductance and must have low noise at a signal frequency of 1.6mHz. The well coupled, single transformer, washer type SQUIDs (WCS) we have studied here often display undesirable parasitic LC resonances in their characteristics. Several approaches aim to reduce the effects of these resonant structures through the use of damping [2,3,4,5]. A quantitative study of the effect of parasitic capacitance on dc SQUID performance was made by Ryhänen, *et al.* [6]. We investigated several damping configurations and found that the addition of a resistor between the SQUID's input coil and SQUID loop adequately controls the most significant resonances. SQUIDs often exhibit excess low frequency noise whose power spectral density scales as $f^{-\alpha}$ where α is commonly near unity. For the gyroscope experiment's 1.6mHz signal, the SQUID's $f^{-\alpha}$ noise far exceeds the white noise. To better understand the low frequency noise, we have studied both WCS devices of the design as described above and weakly coupled, stripline SQUID (SLS) devices, previously characterized by Huber and Cromar, with $f^{-\alpha}$ noise extending to relatively high frequencies [7].

---

* Supported by NASA under contract NAS8-36125

† Contribution of the U.S. Government, not subject to copyright in the United States

Springer Proceedings in Physics, Vol. 64 Superconducting Devices and Their Applications
Editors: H. Koch, H. Lübbig

2. Experimental Apparatus

The SLS test probe is shown in Fig. 1. Extensive electrostatic shielding was used as well as dual π-section filters on all leads. Ferromagnetic shielding reduced the dc magnetic field at the SLS to less than 0.5μT and superconductive enclosures provided ac magnetic shielding. An opto-isolated data acquisition system allowed automated collection of I-V and V-Φ characteristics. The SLS was operated in vacuum allowing a commercial ac conductance bridge/PID system to control temperature. We report data taken at 4.2K.

SLS noise is measured by coupling the device to a following commercial dc SQUID through a transformer-filter network which provides a low frequency current step-up of 4X.

Experiments on the WCS were made in similar probes at Stanford and at NIST (without the vacuum and temperature control feature). Flux locking was accomplished using an ultralow noise (0.33nV/$\sqrt{\text{Hz}}$) amplifier [8] with 100kHz modulation and two-pole integrator electronics similar to the design of Wellstood, *et al.* [9].

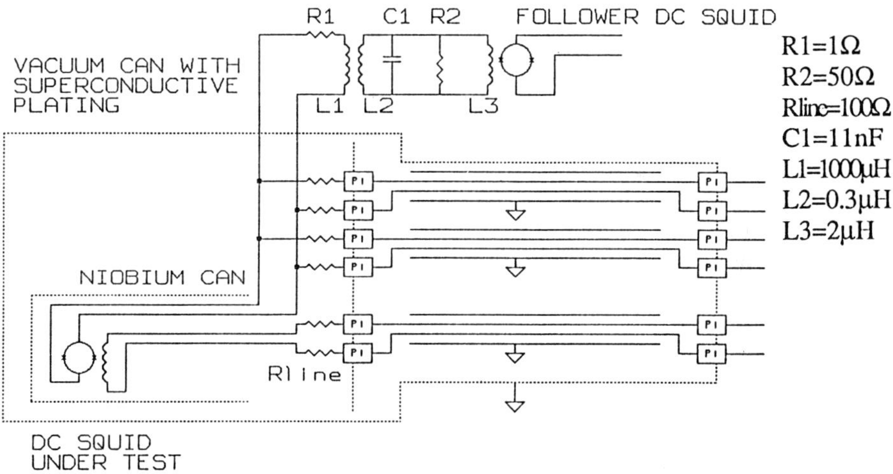

Figure 1: SLS test probe.

3. Device Fabrication

The SQUIDs were fabricated at NIST using a superconducting integrated circuit process with 2.5μm minimum feature size and based on Josephson junctions formed from Nb/Al-oxide-Al/Nb trilayers. The trilayers were formed on 51mm Si wafers whose temperature during deposition was maintained below 70°C. The load locked deposition system was cryopumped to its base pressure of 5.3 x 10^{-6}Pa and the lower Nb electrode (200nm thick) was sputtered in 1.07Pa of Ar at a rate of 2.5nm/s. Next, the Al layer (7nm thick) was sputtered in 0.67Pa of Ar at a rate of 0.1nm/s.

The wafer was then transferred to the load-lock and the Al was oxidized in O_2 for 30 minutes to form the tunneling barrier. The wafer was returned to the deposition chamber and the upper Nb electrode (50nm thick) was sputtered in 1.07Pa of Ar at a rate of 1.5nm/s. The resulting trilayers were patterned by reactive ion etching in CF_4 and O_2. The junction shunting and resonance damping resistors are thermally evaporated $AuIn_2$ with sheet resistivity of approximately 1Ω/□. Insulating layers are thermally evaporated SiO and the SQUID washer and wiring layers are composed of thermally evaporated PbInAu alloy. All layers except the trilayer were patterned by photoresist lift-off.

4. Results

The I-V and V-Φ characteristics for the SLS were acquired with the following SQUID removed; the latter is shown in Fig. 2. The corresponding electrical parameters are given in Table 1. The junction shunt resistance, R_S, and the minimum and maximum critical currents (I_{CMIN} and I_{CMAX}) were inferred from the I-V curve. The SQUID's self inductance L_S was obtained from I_{CMIN} and I_{CMAX} using the work of Peterson and Hamilton [10]. The capacitance and the hysteresis parameter β_C were inferred from the area of the junction and its expected capacitance per unit area. The voltage noise spectral density was measured at the output of the following SQUID and normalized with the responsivity to yield $\sqrt{S_\Phi}$ (Fig 3). This verifies previous measurements taken from 1-1000Hz at NIST, however, the data shown here has slightly higher values of S_Φ. The difference may be due to helium boiling at 4.0K at NIST and 4.2K at Stanford. The origin of the spikes in Fig. 3 is not known.

The characteristics for the WCS without damping showed much more structure than for the SLS. The WCS V-Φ characteristics shown in Fig. 4 illustrate the improvement achieved when a damping resistor joins the input coil to the SQUID loop. The device parameters for a similar WCS are given in Table 1. The WCS coupling coefficient, k, was determined from a measurement of the mutual inductance between the SQUID and input coil, M_{SI}, assuming $k=M_{SI}/(nL_S)$.

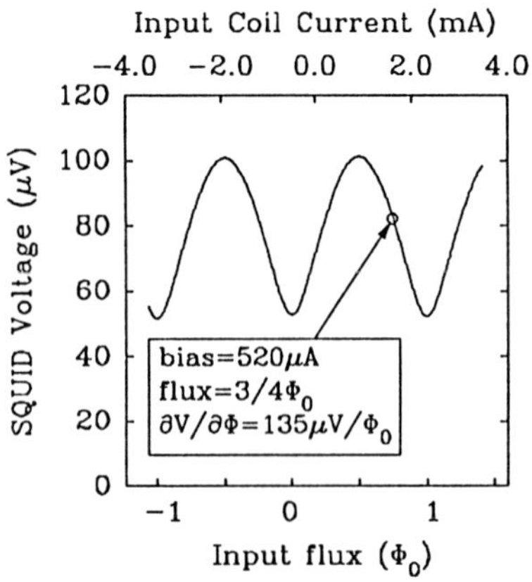

Figure 2: V-Φ curve for SLS.

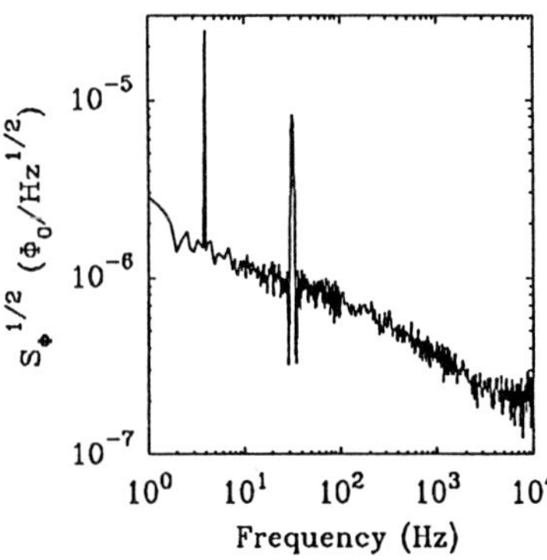

Figure 3: $\sqrt{S_\Phi}$ for SLS at state in Fig. 2.

Table 1: dc SQUID Parameters

	SLS	WCS
Critical Current ($2I_{CMIN}/2I_{CMAX}$)	220μA/440μA	30μA/45μA
$\beta_L = 2L_S I_{CMAX}/\Phi_0$	0.95	1.8
$\beta_C = 2\pi I_{CMAX} R_S^2 C/\Phi_0$ (calculated)	0.3	0.05
L_S	4.5pH	83pH
L_I (calculated)	-	0.16μH
Washer to Input Resistance	Open	29.2Ω
R_S (junction shunt resistance)	0.66Ω	2.8Ω
M_{SI} (SQUID to input coil)	-	3.80nH
M_{SM} (SQUID to modulation coil)	0.73pH	49pH
k	-	0.96
n (input coil turns ratio)	-	44
Typical maximum responsivity	150μV/Φ_0	120μV/Φ_0

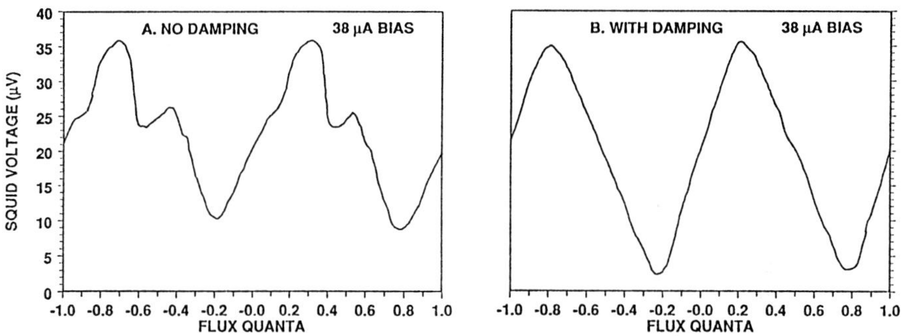

Figure 4: V-Φ curves for WCS without (A) and with damping (B).

For flux-locking measurements on the WCS, a 0.2μH superconductive inductor was placed across the SQUID input coil to simulate a typical source. The SQUID was biased at 43μA and the dc flux was adjusted to a minimum in the V-Φ characteristic. A 100kHz rounded square wave was then injected into the modulation coil. The measured white noise of the flux-locked SQUID (5μ$\Phi_0/\sqrt{\text{Hz}}$ between 50-2000Hz) was dominated by amplifier noise (no cold transformer used so far). The low frequency noise was 36μ$\Phi_0/\sqrt{f}$ for f between 1 and 20Hz.

5. Conclusion

The stripline SQUID exhibited smooth characteristics and a white noise level of $2 \times 10^{-7}\ \Phi_0/\sqrt{\text{Hz}}$. Below 4kHz the noise began to rise, yielding a power spectral density which scaled as approximately $f^{-0.5}$. The undamped WCS showed considerable structure in its I-V and V-Φ characteristics, presumably due to resonance effects. The addition of a damping resistor between the input coil and the SQUID loop appeared to control the dominant resonances, as evidenced by the smooth V-Φ curve. When flux-locked, the latter SQUID showed a white noise level dominated by electronics noise ($5 \times 10^{-6}\ \Phi_0/\sqrt{Hz}$) and low frequency noise which began to rise at about 20Hz with a $f^{-1.0}$ power spectral density. Further work is underway to determine the ultimate flux-locked noise performance of the damped WCS described here.

References

1. J.M. Lockhart, Proc. SPIE **619**, 148 (1986).
2. K. Enpuku, K. Yoshida, and S. Kohjiro, J. Appl. Phys. **60**, 4218 (1986).
3. J. Knuutila, C. Ahonen, and C. Tesche, J. Low Temp. Phys. **68**, 269 (1987).
4. S. Kohjiro, K. Enpuku, K. Yoshida, and K. Yamafuji, Jpn. J. Appl. Phys. **27**, 1871 (1988).
5. V. Foglietti, W.J. Gallagher, M.B. Ketchen, A.W. Kleinsasser, R.H. Koch, and R.L. Sandstrom, Appl. Phys. Lett. **55**, 1451 (1989).
6. T. Ryhänen, R. Cantor, D. Drung, and H. Koch, IEEE Trans. Magn. **27**, 3013 (1991).
7. M.E. Huber and M.W. Cromar, Physica B **165&166**, 77-78 (1990).
8. G.M. Gutt, J.S. Kim, M.R. Condron II, J.M. Lockhart, and B. Muhlfelder, "An Ultralow Noise Amplifier for Superconductive Detectors." To appear in *Superconductor Science and Technology*.
9. F. Wellstood, C. Heiden, and J. Clarke, Rev. Sci. Instrum. **55**, 952 (1984).
10. R.L. Peterson and C.A. Hamilton, J. Appl. Phys. **50**, 8135 (1979).

Noise Properties of NbN-MgO-NbN SQUIDs

O. Dössel[1], *B. David*[1], *D. Grundler*[2], *R. Kobs*[1], *and K.-M. Lüdeke*[1]

[1]Philips GmbH, Forschungslaboratorium Hamburg,
Postfach 540840, W-2000 Hamburg 54, Fed. Rep. of Germany
[2]Institut für Angewandte Physik, Universität Hamburg,
Jungiusstr. 11, W-2000 Hamburg 36, Fed. Rep. of Germany

Abstract. DC-SQUIDs made of refractive materials NbN-MgO-NbN were prepared using thin film preparation techniques. The noise properties of these SQUIDs were investigated. The white energy sensitivity of typical devices is 600h (8mF0/ÀHz). The measured values are compared with theoretical predictions and the standard optimization rules. The low frequency noise was measured directly and using different modulation schemes. The SQUIDs show a relatively high 1/f noise onset in the range of 100Hz. The main contribution comes from out-of-phase critical current fluctuations and can be effectively suppressed by bias modulation.

1. Introduction

All refractory Josephson junctions (Nb,NbN) with barriers made of artificial oxides are more and more the basis of SQUIDs used for biomagnetic measurements where low white noise and an onset of 1/f noise below 1 Hz is of utmost importance.

There are at least two separate sources of 1/f noise in DC-SQUIDs: critical current fluctuations of the Josephson junctions and the motion of flux lines trapped in the body of the SQUID. While the critical current noise can be suppressed by a suitable modulation scheme, the flux noise cannot [1]. The critical current fluctuations can be separated into an in-phase and an out-of-phase component. In a SQUID biased by constant current and constant flux, both modes contribute to the noise. The standard flux modulation technique zeros out the in-phase contribution. Additional bias current modulation suppresses the out-of-phase contributions. Measurements using different modulation schemes allow to distinguish between the different sources of 1/f noise [2].

Although a trend towards Nb-Al_2O_3-Nb junctions can be observed today the advantages of the NbN-MgO-NbN system are still obvious: Together with the recognized compatibility of MgO and NbN this system proves to be extremely stable during thermal cycling and storage. However very little is known about the noise properties of NbN-MgO-NbN based DC-SQUIDs.

2. SQUID preparation

The DC-SQUIDs are based on NbN-MgO-NbN tunnel junctions and are completely fabricated by thin film technology [3]. The SQUID layout is similiar to the square

Springer Proceedings in Physics, Vol. 64 Superconducting Devices and Their Applications
Editors: H. Koch, H. Lübbig

Table I: DC-SQUID parameters (optimum values)

critical current per junction	I_0	9 μA
shunt resistance per junction	R	2.7 Ω
capacitance per junction	C	0.09 pF
junction area	A	9 μm^2
SQUID inductance	L_{sq}	360 pH
input coil inductance	L_{coup}	900 nH
modulation depth parameter	β_L	3
hysteresis parameter	β_C	0.03
white flux noise	$\sqrt{S_\Phi}$	6 $\mu\Phi_0/\sqrt{Hz}$
white energy sensitivity	ε_w	330 h

washer design developed by Ketchen and Jaycox [4]. The two Josephson junctions are located outside of the square washer. The slit separating the junctions and leading to the center hole has a width of 5µm. A 50-turn spiral input coil with a linewidth of 5µm is wound over the top surface of the washer. The center hole of the SQUID is 200µm*200µm in size . The relevant SQUID parameters are listed in Table I.

3. Noise measurements

The white noise was determined at 20kHz with a 1:10 transformer and a low noise operational amplifier (noise voltage: 180pV/$\sqrt{Hz}$ above 200Hz). The low frequency noise was measured by three different methods: A direct measurement using a low noise FET amplifier (noise voltage: 700pV/$\sqrt{Hz}$ above 20Hz), a standard flux-locked loop scheme and a new readout scheme with an additional bias modulation [5]. During the measurement, the SQUIDs were surrounded by superconducting shields (niobium and lead). The noise measurements were carried out on SQUIDs from different production cycles with varying parameters like I_0 and R but similar values of L_{sq} and C.

4. White noise

The dependence of the measured white noise levels on the hysteresis parameter β_C and the modulation depth parameter β_L are shown in Figure 1 and 2. In order to compare SQUIDs with different I_0R products the data for Figure 2 were normalized and show an effective noise temperature. The experimental data can be compared with theoretical predictions for the energy sensitivity given in [6,7]

$$\varepsilon_w = \frac{kT\Phi_O}{I_O \cdot R} * f(\beta_L)\ ; \qquad T_N = T * f(\beta_L)\ ; \qquad \varepsilon_w = 16\, kT \sqrt{(LC/\beta_C)}$$

The predicted value for our SQUIDs would be 50h. Our best measured value with a β_C parameter of 0.03 amounts to 330h. Such an increased noise value is typical

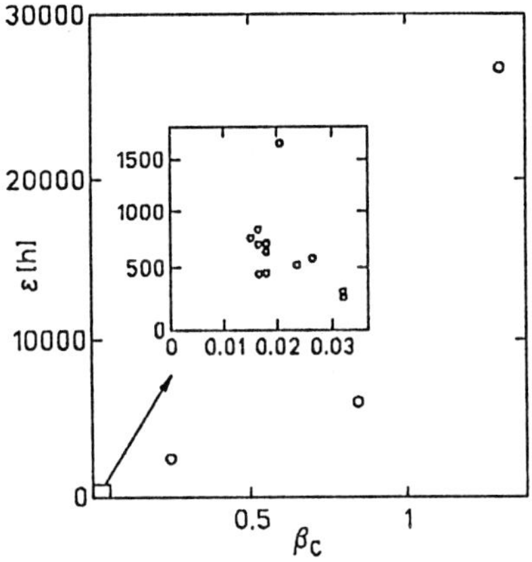

Figure 1: Dependence of the energy sensitivity on the β_C parameter

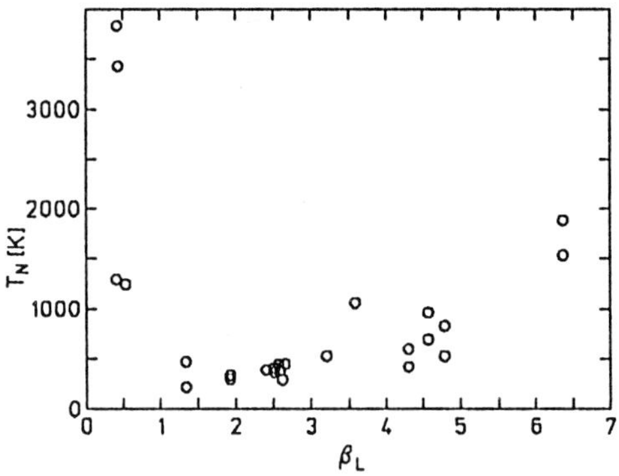

Figure 2: Dependence of the noise temperature on the β_L parameter

for SQUIDs with integrated large inductance input coils. In contrast to standard optimization rules the noise drastically increases for β_C approaching 1. We assume, that parasitic resonances reduce the SQUID performance. A significantly better SQUID performance can be expected from adequate damping. On the other hand the optimization rule $\beta_L=1$ proves to be valid. Although the absolute noise values are about a factor of 10 higher then expected, the minimum is found near $\beta_L=1$.

5. 1/f noise

In order to accentuate the effect of using modulation techniques in Figure 3 an example with untypical high 1/f noise was chosen.

As can be seen in Table II the white noise level is slightly increased by using modulation techniques and the noise level at 1Hz is drastically decreased. Using formula (3) and (4) given in [1] the critical current fluctuation $\sqrt{S_{I0}}/I_0$ can be estimated. Due to the fact that these estimates according to the direct method and the flux modulation method are nearly identical and since the 1/f noise can be completely suppressed by the bias modulation technique, it is clear that critical current fluctuations are the origin of the observed 1/f noise. We assume, that electron traps and resonant tunneling sites in the barrier are responsible for the critical current fluctua-

Table II: Low frequency noise of a NbN-MgO-NbN SQUID

	$\sqrt{S_\Phi}$(1kHz) [$\mu\Phi_0/\sqrt{Hz}$]	$\sqrt{S_\Phi}$(1Hz) [$\mu\Phi_0/\sqrt{Hz}$]	$\sqrt{S_{I0}}/I_0$(1Hz) [ppm/$\sqrt{Hz}$]
direct measurement	10	110(extrapol)	60
flux modulation	20	90	58
flux & bias modulation	22	22	-

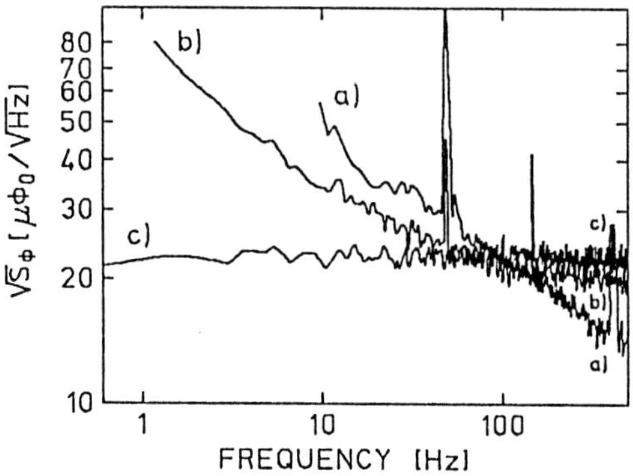

Figure 3: Noise spectra measured directly (a), with flux modulation (b) and with flux & bias modulation (c)

tions. Improved barrier preparation techniques are expected to reduce the low frequency noise.

Acknowledgements

Substantial contributions from G. Kürsten and G. Rabe are gratefully acknowledged.

References

[1] R.H. Koch, J. Clarke, W.M. Goubau, J.M. Martinis, C.M. Pegrum, D.J. van Harlingen, J. Low Temp. Phys. 51, 207-224 (1983)
[2] V. Foglietti, W.J. Gallagher, M.B. Ketchen, A.W. Kleinsasser, R.H. Koch, S.I. Raider, R.L. Sandstrom, Appl. Phys. Lett. 49, 1393-1395 (1986)
[3] B. David, O. Dössel, G. Kürsten, G. Rabe, Proceedings of the ISEC 91, Glasgow, to be published (1991)
[4] J.M. Jaycox, M.B. Ketchen, IEEE Trans. Magn. MAG-17, 400-403 (1981)
[5] O. Dössel, B. David, M.Fuchs, W.H. Kullmann, K.-M. Lüdeke, IEEE Trans. Magn. MAG-27, 2797-2800 (1990)
[6] C.D. Tesche, J. Clarke, J. Low Temp. Phys. 29, 301-331 (1977)
[7] J.J.P. Bruines, V.J. de Waal, J.E. Mooij, J. Low Temp. Phys. 46, 383-386 (1982)

Noise Studies of Uncoupled DC-SQUIDs

T. Ryhänen[1], *H. Seppä*[1], *R. Cantor*[2], *D. Drung*[2], *H. Koch*[2], *and D. Veldhuis*[3]

[1]Metrology Research Institute, Helsinki University of Technology and Technical Research Centre of Finland, Otakaari 7 B, SF-02150 Espoo, Finland
[2]Physikalisch-Technische Bundesanstalt, Institut Berlin, Abbestr. 2–12, W-1000 Berlin 10, Fed. Rep. of Germany
[3]University of Twente, Faculty of Applied Physics, P.O. Box 217, NL-7500 AE Enschede, The Netherlands

Abstract. We have performed noise measurements of uncoupled dc SQUIDs with similar square washer geometries. Mappings consisting of hundreds of flux noise measurements at different points of operation have been carried out using a dc SQUID preamplifier and a computer controlled measurement system. The white and low frequency noise, dynamic resistance, and the flux-to-voltage transfer function have been measured. The dc SQUID noise performance is described using equivalent current noise sources, and the voltage noise is observed to be linearly proportional to dynamic resistance R_{dyn}.

1. Introduction

The development of novel read-out electronics requires simple lumped-circuit models which describe the input and output impedances, forward transfer parameters, noise performance, etc. Using a dc SQUID preamplifier and a computer controlled measurement system, we have measured and mapped the dynamic resistance R_{dyn}, and the flux-to-voltage transfer function $\partial\bar{V}/\partial\Phi_a$, as well as the white and low-frequency voltage noise of a variety of uncoupled square-washer-type dc SQUIDs.[1,2,3,4] Similar measurements have been published by Koch et al. [5] On the basis of our measurements, an alternative way to model the dc SQUID performance in terms of R_{dyn} is presented. An example of an uncoupled dc SQUID with $L = 41$ nH is characterized using this model.

2. Measurement system

The dc SQUID measurements are performed using another dc SQUID [3] as a preamplifier. The voltage signal of the SQUID to be measured is coupled via a wire-wound coil to the preamplifier dc SQUID. The series resistance of the coil $R = 9.4\ \Omega$ and the mutual inductance between the coupling coil and the preamplifier SQUID $M = 1.2$ nH. The preamplifier dc SQUID is operated in flux-locked loop. The preamplifier SQUID voltage is amplified using a low noise operational amplifier, LT1028, followed by a PI controller. The equivalent flux

Springer Proceedings in Physics, Vol. 64 Superconducting Devices and Their Applications
Editors: H. Koch, H. Lübbig

noise of the preamplifier SQUID is $\langle \Phi_{n,1}^2 \rangle^{1/2} = 1.3 \times 10^{-6} \Phi_0/\sqrt{Hz}$. The total voltage noise is measured by a HP 35660A spectrum analyzer controlled by a computer. The bias current and applied flux for the test SQUID are generated by the computer, and the dc characteristics, R_{dyn}, and $\partial \bar{V}/\partial \Phi_a$ are measured using a second read-out electronics coupled directly to the test SQUID.

The total output voltage noise $V_{n,out}$ can be written as

$$V_{n,out}^2 = \left(\frac{R_f}{M_f}\right)^2 \left\{ \langle \Phi_{n,1}^2 \rangle + \frac{M^2}{(R + R_{dyn})^2} \left[(\partial \bar{V}/\partial \Phi_a)^2 \langle \Phi_{n,2}^2 \rangle + 4k_B T R \right] \right\}, \quad (1)$$

where $\langle \Phi_{n,1}^2 \rangle^{1/2}$ is the flux noise of the preamplifier SQUID in flux-locked loop, $\langle \Phi_{n,2}^2 \rangle^{1/2}$ is the intrinsic flux noise of the dc SQUID under test, $\sqrt{4k_B TR}$ is the Johnson noise of the resistor R, R_f is the feedback resistor, and M_f is the feedback coil mutual inductance. Note that R_{dyn} has to be measured before substracting the noise contributions of the preamplifier SQUID and the coupling resistor.

3. Results

The uncoupled dc SQUID [4] serves as a good reference for studies of coupled devices. The device is fabricated from a Nb/Al$_x$O$_y$/Nb trilayer and structured using four mask levels.[3,6] The SQUID washer is 400 μm wide with a hole of 20 μm. The slit is covered by a Nb film. The 20 μm^2 Josephson junctions each have a critical current of 32 μA and a shunt resistor $R = 5.8\ \Omega$. The SQUID inductance $L = 41$ pH, and the junction capacitance $C = 1.0$ pF. The data presented here consists of 168 independent measurements at different points of operation. The lowest flux noise of $4.2 \times 10^{-7} \Phi_0/\sqrt{Hz}$ was measured for a bias current $I_b = 33 \mu$A and $\Phi_a = 0.27 \Phi_0$.

Figure 1 presents the dependence of $\partial \bar{V}/\partial \Phi_a$ on R_{dyn}. For all dc SQUIDs studied the dependence has been nearly linear. The approximations $R_{dyn} \approx R/\sqrt{2}$ and $\partial \bar{V}/\partial \Phi_a \approx R/2L$ lead to $\partial \bar{V}/\partial \Phi_a \approx R_{dyn}/\sqrt{2}L$. From other measurements on coupled devices, this approximation appears to be valid for these devices as well. The scaling of $\partial \bar{V}/\partial \Phi_a$ with R_{dyn} based on numerical calculations and the discrepancies in measurement results due to resonances have been reported previously by Koch et. al.[5]

The voltage noise across the SQUID, $\langle V_n \rangle$, is proportional to R_{dyn}, as shown in Fig. 2. In the inset of Fig. 2 the SQUID noise is modelled using two non-correlated equivalent noise current sources, $\langle i_{n,1}^2 \rangle = 4k_B T/(R/2)$ and $\langle i_{n,2}^2 \rangle = 4^2 \cdot 4k_B T/(2R)$, which describe noise currents through the device and circulating in the SQUID loop, respectively. The dynamic impedance of the dc SQUID is divided into the normal resistance $R/2$ and the additional resistance $R_{dyn} - R/2$ owing to the nonlinear dynamics of the SQUID. The circulating noise current source $\langle i_{n,2}^2 \rangle^{1/2}$ sees a dynamic resistance that is four times higher, since the two junctions are in series. This is taken into account by multiplying the noise current spectral density by the factor 4^2. Using the model shown in

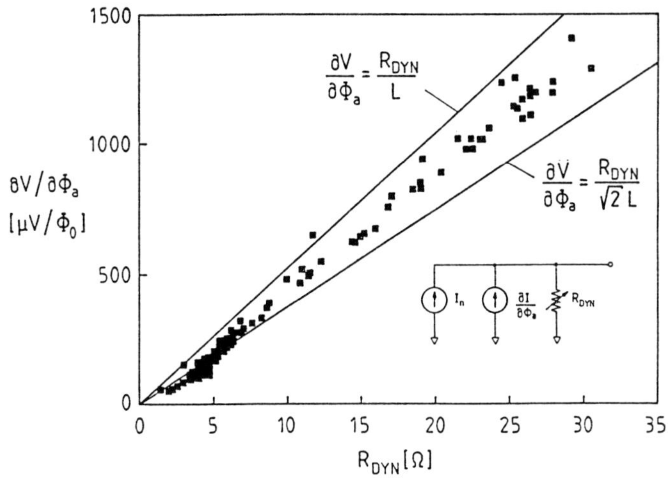

Figure 1. Measured dependence of the flux-to-voltage transfer function $\partial \bar{V}/\partial\Phi_a$ on dynamic resistance R_{dyn}. The inset presents the model used for describing the measured data.

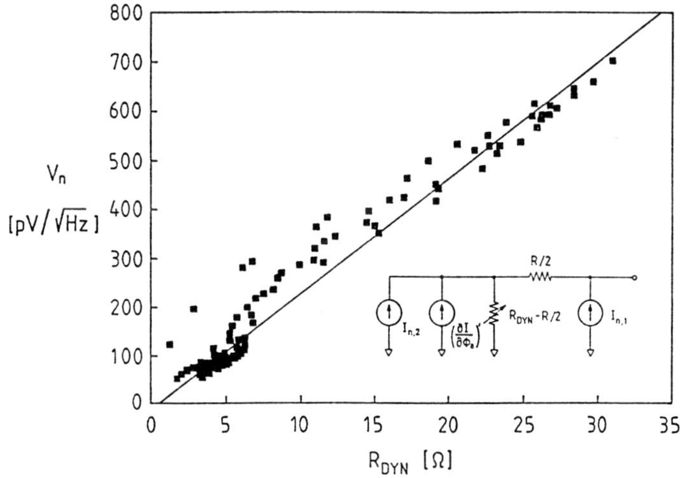

Figure 2. Voltage noise of the dc SQUID as a function of R_{dyn}. The inset presents the equivalent noise model of the dc SQUID.

the inset of Fig. 2 the voltage noise across the SQUID can be written as

$$\langle V_n^2 \rangle \approx \gamma \frac{8k_BT}{R} \{R_{dyn}^2 + 4(R_{dyn} - R/2)^2\} , \tag{2}$$

where γ is a factor taking into account the mixing-down effect of noise. The line in Fig. 2 corresponds to the approximation (2) with the value $\gamma = 3$.

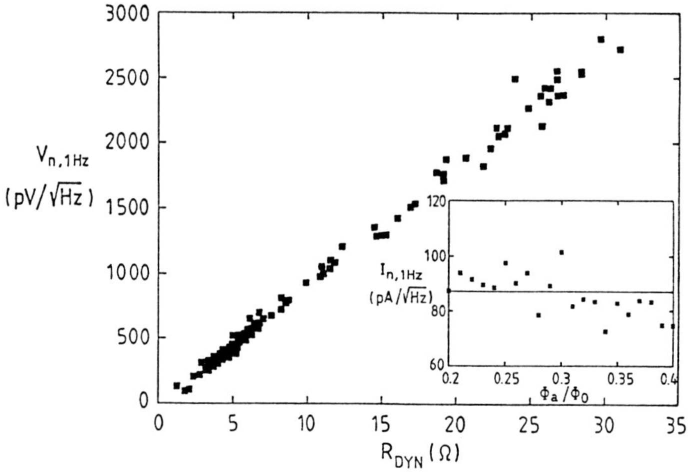

Figure 3. Voltage noise measured at 1 Hz as a function of R_{dyn}. The equivalent current noise $\langle I^2_{n,1\ Hz}\rangle^{1/2} = \langle V^2_{n,1\ Hz}\rangle^{1/2}/R_{dyn}$ as a function of applied flux for a fixed bias current of 35 μA is shown in the inset.

Figure 3 presents the voltage noise $\langle V^2_{n,1\ Hz}\rangle^{1/2}$ at 1 Hz as a function of R_{dyn}. The linear dependence suggests that the equivalent current and flux noise of the SQUID are independent of R_{dyn} and the point of operation. In the inset of Fig. 3, the low frequency current noise $\langle I^2_{n,1\ Hz}\rangle^{1/2} = \langle V^2_{n,1\ Hz}\rangle^{1/2}/R_{dyn}$ at 1 Hz is shown as a function of Φ_a. Only a slight correlation with $\partial\bar{V}/\partial\Phi_a$ is observed.

4. Conclusions

The dynamic resistance is found to be nearly linearly correlated with the flux-to-voltage transfer function. Thus, $\partial I/\partial\Phi_a = (\partial\bar{V}/\partial\Phi_a)/R_{dyn}$ is found to be nearly indepedent of the point of operation. In addition, the voltage noise of the dc SQUID depends linearly on R_{dyn}. The SQUID noise performance can therefore be modelled with an equivalent current noise source.

The low frequency voltage noise is also found to be correlated with R_{dyn}, and the 1/f flux noise is found to be independent of the point of operation. This suggests that, for these small uncoupled dc SQUIDs, the measured noise does not give any possibility of distinguishing the real flux noise from the flux noise arising from critical current fluctuations.

References

[1] J.M. Jaycox and M.B. Ketchen, IEEE Trans. Magn. **MAG-17**, 400 (1981).

[2] M.B. Ketchen and J.M. Jaycox, Appl. Phys. Lett. **40**, 736 (1982).

[3] R. Cantor, T. Ryhänen, D. Drung, H. Koch, and H. Seppä, IEEE Trans. Magn. **MAG-27**, 2927 (1991). The Nb/Al_xO_y/Nb trilayers were provided by D. Veldhuis, University of Twente, Enschede, The Netherlands.
[4] T. Ryhänen, R. Cantor, D. Drung, H. Koch, and H. Seppä, IEEE Trans. Magn. **MAG-27**, 3013 (1991).
[5] R.H. Koch, J. Clarke, W.M. Goubau, J.M. Martinis, C.M. Pegrum, and D.J. Van Harlingen, J. Low Temp. Phys. **51**, 207 (1983).
[6] R. Cantor, D. Drung, M. Peters, and H. Koch, J. Appl. Phys. **67**, 3038 (1990).

The Laser Switch in SQUID Measurements: Fundamental Experiments and Low Frequency Noise Reduction

B. Cabrera

Physics Department, Stanford University, Stanford, CA 94305, USA

Abstract. We are developing a laser-driven superconducting switch. The switch has been used in several fundamental physics experiments which utilize SQUIDs, and in a noise reduction scheme for low-frequency SQUID measurements. The experiments include characterization studies of the laser switch itself, a direct measurement of the pinning force on a single vortex, the measurement of the Cooper pair mass, a study of the Aharonov-Bohm effect, and a measurement of the kinetic inductance of a superconducting thin-film. The noise reduction scheme is based on modulating the input circuit of a SQUID at a frequency above its 1/f knee. The demodulated output is then devoid of the SQUID noise below the modulation frequency. This paper reviews these experiments and reports on their status.

1. The Laser Switch

A number of investigations about the interaction of optical radiation with superconducting microbridges appear in the literature [1], including studies on switching times and relaxation processes. However, the devices described in this paper were the first to utilize the high sensitivity of SQUIDs together with a laser-driven superconducting switch. Using this combination, we have developed new techniques for a number of interesting experiments.

The basic switch (Fig 1) consists of a niobium thin-film microbridge which is illuminated with a laser beam fed through an optical fiber. This fiber is attached to the chip using commercial ruby ferrules which are epoxied to the substrates, typically crystalline silicon or sapphire. We have used two sources for the laser illumination, a 300 mW continuous wave Nd:YAG and several laser diodes.

To characterize the switch performance, we have connected it directly across the input terminals of an rf SQUID [2]. Fig 2a shows a sequence of flux readings taken using the Nd:YAG with a mechanical shutter run in a square wave mode at about 1 Hz. The points along the center of the plot are produced during the laser-on portion of the square wave when the input circuit is open, whereas the wide distribution of quantized points corresponds to flux states obtained during the laser-off portion when the quantum coherence is reestablished around the ring. The quantized flux

Springer Proceedings in Physics, Vol. 64 Superconducting Devices and Their Applications
Editors: H. Koch, H. Lübbig

Fig 1 Schematic of laser switch.

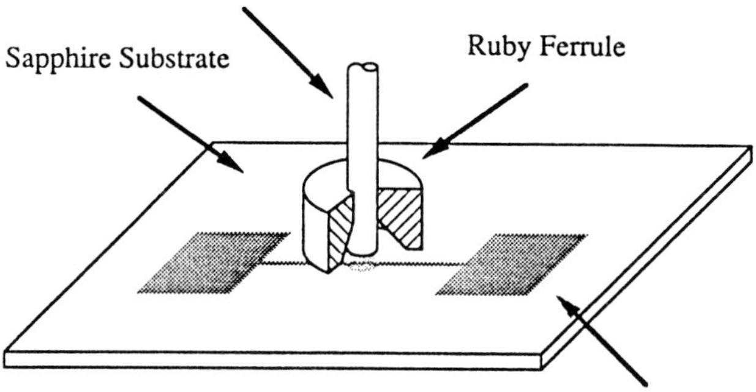

distribution forms a Gaussian which is centered around zero current in the input circuit and has a Boltzmann standard deviation given by $\phi_{rms}{}^2/2L = kT^*/2$ where L (2 μH) is the inductance of the input circuit and T*, usually near the T_c of the film, is the freeze-out temperature below which it is very improbable for vortices to move across the microbridge. The position of each point in the distribution is uncorrelated with respect to the position of the previous point.

Next, we use a pulsed laser diode for the the illumination. Now the pulse duration (laser-on) is about 100 ns and again a quantized Gaussian distribution is formed (Fig 2b). However, successive points in the distribution are now strongly correlated, much like a random walk. These correlations are due to the presence of a 3 ohm shunt resistor which is in parallel with the switch across the input of the SQUID. When the switch remains open for times short compared with the $\tau = L/R$ time constant (.75 μs), the shunt resistor acts like a superconducting short, preventing flux from entering the input coil of the SQUID. The distribution of points formed from a set of values immediately following a particular flux state also are Gaussian, but have a narrower standard deviation given by $\phi_{rms}{}^2 = LkT^*(1 - e^{-2t/\tau})^2$. The centroid of this distribution is near the previous flux state, but shifted toward the zero current position. These properties are described by the conditional probability function [2]

$$P(\phi_{i+1},t;\phi_i) \propto \exp\left(- \frac{(\phi_{i+1} - \phi_i\, e^{-t/\tau})^2}{2LkT^*(1 - e^{-2t/\tau})^2} \right).$$

We performed a careful study of the fit of our data to this function for a range of pulse times from 6 ns to 100 ms. The data fit the function well (Fig 3) and convincingly set an upper limit on the switching times of <6 ns. The thermal relaxation time constant for these films is ≈ 1 ns.

One final point about the switch operation is that in looking closely at the distribution from the laser diode data (Fig 2b), we find non-quantized

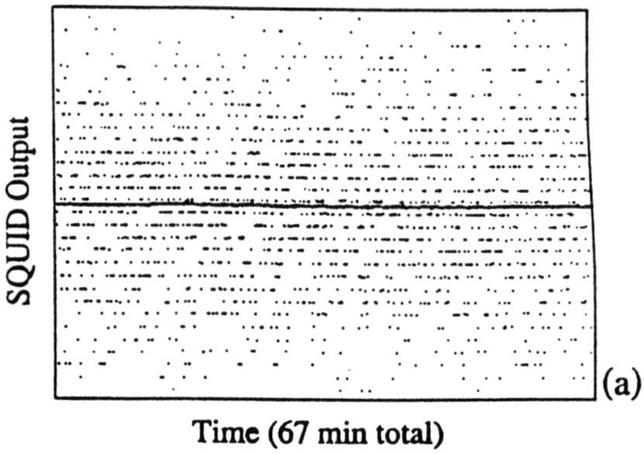

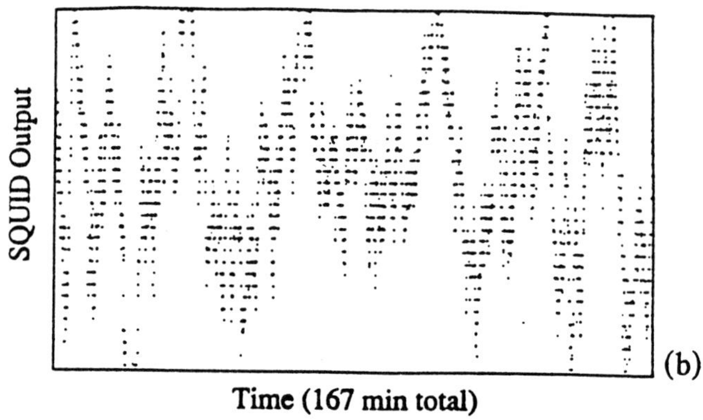

Fig 2 Evolution of flux quantum states (a) using a Nd:YAG laser, and (b) using a laser diode.

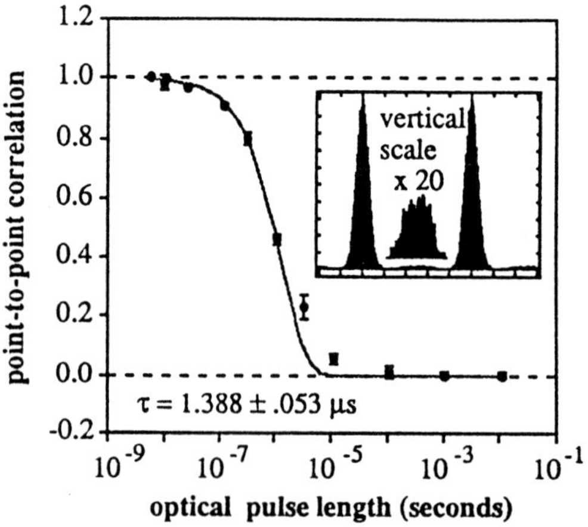

Fig 3 Fit of data to conditional probability function.

flux points which occur about 5% of the time. These correspond to a vortex which is trapped right in the area of the film which was illuminated by the laser. The apparent fractional value in the flux is due to the superposition of two different flux quantum states on opposite sides of the vortex, and its value between the two states is roughly proportional to the position of the vortex along a direction perpendicular to its length. Next, we will utilize this feature to study the vortex pinning force.

2. Vortex Pinning Force

Very recently, we have taken advantage of the flux trapping as a tool to measure the pinning force on a single vortex [3]. We repeatedly open and close the laser switch until a fractional flux value is obtained. Then, we run a current through the 2 μm wide niobium line until the vortex moves to the edge of film, usually all at once, but sometimes in two steps indicating that it fell into a deeper pinning site on its way to the edge and a larger current was needed to free it. The circuit which provides the current has two branches, the first is connected in parallel with the switch and the second is inductively coupled to the SQUID input circuit. An adjustable bridge proportions the drive current in such a way that the SQUID output remains constant unless the vortex moves. Using this technique, we have measured single-vortex pinning forces in the range of 10^{-4} N/m for our 20 nm thick Nb films.

We plan to extend the technique in two ways. First we will use a dc SQUID to improve the signal-to-noise by more than an order of magnitude. With this increased sensitivity, we will examine the elastic region where the vortex moves in its pinning potential but does not escape. It will then be possible to map out the shape of the pinning potential. Second, we will attempt to utilize the technique on high temperature superconductor films such as YBCO. The technique will work, unless we trap many vortices each time we operate the switch. We have been very encouraged by the initial results with low T_c films and feel that this technique will be very useful in pinning studies.

3. Cooper Pair Mass

Originally, we developed the laser switch for an experiment to determine the mass of a Cooper pair [4]. The mass m* is defined through the quantum mechanical relation $m^*\mathbf{v} = (h/2\pi)\, \mathbf{grad}\, \phi - e^*/c\, \mathbf{A}$, where $\mathbf{v}$ is the superfluid velocity and ϕ the phase of the order parameter. For a bulk superconductor at rest in the laboratory, $\mathbf{v}=0$ in the interior so no information can be obtained about m*. In fact it is necessary for $\mathbf{curl\ v} \neq 0$ around a closed path. This condition is easily achieved using a rotating superconductor. From the gauge invariant relation above, it can be shown

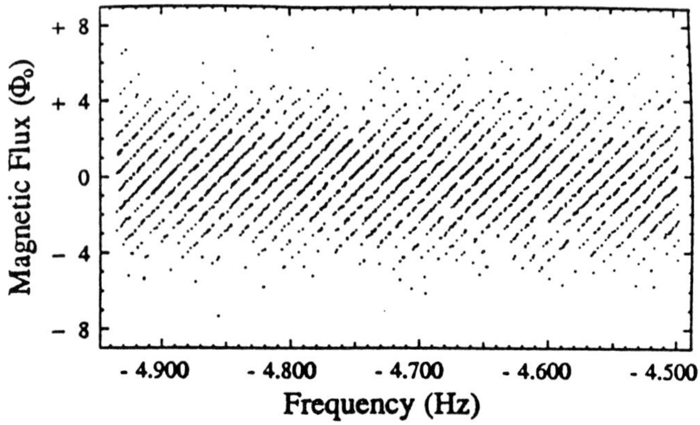

Fig 4 Flux quantum data during slow ramp of spin speed.

that the interior of a rotating superconductor contains a uniform field proportional to the angular velocity ω. This field, $B_L = 2m^*c/e^* \ \omega$ is called the London field.

If we now consider a rotating ring, we can measure this London magnetic flux in terms of the flux quantum $\phi_0 = hc/e^*$. The total flux measured by a stationary pick up loop connected to a SQUID is given by $\phi = n\,\phi_0 - (2m^*c/e^*)\,\omega{\cdot}S$ where S is the area bounded by the ring. This equation corresponds to a set of parallel lines (see data in Fig 4) which are separated by ϕ_0 along the ϕ axis and by $\Delta\omega$ along the frequency axis. Looking at the equation, we see that $h/m^* = 2\,\Delta\omega\,S$. Thus by measuring $\Delta\omega$ and the area S to high precision, we obtain a direct measure of h/m^*. We have made such a measurement with a statistical accuracy of 5 ppm and a systematic uncertainty of 20 ppm.

Measuring an area to such precision required depositing a thin-film superconducting line on the equator of a precision quartz hemisphere, and using laser interferometric methods to measure the equatorial area [4]. The precision for the $\Delta\omega$ determination is made by sweeping the spin speed through about 1000 null crossings. Fig 4 shows a slow spin ramp through 0.5 Hz change. Each point is taken after the coherence of the ring is broken and reestablished using a laser switch positioned to shine on the line. Without the switch we would always remain on the same line and be unable to measure the frequency shift between lines. As with Fig 2a, the distribution of data points along the ϕ axis forms a Gaussian centered around zero current in the ring. Now however, the flux quantum states are tilted due to the additional phase shift of the order parameter produced by the rotation.

Our result for m', the Cooper pair mass with relativistic corrections added is $(m'/2m_e)_{exp} = 1.000\ 084\ (5)\ (20)$. Thus a Cooper pair has very nearly the mass of two free electrons at rest. However, several calculations suggest that $(m'/2m_e)_{theo} = 0.999\ 992\ (4)$, not only smaller in magnitude, but also on the other side of unity. Interestingly, the relativistic corrections contain a kinetic energy and a potential energy term, with $\langle T\rangle/2m_ec^2 = +0.000\ 180$ and $\langle V\rangle/2m_ec^2 = -0.000\ 188$, such that $(m'/2m_e)_{theo} = 1+[\langle T\rangle+\langle V\rangle]/2m_ec^2$. Our measurement suggests that a complete relativistic treatment might result in $m'/2m_e = 1+[\langle T\rangle+{}^1/_2\langle V\rangle]/2m_ec^2 = 1.000\ 086$. A more detailed calculation needs to be performed which includes all first-order relativistic effects in a self-consistent formalism. We also plan to improve the measurements.

4. Absolute Aharonov-Bohm Effect

Our experiment to measure the absolute significance of the magnetic vector potential [5] was motivated by a suggestion that the absolute magnetic flux through superconducting circuits might contain a fractional residue [6]. In this theory, flux quantization through a superconducting circuit would take the form $\int \mathbf{A}\cdot d\mathbf{l} = (n+r)\,\phi_0$ where n is an integer and $0<r<1$ is a residue. The superconducting circuit would trap r at the time the global phase coherence was first established around the circuit. Subsequent changes in the flux threading the circuit, which arise from vortices passing through a Josephson junction or across a microbridge, would change n but not r, since the local phase coherence is always maintained in such microscopic processes with dimensions smaller than the coherence length of the superconductor.

In order to test for the possibility of fractional flux quantization, it was necessary to devise an experimental technique for completely destroying and reestablishing the phase coherence over distances much greater than the coherence length. Repeatedly heating a portion of a macroscopic superconducting circuit does not generate a quantized flux distribution because vortices are trapped in various portions of the superconducting material making up the circuit. Since the locations of these vortices change after each cooling, a continuum of flux values is seen in measurements.

Our laser switch circumvented these experimental difficulties and allowed for the first time a sensitive test of the fractional flux quantization theory. Fig 5a shows a schematic representation of the experiment. The optically switched element is connected directly across the input coil of an rf SQUID sensor. In addition, the circuit threads a toroidal solenoid (Fig 5b) enclosed in a superconducting shield which can couple flux to the circuit while shielding all portions of the superconducting material from the magnetic field. The shield geometry is not a closed toroid, which would prevent any changes in the magnetic flux coupling to the circuit, but rather there is a long overlap region which is electrically isolated.

To test the absolute nature of the Aharonov-Bohm effect, a slowly changing magnetic flux was introduced through the circuit using the toroidal solenoid. Data were taken before, during, and after the ramping (Fig 5c). The centroid of the distribution remains at zero current (as in Fig 2). To calibrate the slope of the ramp portion, an identical run was

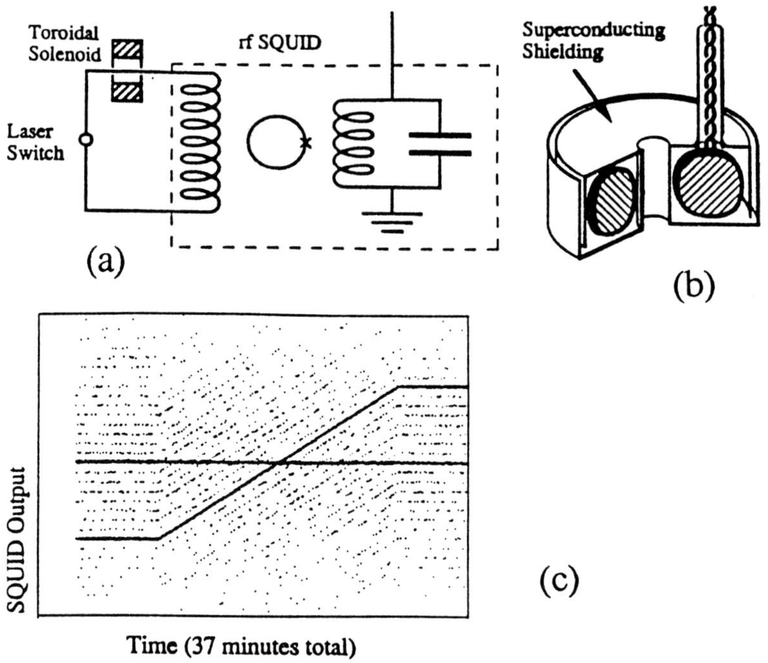

Fig 5. (a) Schematic of superconducting circuit with laser switch and (b) toroidal solenoid. (c) Data run with ramped solenoid current and a superimposed run with the switch superconducting throughout.

performed with the laser intensity turned down so that the switch remained superconducting throughout. These data are superimposed with no dc offset and show that the slopes are the same for the two runs.

To within the experimental uncertainty of about 1%, these SQUID-based data show that flux changes made by a toroidal solenoid when the circuit is open and without global phase coherence are always accounted for in full when the circuit is later allowed to return to the phase-coherent quantum state. There was no evidence for fractional flux quantization in superconducting circuit and the data are completely consistent with the absolute significance of the magnetic vector potential in conventional quantum mechanics.

5. Kinetic Inductance

We have recently completed a detailed study of the kinetic inductance of a thin-film circuit, measuring both its absolute value and its temperature dependence [7]. We find excellent agreement with the predictions of BCS theory in the dirty limit [8].

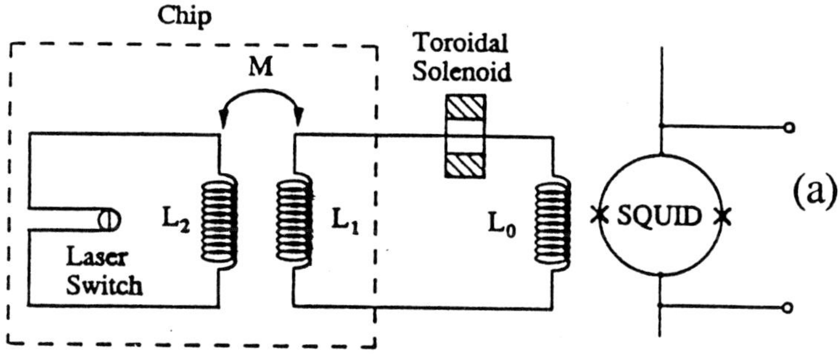

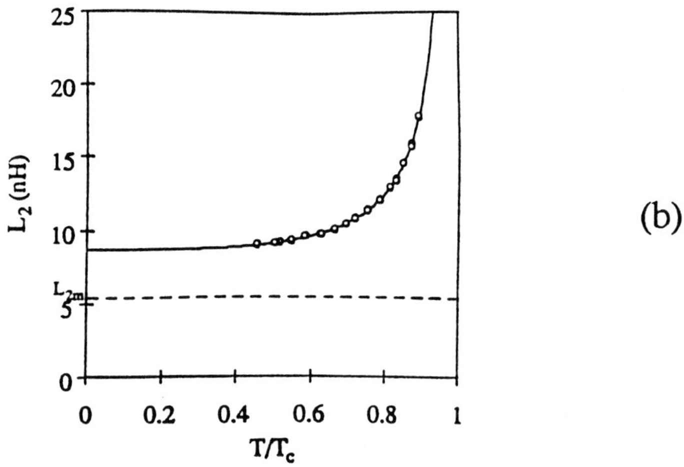

Fig 6 (a) Schematic of kinetic inductance measurement circuit.
(b) Calculated and measured inductance of square loop.

The inductance of a superconductor has two parts, one due to the magnetic field and the other due to the momentum of the current carriers. These can be derived from the fluxoid quantization condition $\phi + (4\pi/c)\,\lambda^2 \int \mathbf{j}_s \cdot dl = n\,\phi_0$ where ϕ is the magnetic flux threading the loop and $\mathbf{j}_s$ is the supercurrent density. We can further obtain a lumped parameter equation by integrating over the cross-sectional area of the wire, so that $\mathbf{j}_s$ becomes the total current I. Then $L_M I + L_K I = n\,\phi_0$, where L_M is the magnetic inductance and L_K the kinetic inductance. For a wire with thickness t and width w both small compared to λ, the current density is approximately uniform across the wire and the kinetic inductance has the simple form $L_K = \mu_0\,\lambda^2\, l/wt$ where l is the length of the circuit. We have transformed from cgs to SI units so that L_K can be easily compared with L_M in henry.

The circuit shown in Fig 6a is patterned from a single 11 nm thick niobium film on a silicon substrate. It consists of a 1 mm square loop which can be interrupted with the laser switch and a 20.5 turn inductively coupled coil which surrounds the square loop. This coil is directly connected to the input circuit of the SQUID after passing through a toroidal solenoid which is used to change the applied flux through the circuit.

Having previously determined the inductance of the SQUID input inductor to be 2.24 μH, the measurement of the total inductance of the square loop is made utilizing three measurements. First the response of the SQUID to a ramped current in the solenoid is measured with the square loop normal. Second the measurement is repeated with the loop superconducting. Finally, a distribution of quantized flux levels is measured in the loop by repeatedly opening and closing switch and the response to a 1 ϕ_0 change is determined. These three independent measurements allow L_2, L_1, and M to be determined. The procedure is repeated over the range of temperatures allowed by the SQUID.

The results for the total inductance of the square loop are plotted in Fig 6b. Also shown is the best fit for the magnetic inductance which is very nearly temperature independent and kinetic inductance which diverges as the temperature approaches the critical temperature. However, the divergence from the BCS dirty limit is significantly slower than the phenomenological relation $1/(1-(T/T_c)^4)$ which is approximately correct for the clean limit but should not be used for this case. A better phenomenological relation is $1/(1-(T/T_c)^4)^{3/2}$. The best fit parameters for the BCS dirty limit are L_{2M} = 5.4 +/- 0.3 nH, $L_{2K}(0)$ = 3.3 +/- 0.2 nH and $\lambda(0)$ = 127 +/- 9 nm, whereas the no free parameter calculated values are L_{2M} = 5.8 +/- 0.1 nH, L_{2K} (0) = 3.1 +/- 0.4, and $\lambda(0)$ = 122 +/- 1 nm. We consider this fit to be excellent and the greatest source of error is due to the uncertainty in the film thickness (11 +/- 1.5 nm).

Note that these values for λ are about 50% larger than usually quoted for Nb films because below about 20 nm in thickness, the resistivity increases and the transition temperature decreases. Also note that for the BCS dirty limit the kinetic inductance can be written as $L_K(0) = (h/2\pi^2)\, R/\Delta$, which depends only on the normal state resistance R and T_c, through $\Delta = 1.76\, kT_c$, and is independent of film thickness and coherence length.

6. Reducing Low-Frequency Noise

In the previous sections, we have discussed a number of SQUID-based fundamental experiments. In all of them, we were forced to use the SQUID at low-frequencies, where the 1/f noise dominates the energy noise spectrum. Much progress has been made on reducing the 1/f noise in dc SQUIDs by making better junctions and by using bias current modulation [9]. However, that technique does not remove the so-called flux noise. In addition, the design for the lowest white noise SQUID, utilizing submicron junctions, often have worse 1/f noise. Instead, we are developing an input coil circuit modulation scheme which is an add-on component between the pickup coil and the SQUID input coil [10].

Fig 7a shows a schematic of the chopping idea. To demonstrate the idea, we have adapted the same circuit used to measure the kinetic

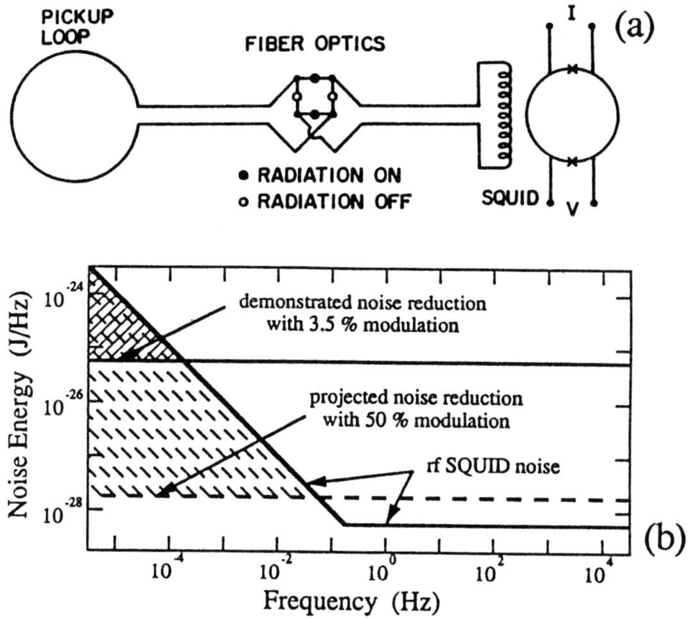

Fig 7 (a) Schematic of input circuit modulation. (b) Noise energy spectra from demonstration run with inductance modulation chip in Fig 6a.

inductance to modulate the SQUID input circuit. Unfortunately, only a 3.5% modulation depth is produced, but the experiment demonstrates improved low-noise performance and proof-of-principle.

Fig 7b shows the noise energy spectral density for our rf SQUID including the 1/f portion (solid line). By forming the difference output between the laser switch on and off SQUID output, we obtain the upper spectral density. Its white noise is $(1/.035)^2 \approx 1000$ times higher, but the low frequency noise is actually lower in the hashed region. We are currently developing a design which uses a dc-SQUID-style washer coil, we expect to obtain modulation depths of over 50%.

7. Conclusions and Acknowledgements

These experiments have demonstrated several interesting techniques utilizing the laser switch in SQUID measurements, and we are confident that these will continue to improve and others will be developed. The low noise reduction scheme has been shown to work, and is a promising way to significantly improve low-frequency or dc SQUID measurements.

The experiments described in this paper were performed by a number of students and colleagues. These include C. E. Cunningham and G. Park currently at Stanford, D. P. Saroff currently at NASA-Ames Research

Center, J. Tate and D. McIntyre currently at Oregon State University, J. T. Anderson currently at Hewlett-Packard, and M. Huber currently at University of Colorado at Denver. This research was funded in part by ONR Contracts N00014-87-K-0135 & N00014-90-J-1528, and NSF Grant DMR 84-05384.

References

[1] L. R. Testardi, *Phys. Rev.* **B4**, 2189 (1971); J. A. Pals, K. Weiss, P. M. T. M. van Attekum, R. E. Horstman, and J. Wolter, *Phys. Rep.* **89**, 323 (1982); and E. M. Gershenzon, M. E. Gershenzon, G. N. Gol'tsman, A. D. Semenov, and A. V. Sergeev. *Sov. Phys. JETP* **59**, 442 (1984).

[2] C. E. Cunningham, G. S. Park, B. Cabrera and M. E. Huber, *Physica* **B165**, 113 (1990).

[3] G. S. Park, C. E. Cunningham, B. Cabrera and M. E. Huber, Stanford Low Temperature Group Preprint BC101-91.

[4] J. Tate, B. Cabrera, S. B. Felch, and J. T. Anderson, *Phys. Rev. Lett.* **62**, 845 (1989).

[5] B. Cabrera, C. E. Cunningham, and D. Saroff, *Phys. Rev. Lett.* **62**, 2040 (1989).

[6] J. Q. Liang and X. X. Ding, *Phys. Rev. Lett.* **60**, 836 (1988).

[7] C. E. Cunningham, G. S. Park, B. Cabrera and M. E. Huber, Stanford Low Temperature Group Preprint BC98-91.

[8] P. B. Miller, *Phys. Rev.* **113**, 1209 (1959)

[9] V. Foglietti, W. J. Gallagher, and R. H. Koch , I*EEE Trans. Magnet.*, **Mag-23**, 1150 (1987).

[10] J. T. Anderson, B. Cabrera, M. A. Taber, S. B. Felch, and J. Tate, *Rev. Sci. Instrum.* **60**, 202 (1989); J. T. Anderson, B. Cabrera, and M. A. Taber, *Rev. Sci. Instrum.* **60**, 209 (1989); and C. E. Cunningham, G. S. Park, B. Cabrera and M. E. Huber, in preparation.

A New DC-SQUID System with 2f-Demodulation

H. Furukawa and K. Shirae

Faculty of Engineering Science, Osaka University,
Toyonaka, Osaka 560, Japan

Abstract. A new dc SQUID system with 2f-demodulation has been devised. The system consists of a main closed circuit and a supplemental closed circuit. The former has an ac amplifier, and responds to the input signal with high frequency. The latter is a novel circuit of the modulation, which utilizes modulation frequency, f and demodulation frequency, 2f. This circuit acts as a dc amplifier, and stabilizes the working point of the ac amplifier in the main circuit. The attained maximum slew rate is 5 x 10^6 $\phi_0 s^{-1}$ at 40 kHz.

1. Introduction

The dc SQUID voltage is a nonlinear function of the magnetic flux threading the dc SQUID ring. Most dc SQUID systems utilize a closed circuit with a modulation technique, called FLL (flux locked loop), to get the linear relation of input versus output. The dc SQUID system is required to have high slew rate in order to accept input signals with any amplitude and frequency. The modulation amplifier, however, limits the slew rate of the system. To overcome this limitation, a dc SQUID system using a two-pole integrator was reported [1]. The system has the maximum slew rate of 3 x 10^6 $\phi_0 s^{-1}$ at 6 kHz. Another type of the dc SQUID system without the FLL was also reported [2]. The system consists of a dc preamplifier and a dc SQUID with the enhanced output voltage by the additional positive feedback. The maximum slew rate is 7.8 x 10^6 $\phi_0 s^{-1}$ at 1.3 kHz.

We have developed a new dc SQUID system to obtain the high slew rate. The system consists of a main closed circuit using an ac amplifier and a supplemental closed circuit. The former responds to the input signal with high frequency and can achieve the high slew rate. The latter is a novel circuit of the modulation, which utilizes modulation frequency, f and demodulation frequency, 2f. The supplemental circuit gives a well stabilized working point of the ac amplifier in the main circuit. This system does not require any dc preamplifier.

Here, the principle of operation on our dc SQUID system and the experimental results are described.

2. Principle of Operation

As shown in Fig.1(a), the dc SQUID has the ϕ-V_{dc} characteristic with a period of ϕ_0 (the flux quantum). The dc SQUID system commonly utilizes the FLL to make

Springer Proceedings in Physics, Vol. 64 Superconducting Devices and Their Applications
Editors: H. Koch, H. Lübbig

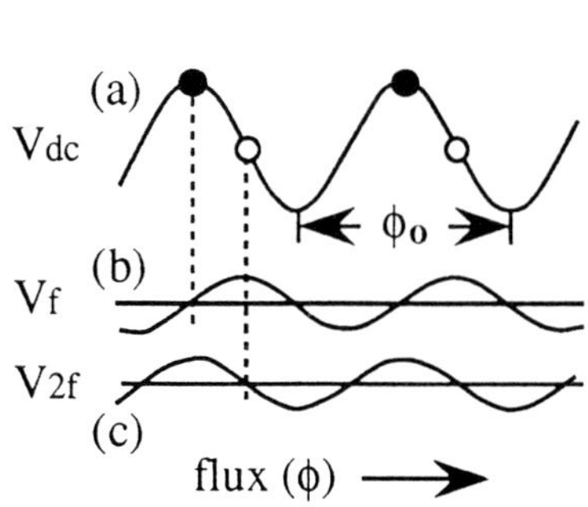

Fig.1. Flux versus voltage characteristic of the dc SQUID. (a) The f component (b) and the 2f component (c) of the dc SQUID voltage which has the AM wave caused by the carrier signal with the frequency, f and the input signal.

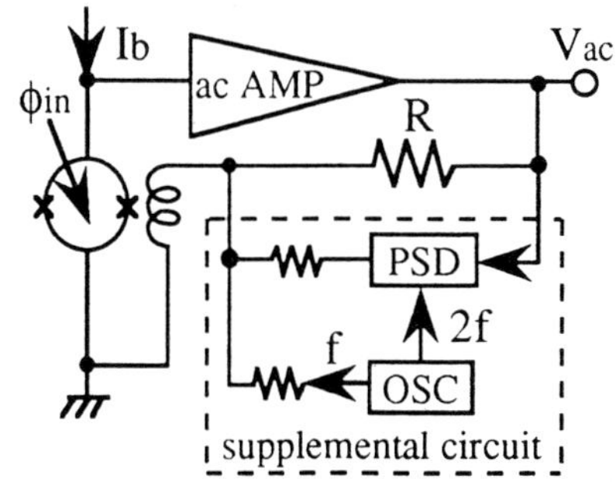

Fig.2. Simplified circuit diagram of the dc SQUID system. The system consists of the main closed circuit through the resistance, R and the supplemental closed circuit through the demodulator, PSD operated at the frequency, 2f.

the proportional output to the input signal. The FLL operates as follows: The nonlinearity of the dc SQUID produces the AM voltage by a modulation flux with high frequency, f and an input signal. The AM voltage includes the amplitude, V_f with the f component shown in (b). The conventional FLL uses V_f, and locks the operating point at one of the solid circles marked in (a).

Now, let us pay attention to the amplitude, V_{2f} with the 2f component in the AM voltage, as shown in (c). To make use of V_{2f}, we developed a circuit shown Fig.2. In the figure, the supplemental circuit has the demodulator, PSD operated at the frequency of 2f, therefore, the operating point of the system is locked at one of the linear regions shown by the open circles in the ϕ-V_{dc} characteristic. The locked point is the working point of the ac amplifier. The ac amplifier and the R make a main closed circuit, and respond to the input signal with high frequency. On the other hand, the ac amplifier and the supplemental circuit make a dc pass, and respond to the low frequency signal.

3. Experimental Results and Discussions

Figure 3 shows the experimental circuit diagram. The main circuit includes three amplifiers, A_{ac}, A_f and A_s, and their passbands are 1 Hz to 30 MHz, dc to 32 kHz and dc to 20 MHz, respectively. The overall voltage gain of the three amplifiers is 90 dB. The supplemental circuit includes the demodulator, PSD operated at 2f and the integrator, A_i with the time constant of 30 ms. The output voltage, V_{out} is fed back to the dc SQUID ring through R.

The dc SQUID is the Nb-VTB type which was fabricated by the double oblique deposition [3]. The pattern is the washer type of 1mm square, and the hole area is 50x50 μm^2 [4]. The SQUID output voltage is $35\mu V/(\phi_0/2)$.

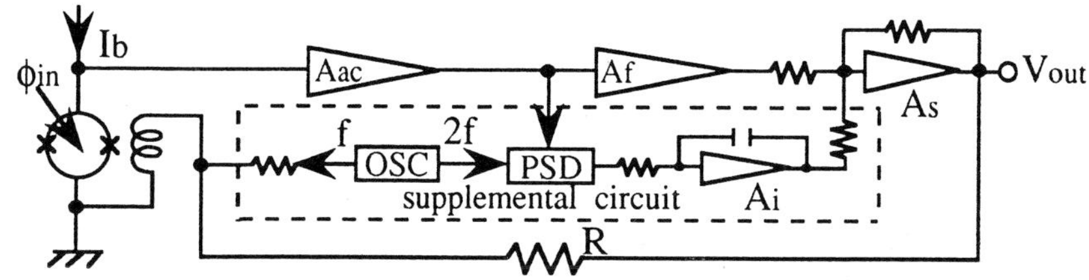

Fig.3. Experimental circuit diagram. The main circuit with three amplifiers of A_{ac}, A_f and A_s can respond to the input signal with high frequency. The supplemental circuit responds to the low frequency signal.

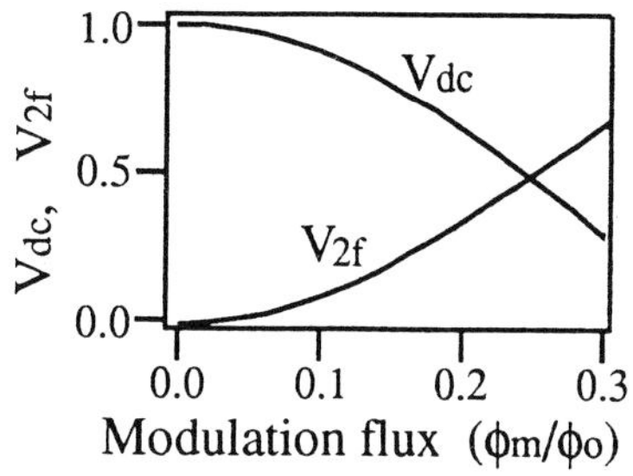

Fig.4. Normalized dc component, V_{dc} and f component, V_{2f} of the dc SQUID voltage with AM wave versus the amplitude of the sinusoidal carrier signal with the frequency, f.

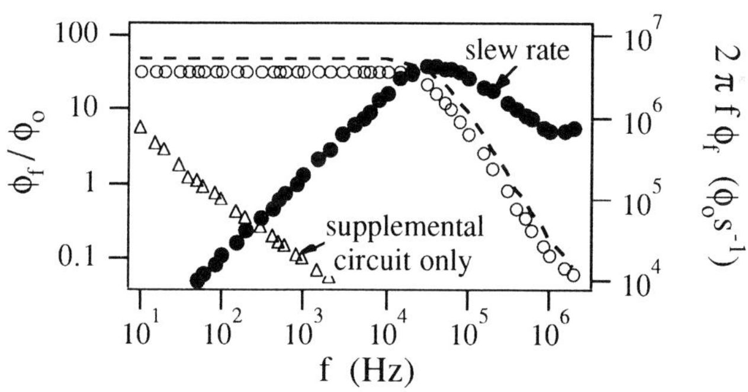

Fig.5. Maximum amplitude of the sinusoidal feedback flux, $\phi_f \sin 2\pi ft$ and the maximum slew rate, $2\pi f\,\phi_f$ versus the frequency, f.

As shown in Fig.4, with the increase of the modulation flux, ϕ_m, V_{dc} decreases and V_{2f} increases. As a compromise, we set 0.11 ϕ_0 to the modulation flux, which corresponds to 90% of the maximum V_{dc}. The frequency of the modulation flux is 11 MHz.

Figure 5 shows the maximum amplitude of the sinusoidal flux, $\phi_f \sin 2\pi ft$ that can be fed back without causing the system to unlock [1] [2]. The triangle marks show the amplitude under the operation of only the supplemental closed circuit (in this case, the gain of A_f is zero). The open circles show the amplitude of the whole system under operation. The dashed line shows the predicted amplitude from the measured open loop gain. The solid circles show the maximum slew rate, $2\pi f\, \phi_f$ [1] [2].

The experimental results show that the main closed circuit can respond to the input signal with high frequency, and the supplemental closed circuit acts as the dc amplifier. The experimental amplitude of the maximum feedback flux, ϕ_f is approximately 70% of the expected value. The maximum slew rate occurs at about 40 kHz, and is approximately $5 \times 10^6\ \phi_0 s^{-1}$. The system noise under the closed loop operation is $3 \times 10^{-5} \phi_0/Hz^{1/2}$ at the white noise region above about 20 Hz. The ac amplifier noise of $2nV/Hz^{1/2}$ is responsible for most of the system noise.

4. Conclusion

A new dc SQUID system with 2f-demodulation has been devised. The system consists of the main closed circuit using the ac amplifier and the supplemental closed circuit of the modulation type using the demodulator operated at 2f. The main closed circuit responds to the input signal with high frequency. The attained maximum slew rate is $5 \times 10^6\ \phi_0 s^{-1}$ at 40 kHz which is highest slew rate ever reported for such a high signal frequency.

References.

[1] Frederick Wellstood, C. Heiden and John Clarke : Rev. Sci. Instrum. 55 (1984) 952

[2] D. Drung, R. Cantor, M. Peters, H.J. Scheer and H. Koch : Appl. Phys. Lett. 57 (1990) 406

[3] H. Furukawa and K. Shirae : Physica B165&166 (1990) 79

[4] M. B. Ketchen : IEEE Trans. on Magnetics, MAG-23 (1987) 1650

Relaxation Oscillating SQUIDs Using Nb/AlO$_x$/Nb Josephson Tunnel Junctions

J. Kawai, G. Uehara, N. Mizutani, Y. Kondo, N. Harada, and H. Kado

Superconducting Sensor Laboratory, 5F Tsukuba Mitsui Bldg., 1-6-1 Takezono, Tsukuba 305, Japan

Abstract. We fabricated Relaxation Oscillating SQUIDs using Nb/AlOx/Nb Josephson tunnel junction techniques and demonstrated that these SQUIDs oscillated at frequencies from 1GHz up to 1.6GHz by means of changing external magnetic flux from zero to $1/2\Phi_0$. A relative linewidth of the spectrum of the order of 10^{-5} was obtained. These results are expected to make the flux-resolution high.

1. Introduction

A dc-SQUID shunted with a resistor and an inductor can function as a Relaxation Oscillating SQUID (ROS), which converts a change in a magnetic field into a frequency deviation relatively easy to digitalize and multiplex. A ROS with nanobridge Josephson junctions has been investigated and the outputs were detected easily by existing FM read-out methods. [1] [2]

We report on the fabrication and characteristics of a ROS composed of Nb/AlOx/Nb tunnel junctions.

2. Design of ROS

The schematic circuit is illustrated in Fig.1. A resistor R_S and an inductor L_S are connected in parallel to a hysteretic dc-SQUID. A ROS can oscillate when a bias current I_B is provided above the critical current of the SQUID I_C. The oscillating frequency of a ROS is given by [3]

$$f = \frac{R_S}{L_S}\left[\ln\left(\frac{I_B - I_m}{I_B - I_C}\right) + \ln\left(\frac{I_B - V_g/R_S - I_C}{I_B - V_g/R_S - I_m}\right)\right]^{-1}$$

where V_g is the gap voltage of the tunnel junction. I_m is the minimum critical current of the SQUID, which is related to I_C as $I_m/I_C=\alpha$. Here α is the McCumber factor, depending on the critical current density. The condition of the bias currents for oscillation becomes broader at a larger α. Since the critical current I_C is modulated with a changing magnetic flux Φ penetrating the SQUID loop, the frequency f also changes with the magnetic flux. The modulation depth of I_C depends on the screening parameter β_L, which is proportional to the products of L_0 and I_C. Here L_0 is the loop inductance of SQUID. In this experiment β_L is designed to be 1, giving a modulation depth of 50%.

Springer Proceedings in Physics, Vol. 64 Superconducting Devices and Their Applications
Editors: H. Koch, H. Lübbig

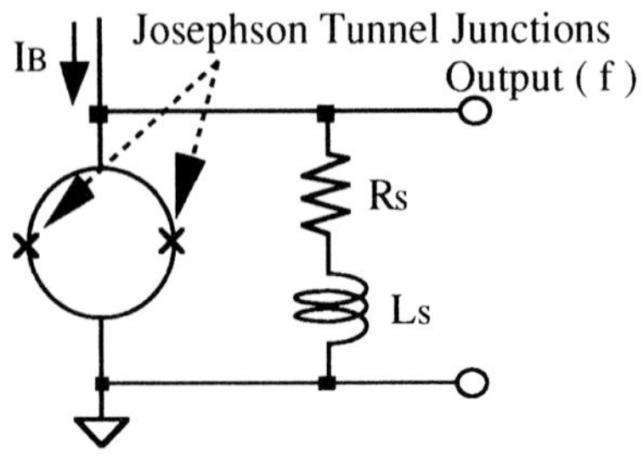

Fig. 1 Schematic circuit of a ROS

3. Fabrication

A schematic cross section of a ROS is shown in Fig. 2. ROS's are fabricated on a Si substrate using a thin-film technique. [4] First, an Nb/AlOx/Nb tri-layer was formed using an underlayer method to improve the I-V characteristics of the tunnel junction. The tunnel barrier AlO_x was formed by oxidizing the surface of Al in 133Pa-O_2 gas for 15minutes at room temperature. The critical current density was held at 2.5kA/cm2. The tunnel junction area was defined to be 4μm2 so that the SQUID had a critical current of 200μA. The junctions, base and counter electrodes, SQUID washers and shunt inductors were patterned by an RIE process with CF_4 gas. The surface of the Nb was nitrogenized with N_2 rf plasma to avoid oxidation. The shunt inductor was 1mm long and its inductance was estimated as ≈1nH. The shunt resisters were formed by patterning the evaporated $AuIn_2$ using a lift-off process. The sheet resistance was held at ≈0.4Ω at a thickness of 200nm by an annealing process.

The I-V characteristics of the tunnel junction is shown in Fig. 3. In Fig. 3 (b) the axis of the current is expanded 10 times so as to show the value of α clearly. In this fabrication α was ≈ 0.1, which is large enough for our experiments.

The device parameters are summarized in Table 1, where the critical current and the shunt resistances were measured values of test elements, and the others were estimated values.

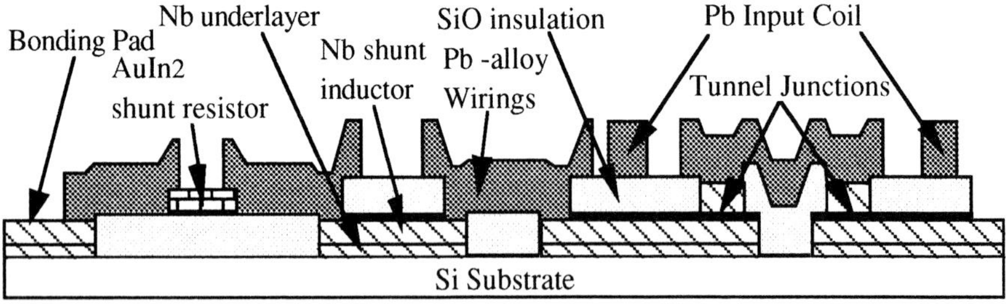

Fig. 2 Schematic cross-section of a ROS chip

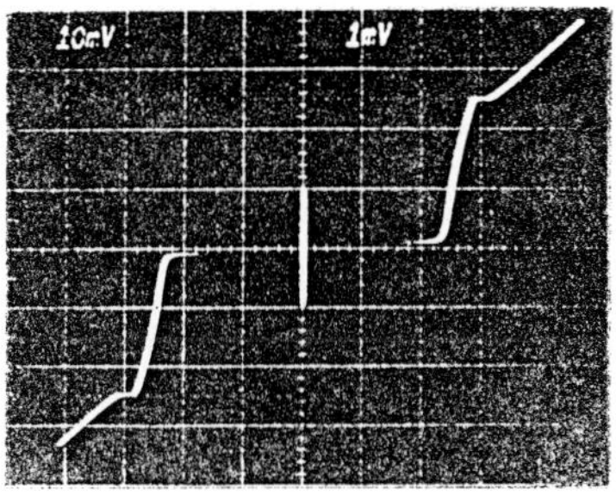

Current (100µA/div.)

Voltage (mV/div.)

(a)

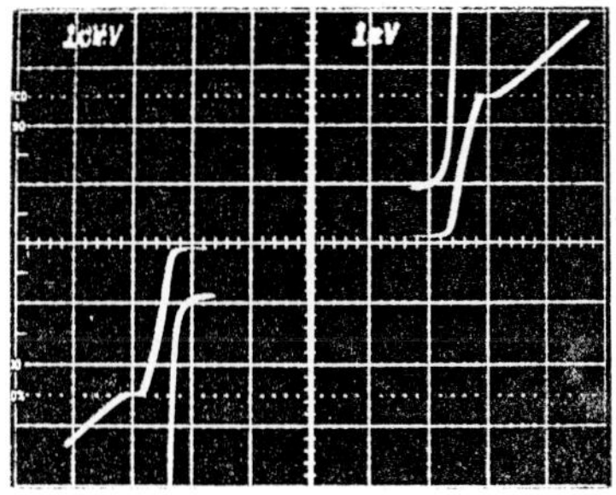

Current (10µA/div.)
Current (100µA/div.)

Voltage (mV/div.)

(b)

Fig. 3 I-V characteristics of the junction

Table 1 Device parameters of a ROS

Parameter	Value
junction size	2µm × 2µm
critical current of SQUID	≈ 200µA
inductance of SQUID	10pH
shunt inductance	1nH
shunt resistance	≈ 0.2, 0.8Ω

4. Experimental Results

A ROS chip was settled on a brass block and connected to a semi-rigid cable through a 50Ω micro-stripline. The output power of the ROS was amplified through a 47dB amplifier and was examined by a spectrum analyzer. The dc-bias current was fed with a regulated voltage supply through a 100kΩ resistor. The frequency spectrum of the ROS oscillating at 1 GHz is shown in Fig. 4. The relative linewidth of the frequency at 3dB below the peak is in the range of 10^{-5}, which is better than any other ROS ever reported.[1] The output power is evaluated as ≈ 1nW.

The frequency changes periodically with every magnetic flux of Φ_0 penetrating the SQUID loop. Fig. 5 shows a fundamental oscillating frequency of 0.499GHz and a maximum frequency of 0.875GHz with $1/2\Phi_0$.

The oscillating frequency increases with an increasing bias current. The dependence of the frequencies on various currents for different magnetic fluxes of zero and $1/2\Phi_0$ are shown in Fig. 6, where (a) and (b) are with different shunt resistances of 0.2Ω and 0.8Ω respectively.

The solid lines are fitting curves. The frequency change with 0.8Ω is from 1.0GHz to 1.6GHz when the bias current is 0.7mA. Since the modulation depth is smaller than the designed value, β_L is larger than 1. Therefore, SQUID loop inductance L_0 is larger than 1nH, which seems to be caused by the parasitic inductances.

Investigating in detail the spectrum and the modulation at the fundamental frequency of 1GHz, we evaluated the flux resolution to be $\approx 10^{-7}\Phi_0/Hz^{1/2}$. [5]

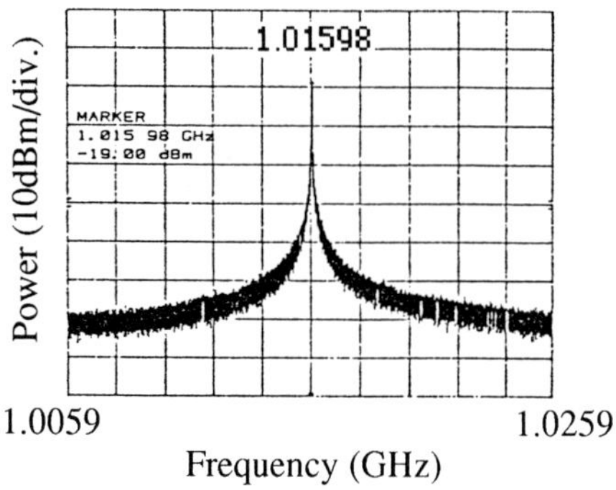

Fig. 4 Frequency spectrum

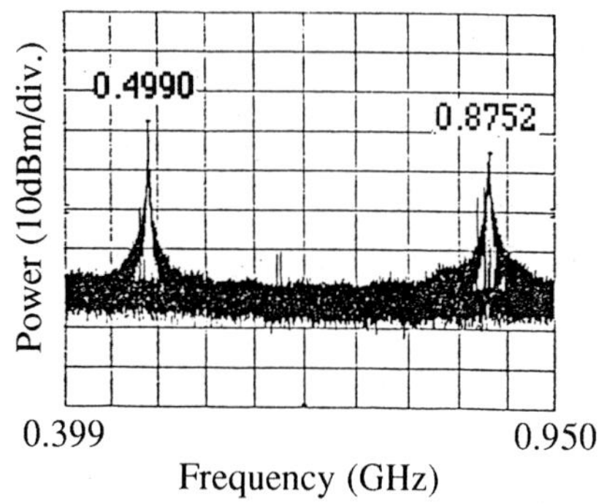

Fig. 5 Frequency spectrum with different magnetic flux

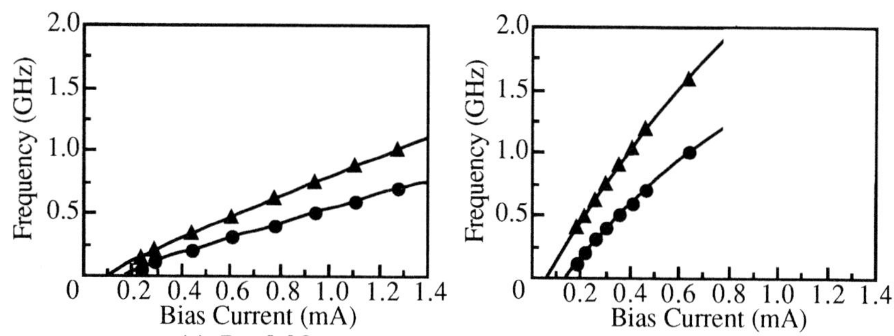

(a) Rs=0.2Ω (b) Rs=0.8Ω

Fig. 6 Dependence of frequency on bias current ● : $\Phi = 0$, ▲ : $\Phi = 1/2\Phi_0$

5. Conclusion

Relaxation oscillation SQUIDs with reliable tunnel junctions of Nb/AlOx/Nb exhibit useful properties like a high flux-resolution and good stability in the oscillating frequency. These properties make the ROS a promising device for a new type of multi-channel SQUID system and other applications. ROS makes a good match to other Josephson devices such as SQUID amplifiers and Josephson counters.

6. Acknowledgement

This work was supported by the Electrotechnical Laboratory. We would like to thank Dr. H. Nakagawa and Dr. S. Takada for their helpful advice on the design and fabrication of ROS.

References

[1] M. Mück, H.Rogalla and H. Heiden, Appl. Phys. A, vol. 46, 97 (1988)
[2] M. Mück and H. Heiden, IEEE Trans. on Mag., vol. 25-2, 1151 (1989)
[3] F. L. Vernon, Jr and R. J. Pedersen, J. Appl. Phys., vol. 39-6, 2661 (1968)
[4] H. Nakagawa and S.Takada, et al., IEEE Trans. on Mag., vol. 27-2 (1991)
[5] Y Kondo and G Uehara, et al., EIC Conference SCE91-4 (1991) (in Japanese)

DC-SQUID Electronics Based on Adaptive Noise Cancellation and a High Open-Loop Gain Controller

H. Seppä

Metrology Research Institute, VTT, Technical Research Centre of Finland, Otakaari 7 B, SF-02150 Espoo, Finland

Abstract. A low-noise SQUID readout electronics with a high slew rate and an automatic gain control feature has been developed. Flux noise levels of $5 \times 10^{-7}\ \Phi_0/\sqrt{Hz}$ at 1 kHz and $2 \times 10^{-6}\ \Phi_0/\sqrt{Hz}$ at 1 Hz have been measured with this readout scheme. The system tolerates sinusoidal disturbances having amplitudes up to 140 Φ_0 at 1 kHz without loosing lock. The electronics utilizes a cooled GaAs FET to control the cancellation of the voltage noise of the room temperature amplifier, a $PI^{3/2}$ controller to provide a high open-loop gain at low frequencies, and a square-wave flux and offset voltage modulation to enable automatic control of the noise reduction. The cutoff frequency of the flux-locked-loop is 300 kHz and the feedback gain is more than 130 dB at 10 Hz.

1. Introduction

For many applications, improving the SQUID readout electronics is as important as reducing the SQUID noise. The field sensitivity of a SQUID magnetometer or gradiometer can be enhanced simply by enlarging the size of the field sensing coil. A good electronics together with a low-noise SQUID magnetometer having smooth characteristics provides a high dynamic range. To increase the slew rate or to minimize the distortion, the feedback gain should be maximized. The most important task of the feedback electronics is to anchor the intrinsic flux tightly at the set value so that the periodic $\Phi - V$ characteristics of the SQUID do not disturb the operation of the magnetic sensor.

Flux modulation offers a simple way to match the low output impedance of the dc SQUID to a low-noise room temperature amplifier. Consequently, the noise contribution of the readout electronics can be avoided. In addition, the phase detection scheme effectively eliminates some sources of low frequency noise. The main disadvantages are the need for a cooled transformer and the bandwidth limitation. The signal bandwidth can be extended by increasing the modulation frequency, but the noise temperature of the preamplifier, unfortunately, tends to increase as well. The use of positive feedback [1,2,3,4] and a directly-coupled amplifier provide an easy way of increasing the bandwidth of the readout electronics without markedly deteriorating the noise performance of the SQUID. The lack of the demodulation circuits also simplifies the electronics, making such an approach very attractive for multi-channel applications.

Springer Proceedings in Physics, Vol. 64 Superconducting Devices and Their Applications
Editors: H. Koch, H. Lübbig

The disadvantage of the positive feedback technique is the depression of the linear region of the SQUID, thereby affecting the slew rate in the flux-locked-loop mode. Fortunately, the improvement in the signal bandwidth compensates the loss in dynamic range [2]; this makes the positive feedback technique very tempting in many applications.

A cooled FET incorporated in the SQUID circuit [3,4] provides control of the positive feedback and thus enables a free choice of the operation point. In addition, a tunable resistor allows the automation of the noise reduction. For the voltage biased SQUID, the positive feedback cannot take place at all, but the contribution of the preamplifier noise can still be reduced [4]. In that mode the feedback coil senses the noise voltage of the amplifier, and the SQUID feeds it back to the amplifier, decreasing the noise. The flux vs. voltage response is not inclined as it is when the positive feedback affects the response, and thus the dynamic range of the SQUID ($\pm\Phi_0/4$) remains unchanged. In addition, the flux-to-current conversion efficiency, $\partial I/\partial\Phi$, is nearly independent of the point of operation, [5] yielding very triangular-shaped $I-\Phi$ characteristics. For the previous reasons we prefer the adaptive noise cancellation method to the positive feedback technique.

2. Electronics

The block diagram of the electronics is shown in Fig. 1. A voltage source, E_s, determines the point of operation, since the operational amplifier zeros the output of the SQUID via a feedback resistor, R_{op}. Consequently, the FET becomes almost voltageless; only the offset voltage and the noise voltage of the amplifier appear across the FET. Next the signal is fed through a $PI^{3/2}$ controller [6] which provides the slope of the gain curve of 30 dB/decade over 5 decades. The flux-locked-loop is completed by feeding the output of the controller into the feedback coil through a resistor R_{nfb}. The flux and the offset voltage are simultaneously modulated. The calculations given in Ref. [4] show that, for a particular value for the feedback resistor R_{pfb}, the effective flux modulation caused by the offset modulation cancels the actual flux modulation. A demodulation circuit, a PI controller, and the FET shown in Fig. 1a drive the system to that point of operation, thus stabilizing the noise reduction efficiency.

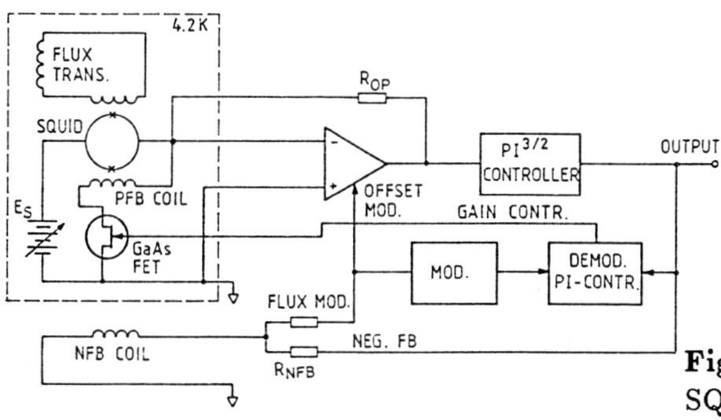

Figure 1. Block diagram of the SQUID readout electronics.

An LT1028 operational amplifier was used as a preamplifier. The $PI^{3/2}$ controller was mainly accomplished by changing the integrating capacitor as a function of frequency so that the slope of the gain curve approached 30 dB/decade. The adaptive gain control circuit was realized with an LT1043 which served as an oscillator and several switches for modulation and demodulation circuits. The whole readout electronics contains no more than four microcircuits, and the size of the circuit board is about 10 cm × 10 cm.

3. Measurement Results

Figure 2a shows the $\Phi - I$ characteristic of the voltage biased SQUID [7] for different gate voltages, V_g . With increasing V_g, the square-wave modulation, superimposed on the response, disappears on one slope and increases on the other. For low values of V_g the system starts to oscillate, which causes the $I - \Phi$ response to bend slightly (not clearly seen in Fig. 2a). Since the FET is voltageless, the flux is not removed, although the channel resistance is varied. This is a very important feature with respect to the automatic control of the noise reduction. The noise spectra of the SQUID [8] with and without the automatic gain control are illustrated in Fig. 2b. The proper adjustment of the gain control makes it possible to reach the intrinsic flux noise of the SQUID. In this case the noise at 1 kHz is about $5 \times 10^{-7}\Phi_0/\sqrt{Hz}$, and $2 \times 10^{-6}\Phi_0/\sqrt{Hz}$ at 1 Hz, both measured in the flux-locked mode. Another SQUID from the same wafer, measured using a dc SQUID as a preamplifier [5], exhibited very much the same features. We conclude that the readout electronics described here do not noticeably deteriorate the noise performance of the SQUID.

Figure 3 summarizes the slew rate, open-loop gain, closed-loop gain, and the noise spectra. The open-loop gain decreases 30 dB/decade from 10 Hz to 30

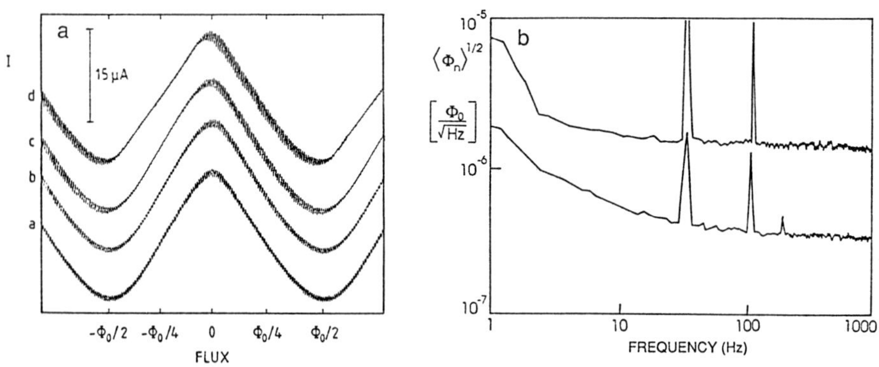

Figure 2. (a) $\Phi - I$ characteristics of the voltage biased SQUID for four different gate voltages a) -1.5 V, b) -0.7 V, c) -0.6 V, d) -0.4 V. The modulation superimposed on the response is caused by the modulation of the offset voltage of the preamplifier. (b) Flux noise spectral density with (lower curve) and without (upper curve) the automatic noise reduction.

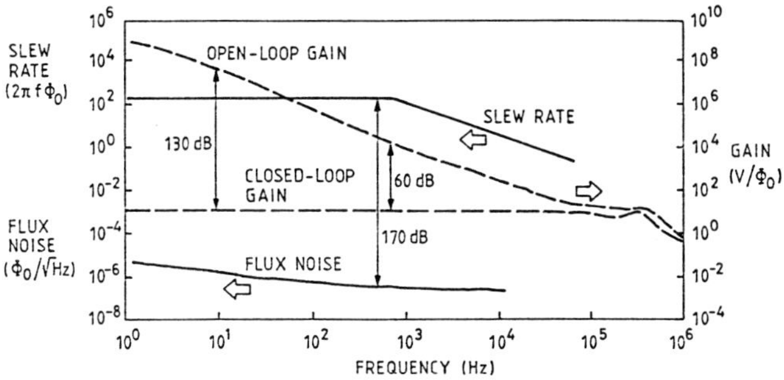

Figure 3. Slew rate, open-loop gain, closed-loop gain, and the flux noise as a function of frequency f. The slope of the open-loop gain curve is approximately 30 dB/decade from 10 Hz to 100 kHz. The cutoff frequency of the closed-loop gain is 300 kHz.

kHz. The slew rate starts to decrease at 1 kHz, at the point where the feedback gain is about 60 dB. The result is very resonable, because in the voltage biased mode, even at 1 kHz, the dynamic range of the SQUID alone is near $\pm\Phi_0/4$. Because of the high slew rate and the low flux noise, the dynamic range of the system can be enormously high, more than 170 dB at 1 kHz.

3. Conclusion

We have shown that an extremely high slew rate and low-noise performance can be achieved by using a SQUID in the voltage biased mode, a $PI^{3/2}$ controller, and a cooled FET as a tunable resistor to control the noise reduction. In addition, the electronics adapts to changes of the SQUID parameters and the environment. The electronics, together with a low-noise SQUID having a sufficiently small pick-up coil, is well suited to be used in noisy environments, e.g., in hospitals.

Acknowledgement

I thank R. Cantor from PTB and L. Grönberg from VTT who fabricated the SQUIDs used to test the electronics. I also express my gratitude to T. Ryhänen and M. Kiviranta for helping me during this work.

References

[1] D. Drung, R. Cantor, M. Peters, H.J. Scheer, and H. Koch, Appl. Phys. Lett. **57**, 406 (1990).

[2] D. Drung, R. Cantor, M. Peters T. Ryhänen, and H. Koch, IEEE Trans. Magn **MAG-27**, 301 (1991).

[3] H. Seppä, A. Ahonen, J. Knuutila, J. Simola, and V. Vilkman, IEEE Trans. Magn. **MAG-27**, 2488 (1991).
[4] H. Seppä, submitted to Rev. Sci. Instrum.
[5] T. Ryhänen, H. Seppä, R. Cantor, D. Drung, H. Koch, and D. Veldhuis, these proceedings.
[6] H. Seppä and H. Sipola, Rev. Sci. Instrum. **61**, 2449 (1990).
[7] R. Cantor, T. Ryhänen, D. Drung, H. Koch, and H. Seppä, IEEE Trans. Magn **MAG-27**, 2927 (1991).
[8] L. Grönberg, H.Seppä, J. Salmi, M. Kiviranta, I. Suni, T. Ryhänen, and R. Cantor, these proceedings.

Investigation of a Double-Loop DC-SQUID Magnetometer with Additional Positive Feedback

D. Drung

Physikalisch-Technische Bundesanstalt (PTB), Abbestr. 10–12, W-1000 Berlin 10, Fed. Rep. of Germany

Abstract. The essential performance parameters (noise, dynamic range, stability) of the PTB direct-coupled magnetometer W5 are presented. The complete magnetometer is integrated on a 7.2×7.2 mm^2 chip. Using a simplified feedback loop without flux modulation, a very low white noise level of typically 2.3 fT/$\sqrt{\text{Hz}}$ is achieved with sufficiently wide margings for the SQUID parameters. Even large preamplifier offset voltage changes of ±4 μV increase the white noise level by only about 20 %. Output signal fluctuations of the order of ±100 fT during periods of 24 hours have been observed with the magnetometer enclosed by a Pb shield.

1. Introduction

A direct-coupled dc SQUID magnetometer (version W5) for biomagnetic multi-channel systems has recently been developed in our laboratory [1]. All cryogenic parts are integrated on a single 7.2×7.2 mm^2 chip fabricated using a four-level Nb/Si_xN_y/Nb process [2]. Additional Positive Feedback (APF) on the sensor chip increases the gradient of the SQUID voltage versus flux (V–Φ) characteristic so that a cooled impedance matching circuit is not required [1]. A Direct Offset Integration Technique (DOIT) without flux modulation greatly simplifies the feedback electronics and eliminates crosstalk and mutual disturbances between SQUID channels caused by the traditional flux modulation [3]. In this paper, it is shown that APF leads to a very low total noise with sufficiently wide margins for the SQUID parameters, even if large preamplifier offset voltage fluctuations of up to ±4 μV are tolerated. Furthermore, it is demonstrated that the long term stability of this SQUID system using DOIT is sufficient for biomagnetic applications.

2. Measurements

The equivalent circuit of the magnetometer and its feedback electronics is depicted in Fig. 1. The circuit is essentially the same as that described in Ref. [1] in detail. Due to the limited space, only the differences between the circuit in Fig. 1 and that in Ref. [1] will be discussed here:

(1) The bias current I_b of the magnetometer is inserted symmetrically into one of the eight pick-up loops instead of asymmetrically into one of the Josephson

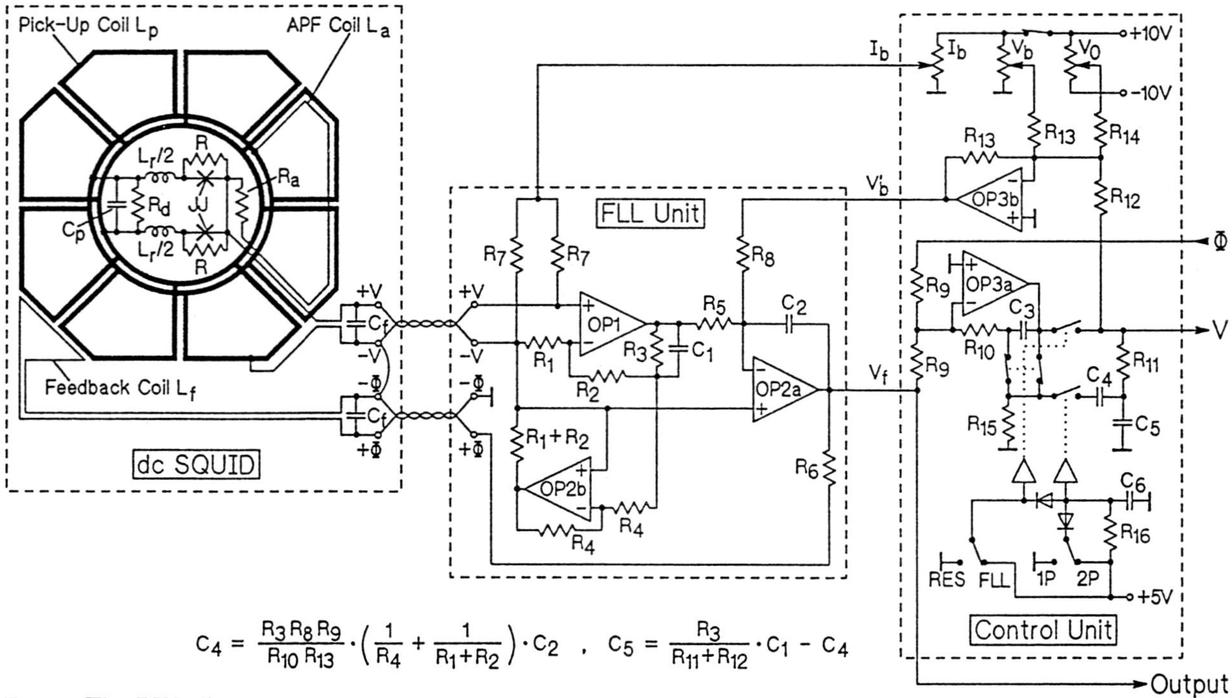

$$C_4 = \frac{R_3 R_8 R_9}{R_{10} R_{13}} \cdot \left(\frac{1}{R_4} + \frac{1}{R_1 + R_2} \right) \cdot C_2 \quad , \quad C_5 = \frac{R_3}{R_{11} + R_{12}} \cdot C_1 - C_4$$

Fig. 1. The PTB direct-coupled magnetometer W5 with feedback electronics. Typical SQUID parameters are: L_a = 14 nH, M_a = 330 pH (mutual inductance L_a,L_P), M_f = 62 pH (mutual inductance L_f,L_P), L_r = 34 pH, $L = L_P + L_r$ = 400 pH, C_f = 100 pF, C_P = 10 pF, C = 1.3 pF (JJ capacitance), I_0 = 6.5 μA (JJ critical current), R_d = R = 10 Ω, R_a = 2×R = 20 Ω. The component values are: R_1 = 4.7 Ω, R_2 = 1 kΩ, R_3 = 18 kΩ, R_4 = 3.9 kΩ, R_5 = 10 kΩ, R_6 = 4.7 kΩ, R_7 = 200 kΩ, R_8 = 100 kΩ, $R_9 = R_{10}$ = 33 kΩ, R_{11} = 47 Ω, R_{12} = 3.6 kΩ, R_{13} = 18 kΩ, R_{14} = 68 kΩ, R_{15} = 10 kΩ, R_{16} = 330 kΩ, C_1 = 47 nF, C_2 = 150 pF, C_3 = 330 pF, C_4 = 18.8 nF, C_5 = 213 nF, C_6 = 100 nF, OP1 = LT 1028, OP2 = AD 708, OP3 = LT 1057.

junctions JJ, and two filter capacitors $C_f = 100$ pF at helium temperature have been added in order to reduce broadband noise disturbing the SQUID via the voltage (±V) and flux (±Φ) leads. In this way, the gradient in the voltage versus flux (V-Φ) characteristic has been increased by typically a factor of two.

(2) The SQUID system in Ref. [1] used a double (two-pole) integrator in order to achieve a high maximum slew rate [4]. Due to the second integrator (preamplifier OP1) the system could only be switched into the flux-locked loop (FLL) mode if the SQUID voltage in the reset (RES) mode was approximately equal to the bias voltage V_b, i.e., if the applied flux was suitably chosen. Feedback loops with a single (one-pole) integrator properly lock with any value of the applied flux. Therefore, the option to use a one-pole (1P) instead of a two-pole (2P) integrator has been realized in Fig. 1, and the feedback loop is automatically switched into the one-pole mode for about 7 ms ($\approx 0.2 \times R_{16}C_6$ for the CMOS switch DG 413 from Siliconix used in Fig. 1) after each reset.

The electronics in Fig. 1 allow the complete characterization of the SQUIDs: The V-Φ characteristics are measured in the RES mode, and the noise and dynamic performance in the FLL mode using the two-pole integrator [1]. For all measurements the magnetometers were immersed in liquid helium (T = 4.2 K) and enclosed by a Pb shield. The sensitivity $B/\Phi = 0.47\ \mathrm{nT}/\Phi_0$ (B is the average flux density in the pick-up coil and Φ_0 is the flux quantum) has been measured without a Pb shield in a magnetically shielded room. The magnetometers have been fabricated using a fast and simple four-level $Nb/Si_xN_y/Nb$ process with Pd shunts and anodized Nb (Nb_2O_5) as insulation layer [1]. A few of the devices (marked by squares in Fig. 2 and by triangles in Figs. 3 and 4) had an additional layer of Si_3N_4 [5] in order to reduce the parasitic capacitance C_P from 10 pF to 2 pF.

A double-loop structure [6] has been chosen for the SQUID which provides an additional working point near the resonance voltage of the small loop (formed by the Josephson junctions JJ, the inductances $L_r/2$, and C_P acting as a short) at a bias voltage $V_b \approx 80\ \mu V$ [1]. At this $L_rC/2$ resonance, the minimum total noise is found for $\beta_L = 2LI_0/\Phi_0$ and $\beta_C = 2\pi I_0 R^2 C/\Phi_0$ ranging between 2 and 3, where L = 400 pH is the SQUID inductance, C = 1.3 pF is the junction capacitance, I_0 is the junction critical current, and R is the shunt resistance (cf. Fig. 2). The noise is minimum for a bias current which maximizes the SQUID voltage swing ΔV, except for large values of β_L where a somewhat larger bias current minimizes the noise (most probably due to intrinsic hysteresis caused by APF for positive feedback gains > 1, cf. Fig. 2c). The system noise is typically 2.3 $\mathrm{fT}/\sqrt{\mathrm{Hz}}$ at 500 Hz, 5 $\mathrm{fT}/\sqrt{\mathrm{Hz}}$ at 1 Hz, and 14 $\mathrm{fT}/\sqrt{\mathrm{Hz}}$ at 0.1 Hz. The feedback range is ±30 nT for frequencies up to about 250 Hz (large signal bandwidth), and is limited by the maximum slew rate for higher frequencies. The small signal bandwidth is typically 80 kHz.

The performance deteriorates if the bias voltage is changed by ΔV_b due to preamplifier offset voltage fluctuations, which are typically within about ±2 μV for several months of operation. Experimentally we found a dependence

$$\sqrt{S_{\Phi i}}/\sqrt{S_{\Phi i}(W)} \approx 1 / \cos(2\times\Delta V_b/\Delta V) \quad , \quad V_\Phi/V_\Phi(W) \approx \cos(3.6\times\Delta V_b/\Delta V) \quad , \quad (1)$$

for $|\Delta V_b/\Delta V| < 0.4$, where $\sqrt{S_{\Phi i}}$ is the intrinsic flux noise of the SQUID, V_Φ =

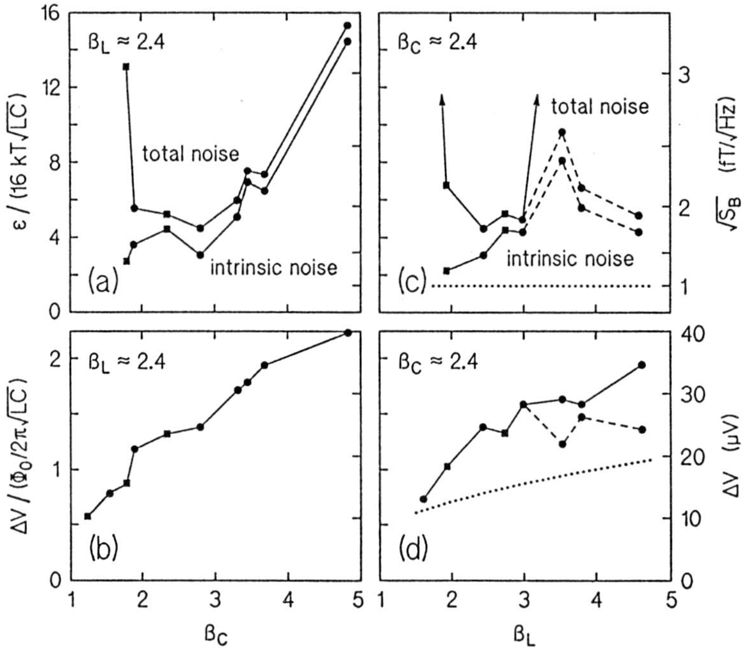

Fig. 2. Normalized white energy sensitivity $\epsilon = S_\Phi/2L$ and normalized voltage swing (a),(b) vs. β_C for constant $\beta_L \approx 2.4$ and (c),(d) vs. β_L for constant $\beta_C \approx 2.4$. For the circles, C = 1.3 pF and C_P = 10 pF, and for the squares, C = 0.85 pF and C_P = 2 pF. The right scales give the flux density noise in fT/$\sqrt{\text{Hz}}$ and the voltage swing in μV for C = 0.85 pF. The SQUIDs were operated at the $L_rC/2$ resonance with a bias current which either maximized the voltage swing (solid lines) or minimized the noise (dashed lines). The dotted lines correspond to the theoretical dependences, $\epsilon \approx 16\,kT\sqrt{LC/\beta_C}$ and $\Delta V \approx I_0R/(2+R/R_a)$, at the traditional working point with a bias current near the SQUID critical current (β_C = 0.5, C_P = 0, [7]).

$\partial V/\partial\Phi$ is the V-Φ gradient, ΔV is the SQUID voltage swing, and (W) denotes the nominal values (cf. Fig. 3). Using a typical value ΔV = 20 μV, a large offset voltage drift ΔV_b = ±4 μV would increase the total white noise by only about 20 % to typically 2.7 fT/$\sqrt{\text{Hz}}$. The large signal and small signal bandwidth would be reduced by about 30 % to typically 175 Hz and 56 kHz, respectively.

The long term stability of the SQUID system has been investigated inside a rf shielded room, where the room temperature was stable within about ±0.5 K. Output signal fluctuations of the order of only ±100 fT during periods of 24 hours have been observed (cf. Fig. 4).

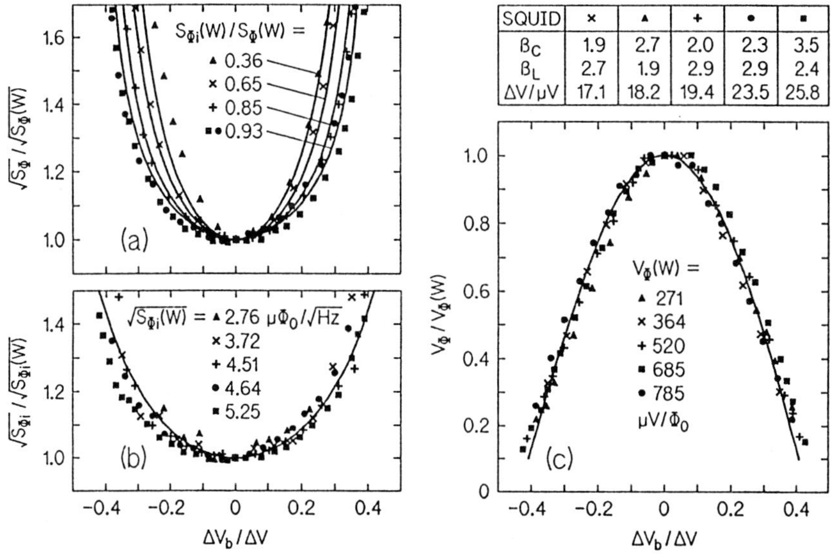

SQUID	×	▲	+	•	▪
β_C	1.9	2.7	2.0	2.3	3.5
β_L	2.7	1.9	2.9	2.9	2.4
ΔV/μV	17.1	18.2	19.4	23.5	25.8

Fig. 3. Normalized (a) total white flux noise, (b) intrinsic white flux noise, and (c) V-Φ gradient vs. normalized bias voltage offset for five representative SQUIDs operated at the $L_rC/2$ resonance. Here C = 1.3 pF and C_P = 10 pF, except for SQUID ▲ which has C = 0.85 pF and C_P = 2 pF. (W) denotes the measured values at the nominal working points. The solid lines are calculated with Equ. (1).

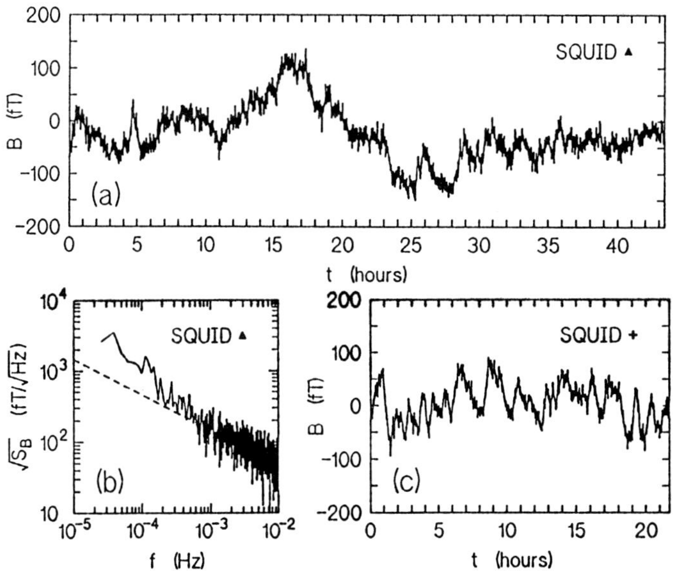

Fig. 4. (a) Flux density noise vs. time for SQUID ▲ in Fig. 3. The bandwidth is 0 - 0.01 Hz. The room temperature was constant within about ±0.5 K. (b) Noise spectrum calculated from curve (a). The dashed line is extrapolated from the 0.1 Hz noise level assuming $S_B \sim 1/f$. (c) Same as (a), but for SQUID + in Fig. 3.

3. Conclusions

It is shown that the PTB direct-coupled magnetometer W5 allows a stable and low-noise operation in a feedback loop without flux modulation, even if preamplifier offset voltage fluctuations of up to ±4 μV (±20 % of the SQUID voltage swing) are tolerated. Based on the investigations with W5, an improved version W6 has been developed where the junction size and capacitance have been reduced by about a factor of two, and the APF resistor R_a and mutual inductance M_a have been increased by a factor of 2.5 in order to reduce the resistive load of the SQUID [3]. The improved version W6 typically has a white noise level of 1.8 fT/$\sqrt{\text{Hz}}$, a SQUID voltage swing of 30 μV, and tolerates offset voltage fluctuations of up to ±6 μV.

Acknowledgements. I wish to thank D. May for assistance with the preparation of the photolithography masks, and M. Peters for fabrication of the magnetometers.

References

[1] D. Drung, R. Cantor, M. Peters, T. Ryhänen, and H. Koch, IEEE Trans. Magn. **MAG-27**, 3001 (1991).
[2] R. Cantor, D. Drung, M. Peters, and H. Koch, J. Appl. Phys. **67**, 3038 (1990).
[3] D. Drung, in *Extended Abstracts of Third Int. Superconductive Electronics Conf.* (Glasgow, 1991), p. 252 and Supercond. Sci. Technol. (to be published).
[4] F. Wellstood, C. Heiden, and J. Clarke, Rev. Sci. Instrum. **55**, 952 (1984).
[5] R. Cantor, T. Ryhänen, D. Drung, H. Koch, and H. Seppä, IEEE Trans. Magn. **MAG-27**, 2927 (1991).
[6] C.D. Tesche, J. Low Temp. Phys. **47**, 385 (1982).
[7] K. Enpuku, K. Yoshida, and S. Kohjiro, J. Appl. Phys. **60**, 4218 (1986).

Switching Probability and Performance in Single-Chip SQUIDs

N. Fujimaki, K. Gotoh, and S. Hasuo

Fujitsu Laboratories Ltd., 10-1 Morinosato-Wakamiya,
Atsugi 243-01, Japan

Abstract. We measured the switching probability of a two-junction SQUID comparator integrated with a superconducting feedback circuit on one chip for use in biomagnetism. The switching probability, its derivative to input flux, and the flux spectral density measured by varying sinusoidal bias frequency and amplitude agree excellently with the thermal-noise theory. The best sensitivity and cutoff frequency were $6.2 \times 10^{-6}\ \Phi_0/\sqrt{\mathrm{Hz}}$ and 1.1 kHz for a bias frequency at 10 MHz--sufficient for practical biomagnetic applications.

1. Introduction

A single-chip SQUID [1] integrates a two-junction (2J) SQUID comparator and a superconducting feedback circuit on one chip (Fig. 1 (a)). Such integration reduces the number of cables in the multichannel SQUID system.

Unlike another digital SQUID scheme [2], this 2J-SQUID comparator receives input flux directly and produces a pulse sequence. The number of pulses is counted by the on-chip feedback circuit by adding the flux quantum in the superconducting loop. The stored flux is fed back to the 2J-SQUID comparator to cancel the input flux. The maximum number of stored flux quanta was confirmed in experiments to be more than 10^5 [3], sufficient for practical applications.

The sensitivity and cutoff frequency depend on the comparator's switching probability, which is determined by thermal noise when other noise is reduced sufficiently. We fabricated a single-chip SQUID using niobium junctions and measured the switching probability.

2. 2J-SQUID Comparator

The 2J-SQUID comparator has no shunt resistor and is supplied assymetrically with a sinusoidal bias current to produce a pulse sequence. The bias amplitude is adjusted to be slightly lower than the threshold value for the zero input flux (Fig. 1 (b)). The switching probability for positive polarity increases with the input flux, while decreasing for negative polarity.

Springer Proceedings in Physics, Vol. 64 Superconducting Devices and Their Applications
Editors: H. Koch, H. Lübbig

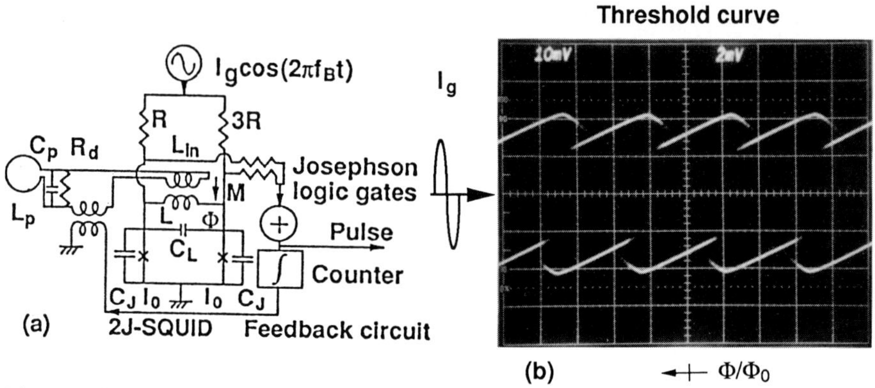

Fig. 1 (a) Single-chip SQUID integrating a 2J-SQUID comparator and superconducting feedback circuit on one chip, and (b) measured threshold curve of the 2J-SQUID comparator.

2.1 Switching probability and sensitivity

Flux spectral density S_ϕ and cutoff frequency f_c are expressed for frequencies lower than the bias frequency as follows [4],

$$S_\phi = \frac{p(1-p)}{f_B(\partial p/\partial\Phi)^2}\,\frac{1}{1+(f/f_c)^2}, \qquad f_c = (\Delta\Phi f_B/\pi)\frac{\partial p}{\partial\Phi} \tag{1}$$

where p is the switching probability per bias clock and $\Delta\Phi$ represents a quantized feedback flux per pulse. The cutoff frequency comes from the counting of the feedback loop, equivalent to a one-pole low-pass filter.

We measured the flux spectral density and cutoff frequency using a fast Fourier transform (FFT) analyzer. Figure 2 shows the result measured for the bias frequency at 10 MHz.

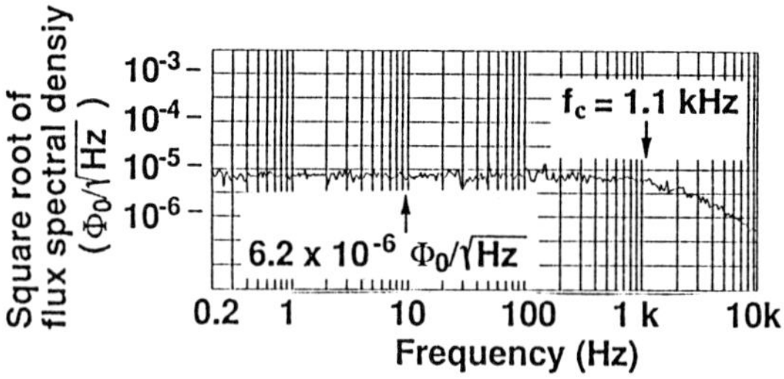

Fig. 2 Measured flux spectral density of the single-chip SQUID for the bias frequency at 10 MHz. The bias amplitude was adjusted to make the switching probability 0.47. Feedback was made by the on-chip superconducting feedback circuit.

2.2 Thermal noise in comparator

In reference [4], the switching probability per bias clock, p, was calculated for a rectangular pulse bias. In practical operation, a sinusoidal bias is easier to suppy.

p is expressed by

$$p=1-\exp\left(-\int_{-T/4}^{T/4} p_0 dt\right), \tag{2}$$

where T is $1/f_B$. p_0 is the switching probability per unit time and is obtained by integrating the escape rate along the ridge in 2J-SQUID potential U (Fig. 3). Because the 2J-SQUID comparator has no shunt resistor, we assume Boltzmann distribution function f, similar to the transition-state method effective for lossless cases [5], and obtained the following equations,

$$p_0=\int_C d\xi_2 \int_0^{\infty} d\dot{\xi}_1 \int_{-\infty}^{\infty} d\dot{\xi}_2(\dot{\xi}_1 f), \tag{3}$$

$$f=C_0\exp\left(-\frac{E}{kT}\right), \quad E=\frac{C_J}{2}(\dot{\phi}^+)^2+\frac{C_J+2C_L}{2}(\dot{\phi}^-)^2+U(\phi_1,\phi_2,I_g), \quad \phi^{\pm}=\frac{1}{\sqrt{2}}(\phi_1\pm\phi_2),$$

where ϕ_1 and ϕ_2 are the phase differences of two Josephson junctions and ξ_1 and ξ_2 are the coordinates normal and parallel to the ridge in the potential. C_0 is a normalization constant, C_J junction capacitance, and C_L the sum of the parasitic capacitance of SQUID inductor and a dummy inductor integrated in the chip, instead of a practical pickup coil. The estimated values for C_J and C_L are 0.29 pF and 11 pF.

Measured curves of p, $\partial p/\partial\Phi$ obtained from f_c through Eq. 1, and $\sqrt{S_\Phi}$ vs. the bias amplitude I_g agree well with the theoretical curve at 10 kHz. As f_B increased, however, measured curves became looser than theoretical curves. Figure 4 compares the measured curves for f_B at 10 MHz and theoretical curves calculated from Eqs. 2 and 3. At the peak of

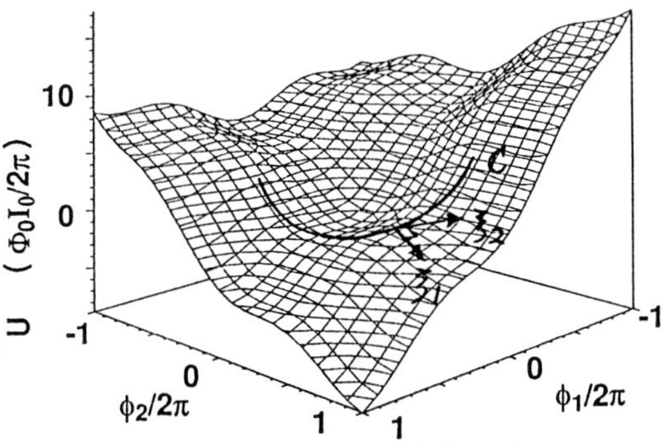

Fig. 3 Potential profile of 2J-SQUID comparator for a normalized bias current I_g/I_0 of 1.4. Measured parameters are equivalent inductance, $L-M^2/(L_p+L_{in})$, at 20 pH and I_0 at 0.1 mA.

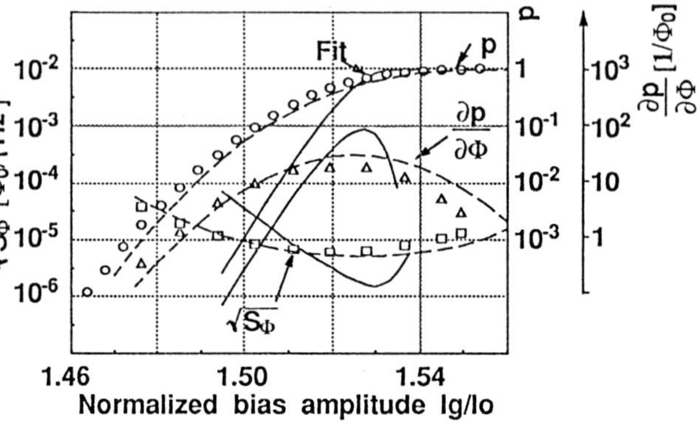

Fig. 4 Measured and theoretical p, $\partial p/\partial\Phi$, and $\sqrt{S_\Phi}$ vs. sinusoidal bias amplitude I_g/I_0 for a bias frequency at 10 MHz. Circles, triangles, and rectangles denote measured values. Solid curves represent theoretical curves calculated from only the thermal noise of the 2J-SQUID comparator through Eqs. 2 and 3. Dashed curves were calculated from both the thermal noise of the 2J-SQUID comparator and superimposed thermal noise flux due to the damping resistor. The horizontal axis of the measured curve was fitted at one point with a p of 0.5.

$\partial p/\partial\Phi$ and $\sqrt{S_\Phi}$, the difference is a factor of 4, because of thermal noise flux from the damping resistor, described in the section that follows.

2.3 Thermal noise flux from damping resistor

Damping resistor R_d is attached to the input inductance of the 2J-SQUID comparator to damp the oscillation induced by switching in a preceding bias clock. Without damping, the switching probability depends on the preceding bias clock probability and, in extreme cases, becomes insensitive to input flux.

The damping resistor, however, produces thermal noise. Its mean square flux noise coupled to the 2J-SQUID comparator was

$$\overline{\Phi_n^2} = \frac{4kT}{R_d}\left(\frac{L_pM}{L_p+L_{in}}\right)^2 \int_0^{f_m} \frac{df}{[1+(f/f_0)^2]^2}, \qquad f_0 = 1/(2\pi\sqrt{\frac{L_pL_{in}}{L_p+L_{in}}C_P}). \tag{4}$$

The upper bound of integral f_m is given roughly by $1/2\tau$, where τ is the width of an equivalent pulse bias that gives the same probability as the sinusoidal bias of frequency f_B and is given by $1/50f_B$ from numerical calculations. By substituting the measured parameters, $L_{in}=0.5\ \mu H$, $L_p=0.6\ \mu H$, $M=3.9$ nH, $T=4.2$ K, $R_d=80\ \Omega$, and a speculated value of $C_p=12$ pF, we obtained $\sqrt{\overline{\Phi_n^2}}$ as, $8.5 \times 10^{-4}\Phi_0$, $2.7 \times 10^{-3}\Phi_0$, $8.3 \times 10^{-3}\Phi_0$, $1.4 \times 10^{-2}\Phi_0$, for f_B of 10 kHz, 100 kHz, 1 MHz, and 10 MHz.

This flux noise makes the switching probability curve looser for a higher f_B. The calculated p, $\partial p/\partial\Phi$, and $\sqrt{S_\Phi}$, including the above flux noise whose amplitude we assume a normal distribution, agree excellently with curves measured for bias frequency f_B from 10 kHz to 10 MHz. Figure 4 shows a comparison for f_B at 10 MHz.

3. Performance of Single-Chip SQUID

As shown in Fig. 2, the flat level of $\sqrt{S_\Phi}$ was 6.2 x 10^{-6} $\Phi_0/\sqrt{Hz}$ and the cutoff frequency, 1.1 kHz for the bias frequency at 10 MHz. The flat level corresponds to a field sensitivity of 8 $fT/\sqrt{Hz}$, when a typical second-order gradiometer pickup coil 2.4 cm in diameter is attached to the chip. This data is sufficient for practical biomagnetism.

If thermal noise from the damping resistor is decreased sufficiently by optimizing design, sensitivity will be further improved by a factor of 4 at a maximum. Sensitivity limited by the thermal noise of the 2J-SQUID comparator can be made higher by increasing the bias frequency above 10 MHz.

4. Conclusions

The measured switching probability and performance agreed well with the theory calculated from the thermal noise of the 2J-SQUID comparator and the damping resistor. Performance is sufficient for practical biomagnetism. Single-chip SQUIDs and a Josephson digital multiplexer are now combined [6] to make a multichannel SQUID.

References

[1] N. Fujimaki, H. Tamura, T. Imamura, and S. Hasuo, IEEE Trans. Electron Devices, **35**, 2412 (1988).

[2] H. Matz, D. Drung, E. Crocoll, R. Herwig, G. Kramer, M. Neuhaus, and W. Jutzi, IEEE Trans. Magn. **27**, 2979 (1991).

[3] N. Fujimaki, K. Gotoh, and S. Hasuo, EIC (The Institute of Electronics, Information, and Communication Engineers) Conference **SCE91-4** (1991) (in Japanese).

[4] N. Fujimaki, H. Tamura, T. Imamura, and S. Hasuo, J. Appl. Phys., **65**, 1626 (1989).

[5] H. A. Kramers, Physica, **VII**, 284 (1940).

[6] K. Gotoh, N. Fujimaki, H. Suzuki, T. Imamura, S. Hasuo, and A. Shibatomi, Extended Abstract of Third International Superconductive Electronics Conference (ISEC'91), 148 (1991).

Digital CMOS Circuits Below 100 K

E. Crocoll, H. Matz, M. Reick, and W. Jutzi

Universität Karlsruhe, Institut für Elektrotechnische Grundlagen der Informatik, Hertzstr. 16, W-7500 Karlsruhe 21, Fed. Rep. of Germany

Abstract. A feedback circuit for a biomagnetic system comprising preamplifier, 12 bit counter and D/A converter with conventional CMOS circuits and clock frequencies between 10 and 20 MHz has been developed for an operation at 77 K and 4.2 K.

1. Introduction

Magnetometers for multi-channel biomagnetic systems should have a resolution below $\sqrt{S_B}$ = 10 fT/$\sqrt{Hz}$ for larger frequencies than 1 Hz, a large dynamic range of D ≥ 2^{12}, slew rates s_{max} > 4000 Φ_0/s at f_S ≈ 1 KHz, a high degree of linearity below s_{max} and a small number of lines between room and cryogenic temperatures.

The number of lines between low and high temperatures could be reduced and the reliability of the system could be improved with a flux-locked loop operating at cryogenic temperatures according to Fig. 1. A system with semiconductor feedback circuits at 300 K and

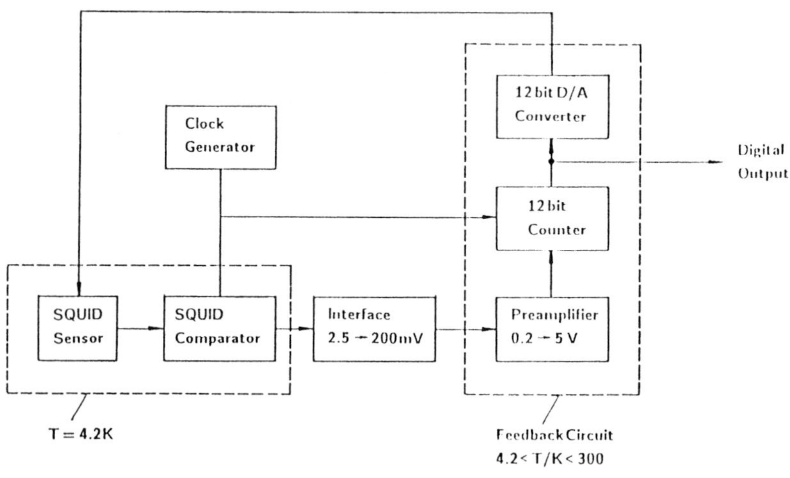

Fig. 1: Scheme of a magnetometer system with digital output.

Springer Proceedings in Physics, Vol. 64 Superconducting Devices and Their Applications
Editors: H. Koch, H. Lübbig

a pick-up coil, a sensor and comparator SQUID integrated on a single chip at 4.2 K has already been tested successfully .[1] It fulfills the above mentioned requirements for biomagnetic applications. Feedback circuits with only superconducting devices have been developed yielding interesting performance parameters. However, the slew rate should be larger. [2]

In this paper the operation of semiconductor feedback circuits according to Fig. 1 is described in a wide temperature range between 4.2 K and 300 K.

2. Interface and preamplifier

The interface between the output voltage of superconducting devices with metallic superconductors in the order of 2.5 mV and the input voltage of semiconductor circuits of 200 mV can be implemented either with two series of Josephson junctions [3] or with semiconductor circuits.

The preamplifier in Fig. 1 increases the pulse amplitude from 0.2V to 5 V, i. e. by a factor 25. It is implemented with CMOS inverters. In Fig. 2 the measured transfer characteristic $U_Q = f(U_I)$ is plotted with the coolant temperature as parameter. The amplification, the slope of the transfer characteristic increases with decreasing temperature. The propagation delay and the rise time of the amplified pulses decreases almost by a factor two between 300 K and 4.2 K, as summarized in Table I. For constant supply voltage the power dissipation decreases with increasing temperature, since the conductances decrease. The power dissipation could be greatly reduced with an appropriately designed layout of the inverter.

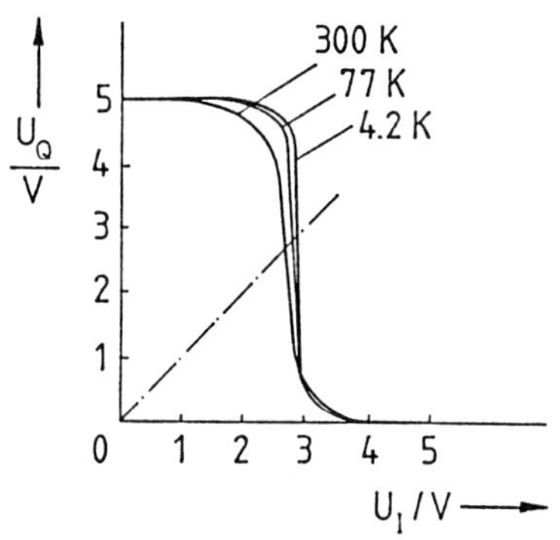

Fig. 2: Measured transfer characteristic of a CMOS inverter with the coolant temperature as parameter.

Table I: Preamplifier characteristics for input pulses with 2 ns rise time.

T/K	4,2	77	300
t_{pd}/ns	≈11	≈11	18
t_r/ns	1,5	1,5	2,8
P_v/mW at 20 MHz	237	198	143

3. Test results of the feedback circuit

The 12 bit counter and D/A converter in Fig. 1 have been assembled with conventional CMOS integrated circuits and metallic resistors. If the comparator generates a pulse, the counter goes down by one step and if there is no comparator pulse, it goes up by one step. The feedback circuits comprising a preamplifier, a 12 bit counter, a D/A converter and a RC-filter to cut down the influence of glitches has been tested with 200 mV pulses at the input of the feedback circuit for various clock frequencies and coolant temperatures.

The large signal behaviour has been investigated in applying a continuous string of pulses at a clock frequency f_C = 10 MHz. After 2^{12} pulses the counter has reached 0 and is reset to its maximum value. The rise time of the expected saw tooth transient is increased owing to the RC-filter. The transients change very little as a function of temperature. For T = 4.2 K it is represented in Fig. 3.

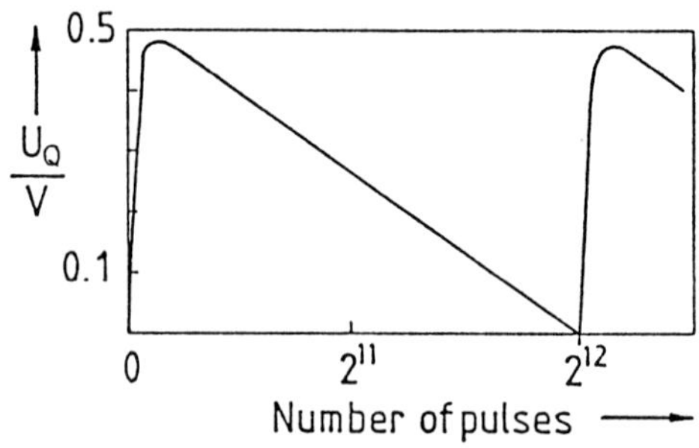

Fig. 3: Large signal response of the feedback circuit with a continuous string of 200 mV input pulses for a 10 MHz clock frequency.

4. Flux locked loop operation

A flux locked loop operation has not yet been tried at 77 K or 4.2 K although there are good chances for success, even at clock frequencies of 20 MHz. The power dissipation of the feedback circuit at 77 K and 20 MHz is 320 mW. It should be further reduced by a proper design of integrated semiconductor circuits in order to minimize the magnetic fields of currents in the feedback circuits at the pick-up coils and in order to reduce the coolant consumption.

5. References

[1] D. Drung, E. Crocoll, R. Herwig, A. Knüttel and M. Neuhaus, "Experimental Gradiometer with a Digital Feedback Loop", Proc. of the Int. Superconductivity Electronics Conf. in Tokyo, Japan, pp. 21-24, August 1987

[2] N. Fujimaki, K. Gotoh, S. Hasuo, J. Atsugi, "Switching Probability and Performance in Single-Chip SQUIDs", this conference

[3] H. Suzuki, A. Inoue, T. Imamura and S. Hasuo, "A Josephson Driver To Interface Josephson Junctions To Semiconductor Transistors", IDEM 88, pp. 290-293

Part III

Applications

Fluxons in Josephson Transmission Lines

N.F. Pedersen

Physics Laboratory I, The Technical University of Denmark,
DK-2800 Lyngby, Denmark

Abstract. The present paper gives a tutorial description of the fundamental properties of fluxons in long Josephson junctions. Expressions for the fluxon velocity, junction voltage, dynamic properties etc. are derived for various conditions of junction geometry and external perturbations. A few key experiments on the fundamental properties will be discussed. An understanding of the nature of fluxon behaviour is essential for new applications such as the flux flow transistor and the phase locked microwave oscillator also described in this volume.

1. Introduction

One of the physical systems where single fluxon properties is most easily accessible for direct experimental measurements is the long Josephson junction, or the Josephson transmission line (JTL).[1].

Fluxons contain one quantum of magnetic flux $\Phi_0 = h/2e = 2.068 \times 10^{-15}$ Vs and have solitonic properties; the possible applications of those unique properties has stimulated a lot of recent research activity. As pointed out in an early experimental paper by Fulton and Dynes [2] moving fluxons in the Josephson Transmission lines (JTL) manifest themselves as the socalled zero field steps (ZFS) in the dc current-voltage characteristic of the Josephson junction. Due to improvements in electronics it has now become possible to measure also the voltage waveform of the fluxon directly, and a comparison with theories and numerical calculations can be made.

2. Physics and the wave equation for the JTL

Figure 1 shows the equivalent diagram for a JTL. In a Josephson junction currents arise from Cooper pairs and normal electrons. The unique non-linear properties are due to the pair current, which may be expressed as $I_0 \sin\phi$. ϕ and the voltage are related through the Josephson relation - $d\phi/dt = 4\pi eV/h$. Here ϕ is the pair phase difference across the junction, and I_0 is the maximum pair current. In addition, a shunt resistance R carries a normal electron current (V/R), and a capacitance C carries a capacitive current C(dV/dt). The physical origin of the inductance L' shown in Fig. 1 is pair currents within the London penetration layer. A possible resistance, R', due to a flow of normal electrons in the same layer is also shown.

Taking all these circuit elements into account we obtain the perturbed sine-Gordon equation [1,3]

$$-\phi_{xx} + \phi_{tt} + \sin\phi = \eta - \alpha\phi_t + \beta\phi_{xxt} \ . \qquad (1)$$

Springer Proceedings in Physics, Vol. 64 Superconducting Devices and Their Applications
Editors: H. Koch, H. Lübbig

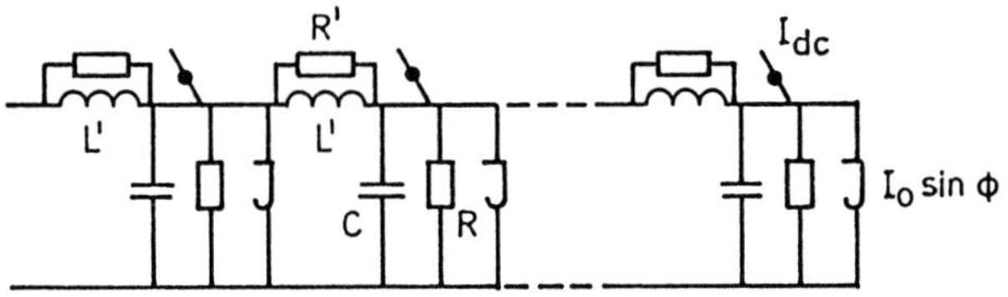

Fig. 1 Equivalent diagram for a Josephson transmission line.

The normalisations used are as follows: Time t is measured in units of $1/\omega_o$, where ω_o, the plasma frequency, is given by $\omega_o = (e4\pi I_o/hC)^{1/2}$. η is the normalized bias current, defined through $\eta = I_{dc}/I_o$. Length is measured in units of the Josephson penetration depth $\lambda_J = (h/4\pi ed\mu_o J)^{1/2}$. α and β are damping parameters given by $\alpha = (\omega_o RC)^{-1/2}$ and $\beta = L'\omega_o/R'$, respectively. J is the current density and d is the magnetic thickness of the junction. The junction length, L, in the x-direction is assumed large, and the width, W, in the transverse direction is assumed small compared to the Josephson penetration depth. Finally, velocities become normalized to the velocity of light in the barrier, $\bar{c} = c(t/d)^{1/2}$, where c' is the velocity of light in vacuum. For typical experimental junctions $\bar{c}$ is a few percent of c.

3. Perturbation calculation for the infinite line

The methods in this section are largely based on the work of McLaughlin and Scott [3]. In Eq. (1), the right hand side, consisting of loss and bias terms, is considered a perturbation to the sine-Gordon equation. The unperturbed sine-Gordon equation has the well known single soliton solution

$$\phi_S = 4\tan^{-1}\exp\xi \quad , \tag{2}$$

where $\xi = (x - ut)\,\gamma(u)$ and $\gamma(u) = (1 - u^2)^{-1/2}$ is the Lorentz factor. Note that the form of the solution - a traveling wave in the parameter $\xi = \gamma(u)(x - ut)$ - is a consequence of the Lorentz invariance of the sine-Gordon equation. Figure 2 shows a schematic drawing of a sine-Gordon soliton, ϕ_S, and its time derivative, which appears as a voltage pulse. The sine-Gordon soliton behaves very much like a relativistic particle with energy, H, and momentum, P, given by

$$H = 8\gamma(u),\ P = 8u\gamma(u) \ . \tag{3}$$

With the normalisations used here the rest mass of the soliton is 8. In Eq. (2) the velocity u is a free parameter. As shown in [3] it is possible to include the perturbation terms by assuming a solution of the same form as that in Eq. (2), but with u to be determined by a power balance equation. Requiring the Hamiltonian or the momentum to be independent of time one finds the velocity to be determined by

$$\pi\eta/4 = u\gamma(u)(\alpha + \beta/3(1-u^2)) \ . \tag{4}$$

In most of what follows we will assume $\beta = 0$ and take only shunt losses into account. In that case the velocity may be found explicitly as

$$u = (1 + (4\ \alpha/\pi\eta)^2)^{-1/2} \ . \tag{5}$$

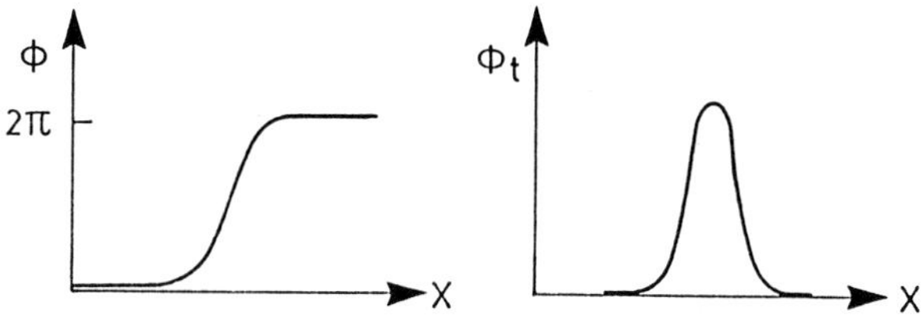

Fig. 2 The sine-Gordon soliton (schematically)

Because Eqs. (4,5) are derived by a perturbational approach, they are not expected to be valid if the perturbing terms are large, i.e. if the bias term approaches one and/or the system is heavily damped.

Another solution to the sine-Gordon equation that may be perturbed under the influence of bias and losses is the soliton-antisoliton solution that may be written [3,4]

$$\phi_{s\bar{s}} = 4\tan^{-1}(\sinh T/u\cosh X) \quad , \tag{6}$$

where $T = u\gamma(u)t$ and $X = \gamma(u)x$. Assuming only shunt-losses (i.e. $\beta = 0$) it is possible in a similar way as for the single soliton case to perform a power balance calculation by requiring the time rate of change of the energy, H, to be zero, i.e. calculating the intergral [4]

$$dH/dt = \int_{-\infty}^{\infty} (\eta\phi_t - \alpha\phi_t^2)dx \tag{7}$$

with ϕ inserted from Eq. (6). For this case a qualitatively new phenomenon occurs. For high incident energies the soliton and antisoliton will pass through each other with a phaseshift δ (spatial advance) given by [3,4]

$$\delta = -2(1-u^2)^{1/2} \ln u \quad . \tag{8}$$

For bias below a certain threshold, η_{th}, to be calculated below, the soliton and antisoliton will annihilate each other, create a breather, and eventually die out as small amplitude, damped plasma oscillations. In evaluating Eq. (7) one finds after rather lengthy calculations [4] that there is an energy loss, Δ H, given by

$$\Delta H \approx 4\pi^2\alpha \tag{9}$$

during the collision. A part of this energy is dissipated in propagating but decaying oscillations of the line. The annihilation threshold may be found by requiring that the total energy of the soliton and antisoliton before the collision, $H = 16\gamma(u)$, is equal to the energy loss plus the rest energy of a stationary soliton and antisoliton, i.e.

$$16\gamma(u) \approx 4\pi^2\alpha + 16\gamma(0) \quad . \tag{10}$$

Eq. (9) together with Eq. (5) leads to [4]

$$\eta_{th} \approx (2\alpha)^{3/2} \quad . \tag{11}$$

This expression for the annihilation threshold has been demonstrated in numerical simulations to be a very good approximation [4].

4. Boundary conditions

The question of boundary conditions becomes important for a comparison with experiments. Figure 3 show three different experimental configurations, the annular JTL (Fig. 3a), the overlap JTL (Fig. 3b) , and the inline JTL (Fig. 3c). For all geometries it is required that the normalized length $l = L/\lambda_J >> 1$ since from Eq. (2) the dimension of a single soliton is typically of the order λ_J. A further assumption is that the normalized width satisfies $w = W/\lambda_J >> 1$ in order to have a one dimensional problem. The introduction of boundary conditions is of course a complication to the infinite line results of the previous section. However it has been shown [5] that some typically encountered boundary conditions are mathematically equivalent to collisions on the infinite line, which were treated in the previous section.

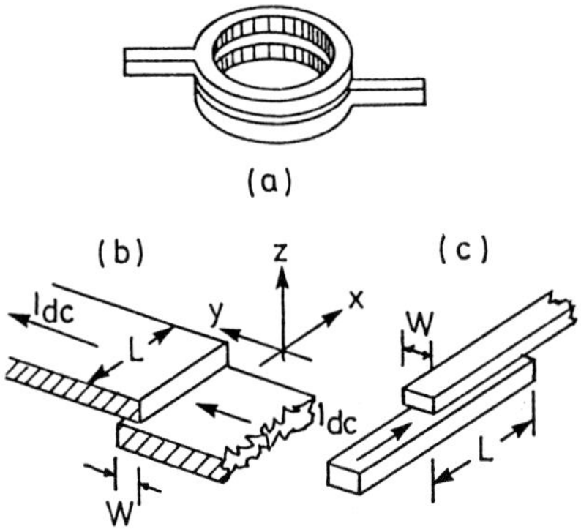

Fig. 3 Geometrical configurations of the JTL.

4.1. The annular JTL

This geometry is the simplest to treat mathematically [6,7]; the periodic boundary conditions are

$$\phi(o,t) = \phi(l,t) + 2p\pi \; . \tag{12}$$

Here p gives the number of full phase rotations along the line. For topological reasons p is a conserved number; for example in experiments it may only be changed by taking the junction through the transition temperature. The simplest case to consider is $p = 1$, i.e. a single fluxon on the circular line. A remark on experiments is appropriate here. Normally the current-voltage characteristic of a long Josephson junction is characterized by a zero-voltage current I_0. The case of the annular junction with $p=1$ is different. As soon as a uniform bias current is applied the fluxon starts moving with a velocity u determined by Eq. (5). Due to the moving fluxon a phaseshift takes place, and - through the Josephson relation - a (normalized) dc voltage, v_1, given by

$$v_1 = (2\pi/l)u \tag{13}$$

occurs. Hence for the annular junction with one trapped fluxon no supercurrent exists and the dc voltage is a direct measurement of the velocity of the single soliton. The IV-curve based on Eq. (5) and Eq. (13) is shown qualitatively in Fig. 4. In addition to the single soliton further solitons may be created only by introducing soliton-antisoliton pairs (for

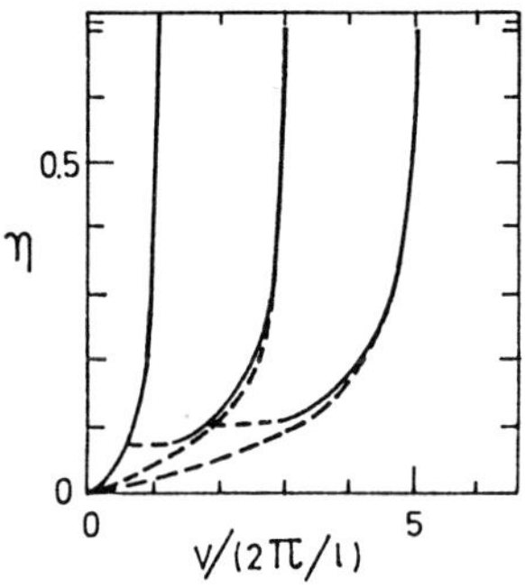

Fig. 4 Qualitative IV-curve of the annular JTL with one trapped soliton. Details of the curve are discussed in the text.

topological reasons). In that case the IV-curves are more difficult to interpret because the effects of collisions must be included. Disregarding the collisions the voltage of these different configurations is given by multiplying the voltage in Eq. (13) with the total number, n, of fluxons and antifluxons, i.e., voltage steps are to be expected at voltage v_1, $v_3 = 3v_1$, $v_5 = 5v_1$ etc. (shown as the dashed curves in Fig. 4). For the higher order branches, v_3, v_5, ... the qualitative effect of the collisions is to lower the average voltage somewhat compared to nv_1. Also a lower bias treshold, η_{th} where a fluxon and a antifluxon annihilate each other (c.f. Eq. (11)) is to be expected. Figure 4 shows qualitatively the higher order steps based on these arguments (full curves). Experimental measurements on a trapped soliton in the annular junction [6,7] are in a very good qualitative agreement with the theory as discussed above. The long annular junction experiment demonstrates in a very clean way the existence of a topological sine-Gordon soliton with a phase change of 2π.

4.2. The overlap and the inline JTL

The IV curve of overlap junction, Fig. 3b, has some features in common with the annular junction discussed above. Here the boundary conditions are

$$\phi_x(0,t) = \phi_x(l,t) = 0 \tag{14}$$

which physically implies that no current flow out at the junction ends. By symmetry the bias current is uniformly distributed over the junction length, and $\eta = I_{dc}/JWL$ may be assumed. There are no topological constraints on the number of fluxons and antifluxons. The mechanism for the first step, n = 1, is a fluxon that moves along the junction and is reflected at the boundary as an antifluxon. The reflection at, say x = l, is at least for small losses [5] equivalent to a collision with a virtual antifluxon at x = l, and the problem may be treated in the framework of Eq. (6) and Eq. (7). If the junction length, l, is very large, the details at the boundaries play only a minor role, and the voltage of the first step is given be Eq. (13). If the junction length is smaller, the energy loss, Eq. (9), and the phaseshift, Eq. (8), must be taken into account. All this leads to IV curves qualitatively simular to those of figure 4, except that all steps now exist. The geometry of the inline JTL is shown in Fig. 3c. The dynamics of the fluxon motion here [1] is quite different from that of the overlap JTL. The bias current flows only within a distance of approximetely 2 (in normalized units) from the ends and is screened from the inside if the junction by the magnetic field created by the bias current itself. To a good approximation this may be described by using $\eta = 0$ in Eq. (1) and the boundary conditions

$$\phi_x(o,t) = -\phi_x(l,t) = I_{dc}/2\lambda_J WJ \equiv \kappa_{dc} \tag{15}$$

such a boundary condition gives an energy input to the fluxon in each junction end. In the interior of the junction it moves in a ballistic trajectory subject to the damping of the line [8,1]

An external magnetic field enter in a somewhat similar way as in Eq. 15. The main difference is that ϕ_x has the same sign in both ends - leading to an energy input in one end and an energy absorption in the other [1].

5. Dynamics in the JTL; a few basic experiments

There are several ways to experimentally observe the phenomena discussed above. The simplest experiment to perform with fluxons on the JTL is to measure its IV-curve, as discussed in the introduction. Such measurements determine the average velocity of the fluxons c.f. Eq. (1). More direct measurements of the shape of the fluxons have recently become possible with the development of Josephson junction sampling circuits (or even traditional electronics with microprocessor data averaging and analysis). Examples of basic studies using these remarkable measurement techniques are found in [9-12]. In [10] there was an experimental observation of a fluxon reflection. Even a direct measurement of a fluxon-antifluxon collision on a JTL has been measured by a Josephson sampling circuit [11]. In [12] both the annihilation case ($\eta < \eta_{th}$) and the passing through case ($\eta > \eta_{th}$) was observed directly. (Stationary) fluxons have also been measured by STM techniques [13] and electron-holographic interferometry [14].

Even in the relatively simple dc - IV - curve measurements [7,15] and in microwave emission measurements [16] can the qualitative behaviour of fluxons together with several details of the dymanics be observed. Methods to move and stop fluxons on the JTL by applying an external control current have been devised, and circuits demonstrating such controlled behaviour have been built and tested [17].

To briefly summarize, the nature and fundamental dynamical properties of fluxons on the JTL are well understood both theoretically and experimentally.

References

[1] N.F. Pedersen in "Solitons". Eds. A.A. Maradudin and V.H. Agranowich North Holland 1986, p 469

[2] T.A. Fulton, and R.C. Dynes, Sol.St.Comm. 12, 57 (1973)

[3] D.W. McLaughlin, and A.C. Scott, Phys.Rev. A18, 1652 (1978)

[4] N.F. Pedersen, M.R. Samuelsen and D. Welner, Phys.Rev. B30, 4057 (1984)

[5] R.M. DeLeonardis, S.E. Trullinger, and R.F. Wallis, J.Appl.Phys.51, 1211 (1980)

[6] A. Davidson, B. Dueholm, B. Kryger, and N.F. Pedersen Phys.Rev.Lett. 55, 2059 (1985)

[7] A. Davidson, B. Dueholm, and N.F. Pedersen, J.App.Phys. 60, 1447 (1986)

[8] D.A. Levring, N.F. Pedersen, and M.R. Samuelsen, J.Appl.Phys. 54, 987 (1983)

[9] J. Nitta and A. Matsuda, Phys.Rev. B35, 4764 (1987)

[10] H. Akoh, S. Sakai and S. Takada, Phys.Rev. B35, 5357 (1987)

[11] A. Fujimaki, K. Nakajima and Y. Sawada, Phys.Rev.Lett. 59, 2895 (1987)

[12] K. Nakajima, H. Mizusawa, H. Akoh, S. Takada and Y. Sawada, Preprint (1990)

[13] H.F. Hess et al., Phys.Rev.Lett. 62, 214, (1989)

[14] S. Hasegawa et al. Phys.Rev. B43, 7631 (1991).

[15] N.F. Pedersen and D. Welner, Phys.Rev. B29, 2551 (1984)

[16] B. Dueholm, O.A. Levring, J. Mygind, N.F. Pedersen, O.H. Sørensen and M. Cirillo, Phys.Rev.Lett. 46, 1299 (1981)

[17] H. Akoh, S. Sakai, A. Yagi and H. Hayakawa, IEEE Trans.Magn. Mag-21, p.737 (1985)

Phase Locking in Long Josephson Junctions

R.D. Parmentier

Dipartimento di Fisica, Università di Salerno, I-84081 Baronissi (SA), Italy

Abstract. Application of the classic McLaughlin–Scott soliton perturbation theory to a fluxon propagating in a long Josephson tunnel junction oscillator reduces the fluxon dynamics to a simplified particle-map. This formalism, applied to a single fluxon oscillator in an external microwave field and to an array of such oscillators mutually coupled through a linear resonator, gives a remarkably detailed agreement with full simulations of partial differential equation models, with a substantial saving — as much as three orders of magnitude — in computer time, and it provides encouragingly accurate descriptions of experimental results.

1. Introduction

The study of phase locking of nonlinear oscillators has a long scientific history; a few of the illustrious names associated with the field are those of Huygens, Lord Rayleigh and van der Pol [1]. The basic physical phenomenon finds numerous practical applications, which helps to explain the widespread interest in the topic. Potential electronic applications surely provide the major motivation for studying phase locking of fluxon oscillations in long Josephson tunnel junctions. For example, series-biased arrays of long junctions operating in the resonant fluxon oscillation mode appear to be serious candidates for employment as local oscillators in integrated superconducting heterodyne receivers for radio astronomy applications [2]; operating such an array in a coherent, phase locked mode offers increased output power and reduced oscillator linewidth, approaching practically useful values, with respect to the figures attainable from a single junction [3].

In order to design a working device, we need an efficient computational scheme to calculate the dynamics of the device model. For phase locking in long junctions, a particularly convenient com-

Springer Proceedings in Physics, Vol. 64 Superconducting Devices and Their Applications
Editors: H. Koch, H. Lübbig

putational scheme is provided by the classic McLaughlin-Scott soliton perturbation theory [4], which reduces the description of a fluxon in a long junction to that of a relativistic particle in a box subjected to energy input and dissipation in a form depending on the details of the model under study. For the basic ideas of this scheme the reader may consult the paper by Pedersen [5]; I will apply the scheme to several specific cases of practical interest, comparing the results with those obtained via full simulations of partial differential equation (pde) models, and with experiments.

2. In-line Geometry Junction in Fixed External Field

The simplest case to study is that of a single in-line geometry junction [6] in an external microwave field that interacts with the junction through the model boundary conditions. The corresponding relativistic particle in this case undergoes ballistic motion in the interior of the box, and exchanges energy with the field through the vibrating walls of the box. The appropriate normalized pde model here is

$$\varphi_{xx} - \varphi_{tt} - \sin\varphi = \alpha\varphi_t - \beta\varphi_{xxt} , \tag{1}$$

with boundary conditions

$$\varphi_x(0,t) + \beta\varphi_{xt}(0,t) = -\varkappa + \eta \ ; \tag{2a}$$

$$\varphi_x(L,t) + \beta\varphi_{xt}(L,t) = +\varkappa + \eta \ . \tag{2b}$$

The terms in Eqs. (1) and (2) have their usual meanings: φ is the phase difference between the junction electrodes, x is the spatial coordinate normalized to the Josephson penetration length, t is time normalized to the inverse of the maximum Josephson angular plasma frequency, the term in α represents shunt dissipation caused by quasiparticle tunneling, the term in β represents dissipation caused by the surface resistance of the junction electrodes, $\varkappa$ is the normalized bias current, η is the normalized magnetic field in the plane of the junction and perpendicular to its long dimension, L is the normalized junction length, and subscripts denote partial derivatives.

A sinusoidal driving term, representing the external microwave field, may be introduced either through the term $\varkappa$ or through the term η. Salerno et al. [7], who studied this model in detail for the case $\beta = 0$, called the former case electric coupling and the latter case magnetic coupling. Salerno and Samuelsen [8] also extended the analysis to cases of mixed electric/magnetic coupling, and Filatrella et al. [9] showed how to incorporate a non-zero β term into the formalism. Phase locking of fluxon oscillations in this model is most readily manifested as the appearance of a constant-voltage current step in the current-voltage characteristic of the first zero-field step of the junction, at a voltage $V = (2n/m)\omega$, where ω is the angular frequency of the applied field, and n and m are integers.

Using the McLaughlin-Scott perturbation theory [4,5] Salerno et al. [7] reduced the pde model of Eqs. (1) and (2) to the following equation for the fluxon momentum (for $\beta = 0$)

$$\frac{dP}{dt} = -\alpha P + 4\pi \sum_{k=0}^{\infty} \left[\varkappa(t) + (-1)^{k+1} \eta(t) \right] \delta\left(X(t) - kL \right) , \tag{3}$$

where P is the "particle" momentum, X(t) is its temporal trajectory, and $\delta(\,)$ is the Dirac delta. They further showed how to manipulate Eq. (3) into the form of an explicit, two-dimensional functional map in terms of the variables t_k, the time, modulo the period of the applied microwave field, of the k'th boundary reflection, and E_k, the fluxon energy at the k'th reflection.

Reduction to a map is particularly convenient inasmuch as one can use standard techniques to study the existence and stability of fixed points, which correspond to phase-locked states of the fluxon dynamics. The salient features that emerge from such an analysis are [7]: (a) For both electric and magnetic coupling, phase-locked steps of height twice the normalized field amplitude exist; for magnetic coupling they exist at the fundamental frequency (n = m = 1) and at all odd subharmonics (n = 1; m = 1, 3, 5, ...) whereas for electric coupling they exist only at the even subharmonics (n = 1; m = 2, 4, 6, ...). (b) The stability range in field amplitude for all steps decreases rapidly with increasing subharmonic order. (c) Instability appears first at the center of a step. (d) For field amplitude values beyond the stability limit, a Feigenbaum-like bifurcation cascade leading to chaos in the fluxon times of flight across the junction is observed. (e) For appropriate model parameters the phase locking steps can cross the zero-current axis, extending to negative current values; such zero-crossing steps are normally associated with a hysteresis in the step amplitude.

To check the validity of the particle-map formalism, it is useful to compare results with those obtained from a full simulation of the original pde model, which has a fairly long and well-tested history of accounting for experimental results. Such a comparison shows that the strongest limitation of the particle-map approach stems from the assumption of single fluxon dynamics. Physically, sufficiently large or sufficiently small energy exchange terms can give rise to either the creation or the annhilation of fluxons within the junction; such effects are clearly manifested in pde simulations [10], but they are not contained in the particle-map model. Quantitative discrepancies can often be attributed to the fact that a fluxon in the particle-map formalism is a point particle with no spatial extension, whereas a pde fluxon has a non-zero width. Consequently, whereas energy exchange at the junction boundaries occurs instantaneously in the map, the exchange interaction is smeared over a certain time interval in the pde. This difference tends to disappear for junction lengths $L \gg 1$, but it may be significant for length values of practical interest. However, within its range of validity, the map predicts results to a surprising level of detail and accuracy. This extends even to fine details such as the hysteresis in zero-crossing steps [9] and the structure of bifurcation cascades [11] and chaotic strange attractors [12] in the instability region.

3. Overlap Geometry Junction in Fixed External Field

Another standard geometry for long Josephson junctions is the overlap geometry [6]. The normalized pde model in this case is

$$\varphi_{xx} - \varphi_{tt} - \sin\varphi = \alpha\varphi_t - \beta\varphi_{xxt} - \gamma \ , \tag{4}$$

with boundary conditions

$$\varphi_x(0,t) + \beta\varphi_{xt}(0,t) = \varphi_x(L,t) + \beta\varphi_{xt}(L,t) = \eta \ . \tag{5a,b}$$

The essential difference with respect to the in-line case is that the bias current, γ, now enters the problem through the pde rather than through the boundary conditions. For the overlap geometry, the interaction with the microwave field can be assumed to occur either through the term γ or through the term η (or possibly both). In what follows, I shall consider only the latter case; the former case is discussed by Filatrella et al. [12] and references therein.

This latter case is evidently fairly similar to the in-line case discussed in the previous section, inasmuch as the interaction with the microwave field takes place only through magnetic field boundary conditions. Using the same formalism as before, it is possible to write a momentum equation corresponding to Eq. (3) for the in-line case. With $\eta = \eta_o \sin(\omega t + \vartheta)$, and for $\beta = 0$, this equation is

$$\frac{dP}{dt} = -\alpha P + 2\pi\gamma + 4\pi\sum_{k=0}^{\infty}(-1)^{k+1}\eta_o \sin(\omega t + \vartheta)\,\delta\big(X(t) - kL\big)\,. \quad (6)$$

The difference between Eqs. (3) and (6) may appear to be trivial, but it has one important consequence: it is no longer possible to manipulate Eq. (6) into a functional map in explicit form. The basic character of a map is still implicitly present, but it is now necessary to perform a numerical inversion of a functional equation at each iteration step in order to find the successive times of flight of the fluxon across the junction. From a practical computational point of view this difference is perhaps not highly significant, since the numerical iteration procedure becomes now only slightly more complicated [7], but it does mean that we no longer have available simple, exact analytic expressions for the ranges of existence and stability of the fixed points.

In spite of the lack of an explicit representation, it is nonetheless possible to obtain some approximate analytic results from the implicit map. For example, it can be shown [7] that the range in bias current for the existence of a phase-locked state at the fundamental frequency or an odd subharmonic is given approximately by $\Delta\gamma = 4\eta_o/L$, once again centered about the unperturbed zero-field step, as compared with $\Delta\varkappa = 2\eta_o$ for the in-line case with magnetic coupling. Direct numerical iteration of the maps gives results that are qualitatively very similar in the two cases, in particular as regards the instability scenario and the existence of zero-crossing steps.

Comparisons of map and pde results for the overlap case show essentially the same level of agreement, and discrepancies, as for the in-line case [10]. Of particular interest is the recent work of Costabile et al. [13], which illustrates zero-crossing steps, accompanied by hysteresis, obtained from experimental measurements and via pde simulations; their results are very similar to those obtained using the particle-map approach.

4. Series-biased Arrays of Coupled Junctions

Experimental studies of phase locking in long-junction arrays go back several years. The pioneering work in this direction was that of Finnegan and Wahlsten [14], who considered locking phenomena in a system of two coupled junctions. Further experimental and numerical results on two-junction arrays, having different geometries, were reported by Cirillo and Lloyd [15] and by Holst et al. [16]. Experimental measurements on arrays of 8, 20, and 100 junctions were described by Pagano et al. [17].

The recent experimental works by Monaco et al. [3] and by Davidson and Pedersen [18] have clearly demonstrated that long-junction fluxon oscillators in a series-biased array can be locked together at a very sharp frequency by means of mutual coupling to a high-Q linear resonant cavity. The experimental device of Monaco et al. [3] consists of a series-biased array of overlap-geometry junctions located near one end of a 50-ohm microstrip resonator which is one half-wavelength long at the operating frequency (~ 10 GHz); the device geometry is essentially that of the 20-junction device described by Pagano et al. [17]. The device of Davidson and Pedersen [18], instead, has the junction array embedded in the side electrodes of a coplanar slotline resonator whose half-wavelength frequency is ~ 35 GHz.

Monaco et al. [3] convincingly demonstrated that the first-order behavior of their device can be described by a simple model in which the physical, distributed resonator is replaced by a lumped, linear tank circuit. In this model each junction is coupled to the tank through a capacitance which also comprises a part of the resonator, and the dynamics of the oscillating fluxon in each junction is described using the particle-map approach appropriate to overlap-geometry junctions, as outlined in the previous section. It seems quite reasonable to suppose that this simple model might also be applicable to the device of Davidson and Pedersen [18]. In this model each junction in the array is described by a pde of the form of Eq. (4), and the boundary conditions for the k'th junction are

$$\varphi_x^{(k)}(0,t) + \beta^{(k)} \varphi_{xt}^{(k)}(0,t) = 0 \quad ; \tag{7a}$$

$$\varphi_x^{(k)}(L^{(k)},t) + \beta^{(k)}\varphi_{xt}^{(k)}(L^{(k)},t) = I^{(k)}(t) \quad , \tag{7b}$$

where $I^{(k)}(t)$ is the normalized current supplied to that junction by the tank. In principle, all of the parameters α, β, γ, and L may vary from junction to junction, due to manufacturing tolerances; to simplify the calculations, Monaco et al. [3] set $\beta = 0$ and considered only variations in the length parameter L.

The current supplied to the k'th junction by the tank may be calculated as

$$I^{(k)}(t) = C^{(k)}\left[Rq_{tt} + \mathcal{L}q_{ttt} - \varphi_{tt}^{(k)}(L^{(k)},t)\right] \quad , \tag{8}$$

where $C^{(k)}$ is the k'th coupling capacitance, R and $\mathcal{L}$ are, respectively, the resistance and inductance of the tank, and q(t) is the total tank charge ($q_t(t)$ is the current in $\mathcal{L}$), which in turn is determined by the equation

$$q_{tt} + 2aq_t + \Omega^2 q = \Omega^2\sum_{k=1}^{N} C^{(k)}\varphi_t^{(k)}(L^{(k)},t) \quad , \tag{9}$$

where $a \equiv R/2\mathcal{L}$, $\Omega \equiv (\mathcal{L}C_{tot})^{-1/2}$, $C_{tot} \equiv \sum_k C^{(k)}$, and N is the total number of junctions in the array.

To construct the corresponding particle-map model it is necessary to calculate the effect of the junctions on the tank. To do this one may assume, consistently with the assumed particle nature of the fluxons, that each fluxon reflecting at the tank boundary can be described as a delta function voltage pulse of weight 4π. In this way the tank voltage may be calculated analytically as a sum of impulse responses [3].

The salient features that emerge from the particle-map model are: (a) The locking range in junction bias current varies linearly with the number of active junctions [3]. (b) The tank voltage varies linearly with the number of locked junctions [3]; this implies that the output power from the array varies as N^2. (c) Junctions can also lock to subharmonics of the tank resonant frequency [12]. (d) For appropriate system parameters, bifurcations and chaos can be observed on subharmonic steps [12].

Once again, a comparison between map and pde results shows quite reasonable agreement [19]. As before, qualitative discrepancies are attributable to the creation or annihilation of fluxons, a mechanism not contained in the map formalism, and quantitative discrepancies are attributable to the assumed point-particle nature of the map fluxons.

5. Conclusions

Phase-locked arrays of long Josephson junction fluxon oscillators seem to be arriving ever closer to the point of offering practical applications in high performance electronic circuits. The pde models needed to describe such devices have become progressively more costly in terms of computer time. In all cases studied so far, the particle-map approach has given a remarkably detailed agreement with full simulations of the pde models, with a substantial saving — as much as three orders of magnitude — in computer time, and it has provided encouragingly accurate descriptions of experimental results.

Acknowledgements

The work described herein is the fruit of a long-standing and felicitous collaboration between the Josephson junction groups working at the University of Salerno, the Cybernetics Institute of the Italian CNR, and the Technical University of Denmark. Special thanks go to my colleagues and friends in those groups. Financial support from MURST and from the Progetto Finalizzato "Tecnologie Superconduttive e Criogeniche" del CNR is gratefully acknowledged.

References

[1] N. Minorsky, "Nonlinear Oscillations" (Van Nostrand, New York, 1962), p. 438.

[2] D.-G. Crete, P. Feautrier, M. Hanus, R. Monaco, P. J. Encrenaz, in: "Stimulated Effects in Josephson Devices", M. Russo and G. Costabile, eds. (World Scientific, Singapore, 1990), pp. 87-98.

[3] R. Monaco, N. Grønbech-Jensen, R. D. Parmentier, Phys. Lett. A **151**, 195 (1990).

[4] D. W. McLaughlin, A. C. Scott, Phys. Rev. A **18**, 1652 (1978).

[5] N. F. Pedersen, *this volume.*

[6] A. Barone, G. Paternò, "Physics and Applications of the Josephson Effect", (Wiley, New York, 1982), Chap. 5.

[7] M. Salerno, M. R. Samuelsen, G. Filatrella, S. Pagano, R. D. Parmentier, Phys. Rev. B **41**, 6641 (1990).

[8] M. Salerno, M. R. Samuelsen, Phys. Lett. A **156**, 293 (1991).

[9] G. Filatrella, G. Rotoli, R. D. Parmentier, Phys. Lett. A **148**, 122 (1990).

[10] N. F. Pedersen, A. Davidson, Phys. Rev. B **41**, 178 (1990).

[11] G. Rotoli, G. Filatrella, Phys. Lett. A **156**, 211 (1991).

[12] G. Filatrella, N. Grønbech-Jensen, R. Monaco, S. Pagano, R. D. Parmentier, N. F. Pedersen, G. Rotoli, M. Salerno, M. R. Samuelsen, in: "Nonlinear Superconductive Electronics and Josephson Devices", G. Costabile, S. Pagano, N.F. Pedersen, and M. Russo, eds. (Plenum, London), *in press.*

[13] G. Costabile, R. Monaco, S. Pagano, G. Rotoli, Phys. Rev. B **42**, 2651 (1990).

[14] T. F. Finnegan, S. Wahlsten, in: "Low Temperature Physics — LT-13", Vol. 3, K. D. Timmerhaus, W. J. O'Sullivan, and E. F. Hammel, eds. (Plenum, New York, 1974), pp. 272-275.

[15] M. Cirillo, F. L. Lloyd, J. Appl. Phys. **61**, 2581 (1987).

[16] T. Holst, J. Bindslev Hansen, N. Grønbech-Jensen, J. A. Blackburn, Phys. Rev. B **42**, 127 (1990).

[17] S. Pagano, R. Monaco, G. Costabile, IEEE Trans. Magn. **25**, 1080 (1989).

[18] A. Davidson, N. F. Pedersen, Appl. Phys. Lett., *in press.*

[19] G. Filatrella, G. Rotoli, N. Grønbech-Jensen, R. D. Parmentier, N. F. Pedersen, *preprint, 1991.*

Soliton Chains in Annular Josephson Junctions: Experiments

A.V. Ustinov[2], *T. Doderer*[1], *B. Mayer*[1], *R.P. Huebener*[1], *I.V. Vernik*[2], *and V.A. Oboznov*[2]

[1]Lehrstuhl Experimentalphysik II, Universität Tübingen, W-7400 Tübingen, Fed. Rep. of Germany
[2]Institute of Solid State Physics, Academy of Sciences, 142432 Chernogolovka, Russia

Abstract. We report for the first time the trapping of several solitons (flux quanta) in annular Josephson junctions. We have realized experimentally two different techniques for the magnetic flux trapping in the junctions. First, using low temperature scanning electron microscopy (LTSEM) we applied the electron beam producing a local heating of the top superconducting film and trapped one by one several magnetic flux quanta in the tunnel barrier area of the Josephson junction. Second, by applying the external monopole–like magnetic field by means of a special coil we were also able to trap the flux in the junction at temperatures near the critical temperature of the superconducting electrodes. We have studied the I-V curves of the junction with several trapped solitons and compared the velocity of the soliton chain with one soliton velocity and the perturbation theory. We link the deviations of the soliton chain velocity from the one soliton velocity with the effects of the interaction between solitons and their bunching at high velocity.

1. Introduction

The annular geometry of long Josephson junction allows for the study of soliton motion without the influence of boundaries, thereby representing the experimental realization of a sine-Gordon system with periodic boundary conditions. Davidson et al. [1, 2] have successfully trapped one soliton (flux quantum) in a ring–shaped annular junction and studied its dynamics. However, with that technique it was not possible to trap more than one soliton in the ring. Still, the study of multi–soliton dynamics and soliton–soliton interaction in a boundaryless system is of a great interest. In our experiments we have studied $Nb - NbO_x - Pb$ junctions with the same geometry as in [1] (Fig. 1) and developed a new technique of flux trapping in these junctions.

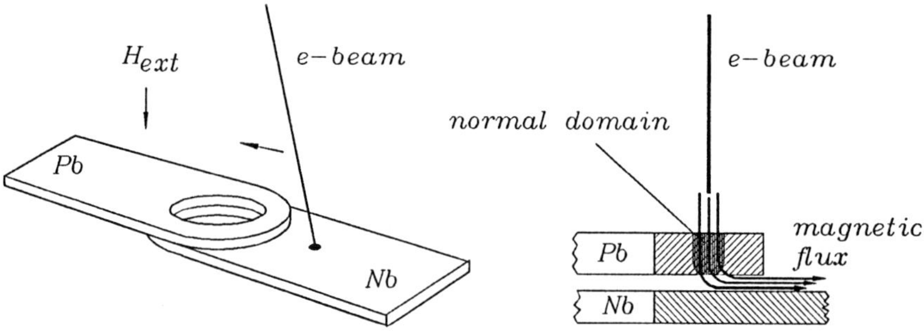

Figure 1: The magnetic flux trapping procedure with an annular junction using LTSEM and the cross–section of the junction at the moment of crossing by the electron beam.

2. LTSEM experiment

First, we have used LTSEM (low temperature scanning electron microscopy) [3] in order to trap magnetic flux in our ring–shaped junction. The electron beam was producing a local heating of the superconducting film. At the presence of a small external magnetic field H_{ext} perpendicular to the film the beam was placed on the *Nb* film outside the junction and moved towards the junction. After it crossed the junction area the trapping of one or more flux quanta in the junction occurred. In fact, the magnetic field was penetrating the locally heated area of the *Pb* film, as shown in Fig. 1b. By repeating of this procedure we trapped one by one several magnetic flux quanta in the tunnel barrier area.

The number n of trapped solitons was easy to detect by measuring the I-V curve. The voltage of the flux–flow step was proportional to n, as expected. Fig. 2 shows the I-V curves for two different trapping procedures (D and E) with $n = 9$ solitons in the annular junction. The trapping was quite well reproducible: two curves are very close to each other especially in the high–voltage range, where the soliton energy is also high.

The comparison of the 1–soliton step with the scaled 3–soliton one (its voltage was divided by $n = 3$) is also presented in Fig. 2. An interesting fact we note from this data is that in the relativistic region the velocity of the chain of solitons is a little higher (branch $B2$), than the velocity of the 1–soliton step at the same level of bias current. As known from numerical simulations [4], the dissipation at high velocity creates an effective attraction between unipolar solitons and their bunching in a localized soliton train moving in the junction. We suppose, that the effect observed can be due to the bunching of solitons, so it deals with

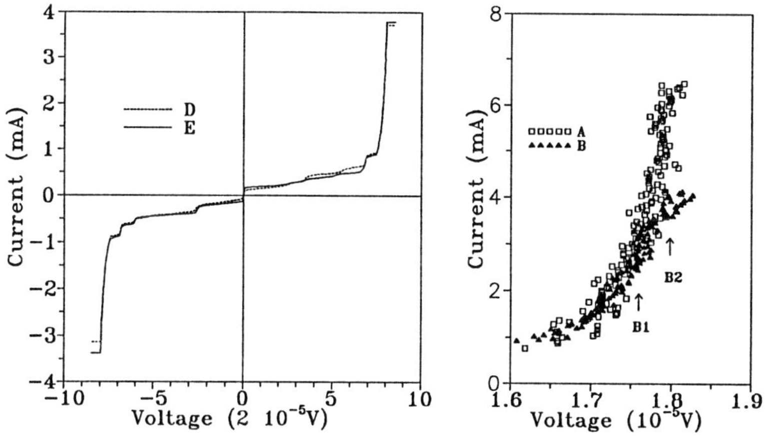

Figure 2: The I-V curves of the annular junction with $n = 9$ trapped solitons (D and E), and the comparison of the steps with $n = 1$ (A) and $n = 3$ (B), the latter scaled by n.

the symmetry breaking in this annular system. The detailed study of this effect now in progress.

3. Monopole field experiment

Fig. 3 illustrates another technique which we have used [5]. A one–turn Nb coil with the same dimensions as the junction ring was prepared on a separate substrate and placed at the distance of a few μm above the junction. The sample was heated above T_c of Pb and then cooled slowly at the presence of the current I_{ext} in the coil. The magnetic field produced by this current had a certain radial component H_{ext} in the plane of the tunnel barrier. Using this method several unipolar solitons in the junction during the cooling process. The I-V curves obtained were similar to those obtained by LTSEM but with much larger critical current and not so well reproducible. We think that the reason for that is the trapped flux in the area of the superconducting films, where the current lines are connected with the coil. Applying I_{ext} at $T < T_c$ we were also able to inject magnetic flux in the tunnel barrier area as a result of the field penetration through film itself (film thickness was comparable with λ_L). In this case the junction behavior was analogous to those one of a non–hysteretic SQUID.

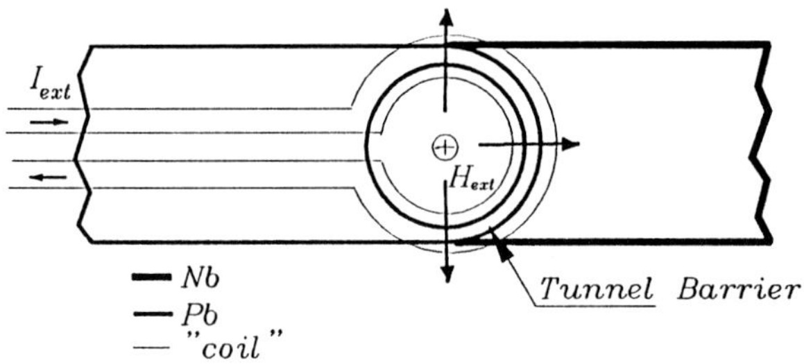

Figure 3: A top view of the junction with the coil producing the monopole–like field component in the tunnel barrier plane.

References

[1] A. Davidson, B. Dueholm, B. Kryger, and N. F. Pedersen, Phys. Rev. Lett. **55**, 2059 (1985).

[2] A. Davidson, B. Dueholm, and N. F. Pedersen, J. Appl. Phys. **60**, 1447 (1986).

[3] R. P. Huebener, Rep. Prog. Phys. **47**, 175 (1984).

[4] P. S. Lomdahl, O. H. Soerensen, and P. L. Christiansen, Phys. Rev. B **25**, 5737 (1982).

[5] The idea of the second experiment was proposed by V. P. Koshelets.

Coupled Josephson Soliton Oscillators

T. Holst[1], L.E. Guerrero[2], N. Grønbech-Jensen[1], J.A. Blackburn[3], and J. Bindslev-Hansen[1]

[1]Physics Laboratory I/MIDIT, The Technical University of Denmark, DK-2800 Lyngby, Denmark
[2]Centro de Fisica, IVIC, Apt. 21827, 1010A Caracas, Venezuela
[3]Department of Physics and Computing, Wilfrid Laurier University, Waterloo, Ontario, Canada N2L 3C5

Abstract. We report on phase-locking phenomena in systems of two closely spaced quasi-one-dimensional Josephson transmission lines, JTLs, biased in resonant fluxon modes where the junctions work as soliton oscillators with a fundamental frequency of 10 GHz. The coupling which is *broad-band* is seen both in the I-V curves and in the microwave signals from the JTLs. Enhanced microwave emission is observed experimentally. A theoretical model for a magnetic dipole coupling mechanism is proposed. Numerical simulations based on the model reproduce the phase-locking phenomena but not the enhanced emission.

1. Introduction

Recently, the non-linear behavior of long and narrow, quasi-one-dimensional Josephson tunnel junctions has been attracting much attention. The junctions act as non-linear Josephson transmission lines (JTLs) in which high-Q resonance modes may be excited by the internal Josephson oscillation [1]. The present study was aimed at achieving coherent radiation from pairs of coupled JTLs. Phase-locked arrays of Josephson soliton oscillators are of technological interest due to their larger power and smaller linewidth [2].

For a narrow junction of length L longer than the Josephson penetration depth λ_J and width $W < \lambda_J$, the travelling wave on the line has soliton character. The soliton/antisoliton is a $\pm 2\pi$ kink/antikink in the Josephson phase, ϕ, across the tunnel barrier between the two superconducting electrodes of the junction.

There is a quantum of magnetic flux, a fluxon/antifluxon, associated with the phase kink/antikink. The equation of motion for the Josephson phase is the much studied perturbed sine-Gordon equation, see [3].

The Lorentz force exerted on the fluxon/antifluxon by a bias current through the junction accelerates the soliton until it reaches a steady-state condition where the total loss balances the input power from the bias. The maximum velocity of the soliton is $u_\infty = \bar{c}$, the Swihart velocity of light in the barrier (for Nb-Nb$_x$O$_y$-Pb junctions, $\bar{c} = 0.025\ c_{vacuo}$). The motion generates a maximum voltage of $V = -d\Phi/dt = -2\Phi_o u_\infty/2L = -\Phi_o \bar{c}/L$. For a junction of length 400 μm this

Springer Proceedings in Physics, Vol. 64 Superconducting Devices and Their Applications
Editors: H. Koch, H. Lübbig

corresponds to a voltage around 40 μV giving a Josephson frequency of about 20 GHz. If N fluxons/antifluxons are threading the barrier, the voltage of the *resonant zero-field steps*, ZFSs, will be proportionally larger, $V_{ZFS*N} = \mp N\Phi_0\bar{c}/L$.

Another type of steps, Fiske steps (FS) are observed when the boundary conditions are determined by an applied magnetic field. They correspond to alternate motion of fluxons and plasma waves along the junction barrier (fluxons are created in one end of the JTL and annihilated in the other end). In the first Fiske mode, on the average only half a fluxon is present per cycle. On the Mth Fiske step the dc voltage is $V_{FS*M} = \mp M\Phi_0\bar{c}/2L$.

When a soliton arrives at the end of the JTL it is reflected and a fraction of the energy in the fluxon is emitted as a pulse. The frequency detected at *one* end of the JTL is half of the Josephson frequency, $f = f_J/2 = eV/h$, in the case of a 400 μm long junctions f is about 10 GHz.

In the following we report on one way of coupling two or more of the non-linear JTLs together in order to study the interaction between them and to increase the output power level.

As a measure of the in-phase coherence between two locked oscillators we use the enhancement factor, F, defined as $F = (P_{12} - P_1 - P_2)/2\sqrt{P_1 P_2}$ which has a maximum value of 1 for two point-like oscillators running totally in-phase. When F exceeds 1 we say that the system exhibits enhanced emission of radiation.

2. Experimental Technique

Figure 1 shows the sample geometry with two long and narrow, 400×20 μm^2 overlap Nb-Nb$_x$O$_y$-Pb tunnel junctions spaced by 35 or 75 μm and sharing the Pb top

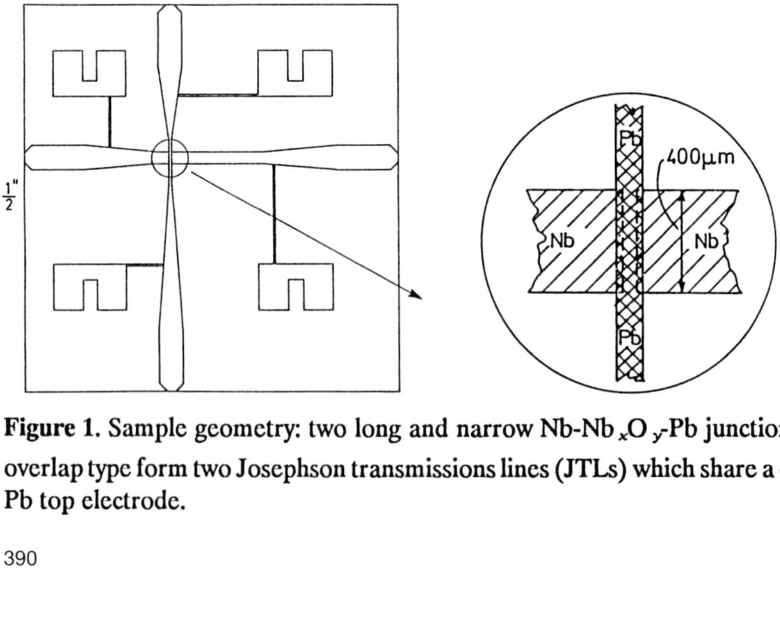

Figure 1. Sample geometry: two long and narrow Nb-Nb$_x$O$_y$-Pb junctions of the overlap type form two Josephson transmissions lines (JTLs) which share a common Pb top electrode.

electrode. The substrates were 1"×1" 0.5 mm thick Corning 7059 glass with low microwave loss. Typical junction parameters were: critical current density $J_c = 3$ - $16\ A/cm^2$, $\lambda_J = 100$ - $250\ \mu m$ (i.e., $L/\lambda_J = 4$ - 1.6), and normalized Swihart velocity $\bar{c}/c = 0.025$. Two independent current sources were used: *a) either* to bias the two JTLs individually *or b)* to bias both of them with one current source (I_{COMMON}) using the second source to provide an additional current (I_{TRIM}). The total voltage across the two junctions was measured. Two current configurations were used: *series aiding* and *series opposed*, given by the relative directions of the two bias currents. In the following the notation +/+ (-/-) and +/- (-/+) will be used to denote the two configurations. For further details about the experimental set-up we refer to Ref. [4].

3. Experimental Results

The experimental results reported here are based on measurements on four samples. The dc I-V curves and the microwave spectra were recorded for the ZFSs and the FSs for the JTLs individually and for both of them running at the same time.

In Fig. 2 we present a typical observation of the frequency-pulling, -mixing and -locking behavior observed when two soliton oscillators both biased in the ZFS1

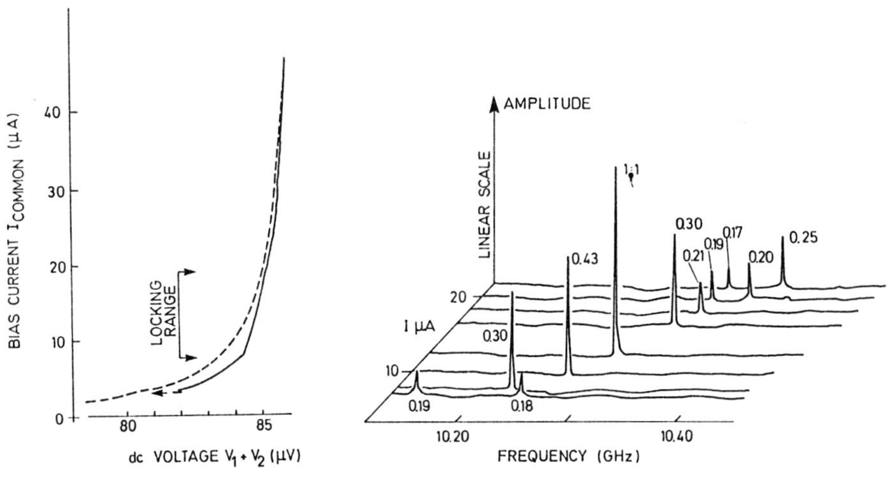

Figure 2. Experimental observation of phase locking, both in the I-V curve and in the microwave spectra, between two JTLs biased in the series aiding mode ZFS+1/ZFS+1 at 4.2 K (S-6006/2). Voltage-pulling and -locking were observed when a trim current, I_{TRIM}, was applied to one of the junctions (to compensate for the difference in junction parameters). The full curve was recorded for $I_{TRIM} = 23$ μA, the dashed curve for $I_{TRIM} = 0$. The corresponding microwave spectra are shown as voltage amplitude vs. frequency for a number of bias points inside and outside the locking range. The total integrated power received is indicated for each peak.

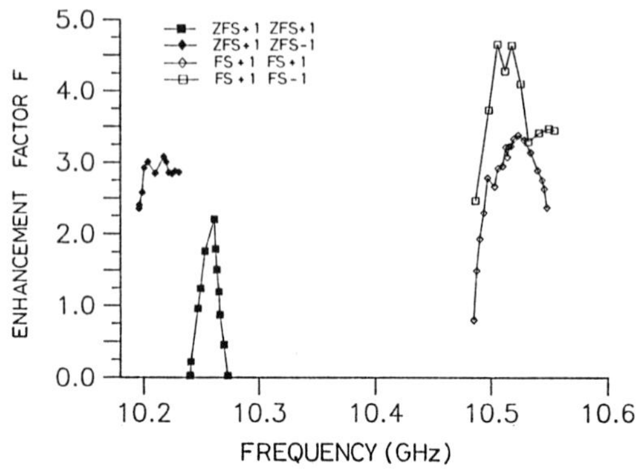

Figure 3. Enhancement factor, $F = (P_{12} - P_1 - P_2)/2\sqrt{P_1 P_2}$ vs. frequency over locking ranges for four different values of bias currents (I_2 swept, I_1 constant), chosen to give phase-locking for all four possible combinations of ZFS±1 and FS±1 in the two Josephson soliton oscillators (sample S-6006/2).

mode were tuned into a phase-locked state. The "3D" plot shows the spectra of detected voltage (linear scale) vs. frequency taken for different values of the common bias current (I_{TRIM} = 23 μA). The integrated power is indicated for each spectrum. It is calculated by multiplying the peak *power* by the *full linewidth at half maximum power (FWHM)*.

From the measured values of P_{12}, P_1 and P_2 the enhancement factor F was found. By sweeping through the locking range by varying the current through one JTL while keeping the other one constant, the variation of F over the locking region was determined. Figure 3 shows such a measurement of the enhancement factor F vs. frequency. The most surprising observation was that F in the middle of the locking region exceeded one by up to about a factor of five. We note that there was a tendency toward stronger interaction (i.e., stronger frequency-pulling) and also higher values of F for the case of two interacting Fiske modes than for interacting zero-field step modes. The emission enhancement was stronger in the series opposed current configurations, ZFS+1/ZFS-1 and FS+1/FS-1, than in the series aiding ones, ZFS+1/ZFS+1 and FS+1/FS+1. The coupling phenomena were observed over all overlapping voltage ranges and for all combinations of steps, e.g., ZFS+2/ZFS+1.

4. Theoretical Model and Numerical Simulations

A number of theoretical models have been proposed and explored for the interaction between Josephson transmission lines, based on capacitive, inductive or resonant coupling [5]. Provided realistic parameters are used, numerical simulations predict phase-locking phenomena for all these coupling models.

For our geometry with two closely spaced, parallel overlap junctions, the model based on the equivalent circuit with broad-band, mutual inductive coupling seems most appropriate. It was investigated in Ref. [4]. Numerical simulations of the coupled sine-Gordon equations reproduced the experimentally observed features of the current-voltage characteristics, most notably the locking regions.

Using the inductive coupling model we have investigated possible origins of the observed large enhancement factors by computing the time-dependent voltages at both ends of the long junctions. The important point to note is that F never exceeds 1, in contrast with experimental findings.

In an attempt to resolve this discrepancy, we have investigated the effect of non-linear damping on the coupling. A two-segment model for the quasi-particle I-V characteristics of the Josephson junction elements was used, the dissipation coefficient α thus depended upon the local instantaneous value of ϕ_t. These simulations also generated maximum F values of 1.

5. Discussion and Conclusions

The experimental observations of enhanced emission ($F > 1$) remains unexplained and the source of this effect is unknown. One possible important mechanism which is yet to be investigated concerns the dynamics of fluxon flipping at the ends of the junctions. It is generally assumed that the radiation originates in this specific stage of the fluxon motion. Simply averaging ϕ_t at the junction boundaries ($x = 0, L$) may not take into account the important acceleration and decelerations which are occurring during reflection. In the frequency regime, such changes in the soliton phase and waveform correspond to conversion of power between harmonics in the soliton spectrum. In the experiments only the power emitted at the fundamental frequency of the fluxon motion (around 10 GHz) is detected. Future experiments will be aimed at capturing more information about the pulse-like microwave emission from single and coupled Josephson soliton oscillators.

Acknowledgements

We thank A. Davidson, M.R. Samuelsen, M. Salerno, M. Devoret, N.F. Pedersen and R.D. Parmentier for fruitful discussions, G. Friis Eriksen for technical assistance, and I. Rasmussen and S. Hjorth for the sample fabrication. This work was supported by the Danish Natural Science Research Council, The Hartmann Foundation, The Danish Research Academy and the Natural Sciences and Engineering Research Council of Canada.

References

1. T.F. Fulton and R.C. Dynes, Solid State Commun. **12**, 57 (1973); D. W. Laughlin and A.C. Scott, Phys. Rev. A **18**, 1652 (1978).
2. A.K. Jain, K.K. Likharev, J.E. Lukens and J. Sauvageau, Phys. Rep. **109**, 309 (1984); J. Bindslev Hansen and P.E. Lindelof, Rev. Mod. Phys. **56**, 431 (1984).
3. N.F. Pedersen, these proceedings.
4. T. Holst, J. Bindslev Hansen, N. Grønbech-Jensen and J.A. Blackburn, Phys. Rev. B **42**, 127 (1990).
5. M. Cirillo and F.L. Lloyd, J. Appl. Phys. **61**, 2581 (1987); M. Cirillo, A.R. Bishop and P.S. Lomdahl, Phys. Rev. B **39**, 4804 (1989); M.B. Mineev, G.S. Mkrtchyan and V.V. Schmidt, J. Low Temp. Phys. **45**, 497 (1981); Y.S. Kivshar and B.A. Malomed, Phys. Rev. B **37**, 9325 (1988); N. Grønbech-Jensen, R.D. Parmentier and N.F. Pedersen, Phys. Lett. A **142**, 427 (1989); and R. Monaco, N. Grønbech-Jensen and R.D. Parmentier, Phys. Lett. A **151**, 195 (1990).

A 500 GHz Quasi-Optical Slot Antenna SIS Mixer

J. Zmuidzinas[1] *and H.G. LeDuc*[2]

[1]Downs Laboratory of Physics 320-47, California Institute of Technology, Pasadena, CA 91125, USA
[2]Jet Propulsion Laboratory 302-231, Pasadena, CA 91109, USA

Abstract - We describe a new quasi-optical SIS mixer design which attempts to obtain efficient radiation coupling into the SIS tunnel junction. Devices were fabricated at the Jet Propulsion Laboratory (JPL) using Nb/Al-oxide/Nb junctions, and a receiver noise temperature of 420 K (DSB) was measured at 490 GHz.

1. Introduction

Superconducting tunnel junction (SIS) mixers are predicted to have sensitivities approaching the quantum limit at frequencies perhaps as high as twice the gap frequency of the superconductor. The major difficulty in the construction of sensitive SIS mixers operating at submillimeter wavelengths is the efficient coupling of radiation into the tunnel junction. A promising approach is to use lenses to focus the radiation from a free-space beam onto an antenna which is integrated with the junction on a dielectric substrate (e.g. [1]). These quasi-optical designs avoid the fabrication problems associated with waveguides at submillimeter wavelengths, and would allow imaging mixer arrays to be constructed. However, present quasi-optical mixers do not couple the radiation into the tunnel junction as efficiently as waveguide designs. The two critical factors are the antenna efficiency and the impedance match between the junction and the antenna. The slot antenna mixer design is our first attempt at dealing with both of these issues.

Our mixer design consists of a twin-slot planar feed antenna, a Nb/Al-oxide/Nb or NbN/MgO/NbN SIS junction, and a tapered microstrip transmission line joining the junction to the two slot antennas (see Fig. 1). A hyperhemispherical lens is used to focus the incident radiation onto the twin-slot antenna (Fig. 2).

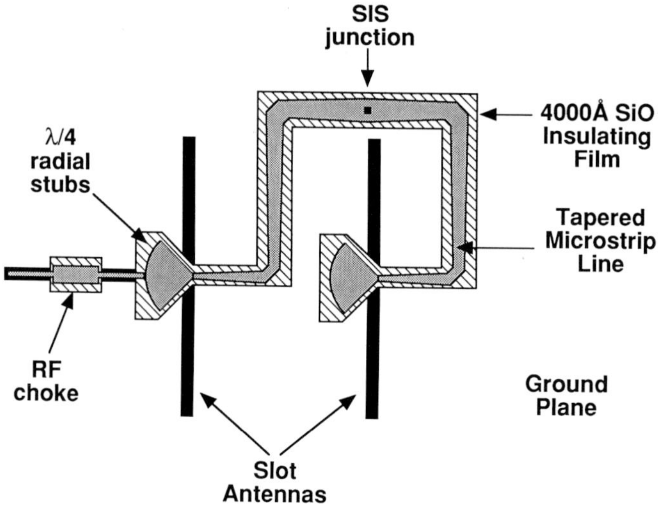

Figure 1. Diagram of the slot-antenna mixer.

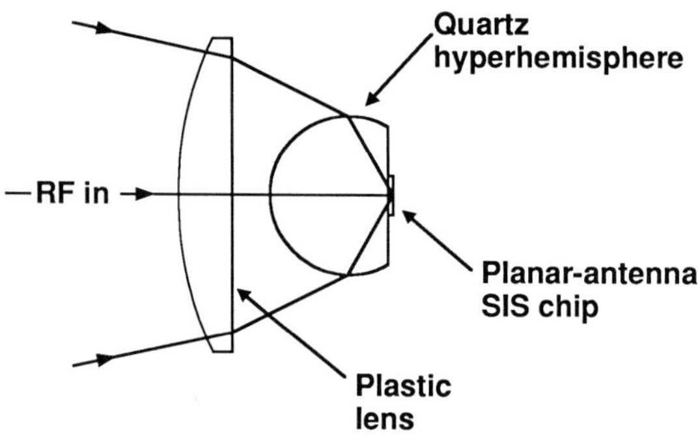

Figure 2. Optical layout for slot-antenna mixer.

2. Slot Antennas on Dielectric Substrates

A twin-slot antenna on a semi-infinite dielectric substrate has a number of desirable properties, including a low impedance and a symmetric antenna pattern. The antenna properties were computed rigorously using a moment-method technique [2]. On quartz ($\epsilon_r = 4.53$), a slot length of $L = 0.5\lambda$ (λ is the wavelength in free space) and width $W/L = 0.04$ gives a real impedance of $Z_a = 35 + j0\,\Omega$. A slot antenna with these dimensions has a one octave impedance bandwidth. Detailed circuit simulations of the mixer indicate an overall bandwidth of 1.6:1.

The moment-method calculation also predicts the radiation pattern of the twin-slot antenna. The separation between the slots ($S = 0.29\ \lambda$) was adjusted to produce a symmetric pattern into the dielectric substrate with a half-power beam width of

47°. The main beam efficiency is calculated to be $\eta_{MB} = 70\%$. The remaining power is radiated into small E-plane sidelobes (4%) and backwards into "air" (26%). The phase and polarization characteristics of the antenna are predicted to be excellent.

3. Impedance Matching

As compared to other designs, the slot-antenna mixer uses an SIS junction with a fairly large area. The area is scaled inversely with frequency; the design rule is that $1/\omega C = 4\,\Omega$ (see below). A 500 GHz mixer requires a junction area of 0.9 μm^2, assuming a specific capacitance of 85 fF μm^{-2} appropriate for Nb/Al–Oxide/Nb junctions with J_c = 10 kA cm^{-2} [3]. Junctions with larger areas have a higher saturation power, require a smaller magnetic field to suppress the Josephson effect, are easier to fabricate, and are electrically more robust. The disadvantage is that a larger junction has a lower impedance which may be more difficult to match. Our mixer design does not attempt to tune out the junction capacitance. Rather, we transform the antenna impedance down to $1/\omega C = 4\,\Omega$ using the tapered microstrip line. We have intentionally avoided using resonant circuits in this initial design because accurately predicting and fabricating such circuits at high frequencies is difficult. However, our failure to tune out the junction capacitance reduces the coupling efficiency by roughly a factor of 3 if $\omega R_N C = 5$.

The microstrip lines are tapered from a characteristic impedance of 35 Ω at each slot to an impedance of 8 Ω at the junction, so the impedance seen by the junction is 4 Ω since the two lines are in parallel. The characteristics of the microstrip lines were calculated using an adaptation of the method described by Whitaker et al. [4], but since the design uses a tapered line, it is fairly insensitive to the actual phase velocity and characteristic impedance.

4. Device Fabrication and Testing

The devices were fabricated at JPL. The mixers tested had junction areas of about 2.3 μm^2, normal-state resistances of 9 Ω, and current densities of 10 kA cm^{-2}, which give $\omega R_N C = 5.3$ at 500 GHz. The typical normal to subgap resistance ratio was $R_{sg}/R_N = 10$.

The results of a measurement of the receiver sensitivity at 490 GHz is shown in Figure 3. A magnetic field was used to suppress the Josephson effect. Structure is still visible in Fig. 3 in the vicinity of the second Shapiro step at 2 mV, but is suppressed to a remarkable degree compared to the case with no magnetic field. The receiver noise temperature was measured to be 420 K (DSB). Estimates of the conversion loss and mixer noise temperature are $L_c = 10\,dB$ and $T_m = 240K$.

We fully expect that a substantial improvement of the mixer sensitivity can be achieved, and that receiver noise temperatures at 500 GHz in the 200 K range will be obtained through a combination of better impedance matching and a reduction of the optical losses. In the future, we also expect to measure the frequency response of the mixer and to map the receiver beam pattern.

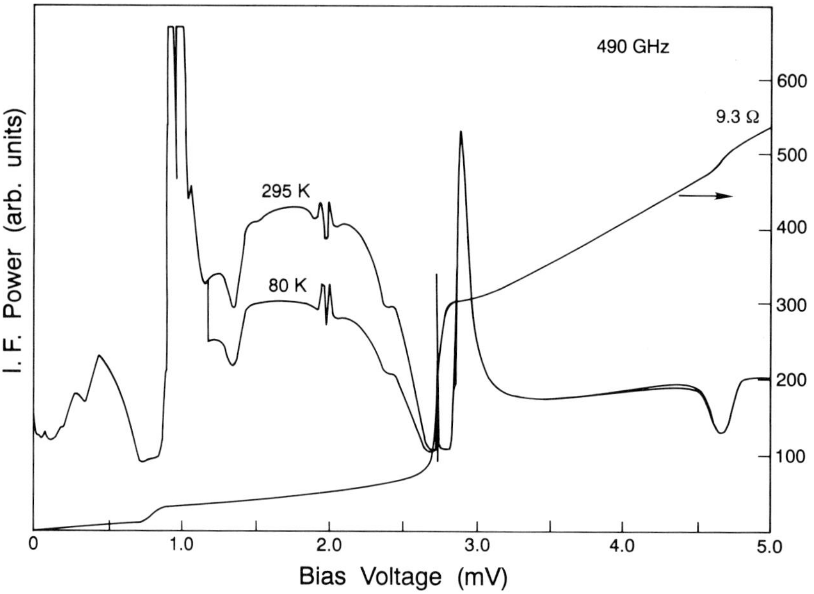

Figure 3. The I.F. output power of the mixer as a function of bias voltage measured with hot (295 K) and cold (80 K) loads at the receiver input. Also shown is the pumped I-V curve.

Acknowledgement

This work was supported in part by NASA grant NAGW-107 and NSF grants DPP88-18384 (to the University of Illinois) and AST-9057123.

References

[1] T. H. Büttgenbach, R. E. Miller, M. J. Wengler, D. M. Watson, and T. G. Phillips, IEEE Trans. Microwave Theory Tech., **MTT- 36**, 1720 (1988).

[2] M. Kominami, D. M. Pozar, and D. H. Schaubert, IEEE Trans. Antennas Propagat., **AP-29**, 600 (1985).

[3] J. Stern, private communication, 1991.

[4] J. F. Whitaker, R. Sobolewski, D. R. Dykaar, T. Y. Hsiang, and G. A. Mourou, IEEE Trans. Microwave Theory Tech., **MTT-36**, 277 (1988).

Numerical and Experimental Results on Josephson Junctions Irradiated by a Biharmonic Drive

D. Andreone, V. Lacquaniti, and S. Maggi

Istituto Elettrotecnico Nazionale "G. Ferraris",
Strada delle Cacce 91, I-10135 Torino, Italy

Abstract. We report some numerical results regarding the characteristics of rf-induced constant voltage steps in a voltage-biased Josephson junction irradiated by a biharmonic drive, namely a rf signal obtained by combining two microwaves of frequency ω and 2ω. The maximization of step amplitudes depends strongly on a careful choice of biharmonic drive parameters. We have made experiments on single, large, Josephson junctions, using a biharmonic signal with a fundamental frequency in the range 8 – 12 GHz. Under these conditions, only a small enhancement of the amplitude of the quantized steps with respect to the single drive case has been observed.

1. Introduction

Josephson voltage standard is at present realized at 1 and 10 V levels by means of large arrays of Josephson junctions coupled to a 70 or 90 GHz microwave source [1]. Another device with a small number of junctions in series (20 with the present design), working at 10 GHz, can achieve quantized voltages up to 100 mV [2]. A major problem of this device is the narrow range of constraints that must be fulfilled by the junction parameters. In fact, using high order steps ($n \approx 200$), stability becomes a hard compromise involving the junction dimensions and current density [3] and, in any case, for a voltage biased junction, the maximum semiamplitude of the n-th step above (below) the quasiparticle I-V characteristic, normalized to I_C, is $i_{n,M} = |\,\mathrm{J}_n(a)\,|$, where a is proportional to the rf voltage.

Recently R. Monaco has demonstrated that, in the limit of a voltage biased junction, if the driving microwave signal is rich in harmonics, steps with a maximum amplitude greater than that given by the above equation can be obtained [4]. In particular, when a rf signal composed of both the fundamental signal and its 2nd harmonic is considered,

$$i_n = \sum_{j=-\infty}^{\infty} (-1)^n \, \mathrm{J}_{n+2j}(a)\, \mathrm{J}_j\!\left(\frac{\rho a}{2}\right) \sin\left[-(n+2j)\theta + \tilde{\phi}\right], \qquad (1)$$

where i_n, in units of I_C, is the current on the n-th step above (below) the quasiparticle line, $a = 2eV'/\hbar\omega$ is the normalized amplitude of the fundamental rf signal, ρ is the amplitude ratio and θ the phase difference of the 2nd harmonic and the fundamental signal. It is thus possible, in principle, to increase the array stability with less stringent requirements on the junction design. In order to apply this result to the experimental situation, we computed the predicted effect for large order steps. The experiments have been performed on single junctions to analyze the effects of a biharmonic drive in a non-ideal case, i.e., for long junctions that cannot be considered to be voltage biased.

2. Numerical Results

Eq. (1) has been used to compute the semiamplitude of the n-th biharmonic step up to $n = 200$. The dependence of the amplitude of the fundamental rf signal and of the amplitude ratio of the biharmonic signals, that maximize the steps, on the step order is shown in Fig. 1. The numerical data have been fitted respectively as $a_M = 1.4n$ and $\rho_M = 0.78/n^{0.2}$. The advantage of using a biharmonic drive seems to be paid by the necessity of using a much larger fundamental rf signal than in the single drive case, where $a_M \sim n$ (Fig. 1a). It should be emphasized that not only the maximum but any increase of the biharmonic steps over the Bessel behaviour leads to this result. On the other hand, when the ρ_M is considered, it decreases by increasing the step order (Fig. 1b). This, compared with the previous result, indicates how untrivial is the role of the 2nd harmonic: it should be present but its amount is less affecting the step amplitude than the fundamental signal at large quantized voltages.

The effect of the phase difference θ has also been evaluated. Its general features, for odd and even steps, are: (a) the step semiamplitude is always maximized for $\theta = 0$; (b) when $n = odd$ the step semiamplitude is zero for $\theta = \pi/2$, (c) when $n = even$ and $\theta = \pi/2$ the pattern is specular to that for $\theta = 0$ and π.

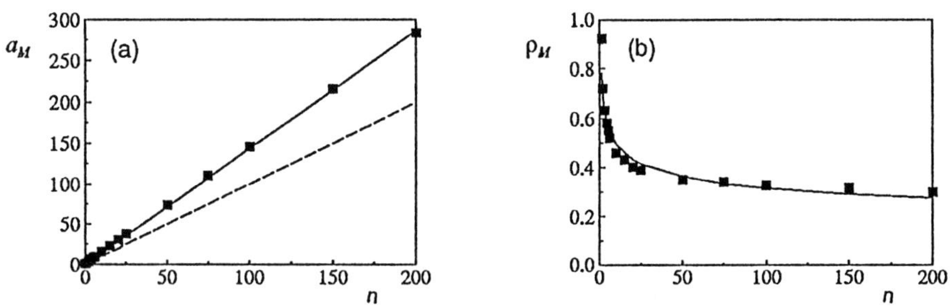

FIG. 1. Normalized amplitude of the fundamental rf signal (a) and amplitude ratio (b), that maximize the biharmonic steps, versus the step order n. The continuous lines are best fit of numerical data. The dashed line in (a) is a_M versus n for the single drive case.

3. Experimental Results

The measurements have been carried out on single Josephson junctions using a biharmonic signal originated by a 8 – 12 GHz klystron and a frequency doubler. The microwave circuit is described elsewhere [5]. Two types of junctions have been used,

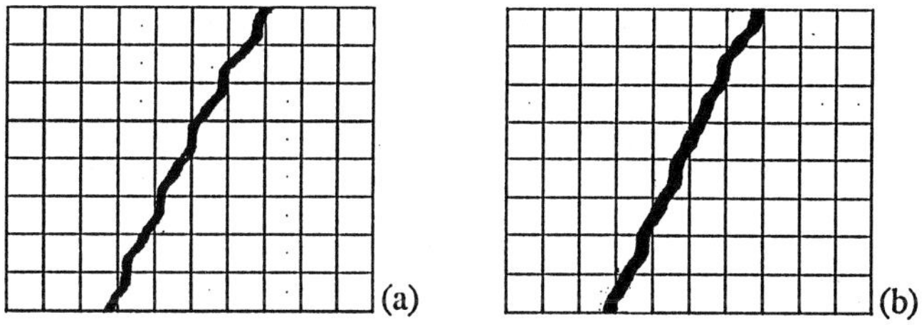

FIG. 2. Effect of the phase difference on the biharmonic steps: (a) $\theta = 0$, (b) $\theta = 150°$. Scales: X=50 μV/div, Y=200 μA/div. The center step is at 150 μV.

TABLE I. Experimental results with a Nb/Pb junction. The step semiamplitudes are in units of I_C.

Step order	Computed step semiamplitude			Measured step semiamplitude		
	single drive	*biharm. drive*	*enhancem. (%)*	*single drive*	*biharm. drive*	*enhancem. (%)*
8	0.324	0.541	67	0.107	0.13	22
12	0.286	0.505	77	0.083	0.097	17
14	0.272	0.492	81	0.077	0.083	9

whose main parameters are: (a) for Nb/Pb junctions, $530 \times 200\ \mu m^2$, $J_C = 0.5 \div 3\ A/cm^2$, $\lambda_J = 250 \div 580\ \mu m$, $f_P = 3 \div 7$ GHz, and (b) for Nb/Nb junctions, $75 \times 20\ \mu m^2$, $J_C = 70 \div 250\ A/cm^2$, $\lambda_J = 25 \div 50\ \mu m$, $f_P = 20 \div 40$ GHz. Both types of junctions are large with respect to λ_J ($L/\lambda_J = 2 \div 3$), and, for the Nb/Nb junctions, the high plasma frequency indicates a chaotic behaviour. Hence these junctions cannot be considered to be voltage biased: the steps induced by the single drive are not those predicted and the effect of the biharmonic drive is different from what has been evaluated by means of Eq. (1). The main results obtained with a Nb/Pb junction are given in Table I. The enhancement of the step amplitude was almost independent of the power of the fundamental signal. Similar results were obtained with Nb/Nb junctions. The effect of the phase difference is shown in Fig. 2. In this case, also, the behaviour predicted by the calculations was not strictly observed.

4. Conclusions

The application of a biharmonic signal to an ideal, voltage biased Josephson junction results in an increase of the step amplitude depending on the amplitude and phase of both the fundamental and the 2nd harmonic. The numerical calculations show that the enhancement of the step height due to the biharmonic drive increases with the step order [5]. However, for a given step, the fundamental rf amplitude needed to maximize the step is much higher than in the single drive case.

The results of the theory have been experimentally verified up to $n = 10$ in a small junction [4]. Our experimental results on large junctions show a different behaviour: the effect is still present but the enhancement is much lower. A possible reason of this behaviour is that our junctions should be considered bidimensional and this leads, as reported by Klein *et al.* [6] for a resonant junction driven by a single rf signal, to a reduction of the step amplitudes with respect to the Bessel behaviour and to their independence on the rf power. The extension of the calculations to bidimensional, current biased junctions is thus needed.

Acknowledgements

The authors wish to thank CSELT, Torino, and M. Negro, IEN, for the support in the experimental work. This work has been partially supported by National Research Council, under the Progetto Finalizzato "Tecnologie Superconduttive e Criogeniche".

References

[1] J. Niemeyer, R. Poepel, L. Grimm, Final Report on BCR Project, Contract 3108.
[2] D. Andreone, V. Lacquaniti, G. Costabile, M. Lepore, R. Monaco, S. Pagano, in Proc. of the Workshop on Stimulated Effects in Josephson Devices, (World Scientific Press, 1990).
[3] R.L. Kautz, J. Appl. Phys. **62**, 199 (1987).
[4] R. Monaco, J. Appl. Phys. **68**, 679 (1990).
[5] D. Andreone, V. Lacquaniti, S. Maggi, to be published in Proc. of NATO Advanced Research Workshop on Superconducting Electronics, (Capri, Italy, 1990).
[6] U. Klein, S. Schroeder, J.H. Hinken, IEEE Trans. Instrum. Meas. **40**, 291 (1991).

Phaselocked States in Josephson Junction Resonator Systems

J. Mygind[1], *H. Dalsgaard-Jensen*[2], *and A. Larsen*[3]

[1]Physics Laboratory I, Technical University of Denmark, DK-2800 Lyngby, Denmark
[2]Danish Institute of Fundamental Metrology, DK-2800 Lyngby, Denmark
[3]Telecommunications Research Laboratory, DK-2970 Hørsholm, Denmark

Abstract. Mutual phaselocking has been observed between small Josephson tunnel junctions integrated in superconducting microstrip resonators. With the junctions individually dc biased on self-induced current steps, we observe phaselocked microwave emission. Locking range and generated power is reported.

1. Introduction

In order to increase the available power of Josephson oscillators one may add the power from several junctions and most efficiently bring these junctions into a phaselocked state[1]. For a series connected array with N junctions the power transferred to an unmatched load scales with the square of N, the linewidth of the radiation decreases with N and the impedance increases with N. Recently progress have been made using long Josephson junctions operating as fluxon oscillators and output levels of 0.1 μW have been reported[2].

A Josephson junction weakly coupled to a resonant circuit exhibits self-pumped current singularities in the dc IV-characteristic at voltages corresponding to the resonance frequencies of the loaded resonators[3]. The appearance of the steps can be understood by calculating the effective admittance load, $Y_\Sigma(\omega)$ on the ideal Josephson element described by the two Josephson relations, $I_S = I_c \sin\varphi$ and $V_J = (\hbar/2e)d\varphi/dt$, where φ is the phase difference of the macroscopic wavefunctions of the superconducting electrodes and I_c is the maximum supercurrent. I_S is the supercurrent, and V_J is the voltage across the element.

In the case of a small tunnel junction integrated in the middle of a long open-ended microstrip resonator, $Y_\Sigma(\omega)$ consists of the sum of the admittance of the Josephson junction and the admittance of the two microstrip halfs transformed to the junction position, and hence given by $Y_e = 2Y_0 \tanh \gamma\ell$, where Y_0 is the characteristic admittance of the microstrip. $\gamma = \alpha + j\beta$ is the propagation constant of the microstrip and ℓ is the length of the microstrip half. In the simple RSJ-model the passive admittance of the junction is given by $Y_J = G_J + j\omega C_J$, where G_J is a linear conductance and C_J is the capacitance of the junction. Assuming a low-loss microstrip ($\alpha\ell \ll 1$) the total load admittance is

$$Y_\Sigma(f) = G_J + 2Y_0\,\alpha l(1 + \tan^2 \pi\frac{f}{f_e}) + j\,(2\pi f C_J + 2\,Y_0\,\tan \pi\frac{f}{f_e})\ , \qquad (1)$$

where $f_e = v_{ph}/2\ell$, and v_{ph} is the phase velocity in the microstrip.

Springer Proceedings in Physics, Vol. 64 Superconducting Devices and Their Applications
Editors: H. Koch, H. Lübbig

When $\mathrm{Im}Y_\Sigma = 0$, with solutions denoted $f = f_n$, the loading circuit exhibits a resonance, and the voltage oscillation amplitude and hence the amplitude of $d\varphi/dt$ will at these frequencies be significantly changed. Due to the nonlinear interaction with the supercurrent $I_c \sin\varphi$, a large nonzero mean current can be carried by this channel and a step structure appears in the dc IV-characteristic.

With the junction biased at a dc voltage corresponding to the k'th harmonic of the resonance frequency f_n, parametrically excited steps may occur at these voltages too. Here we only discuss excitation of the fundamental mode f_1. To distinguish between the two types of steps we denote the steps originating from different solutions, f_n, to the resonance condition, mode steps (MSn), and the parametrically induced steps, parametrically generated mode steps (PMSk). In a recent paper[4] we have described this in more detail.

2. Mutual locking

We have observed mutual locking in a number of samples. For clarity the results reported here are from one particular sample (H5-2) at 4.2 K. The two junctions, A and B, on this sample were nearly identical with a maximum critical current of $I_c = 770$ μA. From the voltage of the first Fiske step $V_{FS1} = 298$ μV and the magnetic field necessary to suppress the supercurrent $B_0 = 305$ μT we find $C_J = 28$ pF, the magnetic thickness $d = 226$ nm and hence the Josephson penetration depth $\lambda_J = 30$ μm (normalized length $L/\lambda_J \approx 1$). The microstrip resonators (width $w = 200$ μm and length $2\ell = 1100$ μm) are placed parallel to each other with a separation of 4 μm. With the magnetic field applied (220–240 μT) $I_c \approx 250$ μA.

For both junctions the first mode step MS1 is at $V_{MS1} = 17$ μV, and when dc biased on this step, we observe radiation in the range from just below 8 GHz at the bottom to a maximum value of approximately 8.36 GHz at the top of the step. We also observe steps at voltages corresponding to integral values of V_{MS1}. These steps are the PMSk's. Biased on these steps we also observe radiation near 8 GHz. Focusing now on PMS2, we also find MS2 at a voltage about 0.1 μV higher than that of PMS2. On the PMS2 we observe emitted microwave radiation not only at 16 GHz as expected from the voltage on PMS2, but also at the fundamental frequency 8 GHz. On MS2, however, we do not observe any microwave signal at 8 GHz, but only at 16 GHz.

We have previously shown[5] the locking of states on MS1 in junction A to states on MS1 in junction B. Locking, however, is equally possible when the junctions are biased on higher order steps. In Fig. 1 is shown a reproduction of IV-traces of PMS2 and MS2 for junction A, labelled A8 and A16, respectively. The three pairs of traces are obtained when junction B is biased (a) in the zero-voltage state (B÷), (b) on PMS2 (B8), and (c) on MS2 (B16). The current bias through junction B is held constant in all traces. The voltage of the pairs are offset for clarity. The dots indicate the begin and end of the locking ranges.

The locking ranges in the different configurations are largest when the junctions are biased on the same type of step (A8-B8 and A16-B16). In the

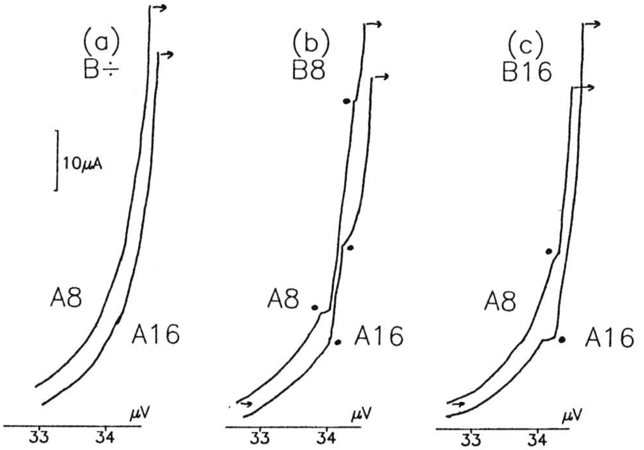

Figure 1: *Reproduction of IV-traces of PMS2 (A8) and MS2 (A16) of junction A, when junction B is biased (a) in the zero-voltage state (B÷), (b) on PMS2 (B8), and (c) on MS2 (B16).*

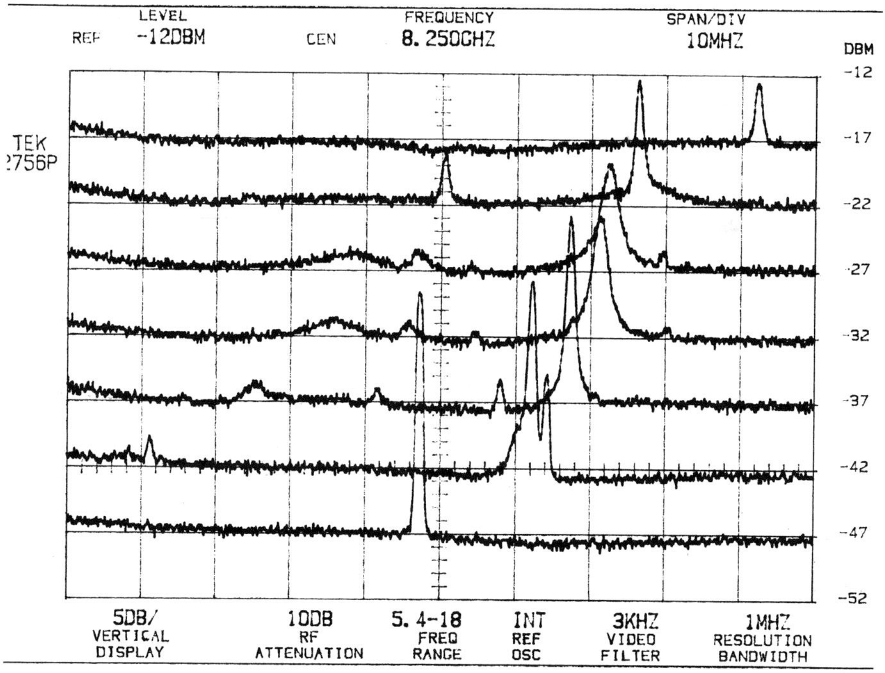

Figure 2: *Spectra of the locking process in the A8-B8 configuration when the bias current in junction A is increased. Also several mixing products are seen.*

latter case the locking range is limited by the step height. When biased in different states (A16-B8 and A8-B16) the locking ranges are generally smaller.

Fig. 2 shows spectra of the locking process in the A8-B8 configuration when the bias current in junction A is increased. From this figure we observe a power of the locked signal of −29 dBm. Based on an evaluation of the cable loss and the total impedance mismatch losses, we estimate an available power of $P_{\Sigma} = 2$ nW.

Acknowledgements. We thank J. Niemeyer, R. Pöpel and M. R. Samuelsen for discussions. This work has been partially funded by the Commission of the European Community and the Danish Research Academy.

References

[1] K. Wan, B. Bi, A. K. Jain L. A. Fetter, S. Han, W. H. Mallison, and J. E. Lukens, IEEE Trans. Magn. **27**, 3339 (1991).

[2] S. Pagano, R. Monaco, and G. Costabile, IEEE Trans. Magn. **25**, 1080 (1989)

[3] N. R. Werthamer and S. Shapiro, Phys. Rev. **164**, 523 (1967).

[4] A. Larsen, H. D. Jensen, and J. Mygind, Phys. Rev. B, **43**, 10179 (1991).

[5] H. D. Jensen, A. Larsen, and J. Mygind, IEEE Trans. Magn. **27**, 3355 (1991).

Microwave Properties of Josephson Junctions Strongly Coupled to Microstrip Resonators

A. Morgenstjerne[1], *J. Mygind*[1], *H. Dalsgaard-Jensen*[2], *and A. Larsen*[3]

[1]Physics Laboratory I, Technical University of Denmark, DK-2800 Lyngby, Denmark
[2]Danish Institute of Fundamental Metrology, DK-2800 Lyngby, Denmark
[3]Telecommunications Research Laboratory, DK-2970 Hørsholm, Denmark

Abstract. *In situ* determination of the losses in a system consisting of a small Josephson junction integrated in the middle of a long superconducting microstrip resonator have been made at the $n = 1$, 2 and 3 mode frequencies. From the quality factor, Q_n, and the junction loss, G_J, the microstrip resonator loss, G_e, is found as function of temperature at 8, 17 and 25 GHz.

1. Introduction

In a recent paper[1] we have shown that exceptionally high coupling factors may be achieved in systems consisting of a small $Nb/Al_2O_3/Nb$ junction integrated in the middle of a resonant section of an open-ended superconducting microstrip transmission line. In this limit we observe a series of large self-induced steps (PMSk, parametrically generated mode steps) excited by parametric pumping of the fundamental resonant mode by the kth subharmonic of the Josephson frequency[2]. The self-induced mode steps (MSn) usually found in the dc IV-characteristic of Josephson junctions interacting weakly[3] with the n'th resonant mode seem to become unstable and to be dominated by the PMSk in the multi-mode system with stronger coupling and large harmonic overlap between the mode resonances.

2. Experimental system

The sample preparation and layout of the small ($L/\lambda_J \approx 1$, λ_J is the Josephson penetration length) tunnel junction integrated in the middle of a long (length ℓ, width w) microstrip resonator has been described in detail elsewhere[1]. The ideal element is loaded by the passive junction admittance $Y_J(f) = G_J + j\,2\pi f C_J$ (representing the junction capacitance C_J and the (in general nonlinear) conductance G_J), and the load admittance Y_e of the two stubs transformed to the position of the junction. The total load admittance is given by

$$Y_\Sigma = G_\Sigma + j\left(2\pi f C_J + 2Y_0 \tan \pi \frac{f}{f_e}\right) \qquad (1)$$

$$G_\Sigma = G_J + 2Y_0 \alpha \ell \left(1 + \tan^2 \pi \frac{f}{f_e}\right)$$

where the approximation is valid for low-loss microstrip lines, i.e. $\alpha\ell \ll 1$. Y_0 is the characteristic admittance of the microstrip.

Springer Proceedings in Physics, Vol. 64 **Superconducting Devices and Their Applications**
Editors: H. Koch, H. Lübbig

The real part of the propagation constant, $\gamma = \alpha + j\beta$, is the sum of two terms: α_c and α_d, the attenuation factors describing surface and dielectric losses, respectively. For the dielectric thickness, d, much smaller than the wavelength the resonator is effectively open-ended along its periphery and radiation losses are negligible. The surface loss factor α_c is connected with the real parts R_a and R_b of the surface impedances $\alpha_c = \frac{R_a+R_b}{2w}Y_0$. α_c depends on frequency through R_a and R_b, but is independent of w. The dielectric loss factor is $\alpha_d = \frac{\pi}{2l}\frac{f}{f_e}\tan\delta$ where $\tan\delta$ is the loss tangent.

We shall neglect the susceptance of the ideal element[1], and hence the mode resonances occur for Im $Y_\Sigma = 0$. This can be rewritten as

$$\tan\pi\frac{f}{f_e} = -\frac{C_J}{C_e}\pi\frac{f}{f_e}; \qquad f_e = \frac{c_0}{2l\sqrt{\epsilon_r\chi_r}} = \frac{\pi f}{\beta l} \tag{2}$$

where $C_e = \epsilon_r\epsilon_0\frac{2lw}{d}$ is the capacitance of the resonator electrodes, and f_e is the first full wavelength resonance frequency of the unloaded microstrip resonator. The nth mode resonance frequency f_n is the solution to Eq. 2 and the quality factor Q_n of the nth mode is

$$Q_n = \pi\frac{f_n}{G_\Sigma}\left(C_J + C_e\left(1+\tan^2\pi\frac{f_n}{f_e}\right)\right) \quad . \tag{3}$$

3. Results

The small Josephson junction can be used as a very sensitive broadband microwave detector when biased at a fixed dc current on the quasi-particle curve in the vicinity of the superconducting gap. In the small signal limit the change, ΔV, in the dc voltage across the junction is proportional to the power of the applied microwave signal. Fig. 1 shows a typical voltage response near the fundamental mode resonance at about 8 GHz. The quadratic power dependence has been carefully checked and the measurements have been carried out at different bias currents. Only the sensitivity changes with the bias point.

Inserting the measured values of f_n and Q_n in Eq. 3, G_Σ is iteratively calculated using Eq. 2, C_e and C_J. C_J is found from measured values of the voltage of the first Fiske resonance and the magnetic field necessary to suppress the supercurrent in the magnetic interference pattern[1].

In Fig. 2 is shown the found values of G_Σ versus the dynamic conductance, G_J, measured in the bias point for the first three resonant modes. The slope of the straight lines is equal to one and as seen from Eq. 1 we are able to determine the damping constant of the microstrip resonator.

The temperature and frequency dependence of α_c has been calculated for the electrodes using Mattis-Bardeen theory. Typically for several junction-resonator systems we find that the dielectric losses dominate at 2 K with a value of 0.26, 0.68, and 1.5 m^{-1} at 8.3, 16.7, and 25.3 GHz, respectively. At 4.2 K the surface losses are of similar magnitude and show the predicted frequency dependence[4].

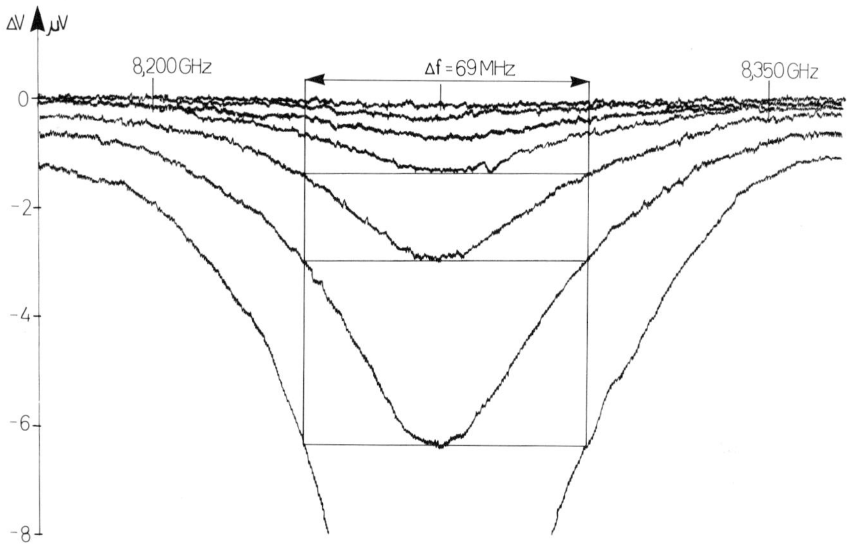

Figure 1: *Typical voltage response, $\Delta V(f)$, near f_1. The junction was biased just below the gap voltage. Sequential curves are recorded with the applied power incremented by 3 dB. The Q_1 value is found to 120 ± 5 at $G_J = 48$ mS, $T = 4.2$ K (Sample H5-2B).*

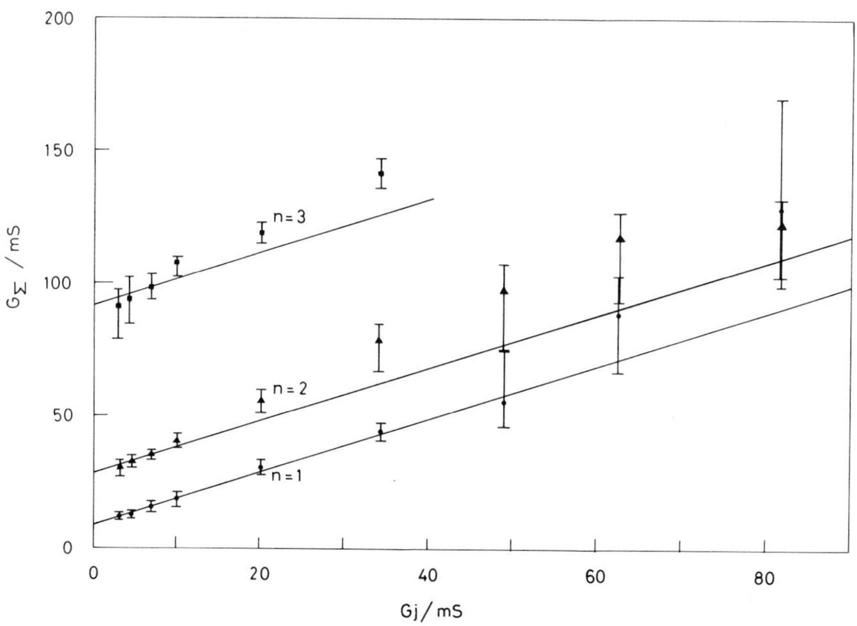

Figure 2: *Measured total load conductance, G_Σ, plotted as function of the dynamic conductance, G_J, for the first three resonant modes at 8.34, 16.68, and 25.31 GHz at $T = 1.9$ K in sample H5-2B.*

Acknowledgements. We thank J. Niemeyer and M. R. Samuelsen for valuable discussions, and R. Pöpel for help with the Mattis-Bardeen calculations. This work was partly funded by the Commission of the European Communities and the Danish Research Academy.

References

[1] A. Larsen, H. Dalsgaard Jensen, and J. Mygind, Phys. Rev. B. **43** 10179 (1991.)

[2] L. Kuzmin, H. K. Olsson, and T. Claeson, in *SQUID-85*, edited by H. D. Hahlbohm and H. Lübbig (W. de Gruyer, Berlin, 1985) pp. 1017-1022.

[3] K. K. Likharev, *Dynamics of Josephson Junctions and Circuits*, (Gordon and Breach, New York, 1986), Chapter 12.

[4] to be published.

Multiloop Integrated DC-SQUID Low Noise RF Amplifiers

M.A. Tarasov, V.Yu. Belitsky, G.V. Prokopenko, and L.V. Filippenko

Institute of Radio Engineering and Electronics, Academy of Sciences, Marx Avenue 18, 103907 Moscow, Russia

Abstract. The noise and signal parameters of several types of RF amplifiers based on different SQUIDs were studied. A new type of the multiloop dc SQUID in the form of a second order gradiometer with integrated input coil and extremely low stray capacitances was designed. An amplifier noise temperature lower than 0.5 K and a power gain of nearly 20 dB were obtained at 430 MHz for 60 MHz bandwidth.

1. Introduction : SQUID–amplifier signal and noise parameters

As mentioned in [1] a SQUID amplifier (SQA) may be viewed as a peculiar type of parametric amplifier. For a parametric up–converter the power gain equals pump to signal frequency ratio. The pump frequency in a dc SQUID with dc bias is the Josephson oscillation and the well known expression $G=\alpha^2 r/L\omega$ may be explained as the ratio of frequency r/L limiting the Josephson current in the loop to the signal frequency. In the case of sufficient capacitance in the loop the limiting resonant frequency will be $(LC)^{-1/2}$. Thus it is useful for gain increase to reduce both inductance and stray capacitance of the SQUID loop.

With the noise temperature of the SQA $T_N=6.5T\omega L/r\alpha^2$, i.e. for $\omega=10^9$, $L=10^{-10}$H one can, according to [2], obtain $T_N=0.04T$, where T – physical temperature, and $R_i^{opt}\simeq 0.3\alpha^2\omega L_i$ for Johnson noise source in the SQUID. In real SQUIDs the T_N may be 2–3 times higher. The quantum limit of SQA at T=0, according to [3], may be calculated by replacing the Johnson noise by zero fluctuations in the expression for the tuned SQA noise. The obtained value $T_N \geq hf/k\ln 2$ is just the same as for any amplifier in the quantum limit.

2. Planar integrated and hybrid single–loop dc SQA studies

At the beginning of our studies of SQA we tried to use three types of SQA, the first and the second contain shunted tunnel junctions and the third has junctions with a semiconductor interlayer. We studied SQUIDs with junctions situated both inside and outside the inductive square loop and used both integrated and hybrid input spiral coils. The measured output noise and voltage gain have some interesting features in the IV curve such as the presence of current steps at voltages corresponding to the resonant frequencies of the SQUID. If the bias point is placed at the maximum of dynamic resistance near one of these steps the amplifier gain and noise temperature may be not worse than for a bias point near the Josephson critical current. It is possible to explain this fact, according to [4], by the presence of broad band Josephson oscillations which give rise to the noise at signal frequency for the bias voltage near the critical current (zero voltage step). This noise will be sufficiently lower for the bias voltage near higher self-resonant voltage steps at $V>100\mu V$.

These results lead to the conclusion that it is possible to rise the gain and reduce the noise if the bias point is located near the resonant step where the dynamic resistance exceeds the normal one and the Josephson frequency exceeds the values obtainable near the critical current. Another method to increase the Josephson frequency is to use external microwave pumping and dc bias near the Shapiro steps [5] and it is much similar to the mode with dc bias near self-resonant steps.

The best results for hybrid SQA with Cu spiral wire-wound coil were G>20 dB, $T_N=1.2\pm1$ K at 100 MHz.

3. Four-loop dc SQUID amplifier studies

The main disadvantages of the previous studied SQA are relatively low amplification frequency and a high influence of external magnetic fields on SQA parameters. For a signal frequency increase with preservation of the low noise temperature and high gain the Josephson frequency in the SQUID loop should be increased. It means that the loop inductance and capacitance should be decreased. In the commonly used integrated SQUID structure the stray capacitance in the loop exceeds 10 pF. The capacitance between the loop and the input coil evaporated above the loop is even larger and this capacitance leads to the significant stray input-output feedback.

To eliminate these disadvantages and increase the input signal frequency and bandwidth we designed a four-loop dc SQUID with an integrated input coil in the form of rectangular turns inside the loops (see Fig.1). The input coil consists of four series connected square turns with 10 μm linewidth made of Nb film. Parallel connection of the partial loops allows to reduce the inductance of the loop and increase the resonant frequency. Series connection of the

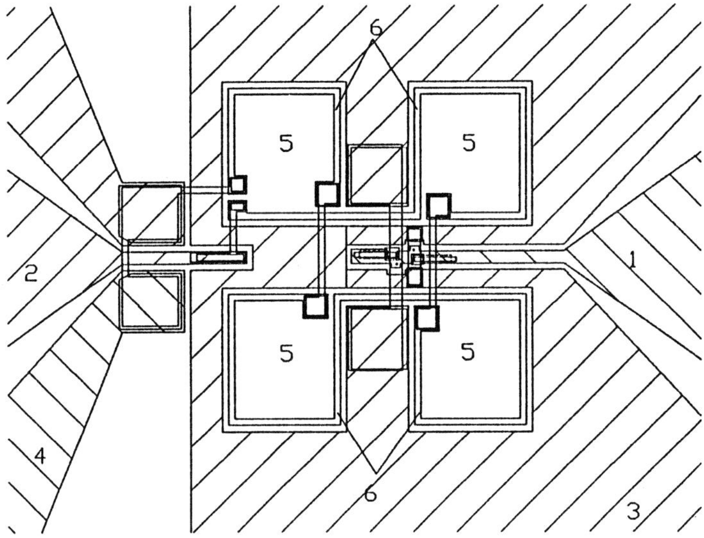

Fig.1 Four-loop planar dc SQUID with integrated input coil. (1,3) and (2,4) - output and input coplanar lines, (5) - parallel connected SQUID loops each 200x200 μm^2, (6) - input coil.

input turns allows to increase the input coil inductance and make the impedance matching with the input 50 Ohm line easier. The position of input turns inside the loops was chosen to reduce the stray input-output capacitance.

The SQUID loop inductance of this configuration is 100 pH, the input coil 1.3 nH, the mutual inductance 300 pH. The stray capacitance in the SQUID loop is 1.8 pF, the junction sum capacitance 0.8 pF, the input coil capacitance 0.3 pF, the loop-coil capacitance 2.2 pF. Nb-AlO_x-Nb shunted tunnel junctions of 2.5x2.5 μm^2 area were used as Josephson junctions. For input signal matching we used a resonant input circuit. At 400 MHz the gain of the SQA was G≃20 dB and the noise temperature was not worse than 0.4 K for 60 MHz bandwidth.

4. Discussion

The nearest competitor of SQA are cooled HEMT amplifiers which have T_N~(1-3) K at f~(1-3) GHz, but have rather high power dissipation and nearly achieved their asymptotic parameters. SQA has more than one order reserve in the margins of gain, noise temperature and signal frequency even on the basis of the present technology. The advantages of SQA are extremely low power dissipation of several picowatt, small size ~1 mm^2 and full electrical and temperature compatibility with superconducting sensitive devices such as Josephson and SIS mixers.

The SQA has good prospects in using high T_c superconductors. According to the above estimations at 77 K HTSC SQA should have noise temperatures lower than 10 K at $f \sim 10^2$ MHz. Concerning HTSC SQA asymptotic margins they are even better than of low-temperature SQA because the energy gap voltage of YBaCuO $V_\Delta \simeq 25$ mV and that of Nb only 1.4 mV, so at the same temperatures the gain may be nearly an order of magnitude higher and noise temperature lower in HTSC SQA. If we compare HTSC SQA at 77 K and LTSC SQA at 4.2 K the noise temperature ratio will be $T^N_L/T^N_H=(T_1/T_2)\bullet(\Delta_H/\Delta_L)\simeq 1$.

5. Conclusion

To improve SQUID amplifier power gain and noise temperature the Josephson oscillation frequency in the loop should be increased by means of a decrease in loop inductance and stray capacitances. To decrease the loop inductance four partial loops of a multiloop gradiometer were connected in parallel. The single input turns situated inside each loop were connected in series to increase the input coil inductance for better impedance matching. The measured noise temperature of such a SQUID amplifier was $T_N<0.5$ K at a signal frequency $f \simeq 430$ MHz for 60 MHz bandwidth.

References

1. J.E.Zimmerman and D.B.Sullivan, Appl. Phys. Lett. 11,N5,360–362 (1977).
2. J.Clarke, C.D.Tesche and R.P.Giffard, J. Low Temp. Phys. 37,n3/4, 405–420 (1979).
3. R.H.Koch, D.J. Van Harlingen and J.Clarke, Appl. Phys. Lett. 38, 380–382 (1981).
4. V.N.Gubankov and M.A.Tarasov, Radiotechn. and Electron. (in Russian) 25,N 2, 381–384 (1980).
5. M.A.Tarasov, V.Yu. Belitsky, G.V.Prokopenko, L.V. Filippenko and V.P. Koshelets, ISEC91, to be published.

Subharmonic Pumping and Noise Properites of the Josephson Parametric Amplifier

R. Movshovich[1], *B. Yurke*[1], *A.D. Smith*[2], *and A.H. Silver*[2]

[1]AT&T Bell Laboratories, Murray Hill, NJ 07974, USA
[2]TRW Space and Technology Group, Redondo Beach, CA 90275, USA

Abstract. We report the observation of substantial amplification due to multiwave (more than four) mixing in a Josephson-parametric amplifier. Amplification was achieved with sub-harmonic pumping at $\frac{2}{3}$, $\frac{2}{5}$, and $\frac{2}{7}$ of the signal frequency near $19GHz$. The processes responsible for the amplification were, correspondingly, five-, seven-, and nine-wave mixing. For the case of five-wave mixing we have observed amplification in excess of $35dB$ and noise referred to input of a few degrees K before the amplifier undergoes a pitchfork instability. The noise rise at the pitchfork bifurcation was also studied and compared with theory.

Josephson-parametric amplifiers have been successfully operated in both three-wave and four-wave mixing modes. However, one is not limited to the three-photon and four-photon processes. The nonlinear Josephson inductance employed in parametric amplification has higher order nonlinear terms that are comparable in magnitude to the second and third order terms responsible for three-wave and four-wave mixing, respectively. In this paper we report the observation of gain from a Josephson parametric amplifier in which these higher order nonlinearities are employed in multi-wave mixing processes involving five or more photons at microwave frequencies. This multi-wave mixing was achieved by pumping the amplifier with a subharmonic of twice the signal carrier frequency. For example, in the case of the five-wave mixing the frequency of the pump was chosen to be $\frac{2}{3}$ of the fundamental frequency (or the frequency of the signal to be amplified). In this case three pump photons combined to produce two signal photons, making a total of five photons taking part in the process.

Early work on Josephson-parametric amplifiers (reviewed in reference 1) was plagued by large excess noise. Lately a number of groups succeeded in producing devices that overcame the large excess noise problem.[2-7] Operated as a five-wave mixer, our device exhibited 35 dB gain with $3K$ noise temperature. We compare our noise measurements with the theoretical calculations[8] for a generic system undergoing a pitchfork bifurcation. Our data also displays a family of closely spaced peaks that emerge in the region of the bifurcation, which is a fingerprint of such a system.[9,10]

The Josephson-parametric amplifier used in this experiment is of the same design as the one that was used to demonstrate thermal[3] and vacuum noise[5] squeezing. The apparatus was similar to the one used in previous investigations[5,6] that employed three-photon mixing. For the current experiment a semi-rigid beryllium-copper coaxial waveguide was used to deliver the pump to the input of the Josephson-parametric amplifier.

Vacuum fluctuations or thermal noise input for the amplifier was provided by a variable temperature microwave termination that could be cycled between $30mK$ and $2K$. The detector system noise temperature was measured to be about $300K$. The amplifier's loss parameter η was measured to be between 0.480 and 0.698, the actual power loss for the signal as it passes through the amplifier being $1 - \eta$.

A probe signal at a frequency about $19GHz$ was coupled to the signal going from the noise source to the input port of the JPA. The output signal from the JPA was amplified further by two high electron mobility (HEMT FET) amplifiers operated at $4.2°K$. The signal was then mixed down with the local oscillator, amplified further, and finally delivered to the input of the HP8566B spectrum analyzer. During a single run the pump power, signal probe power, and the detuning from the fundamental frequency were stepped over the range of interest.

Most of the data for the five-photon process was taken with the pump at a frequency of $\nu_p = 12.580\ GHz$. This frequency is 2/3 of the fundamental frequency of $\nu_0 = 18.87\ GHz$ about which the amplified signal and idler signals appeared as the pump power was increased during a single sweep series. Fig. 1 shows four spectrum analizer traces from one of the runs. The trace in Fig. 1(a) was taken with a rather low pump power. It shows the probe signal at $11MHz$ and an idler at $9MHz$ which is just starting to come up in amplitude. Fig. 1(b) shows the trace when the pump power was increased further and the probe and idler are about the same height. As the pump power is increased even more the spectrum breaks into spikes, as shown in Fig. 1(c). These extra peaks are characteristic of the system operated just below the pitch-fork instability[9,10], and are the result of the higher order mixing effects. Finally, Fig. 1(d) displays a trace taken with the pump power above the critical value for the onset of the pitch-fork bifurcation. The central peak at $10MHz$ is due to the amplifier breaking into oscillation. The linewidth of this peak is instrument limited and the oscillation is phase-locked to the pump. The pattern of spikes still persists, in agreement with theory.[9,10]

Fig. 2 shows the noise temperature referred to input of the JPA as a function of the gain for the probe signal as the pump power was stepped through the region of the pitchfork bifurcation for the same data run. Probe signal of each trace of the data run enable determination of the power gain. The noise at the output of the JPA was then divided by the classical gain times the JPA's loss to determine the noise temperature referred to input. As the plot of Fig. 2 shows, the amplifier is very well behaved until the gain is nearly $30dB$, without showing signs of any gain-dependent excess noise. As the pump power is increased further, the amplifier enters the regime of the pitchfork instability, where both increasing noise and decreasing classical probe gain result in a dramatic increase in the noise temperature referred to the input of the JPA,

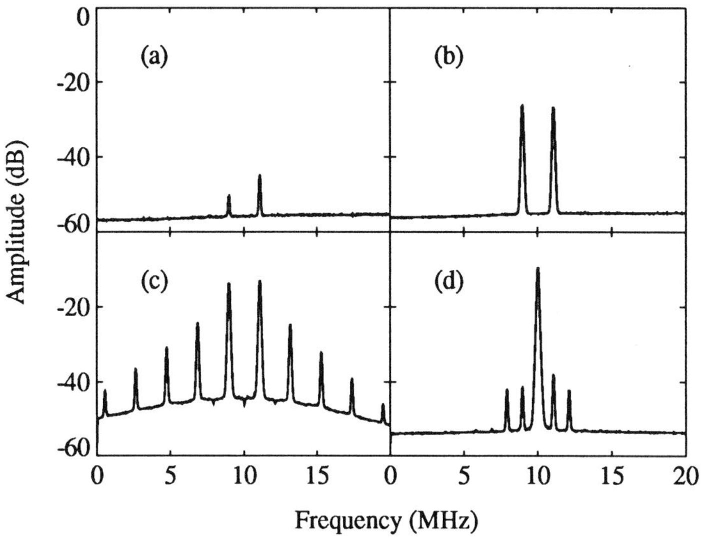

Figure 1: The spectrum analyzer amplitude traces as a function of frequency. Detuning for this run was set to $1MHz$ and the probe amplitude was $-50dBm$. The relative pump powers are: (a): $-6dB$; (b): $-0.9dB$; (c): $-0.3dB$; (d):$0dB$

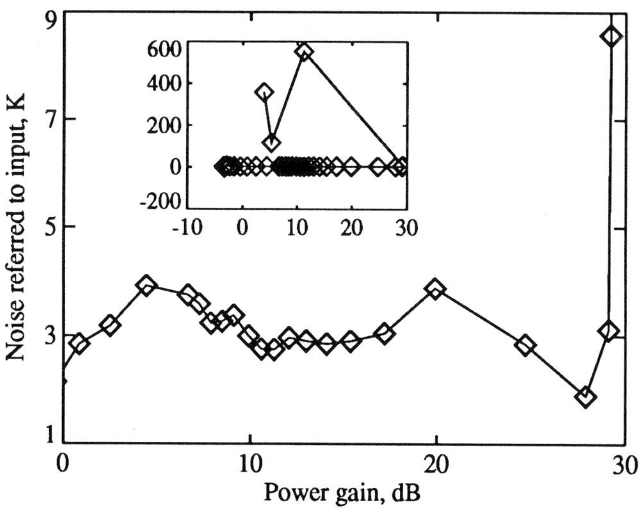

Figure 2: Noise referred to input as a function of the probe gain in the region below the critical value of the pump power for the data displayed in Fig. 1. The line connects consecutive data points as the pump power is increased.

as shown in the inset of Fig. 2. This behavior is qualitatively similar to the one expected for a generic system undergoing a pitchfork bifurcation.[9,10] The average value of the noise referred to the input of the JPA is about $3°K$, which is a factor of two greater than the expected value of $\approx 1.5°K$. We attribute this discrepancy to an additional (probably four-wave) mixing process about the pump frequency that turns on at a lower pump power than the one used to collect the data.

We have also used $\nu_p = \frac{2}{5}\nu_0$ and $\nu_p = \frac{2}{7}\nu_0$, which resulted in seven-wave and nine-wave mixing. For these higher order processes amplification of $\approx 15dB$ resulted when the amplifier underwent a pitchfork bifurcation and nearly $30dB$ was seen when the instability was of the Hopf type.

References

[1] M. J. Feldman and M. T. Levinsen, IEEE Trans. Mag., **MAG-17**, 834 (1981)

[2] H. K. Olsson and T. Claeson, Jpn. J. Appl. Phys., **26**, 1547 (1987)

[3] B. Yurke, P. G. Kaminsky, R. E. Miller, E. A. Whittaker, A. D. Smith, A. H. Silver, and R. W. Simon, Phys. Rev. Lett., **60**, 764 (1988)

[4] B. Yurke, L. R. Corruccini, P. G. Kaminsky, L. W. Rupp, A. D. Smith, A. H. Silver, R. W. Simon, and E. A. Whittaker, Phys. Rev. A , **39**, 2519 (1989)

[5] R. Movshovich, B. Yurke, P. G. Kaminsky, A. D. Smith, A. H. Silver, R. W. Simon, and M. V. Schneider, Phys. Rev. Lett., **65**, 1419 (1990)

[6] P. Bryant, R. Movshovich, and B. Yurke, Phys. Rev. Lett., **66**, 2641 (1991)

[7] B. Yurke, R. Movshovich, P. G. Kaminsky, P. Bryant, A. D. Smith, A. H. Silver, and R. W. Simon, IEEE Trans. Mag., **MAG-27**, 3374 (1991)

[8] P. Bryant and K. Wiesenfeld, Phys. Rev. A, **33**, 2525, (1986)

[9] P. Bryant, K. Wiesenfeld, and B. McNamara, Phys. Rev. B **36**, 752 (1987)

[10] P. Bryant, K. Wiesenfeld, and B. McNamara, J. Appl. Phys., **62**, 2898 (1987)

Imaging of Self-Resonant and RF-Induced States in Josephson Junctions

T. Doderer[1], *R.H. Huebener*[1], *C.A. Krulle*[1], *B. Mayer*[1], *J. Niemeyer*[2], *R. Pöpel*[2], *and D. Quenter*[1]

[1]Physikalisches Institut, Lehrstuhl Experimentalphysik II, Universität Tübingen, Morgenstelle 14, W-7400 Tübingen, Fed. Rep. of Germany

[2]Physikalisch-Technische Bundesanstalt, Bundesallee 100, W-3300 Braunschweig, Fed. Rep. of Germany

Abstract. Using Low Temperature Scanning Electron Micoscopy (LTSEM) we investigated dynamic states of Josephson tunnel junctions: Fiske modes and photon-assisted tunneling. A square junction of intermediate size biased at Fiske steps clearly shows one- and two-dimensional cavity modes depending on the orientation of the external magnetic field. On the other hand, microwave injection into a superconducting tunnel junction results in a standing wave due to the large impedance mismatch between the junction cavity and the surrounding structures. By imaging the spatial distribution of the photon-assisted tunneling current density, we obtained the microwave distribution inside the tunnel junction.

1. Introduction

Acting as nonlinear elements, Josephson tunnel junctions show interesting dynamical behavior [1, 2]. Most of the experimental studies examine the current-voltage characteristic (IVC) of the junctions of various geometries under various experimental conditions. For example, a small external dc-magnetic field in the plane of the junction barrier results in the well known Fiske steps [3, 4]. They can be explained by the excitation of self-induced resonances inside the Josephson transmission line. With junctions smaller than the Josephson penetration depth λ_J these excitations are expected to be standing electromagnetic waves inside the junction cavity [5, 6].

Acting as passive elements, Josephson tunnel junctions represent superconducting transmission lines or open ended cavities. Due to the large impedance mismatch between the junction and the surrounding structures, the injection of microwaves can result in a standing wave.

The IVC contains only spatial averaged information of the various dynamic processes inside the junction. By applying spatially resolved measuring techniques like LTSEM one obtains a deeper insight into the corresponding junction behavior.

In section 2 the basic principles of LTSEM will be discussed. After that, we present the models for the LTSEM signals for the self-resonant states (sec.

3.1) and the rf-induced states (sec. 3.2). The experimental results are shown and discussed in sec. 4. Finally, we summarize our conclusions in sec. 5.

2. Low Temperature Scanning Electron Microscopy (LTSEM)

Using a special cryo stage [7] with a conventional Scanning Electron Microscope (SEM) it is possible to scan a sample kept at about 4 K with the electron beam. For recent reports on LTSEM and its various applications to low temperature physics we refer to [8, 9].

Concerning the studies of Josephson tunnel junctions presented in this paper, the basic principle of LTSEM is the beam-induced local heating effect of the junction. The lateral extension of the thermally disturbed area is given by the thermal healing length

$$\eta = \left(\frac{\varkappa D}{\alpha_{th}}\right)^{1/2} \tag{1}$$

mainly determining the spatial resolution of LTSEM (1 - 3 μm). Here, $\varkappa$ and D are the heat conductivity and the total thickness of the irradiated films, respectively. The heat transfer coefficient α_{th} describes the thermal flux from the thin film into the substrate.

Due to local heating of the superconducting tunnel junction, some of the junction properties are changed locally as follows: Decrease of the critical tunneling conductivity j_c, increase of the London penetration depth of the two superconducting electrodes λ_{L1} and λ_{L2}, increase of the quasi-particle density inside the electrodes and, therefore, increase of the tunneling conductivity of the quasi-particles, etc.

Scanning the electron beam across the sample and recording the desired signal result in imaging of distinct sample properties depending on the experimental conditions. Because of the short thermal relaxation time of the thin film samples (typically 10^{-7} s [10]) and the low scanning speed (typically 100 μm/s), the irradiated sample is nearly in thermal equilibrium at each beam focus position.

With superconducting tunnel junctions the electron beam induced response signal of the junction used for imaging is the change of the total junction current or voltage. To obtain local information of the distinct junction properties from the LTSEM images, specific models for the signal generation are required. Recently, LTSEM was applied to superconducting tunnel junctions to obtain information on the spatial variation of the quasiparticle tunneling current density [11], of the maximum Josephson current density [12, 13], and of the energy gap in the junction electrodes [14]. In this paper, we want to present the latest results of our LTSEM studies of self-resonant [15, 16] and rf-induced states [15, 17] in Josephson tunnel junctions.

3. Models for the LTSEM Signals

3.1 Self-Resonant States (Fiske Modes)

Here, we want to present a qualitative model for explaining the LTSEM signal generation if the Josephson tunnel junction is in a self-resonant dynamic state (Fiske resonance). A detailed analysis of this problem together with the experimental results and numerical simulations with the perturbed sine-Gordon equation can be found in [16]. The following analysis is restricted to small junctions, i. e. the length L and width W of the junction have to be smaller than $2\pi\lambda_J$.

The local temperature increase due to the electron beam irradiation results in an increase of the quasi-particle density within the disturbed area. Therefore, the dynamics of the resonance will be changed as a whole by increasing the junction resonator losses which determine the quality factor Q. Usually, there are two main contributions to the junction quality $Q^{-1} = Q_q^{-1} + Q_s^{-1}$, where Q_q describes the quasi-particle tunneling losses and Q_s the surface losses, respectively. Neglecting effects at the junction boundaries due to the finite size of the perturbed area, the electron beam induced change of Q_q is independent of the beam position across the junction, if the tunneling conductivity is homogeneous. On the other hand, the electron beam induced change of Q_s is expected to be dependent on the beam position for the following reason.

The increase of the quasi-particle density in the irradiated area causes an increase of the real part of the surface impedance R_s and, hence, of the surface losses. Using classical electrodynamics [18], the surface losses are not only proportional to R_s but also to the square of the tangential ac-magnetic field $\mathbf{H}_{\parallel}$ within the cavity. Of course, $\mathbf{H}_{\parallel}$ is spatially dependent for a standing electromagnetic wave excitation in the junction resonator.

Combining both loss mechanisms, in the case of weak electron beam perturbation the change of the junction Q is expected to be

$$\Delta Q = c_1 + c_2 \mathbf{H}_{\parallel}^2 . \tag{2}$$

The spatially independent coefficients c_1 and c_2 correspond to ΔQ_q and ΔQ_s, respectively.

Scanning the electron beam across the junction area, we obtain a spatially modulated change of the junction quality Q proportional to $\mathbf{H}_{\parallel}^2$. As already mentioned, the LTSEM signal used for imaging is either the change of the total junction current or voltage. With a Fiske resonance, a change of Q results in a change of the step structure in the IVC of the junction. For a current biased junction we obtain a decrease $\Delta V_{m;n}$ of the step voltage $V_{m;n}$. Here, the integers m and n denote the two-dimensional cavity mode and for a rectangular junction with length L and width W. The step voltage is given by [19]

$$V_{m;n} \approx \frac{h\bar{c}}{4e}\left(\left(\frac{m}{W}\right)^2 + \left(\frac{n}{L}\right)^2\right)^{1/2} . \tag{3}$$

Here, h is Planck's constant, $\bar{c}$ the Swihart velocity, and e the elementary charge. Using only a small perturbation with the electron beam, we obtain a linear relationship between ΔQ and the detected voltage decrease $\Delta V_{m;n}$. Hence, the voltage image $\Delta V_{m;n}(x, y)$ gives information about the spatial distribution of the ac-magnetic field of the cavity mode. Here, (x,y) denotes the coordinates of the electron beam focus.

3.2 rf-Induced States (Photon-Assisted Tunneling)

Each superconducting tunnel junction represents an open ended superconducting cavity. Injection of microwaves with a frequency ν_e larger than the plasma frequency of the junction ν_p results in the propagation of the rf inside the junction. Due to the large impedance mismatch between the junction and the surrounding structures, a standing wave develops. As already pointed out by Hamilton and Shapiro [20], for junctions not small compared to the wavelength $\bar{\lambda}$ of the rf inside the junction, the photon-assisted tunneling current density [21, 22] will be spatially modulated in accordance with the standing wave.

LTSEM offers the opportunity to measure the spatial distribution of the photon-assisted tunneling current density and, hence, the distribution of the standing microwave inside the junction cavity [15, 17]. Figure 1 shows schematically the electron beam induced transitions in the irradiated part of the current biased

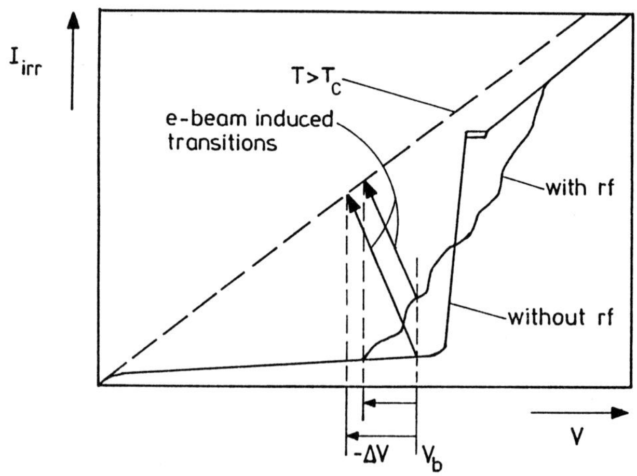

Figure 1: Schematic presentation of the electronic processes within the electron beam irradiated part of the current biased tunnel junction. The IVC is shown with and without rf-input.

junction. Due to the local irradiation this part can be heated above the superconducting transition temperature and the quasi-particle tunneling current will be increased by $\Delta I(x,y)$. Since the total junction is current biased, the nonirradiated part acts as a load resistor and the differential resistance at the bias point $\partial V/\partial I(V_b)$ determines the slope of the transition in Fig. 1 and, hence, the amplitude of the electron beam induced voltage signal $\Delta V(x,y)$. If the irradiated area of the junction is small compared to the nonirradiated area and if the resulting voltage signal is small, we have

$$\Delta V(x,y) = -\frac{\partial V}{\partial I}\left(V_b\right)\cdot\Delta I(x,y) \ . \qquad (4)$$

Here, we use the imaging technique already applied to measure the spatial distribution of the quasi-particle tunneling current density and more details can be found in [11].

From Fig. 1 we expect larger voltage signals $-\Delta V(x,y)$ from areas with a smaller amount of the rf-voltage $V_e(x,y)$ inside the junction and vice versa, if $V_b < 2\Delta/e$ [20, 22]. 2Δ denotes the sum of the energy gaps of the two junction electrodes. On the other hand, if $V_b > 2\Delta/e$ we expect the opposite.

Note, for imaging the spatial distribution of the standing microwave inside a superconducting tunnel junction, this junction serves not only as the resonator but also as the photon detector. Hence, the coupling of the rf-detector to the rf-resonator is perfect.

4. Experimental Results and Discussion

For investigating the self-resonant states we used a $50\,\mu m \times 50\,\mu m$ $Nb/Al_2O_3/Nb$-tunnel junction. The relative length and width of the junction was $L/\lambda_J = W/\lambda_J = 2.0$. A sufficient homogeneous magnetic field parallel to the junction barrier was generated by two crossed pairs of coils. For all values of the magnetic field used for our investigations (up to 10 mT perpendicular to the electron beam) we observed only a constant shift of the SEM pictures but no overall reduction of the LTSEM resolution.

Figure 2 shows the voltage images $-\Delta V_{0;n}(x,y)$ of the first four one-dimensional Fiske modes. The external dc-magnetic field is oriented perpendicular to the scan direction of the electron beam and parallel to one junction edge. The comparison between our model based on the excitation of cavity modes (sec. 3.1) and the experimental results shows good agreement if we take into account the nonvanishing diameter of the area perturbed by the electron beam [16].

S. Meepagala et al. reported results with a laser scanning technique investigating zero-field current steps (ZFS) [23]. With a junction of intermediate length their experimental studies also show an oscillatory spatial distribution of the signal corresponding to a standing-wave excitation.

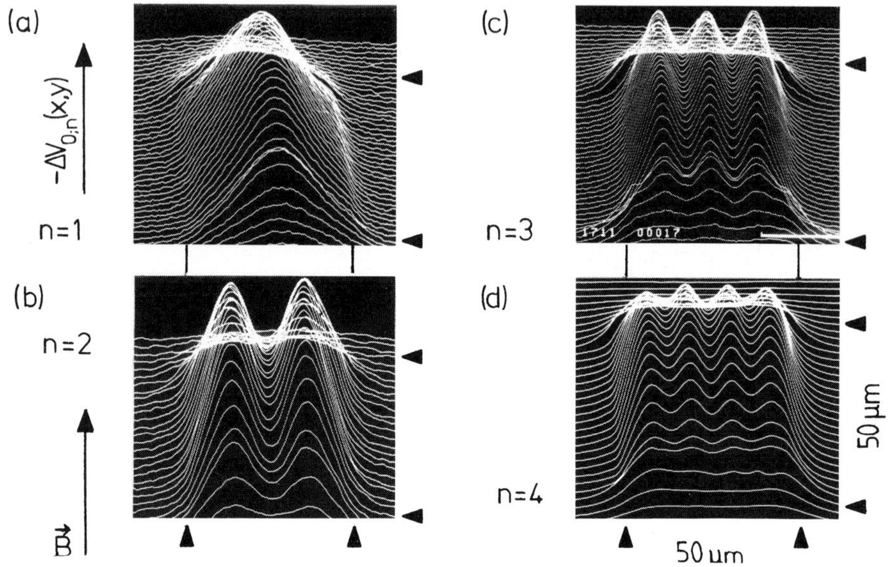

Figure 2: Voltage images $-\Delta V_{0;n}(x,y)$ of the one-dimensional Fiske modes (0;n) with n = 1 - 4. As indicated, the external dc-magnetic field is parallel to one edge of the junction. (a) - (d) correspond to n = 1 - 4. The arrows mark the junction boundaries.

Two-dimensional Fiske resonances are excited for a nearly diagonal orientation of the external dc-magnetic field. Figure 3a shows the voltage image $-\Delta V_{1;1}(x,y)$ of the first two-dimensional Fiske mode. The signal calculated from our model is displayed in Fig. 3b.

For the rf-induced states we used $Nb/Al_2O_3/Nb$-tunnel junctions of different geometries concerning the dimensions of the junction and the rf-injection. The junctions are integrated into a superconducting microstripline well coupled to a receiving antenna and terminated by a lossy stripline. For details concerning the fabrication and the coupling design of the junctions to the stripline we refer to [24]. The frequency of the applied microwave was 70 GHz, whereas the junction plasma frequency was about 29 GHz. Hence, a deep penetration of the rf into the junction is expected.

Figure 4 shows the voltage images of the spatial rf-distribution in three different quasi-onedimensional superconducting tunnel junctions. All of the junctions have a width of 20 μm measured perpendicular to the rf-flow inside the junctions. The lengths are chosen to be about $\bar{\lambda}/2 \approx 62\,\mu m$ (Fig. 4a), $\bar{\lambda} \approx 124\,\mu m$ (Fig. 4b), and $2\bar{\lambda} \approx 250\,\mu m$ (Fig. 4c). Note, in Fig. 4a and b the rf is injected from the left-hand side, whereas in Fig. 4c the rf enters the junction from the right.

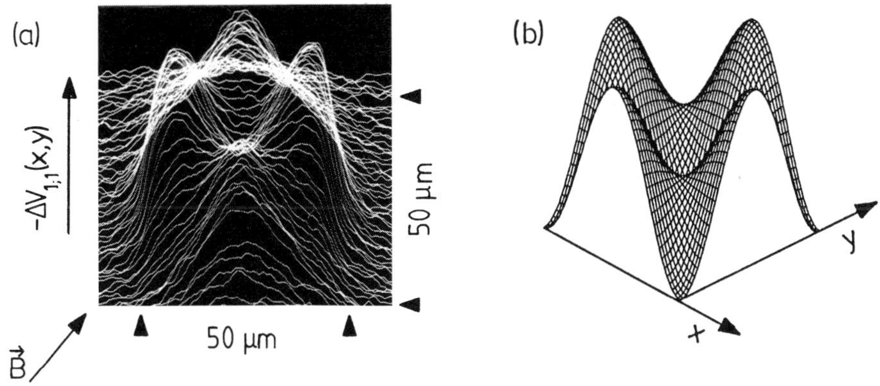

Figure 3: (a) Voltage image $-\Delta V_{1;1}(x, y)$ of the first two-dimensional Fiske mode (1;1). The external dc-magnetic field is oriented nearly diagonally from the lower left to the upper right corner of the junction. (b) Calculated signal for the (1;1) mode. For better visibility, the junction is rotated about 45° relative to (a) and only the spatially modulated part of the signal is shown.

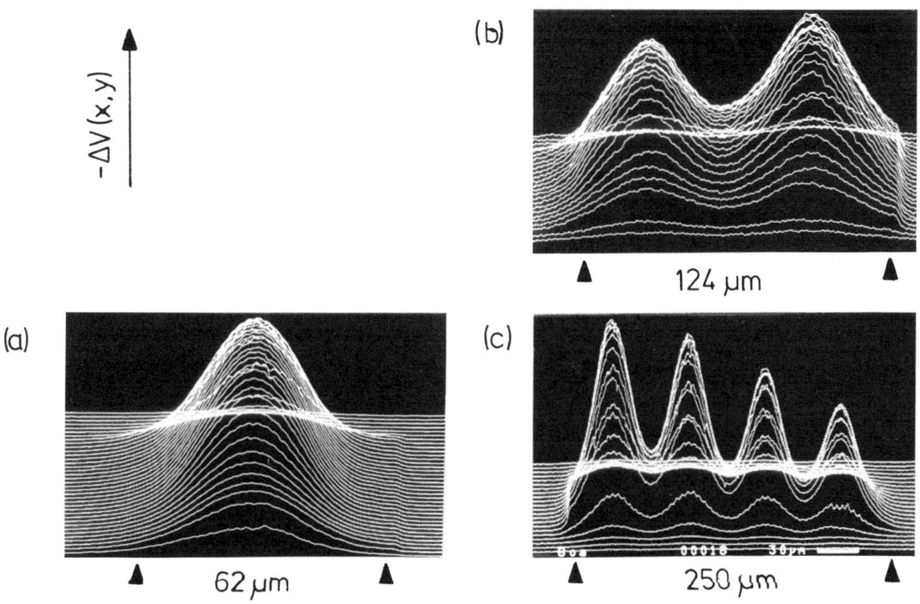

Figure 4: Voltage image $-\Delta V(x, y)$ of the spatial rf-distribution inside three 20 µm wide superconducting tunnel junctions of different length L: (a) L = 62 µm ≈ $\overline{\lambda}/2$, (b) L = 124 µm ≈ $\overline{\lambda}$, (c) L = 250 µm ≈ $2\overline{\lambda}$. In Fig. (a) and (b) the rf (70 GHz) enters from the left-hand side, whereas in Fig. (c) the rf is injected from the right.

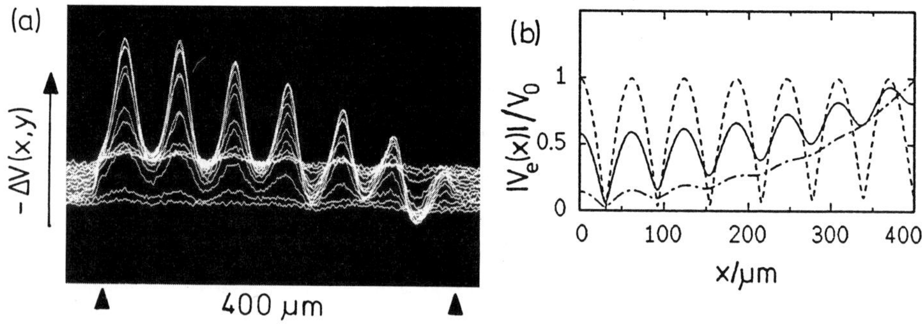

Figure 5: (a) Voltage image $-\Delta V(x,y)$ of the spatial microwave distribution inside a $20\,\mu m$ wide and $400\,\mu m$ long superconducting tunnel junction. The rf (70 GHz) enters from the right-hand side. (b) Calculated rf-voltage distribution inside the junction for three different values of the damping constant α. V_0 denotes the amplitude of the injected rf-voltage.
—·—·—· $\alpha = 1/150\,\mu m$, ———— $\alpha = 1/350\,\mu m$, – – – – $\alpha = 1/3500\,\mu m$

To study the rf-distribution and the damping, we took the voltage image of a junction with a length of $400\,\mu m$ and a width of $20\,\mu m$ (Fig. 5a). For comparison with calculated rf-distributions (Fig. 5b) it is important to point out, that a small rf-voltage $V_e(x,y)$ results in a large LTSEM voltage signal $-\Delta V(x,y)$ and vice versa (sec. 3.2). To obtain Fig. 5b, we assumed a damped rf-propagation with a damping constant α, complete reflection of the microwave at the junction boundaries without any phase shift (open ended cavity), and rf-injection from the right. Since the voltage signal $-\Delta V(x,y)$ in Fig. 5a is not a simple function of the local rf-voltage $V_e(x,y)$, only a qualitative comparison of the experimental results with the calculated curve is possible. From the pronounced undulation and the spatial dependence of the voltage standing-wave ratio (VSWR), it appears that the value $\alpha = (350\,\mu m)^{-1}$ for the damping constant agrees with the experiment. This value for α is also theoretically obtained for this junction at 70 GHz [24, 25], if we take only the surface losses into account. The tunneling losses are expected to be about one order of magnitude less than the surface losses for this junction.

Up to now, we did not measure self-resonant and rf-induced states with the same junction. Nevertheless, we can compare the LTSEM images of the one-dimensional Fiske modes (Fig. 2) with the rf-induced cavity modes in quasi-onedimensional tunnel junctions (Fig. 4). Despite the very different ways of signal generation in both cases (sec. 3) we obtain similar results for the Fiske resonances $n = 1$, 2, and 4 (Fig. 2a, b, and d) and for the rf-induced states with junctions of length $\bar{\lambda}/2$, $\bar{\lambda}$, and $2\bar{\lambda}$ (Fig. 4a, b, and c). Note, in Fig. 2 large signals correspond to a large ac-magnetic field $H_{\parallel}(x,y)$, whereas in Fig. 4 large signals correspond to a low ac-voltage $V_e(x,y)$. However, for a standing wave in a cavity, we expect a large $H_{\parallel}(x,y)$ where $V_e(x,y)$ is low and vice versa.

Obviously, rf-injection into a superconducting open ended cavity results in a standing electromagnetic wave. From the comparison of Fig. 2 and Fig. 4 we conclude, that the observed Fiske modes are standing-wave excitations of electromagnetic waves, too. The most obvious difference between the images of the self-resonant and the rf-induced states is the spatial dependence of the signal maxima along the resonator. Due to the well defined rf injection only at one boundary of the junction in case of the rf-induced states, we observe a damped standing-wave. On the other hand, in case of the self-resonant states we expect no damping along the junction resonator because of the self-excited standing-wave in the Josephson tunnel junction.

5. Conclusions

Low Temperature Scanning Electron Microscopy (LTSEM) offers a well suited tool for the investigation of dynamic states of Josephson tunnel junctions with a spatial resolution of 1 - 3 μm. With this technique it is possible to obtain a deeper insight into the corresponding junction behavior. We probed the rf-field distributions for self-resonant and rf-induced states. In the former case, one- and two-dimensional Fiske modes were imaged with a square tunnel junction (length and width equal to 2.0 λ_J). The Fiske modes in that junction can be attributed to standing-wave excitations of electromagnetic waves. In the latter case, we imaged the distribution of standing microwaves injected into the junction, by measuring the spatial distribution of the photon-assisted tunneling current density. From this, the Swihart wavelength and, hence, the Swihart velocity and the damping constant can be extracted.

For many rf-applications of Josephson tunnel junctions it is highly desirable to know their interaction with the external microwave system and the propagation of the microwaves inside the devices. Recent investigations of rf-driven long tunnel junctions indicate the inverse ac Josephson effect due to phase-lock of the fluxon oscillations and the externally applied rf-oscillation [26]. In most of the theoretical studies the influence of the external microwave is introduced by the boundary conditions or is assumed to be uniform across the whole junction. But in general, the distribution of the rf-field inside the junction is not known. If the frequency of the external applied rf is larger than the plasma frequency of the junction, the electromagnetic field of the microwave deeply penetrates into the junction and a standing-wave develops. For junctions not small compared with the Swihart wavelength, a spatially inhomogeneous and temporally harmonic ansatz for describing the influence of the applied rf is needed.

Acknowledgments

We would like to thank A.V. Ustinov, U. Klein, and P. Dammschneider for valuable discussions. The preparation of the square tunnel junctions by R. Gross and M. Koyanagi at ETL, Tsukuba, Japan is gratefully acknowledged. Financial support of this work was provided by the Deutsche Forschungsgemeinschaft.

References

[1] A. Barone, G. Paternò: Physics and Applications of the Josephson Effect, John Wiley & Sons, New York (1982)
[2] K.K. Likharev: Dynamics of Josephson Junctions and Circuits, Gordon and Breach Science Publishers, New York (1986)
[3] M.D. Fiske: Rev. Mod. Phys. **36**, 221 (1964)
[4] D.D. Coon, M.D. Fiske: Phys. Rev. **138**, A744 (1965)
[5] I.O. Kulik: JETP Lett. **2**, 15 (1965)
[6] R.E. Eck, D.J. Scalapino, B.N. Taylor: Phys. Rev. Lett. **13**, 15 (1964)
[7] T. Doderer, H.-G. Wener, R. Moeck, C. Becker, R.P. Huebener: Cryogenics **30**, 65 (1990)
[8] R.P. Huebener in: Advances in Electronics and Electron Physics, Vol. **70**, P.W. Hawkes, ed., Academic Press, New York, 1-75 (1988)
[9] R. Gross, T. Doderer, R.P. Huebener, F. Kober, D. Koelle, C. Kruelle, J. Mannhart, B. Mayer, D. Quenter, A. Ustinov: Physica B **169**, 415 (1991)
[10] J.R. Clem, R.P. Huebener: J. Appl. Phys. **51**, 2764 (1980)
[11] H. Seifert, R.P. Huebener, P.W. Epperlein: Phys. Lett. **95A**, 326 (1983)
[12] J. Bosch, R. Gross, M. Koyanagi, R.P. Huebener: J. Low Temp. Phys. **68**, 245 (1987)
[13] J. Mannhart, J. Bosch, R. Gross, R.P. Huebener: J. Low Temp. Phys. **70**, 459 (1988)
[14] R. Gross, D.B. Schmid, R.P. Huebener: J. Low Temp. Phys. **62**, 245 (1986)
[15] T. Doderer, D. Quenter, B. Mayer, C.A. Krulle, A.V. Ustinov, R.P. Huebener, J. Niemeyer, R. Fromknecht, R. Poepel, U. Klein, P. Dammschneider, J.H. Hinken in: Nonlinear Superconductive Electronics and Josephson Devices, N.F. Pedersen, M. Russo, A. Davidson, G. Costabile, S. Pagano, eds., Plenum Publishing Corp., New York, to be published
[16] B. Mayer, T. Doderer, R.P. Huebener, A.V. Ustinov, to be published
[17] C.A. Krulle, T. Doderer, D. Quenter, R.P. Huebener, R. Pöpel, J. Niemeyer, to be published
[18] J.D. Jackson: Classical Electrodynamics, ch. 8, John Wiley & Sons, New York (1975)
[19] M.A.H. Nerenberg, P.A. Forsyth, Jr., J.A. Blackburn: J. Appl. Phys. **47**, 4148 (1976)
[20] C.A. Hamilton, S. Shapiro: Phys. Rev. B **2**, 4494 (1970)
[21] A.H. Dayem, R.J. Martin: Phys. Rev. Lett. **8**, 246 (1962)
[22] P.K. Tien, J.P. Gordon: Phys. Rev. **129**, 647 (1963)
[23] S. Meepagala, J.T. Chen, J.-J. Chang: Phys. Rev. B **36**, 809 (1987)

[24] R. Pöpel, J. Niemeyer, R. Fromknecht, W. Meier, L. Grimm: J. Appl. Phys. **68**, 4294 (1990)
[25] R. L. Kautz: J. Appl. Phys. **49**, 308 (1978)
[26] N. F. Pedersen, M. Russo, A. Davidson, G. Costabile, S. Pagano, eds.: Nonlinear Superconductive Electronics and Josephson Devices, Plenum Publishing Corp., New York, to be published

1 V and 10 V Josephson Voltage Standards

J. Niemeyer

Physikalisch-Technische Bundesanstalt, Bundesallee 100,
W-3300 Braunschweig, Fed. Rep. of Germany

Abstract. The operation principle and the design of Josephson series arrays for precise dc voltage measurements up to more than 10 V are described. Properly designed circuits have been successfully operated by several national standards laboratories at rf frequencies between 20 GHz and 100 GHz. Aspects of various circuit fabrication technologies are compared with results obtained in $Nb/Al_2O_3/Nb$ series arrays, which is the technology most commonly used for manufacturing digital Josephson circuits and SQUIDs.

Precision measurements show that the accuracy of a complete, real instrument consisting of electronic components presently available for rf and dc biasing the Josephson circuit is restricted to a few parts in 10^{10}.

1. Introduction

Since 1984 series arrays of Josephson tunnel junctions have been successfully used as precise dc voltage standards [1]. The dc voltages are generated by a microwave phase-locked to each Josephson oscillator of the array. If the external oscillator operates at the frequency f_e, the reference voltages are generated at $V_n=n(h/2e)f_e$ with n=0,1,2,3... [2]. The voltage steps used widely overlap each other (zero current steps) [3]. This allows the series array to be operated at even higher voltages step by step with the fundamental distance V_1 between adjacent steps. This type of constant voltage step reaches up to about 1mV per single junction. This relatively low dc voltage requires the use of series arrays to reach higher output voltages. All arrays developed up to now are based on the idea of arranging the chain of junctions as the upper electrode of a microstripline [1,4]. The complete series array is divided into a number of smaller arrays connected via quarter wavelength transformers in parallel to a microwave antenna which transforms the impedance of a waveguide to that of the circuit stripline. Each parallel microwave path is terminated by a lossy stripline. With respect to the dc connections the parallel paths are series connected. The galvanic separation from the microwave distribution network is also accomplished by the quarter wavelength transformers. There have been recent developments in the fabrication of very large arrays for the generation of reference voltages up to more than 10V in various technologies [5,6] and the use of relatively low frequencies for the external microwave drive [7,8].

The precision and reproducibility of the Josephson series array voltage standard has led the Comité International des Poids et Mesures to define the Josephson constant $K_j=2e/h$ as the value of K_{j-90}=483597.9GHz/V. This makes it possible to transfer the voltage with the accuracy that can be provided by the present Josephson voltage standards and it also ensures that the volt is the same internationally in all calibration services within these limitations.

Springer Proceedings in Physics, Vol. 64 **Superconducting Devices and Their Applications**
Editors: H. Koch, H. Lübbig

2. Single Junction Parameters and Materials

The principle most important in designing the correct junction size is to ensure a stable phase-lock between the external microwave oscillator and the Josephson junction oscillator provided that the constant voltage steps have a maximum current width. The correct design of a single junction is vital for the stable operation of the series array standard containing many junctions. Assuming that a single junction spontaneously and independently switches to another voltage step only once every few hours, one would never reach a stable situation with a large array consisting of more than 20 000 junctions. The requirement of a strictly non-chaotic behaviour leads to restrictions on the junction's critical current density if the external microwave frequency is fixed and also on the geometry parameters of the junction. All these quantities are material-dependent, and the technology used to fabricate the junctions, therefore, influences to a certain degree the optimum junction design. A detailed description and calculation has recently been given by Kautz [9]. To avoid a chaotic reaction of the junction, the external microwave f_e must be smaller than the junction plasma frequency:

$$f_p=(ej_c/\pi h C_s)^{1/2}.$$

According to [9], f_e should not be smaller than $1.8f_p$. If, due to imperfect circuit fabrication, the parameter spread of the array junctions is not small, it is better to choose an optimum value of $f_e=3f_p$ to ensure that all junctions operate on stable voltage steps in the presence of a certain range of microwave powers due to the attenuation along the line. The restriction of the external microwave frequency to a minimum value of $3f_p$ limits the critical current density j_c of the junction to

$$j_{cmax}=f_e^2\pi h C_s/9e.$$

In the case of a single tunnel junction constructed from $Nb/Al_2O_3/Nb$, in experiments we found a stable phase-lock for a sufficiently wide range of microwave powers at critical current densities of less than $j_{cmax}=42A/cm^2$ (Fig.1). To reach the stable phase-lock region the current density of this junction was decreased stepwise by post-fabrication tempering at temperatures between 200 and 250°C. With the specific junction capacitance of $C_s=6\mu F$, $f_e=70$ GHz, and the above

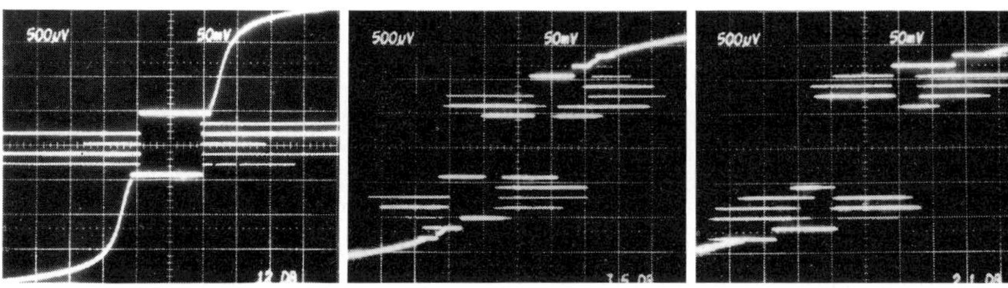

Fig.1 Dc characteristics of a single Josephson tunnel junction ($Nb/Al_2O_3/Nb$; junction area: 20µmx40µm, critical current: 550 µA) under microwave radiation of $f_e=70$ GHz. Constant voltage steps are obtained at all rf power levels. The rf power is increased from the left in three steps: 12 dB; 3.5dB; 2.1dB (arbitrary units) Vertical scale: 500µV/div. Horizontal scale: 50µA/div.

value of j_{cmax} one obtains $f_e=2.3f_p$. This limit was found to be rather sharp. It is also in good agreement with the estimate in [9].

In order to obtain a stable phase-lock between the external and the junction oscillator, the phase difference of the superconducting order parameter across the junction barrier should also be spatially uniform. rf and dc currents must, therefore, be uniformly distributed over the junction area in the presence of large magnetic fields of rf currents [10]. This requirement restricts the junction length (parallel to the current flow) to

$$l_{max}=[3h/\{4\pi^2 f_e C_s V_n e\mu_o(\lambda_{L1}+\lambda_{L2})\}]^{1/2}.$$

$\lambda_{L1,2}$ are the London penetration depths of the superconducting junction electrodes. The thickness of the insulating barrier is small compared with the penetration depths and is therefore negligible.

Standing waves of the microwave inside the junction must also be avoided. This restricts the junction width (perpendicular to the current flow) to

$$w_{max}=(1/2f_e)(2\mu_o C_s[\lambda_{L1}+\lambda_{L2}])^{-1/2}.$$

These effects are also confirmed by direct measurements of the rf current distribution across the area of a $Nb/Al_2O_3/Nb$ junction by means of a low-temperature scanning electron microscope [11].

The maximum current width of the nth constant voltage step can be calculated from

$$\Delta I_{nmax}=2I_{cmax}\,|\,J_n(2eV_{rf}/hf_e)\,|_{max}$$

with $I_{cmax}=j_{cmax}l_{max}w_{max}$. J_n is the Bessel function and V_{rf} the rf voltage amplitude across the junction. In order to obtain j_{cmax}, the maximum of J_n is chosen, corresponding to the experimental situation in which the step size of the desired reference voltage is maximized by adjusting the incident microwave power to the optimum value (Fig.1).

The junction parameters described are dependent on the superconducting and insulating materials used for the junction fabrication. In Tab.1 the optimum junction parameters are listed as calculated for various junction types and an external drive frequency of 70GHz. The material parameters were found in [1,4,9,11].

If the external microwave frequency is fixed, the only parameters that determine the differences in the optimum voltage step size for the various junction types are the penetration depths of the junction electrodes. The smaller the penetration depths, the larger the junction can be made at a given frequency or synonymously, at a fixed value of the current density. Junctions with pure NbN electrodes with very large penetration depths are therefore not optimum for generating broad constant voltage steps, whereas $Nb/Al_2O_3/Nb$ junctions produce the widest step amplitudes. If NbN junction electrodes are backed with Nb, the same step size as for pure Nb electrodes may be reached. The advantages of NbN circuits are the relatively large gap voltage of up to 5 mV and their excellent durability . $Nb/Nb_2O_5/PbBi$ junctions reach a step size two thirds of that at $Nb/Al_2O_3/Nb$ but with a smaller junction size, which may be of advantage in the case of circuits with a very large number of junctions and if a smaller external microwave frequency (see below) is to be used. Lead alloy

Tab.1 Optimum junction parameters for voltage standard design. The various junction types are denoted by their layer sequence. The external microwave frequency is chosen to be 70GHz, V_n=1mV, and $f_p=f_e/3$.

Junction-type / Junction parameter	Pb In Au/ Pb In oxide/ Pb Au	Nb/ Nb_2O_5/ Nb	Nb/ Al_2O_3/ Nb	Nb N/ Mg O/ Nb N	Nb N Nb/ Mg O/ Nb N Nb
C_s ($\mu F/cm^2$)	3	12	6	8	8
j_{cmax} (A/cm^2)	21	84	42	56	56
$\lambda_{L1}/\lambda_{L2}$ (nm/nm)	150/ 150	80/ 140	80/ 80	>200/ >200	>200 to 80/ >200 to 80
l_{max} (μm)	20	12	19	< 11	<11 to 17
w_{max} (μm)	67	40	65	<35	< 35 to 56
j_{cmax} (μA)	280	400	520	<216	< 216 to 533
ΔI_{7max} (μA) (n=7)	190	270	350	<150	< 150 to 362

junctions provide sufficiently large steps for voltage standard purposes [1] but they are not durable. The all-niobium circuits need an additional wiring connection because the junction is left without connections after the patterning process with reactive ion etching. In the presence of very large rf drive currents this connection should retain its superconducting properties up to dc currents of 100mA, otherwise it might be driven into the normal conducting state during the standard operation. To achieve such a good connection to a fresh Nb surface is difficult. In the case of lead alloy and Nb/Nb_2O_5/PbBi circuits this problem does not arise because the junction´s top electrode also forms the wiring layer. All types of junctions need a plasma sputtering process to form the tunneling barrier with the exception of Nb/Al_2O_3/Nb junctions for which a thermal oxidation process is used. This causes a very small junction parameter spread over the whole wafer in the latter case, which is important for manufacturing large circuits with many junctions.

The junction parameters are also dependent on the external microwave frequency. Fig.2 shows this dependence for the most commonly used junction types, Nb/Al_2O_3/Nb and Nb/Nb_2O_5/PbBi. It is assumed that V_n=1mV for all frequencies. If this value is exceeded the constant voltage steps no longer widely overlap because photon-assisted excess currents shift the constant voltage steps away from the voltage axis. It is also assumed that $f_p=f_e/3$, a condition that safely avoids chaotic behaviour even if many junctions with a certain parameter spread are used in a circuit. With decreasing frequency the size of the junction increases rapidly, due in particular to the large increase of the optimum junction width. The maximum step width also decreases due to the fact that V_n is kept at a constant value of 1 mV, an assumption which allows the number of junctions in a standard array with a fixed maximum output voltage to be kept constant,

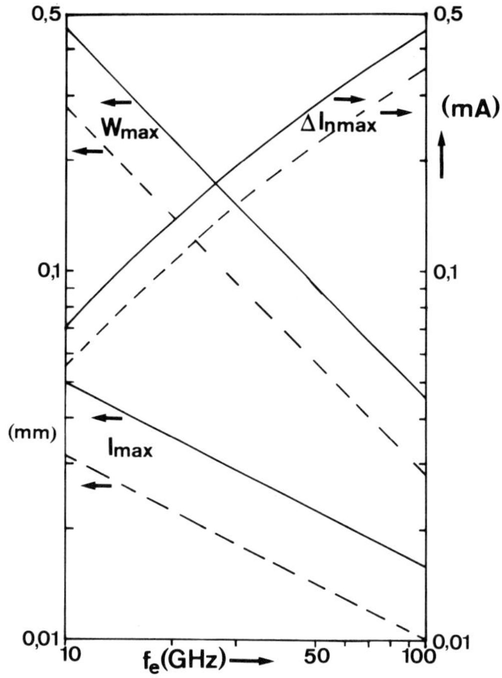

Fig.2 Dependence of the optimum junction parameters on the frequency of the external microwave. It is assumed that $f_p=f_e/3$ and $V_n=1mV$. The latter assumption increases the order of the constant voltage step used for the calculation of the curves with decreasing frequency. The continuous lines denote the values for $Nb/Al_2O_3/Nb$ circuits, the broken lines are for $Nb/Nb_2O_5/PbBi$ circuits.

independent of the external microwave frequency used. From the curves in Fig.2 it can be deduced that for $Nb/Nb_2O_5/PbBi$ junctions down to a drive frequency of 25GHz, the junction size is acceptable even for the fabrication of large circuits. For $Nb/Al_2O_3/Nb$ this is the case only for frequencies down to 45GHz. If even smaller frequencies should be used and the junction size cannot be further increased, one has to accept a decrease in the maximum current width of the step. This effect can be reduced by decreasing V_n. If for instance V_n=0.5mV is used instead of 1mV, the maximum step width will be increased by about a factor of 2.3 but also the number of junctions must be doubled to reach the same output voltage of the array as before. This again would increase the circuit size. From the viewpoint of having convenient and relatively inexpensive microwave equipment, the use of f_e=10GHz would be more desirable than the range of 20 to 40GHz used in present experiments [7,8] or designs [12].

In Tab.2 the optimum junction parameters for various junction types are given for an external microwave frequency of 10 GHz. The data show that the junction width is almost half a mm in all cases, with the exception of the $Nb/Nb_2O_5/PbBi$ junctions. The easiest way to produce at least 1V circuits of a reasonable size would therefore be to use $Nb/Nb_2O_5/PbBi$ technology. With the microwave design of reduced size described in [6] it might be possible to use the other junction types but at least for 10V circuits it would be very difficult. In that case a different microwave distribution might be necessary. Also junction stacking would reduce the size of the circuits. This technique requires all niobium trilayer technology. It should also be noted that these estimates are made for the optimum size of the voltage steps. If a real, large series array is considered, the optimum current amplitude will not be reached because the actual step size is often more than three times smaller due to non-ideal microwave coupling. This

Tab.2 Optimum junction parameters for voltage standard design. The various junction types are denoted by their layer sequence. The external microwave frequency is 10GHz, V_n=1mV, and $f_p=f_e/3$.

Junction parameter \ Junction-type	Pb In Au/ Pb In oxide/ Pb Au	Nb/ Nb_2O_5/ Nb	Nb/ Al_2O_3/ Nb	Nb N/ Mg O/ Nb N	Nb N Nb/ Mg O/ Nb N Nb
$C_s(\mu F/cm^2)$	3	12	6	8	8
$j_{c\,max}$ (A/cm^2)	0.43	1.7	0.86	1.15	1.15
$\lambda_{L1}/\lambda_{L2}$ (nm/nm)	150/ 150	80/ 140	80/ 80	>200/ >200	>200 to 80/ >200 to 80
l_{max} (µm)	53	32	50	<29	<29 to 45
w_{max} (µm)	469	280	455	<245	<245 to 392
$I_{c\,max}$ (µA)	107	152	196	<82	<82 to 203
$\Delta I_{n\,max}$ (µA) (n = 49)	39	55	71	<30	<30 to 73

even increases the difficulties of using very low frequencies without seriously affecting the stability of the reference voltage.

Another interesting case is the use of very high external microwave frequencies in the range of 1000GHz. This would locate the first order step close to the gap voltage. The step size might be even larger than that calculated from the simple Bessel function dependence (ΔI_{nmax}=1.6mA; l_{max}=3.5µm; w_{max}=4.6µm; V_n=2mV; f_e=70GHz; Nb/Al_2O_3/Nb) because of effects due to the Riedel peak. The junction size would also be small but still convenient to fabricate and the number of junctions may be decreased.

3. Series Array Design

1V arrays containing 1500-3000 junctions have been fabricated in lead alloy, all-niobium junction and Nb/Nb_2O_5/PbBi-technology.The external microwave frequencies used range from 70GHz to 100GHz [1,11,13]. In the case of Nb/Nb_2O_5/PbBi circuits 35 GHz [7] and 24GHz [8] designs have been successfully operated. 10V standard chips with up to 20 000 junctions were first realized with Nb/Nb_2O_5/PbBi junctions [5] followed by the development of Nb/Al_2O_3/Nb circuits [6]. The 10V circuits are operated at 70GHz only. All types of circuits are basically designed in the same way and were successfully used to generate the desired reference voltages.

Fig.3 shows a typical design of an integrated microwave circuit for 70GHz with a series array containing 2000 Nb/Al_2O_3/Nb junctions for a 1V standard with the series array distributed over 8 parallel lines and a 10V circuit with a less space-

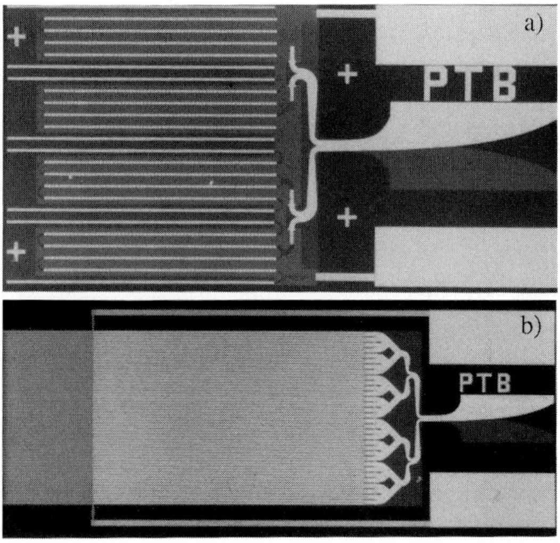

Fig.3 a) 1V standard chip split into 8 microwave paths; chip size: 8mmx16mm.
b) 10V standard chip split into 64 microwave paths; chip size 10mmx27mm.

consuming microwave distribution. This allows the 64 parallel paths with 20 000 junctions to be guided very close together. The reason for distributing the junctions over several parallel microwave paths is to ensure that the coupling of the microwave power to all junctions is sufficiently uniform in the presence of a finite attenuation of the stripline. The number of parallel paths should be kept as small as possible to avoid dividing the microwave power into fractions too small to generate stable constant voltage steps across the single lines. The longer a line can be made without excessive attenuation the smaller the number of parallel paths may be kept. The maximum length of a line is determined by its attenuation per unit length, the junction parameter spread, and the difference between the single junction plasma frequency and the external microwave frequency. The larger the drive frequency compared with the plasma frequency, the more stable especially low-order constant voltage steps become, preferentially generated by low rf power at the end of longer lines. The experimental verification of this effect has been reported in [14].

In order to determine the attenuation of the series array stripline path, the rf properties of this circuit, a section of which together with its lumped microwave circuit model is shown in Fig.4, have been analyzed in detail in [6] for $Nb/Al_2O_3/Nb$ circuits over a Nb groundplane both separated by a SiO dielectric layer. The main result is described in Fig.5. h_1 is the thickness of the main dielectric layer. l_4 denotes the width by which the junction area is larger than the window in the second SiO layer (thickness h_2). Through the window the junction's top electrode is coupled to the connecting wiring layer. An additional wiring layer is needed in the case of all Nb junction circuits because the junction sandwiches are patterned after their complete wafer deposition by a reactive ion etching process that leaves the junctions unconnected. The attenuation of the stripline can be exceedingly large if l_4 is larger than 2μm (Fig.5). This is due to a resonance caused by the shortened parasitic stripline of length l_4. With l_4=2μm,h_1=1μm, and h_2=350nm one obtains α_k=6.23 1/m (54.1dB/m) for

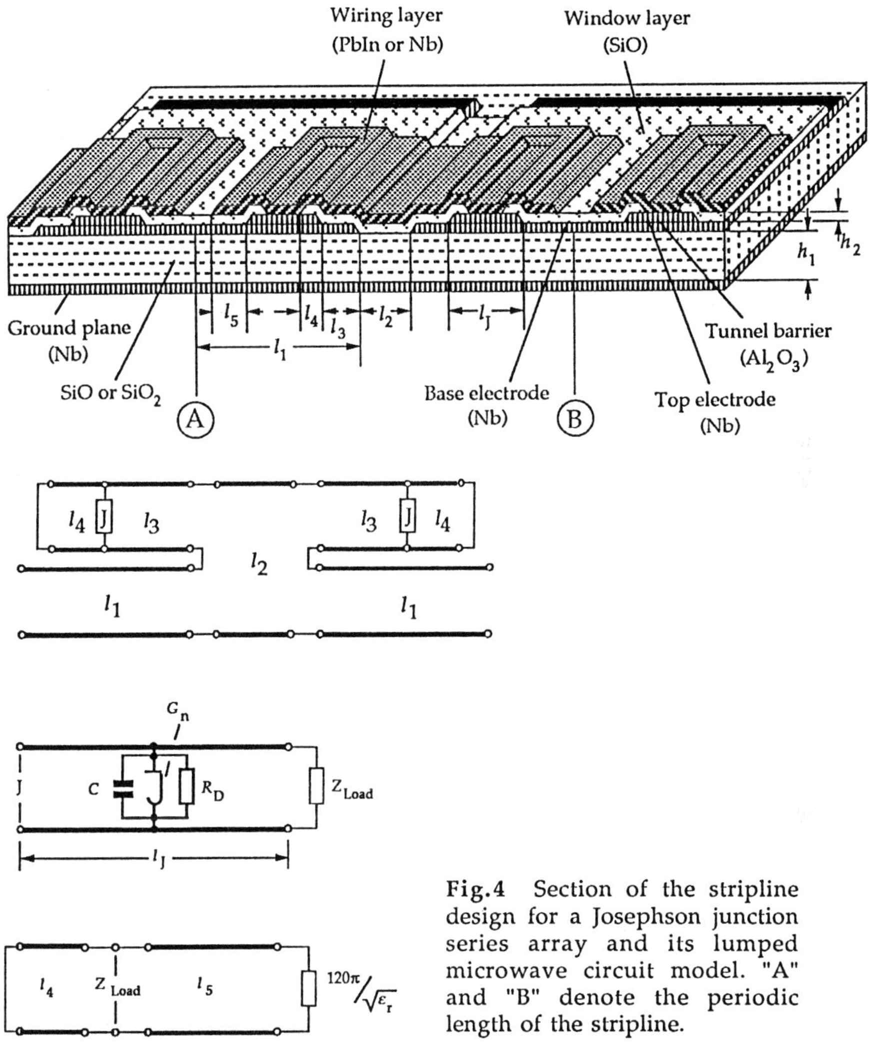

Fig.4 Section of the stripline design for a Josephson junction series array and its lumped microwave circuit model. "A" and "B" denote the periodic length of the stripline.

f_e=70GHz. This value equals an attenuation of 2.7x10^{-3} dB/junction. This value decreases to 0.93x10^{-3} dB/junction if h_1 is increased to 3µm while keeping the other geometry parameters constant. In the case of the lead alloy and the $Nb/Nb_2O_5/PbBi$ circuits l_4 equals zero because the junction´s top electrode also forms the connection to the next junction. The junction area and the window opening are therefore identical in these cases. As l_4 must be rather small an alignment precision of the window mask of at least 1µm is desirable for the production of $Nb/Al_2O_3/Nb$ cicuits with low attenuation.The above calculated values of the stripline attenuation are in good agreement with those calculated in [9] for $Nb/Nb_2O_5/PbBi$ circuits driven at 96GHz. Fig. 1 shows that the microwave power needed to obtain stable zero current steps close to 1 mV for a single

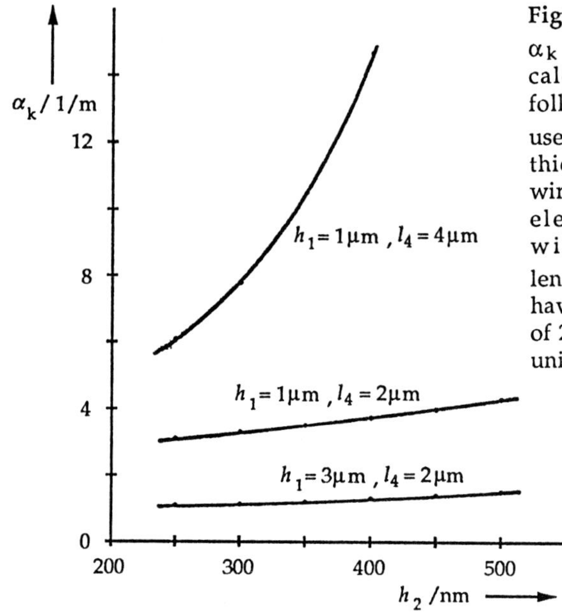

Fig.5 Total attenuation constant α_k of the circuit in Fig.4. To calculate these curves the following numbers have been used: $\varepsilon_{SiO}=5.7$; $\varepsilon_{Al2O3}=10$; barrier thickness=2nm; width of the wiring=62μm; width of the base electrode=70μm; junction width=46μm; junction length=26μm. The values of α_k have to be multiplied by a factor of 20loge=8.686 to obtain them in units of dB/m.

$Nb/Al_2O_3/Nb$ junction at $f_e= 2.3f_p$ ranges from about 3.5dB to 2.1dB. This attenuation should therefore be tolerable for the proper operation of a single series array microstripline without dividing it into parallel microwave paths. This length is equivalent to about 1000 junctions connected in series if h_1 is somewhat smaller than 1μm. This number is valid for $Nb/Al_2O_3/Nb$ circuits but should also be applicable in series arrays made of other materials because the microwave attenuation of all circuit types is similar.

If one accepts lower-order steps at less than 1mV it should be possible to use even much longer single path arrays. In a real circuit the junction parameter spread must also be taken into account. The larger the spread, the shorter the single paths should be. This allows a closer approach to the optimum microwave power for all junctions, even for those which are, with respect to their parameters, close to the condition of chaotic behaviour. In the case of $Nb/Al_2O_3/Nb$ arrays, the parameter spread is the smallest of all the fabrication technologies. It should therefore be possible to divide a 20000 junction array into only 8 or 16 parallel paths; this has been experimentally proved for $Nb/Nb_2O_5/PbBi$ circuits [5]. Beside this a 30 000 junction array should be used instead of one that contains only 20 000 junctions. It allows the power consumption for the generation of a 10V dc reference voltage to be decreased. This is because the generation of low-order steps needs much less power than high-order steps require (cf. Fig.1)

There is another parasitic attenuation mechanism which can occur in all niobium junction circuits. If the wiring layer is not properly connected to the top electrodes of the junction, it will form a Josephson junction that also absorbs microwave power. In the worst case this connection is driven to its normal conducting state and thus prevents the generation of zero resistance voltage steps. To avoid these effects we found by experiment that the critical current of the wiring connections should reach values of 100mA at least. To obtain such high connection currents, the Nb top electrode surface should be passivated by reactive ion nitridation directly after the sandwich is produced, and carefully cleaned before the wiring layer is deposited. The wiring problem cannot occur in $Nb/Nb_2O_5/PbBi$ and lead alloy circuits because the junction top electrodes also form the connection to the adjacent junctions.

4. Precision

The accuracy of a voltage calibration with a Josephson series array standard is generally not determined by the standard chip or its fabrication technology but by the object to be calibrated and the electronic equipment necessary to compare the test object at room temperature with the chip's reference voltage at 4.2K.

Fig. 6 shows a long-term recording of the 10V output of a Zener reference standard measured by means of a Josephson 10V series array standard. The noise signal of 200nV peak to peak at a bandwidth of 0.1Hz is completely determined by the noise of the Zener standard. It limits the accuracy of the 10V calibration to 40nV standard deviation at 0.1Hz bandwidth. To determine the limitation of the accuracy of a complete Josephson array standard (array plus electronic equipment) a direct comparison between a 1V array standard driven by a 70GHz klystron, the frequency of which is stabilized to 1 part in 10^{10} by a traditional phase-lock loop used as reference system, and two other different 1V Josephson array standards was performed. Both systems under test were frequency monitored and controlled by an EIP counter now commonly used in several standards laboratories. The microwave sources consist of a klystron in the first case and a Gunn diode in the second.

The results of this comparison are shown in Fig.7. The comparison of "klystron versus klystron" results in a peak-to-peak noise of about 2 nV at 0.1Hz bandwidth and a common drift of about the same value per hour. This is equivalent to a standard deviation of the complete instrument of 4×10^{-10} at 0.1Hz bandwidth. The noise figure is only slightly larger than the noise of the N1a nanovoltmeter used as zero instrument (see insert of Fig. 7a). If the klystron of the instrument to

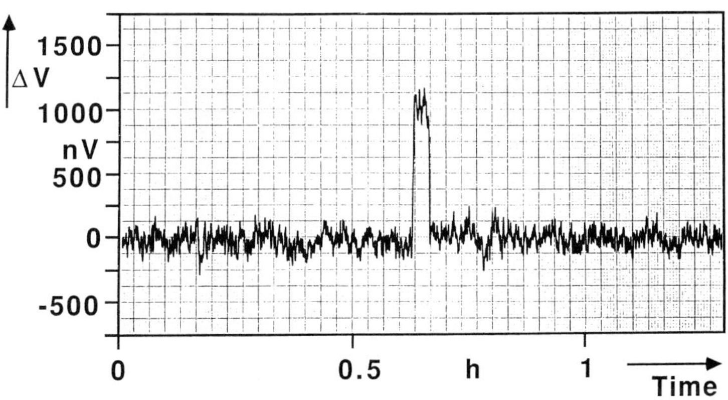

Fig.6 Calibration curve of a Zener reference standard with a 10V Josephson standard. The voltage level ($\Delta V=0$) is 9.999 968 716V. Calibration peak: 1μV

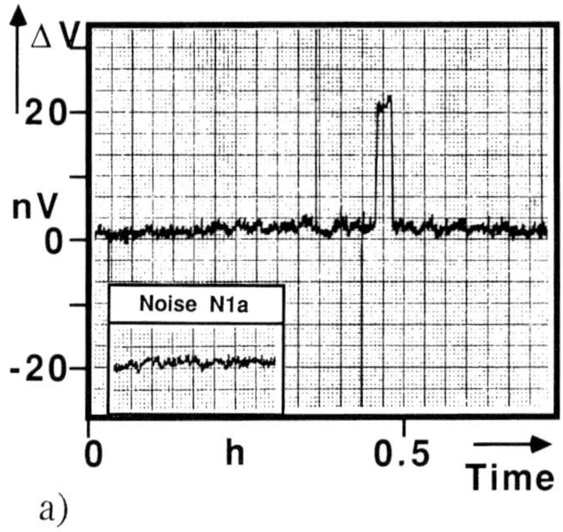

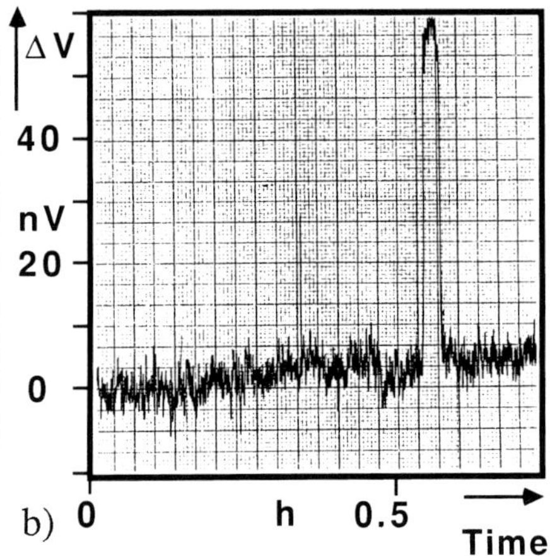

Fig.7 a) Comparison of the output voltages of two Josephson series array standards (klystron vs. klystron. The voltage level is 1V, the bandwidth 0.1Hz. The calibration peak is 20 nV. The insert shows the noise of the nanovoltmeter used as zero instrument.
b) Comparison of the output voltages of two Josephson series array standards (klystron vs. Gunn oscillator). The voltage level is 1V, the bandwidth 0.1Hz.

be compared with the reference system is replaced by a Gunn oscillator, the output noise of the standard increases significantly to 10nV (peak to peak) (Fig.7). The frequency stability of the Gunn oscillator can be improved to 3×10^{-11} by redesigning the electronic circuit that drives the oscillator [15], which is a great improvement.

Acknowledgment. This work was supported by the Commission of the European Communities, Community Bureau of Reference (BCR), Contract No.3108.

References

[1] J. Niemeyer, J.H. Hinken, R.L. Kautz, Appl. Phys. Lett. **45**, 478 (1984);
[2] S. Shapiro, A.R. Janus, S. Holly, Rev. Mod. Phys. **36**, 223 (1964)
[3] M.T. Levinsen, R.Y. Chiao, M.J. Feldman, B.A. Tucker, Appl. Phys. Lett. **32**, 776 (1977)
[4] J. Niemeyer, L. Grimm, W. Meier, J.H. Hinken, E.Vollmer, Appl. Phys. Lett. **47**, 1222 (1985)
[5] C.A. Hamilton, F.L. Lloyd, C. Kao, W.C. Goeke, IEEE Trans. Instrum. Meas. **38**, 314 (1989)
[6] R. Pöpel, J. Niemeyer, R. Fromknecht, W. Meier, L. Grimm, J.Appl. Phys. **68**, 4294 (1990)
[7] F. Müller, H.-J. Köhler, P. Weber, K. Blüthner, H.-G. Meyer, Phys. Stat. Sol. (a) **121**, K73 (1990), J. Appl. Phys. **68** , 4700, (1990)
[8] C.A. Hamilton, R.L. Kautz, M. Stieg, K. Chieh, M.B. Simmonds, W.F. Avrin, Proceedings of the Conference on Precision Electromagnetic Measurements, Ottawa, **40**, (1991)
[9] R.L.Kautz, Proceedings of the International School of Physics, Enrico Fermi, Course CX (1989)

[10] R.L.Kautz, G. Costabile, IEEE Trans. Magn. **Mag-17,** 780 (1981)
[11] J. Niemeyer, Y. Sakamoto, E. Vollmer, J.H. Hinken, A. Shoji, H. Nakagawa, S. Takada, S. Kosaka, Jap. J. Appl. Phys. **25,** L343 (1989)
[12] W.F. Avrin, M.B. Simmonds, C.A. Hamilton, IEEE Trans. Magn. **27,** 3359 (1991)
[13] C.A. Hamilton, R.L. Kautz, R. Steiner, F.L. Lloyd, IEEE Electr. Dev. Lett. EDL-**6,** 623 (1985)
[14] J. Niemeyer, Josephson Series Array Potentiometer, pp. 228-254 in Superconducting Quantum Electronics, V.Kose, Ed., Springer-Verlag, New York, (1989)
[15] H. Yoshida, Y. Sakamoto, T. Endo, to be published in IEEE Trans.Instr. Meas.(1991)

Ratio Standard for DC Resistance Using a Second Generation of Josephson Junction Arrays

J. Kohlmann, P. Gutmann, K. Löhr, T. Weimann, and J. Niemeyer

Physikalisch-Technische Bundesanstalt, Bundesallee 100,
W-3300 Braunschweig, Fed. Rep. of Germany

Abstract. A ratio standard for dc resistance using a three-terminal Josephson junction array is described. It covers the resistance range above 10 kΩ applying ratios from 1:1 to 1:10 with a single device. Some features of the circuit resulting in an outstanding performance of the ratio standard are discussed. Preliminary measurements show a relative reproducibility of $1 \cdot 10^{-7}$, given by the detector noise.

1. Introduction

Planar resistors with the quality of a standard are widely used in commercial products and consumer goods. The industry is interested in resistors of 10 kΩ to 10 MΩ, the resistance of which is traceable to the national standard with relative uncertainties of the order of 10^{-7}. The link between the national resistance standard, represented by the quantum Hall resistance $h/2e^2 = 12.9$ kΩ and the standards, instruments and test equipment to be checked is the ratio standard. A new ratio standard using a Josephson junction array is described, the random and systematic error of which can be made negligible. We have realized and successfully tested a prototype instrument which we do not think is available at present with such excellent data.

2. Principle

The principle of the ratio standard is shown in Fig. 1. The standard is a modified Wheatstone bridge in which two resistors in series are replaced by two voltage sources, resulting in an ideal voltage divider with zero source resistance. The voltage sources are realized by a three-terminal Josephson junction series array with 2000 tunnel junctions. The irradiation of this array with microwaves of frequency f produces constant voltage steps $U_J = j \cdot (h/2e) \cdot f$ due to phase locking of the Josephson oscillator by the external oscillator (j = 1, 2, 3, ...; h denotes the Planck´s constant and e the electron charge). At f = 70 GHz voltage steps appear at distances of $U_J = 145$ µV. A single Josephson junction generates up to seven steps (about 1 mV). To reach voltages $m \cdot U_J$ and $n \cdot U_J$ of the order of 0.5 V every part of the three-terminal array must be biased on about the 3500th step.

The ratio of two given resistors may be measured in a single measurement, "in bridge", balancing to the nearest ratio n/m with an off-balance voltage $U_{ab}/U_J = (m \cdot R_0 - n \cdot R_1)/(R_0+R_1)$, or in two measurements, "by substitution", measuring both resistors one after the other at the same place e.g. R_0, whereas the value of R_1 may be changed to balance the bridge.The integer step control of m and

Springer Proceedings in Physics, Vol. 64 Superconducting Devices and Their Applications
Editors: H. Koch, H. Lübbig

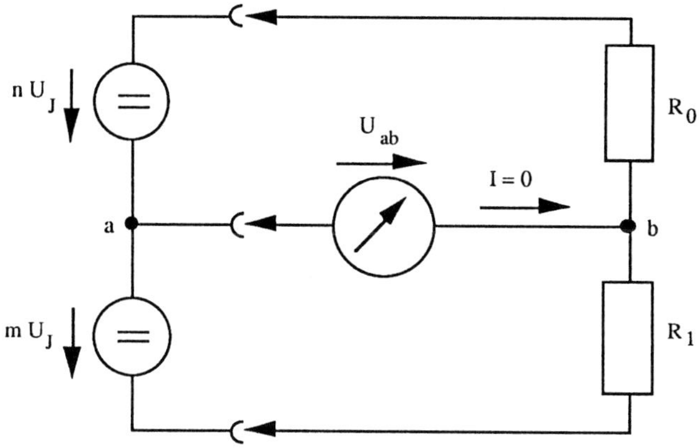

Fig. 1. Principle of the ratio standard

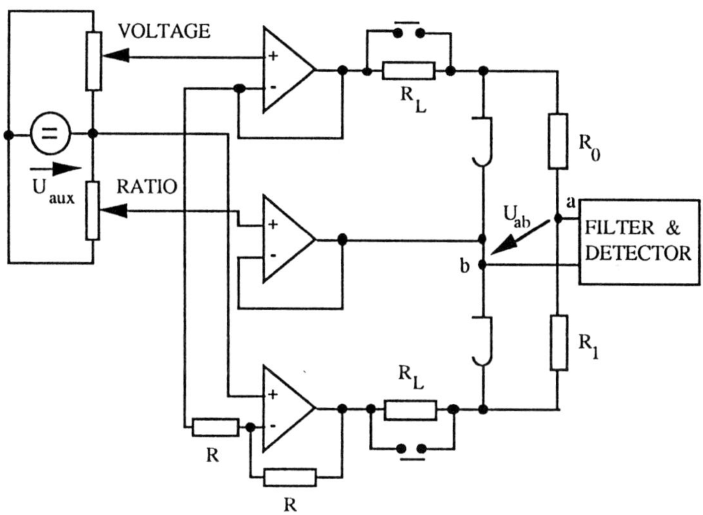

Fig. 2. Simplified circuit of the ratio standard

n reduces the requirements of absolute accuracy by a factor of (m+n) of the detector´s linearity or of the balancing resistor, depending on the measuring procedure.

In Fig. 2 a simplified electrical circuit of the ratio standard is shown. The desired voltage levels are produced by an auxilary voltage source U_{aux} (dc calibration). Suitable voltage steps are produced, applying microwave power (ac calibration), and then stabilized by introducing resistors $R_L = 10\ \Omega$.

3. Technology

The design of the Josephson potentiometer arrays is based on the 1 V Josephson voltage standard of the Physikalisch-Technische Bundesanstalt (PTB), containing

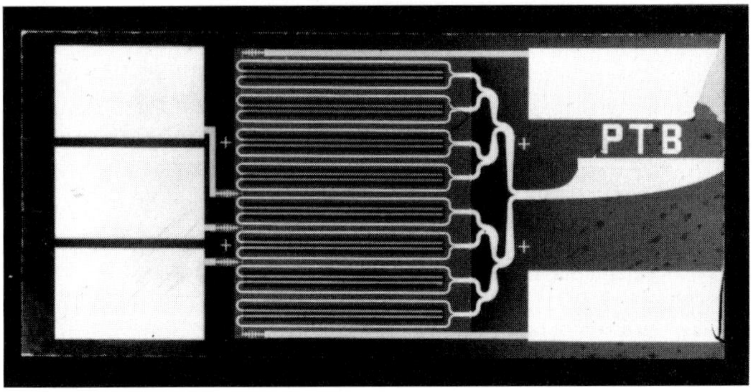

Fig. 3. Photograph of the Josephson potentiometer

2016 tunnel junctions in 16 parallel paths [1]. Additional dc pads are added at the end of some paths to divide the Josephson voltage of the whole array in the ratio of 1:1 (first design) or in the ratios of 1:1, 1:2 and 1:5 (second design; Fig. 3). The two designs allow a comparison of the various ratios with an optimum signal-to-noise ratio.

The Josephson arrays are produced in the standard process for the fabrication of large series arrays with an Nb/AlO_x /Nb tunnel sandwich [1]. The usual photolithographic processes are applied to sequentially mask and pattern the layers of the circuit. Details of the design and fabrication are described elsewhere [1, 2].

4. Features and data

We found some features of our circuit by experimental test, resulting in an outstanding performance of the ratio standard:

* The adjusted ratio m/n is preserved by reversing the auxilary voltage without any dc or ac recalibration
* A symmetric three-terminal device can be used to check ratios ranging between 1:1 and 1:10
* A free-running microwave Gunn oscillator does not impair accuracy
* Independent control of m and n is possible by appropriate dc voltage bias

The straightforward use of these features allows an extremely simple and fast measurement procedure, eliminating offsets and drifts by voltage reversal without any switching of the resistors under test or dc and ac bias adjustment, which can easily be set under computer control.

With voltages $(m+n)\cdot U_J = 1.45$ V we have checked the known ratio of two commercial 10 kΩ resistors in a two-terminal configuration. Though our experimental set-up is in a preliminary stage, the relative systematic error was smaller than $2\cdot10^{-6}$, given by the uncertainty of the temperature of the resistors at our disposal, and the relative reproducibility was $1\cdot10^{-7}$, limited by the detector noise.

5. Conclusion

We have presented a new concept of a ratio standard for dc resistance and dc voltage using a Josephson junction series array potentiometer. Its successful realisation is demonstrated by tests and preliminary results. Further improvements are planned, e.g. lowering the detector noise, reducing dc drifts of thermal EMFs, and virtually increasing leakage resistances by shielding and guarding. A powerful new concept therefore exists to build up the decadic OHM scale from 10 kΩ to several TΩ.

References

[1] R. Pöpel, J. Niemeyer, R. Fromknecht, W. Meier and L. Grimm, J. Appl. Phys. **68**, 4294 (1990)

[2] J. Kohlmann and P. Gutmann (to be published)

Superconducting Instrumentation for Precision Measurement and Control

R.V. Duncan

Sandia National Laboratories, Albuquerque, NM 87185, USA

Abstract. The Josephson series-array voltage standard (JAVS) has dramatically improved the level to which high-accuracy electrical equipment may be monitored and stabilized. For example, a small pressure dependence in the output of some solid-state voltage standards has been measured with the JAVS. The use of two such series-arrays in a superconductive loop promises to be a new method of ultra-small frequency difference measurement. The accuracy of these new measurement configurations is set by the superconductive loop inductance and the Josephson junction parameters.

1. Introduction

Series-arrays [1] of Josephson junctions, all biased by a common microwave source, are in use within approximately twenty laboratories throughout the world as an intrinsic voltage standard. At Sandia National Laboratories (SNL), and similarly at many other laboratories, this standard is routinely used to track the drift, accuracy, and linearity of digital volt meters (DVM), voltage reference standards, and voltage calibrators to an uncertainty better than 0.1 ppm. In collaboration with the National Institute of Standards and Technology (NIST, formerly NBS) we have used this technology to detect small systematic variations in the output of some solid-state voltage standards with elevation. Such refinements in measurement technology would be impossible without this wide-spread use of the JAVS. Recently a powerful new measurement technique utilizing these series-arrays of Josephson junctions has been developed [2]. Here two such arrays have been placed in series-opposition in a superconductive measurement loop which is inductively coupled to a SQUID. We [3] have analyzed this interferometer circuit within the framework of the Steward-McCumber model [4] to determine the measurement accuracy [5]. This circuit, which we refer to as a macroscopic quantum circuit (MQC) due to its superconductive coherence throughout the loop, may prove useful as an ultra-accurate interferometric readout at millimeter wavelengths for such applications as gravity wave detection and Sagnac-effect gyroscopes [6]. The full utilization of a MQC requires

Springer Proceedings in Physics, Vol. 64 Superconducting Devices and Their Applications
Editors: H. Koch, H. Lübbig

the development of a superconductive positioner to fine-tune cavity oscillators used to bias the Josephson junctions and possibly to tune the MQC frequency response directly.

The SNL Josephson voltage standard is similar to that used by NIST [7] except that it incorporates cryogenic (rather than room-temperature) passive filtering [8], and it is cooled continuously by a refrigerated Dewar [9] which maintains the standard at 4K with no helium loss or routine operator intervention. In the near future this refrigerated Dewar will be used to simultaneously support the Josephson array and to re-liquify the ^4He boil-off from a nearby quantum Hall effect resistance standard.

2. Environmental effects on solid-state voltage standards

Solid-state voltage standards (SSVSs) utilizing Zener diode references are in common use today for dc voltage metrology [10]. The JAVS provides an extremely accurate measurement system against which these SSVSs may be critically evaluated. The thermal coefficient of many SSVS have been measured against the JAVS and found to be typically 17 ppm/K. Since the set temperature of the SSVS enclosure is stable to better than 2 mK, and when intentionally over-heated it returns to its set point to within 5 mK, thermal effects create inconsequential errors in all the SSVSs observed to date. A systematic variation in some SSVS with elevation (and hence pressure) has been detected. One such SSVS, designated SSVS#1, was transported between four locations within the United States during 1989. This device was operated at elevations ranging from sea-level to 1,500 meters and no elevation or pressure dependence was observed in its output. SSVS#1 displayed excellent repeatability and stability as measured against the JAVS from location to location (once its drift rate was taken into account, SSVS#1 typically deviated by no more than 0.1 ppm during shipment). No variation of the JAVS has been observed during controlled temperature and pressure changes [11].

In August, 1990 another SSVS (called SSVS#2) was shipped round-trip between SNL in Albuquerque and NIST in Gaithersburg. At each location the device was measured against a Josephson voltage standard. The average of its four 1.018V taps was 0.98 ppm higher at Gaithersburg than at Albuquerque - a factor of ten larger in variation than the SSVS#1 result! A comparison of the value of SSVS#2 after its return from NIST to its value before it was shipped agreed to within 0.15 ppm. In order to check for a pressure dependence in the SSVS#2 output, both SSVS#1 and SSVS#2 were measured in a mobile laboratory at an elevation of 3,260 meters (on the summit of Sandia Crest, which is located 50 km from SNL) in collaboration with S. L.

Kupferman. These measurements were made against a high-accuracy 8.5-digit DVM which was calibrated against the JAVS immediately before and immediately after the mountain measurements. SSVS#2 experienced a -3.40 ppm deviation in the average of its four 1.018V taps with the 1,760 meter elevation gain, while SSVS#1 displayed an inconsequential +0.8 ppm deviation. Given the previous performance of SSVS#1, this deviation was likely due to a systematic change in the DVM gain with elevations. Only the 1.018V taps of SSVS#2 displayed this variation with elevation, indicating that a resistive voltage divider within SSVS#2 was most likely the source of this systematic error. These results serve as a warning that all transportable voltage standards used in critical metrology applications must be individually qualified through their full range of environmental conditions against a JAVS. In the future we will use our JAVS and an automated pressure/ temperature/ humidity-controlled chamber to make these critical environmental qualification measurements. Notice also that these very small systematic effects could not have been detected without the JAVS.

3. Macroscopic Quantum Circuit Measurements

Consider two rf-biased series-arrays connected in series-opposition with superconductive wire, forming the MQC with a total self-inductance L. The supercurrent i_S in the MQC which builds up over time t due to the difference in electric potential ΔV between the two rf-biased devices is given by

$$i_S(t) = \frac{1}{L}\int_0^t \Delta V \, dt'. \tag{1}$$

This supercurrent is then read out by a SQUID which is magnetically coupled to the MQC. This method of ultra-precise potentiometry was developed by Clarke [12], and more recently used by Jain et al. [13] to obtain measurements with a precision of $\Delta V/V \approx 3x10^{-19}$ using single rf-biased junctions operating at about 300 μV. Kautz and Lloyd [2] have used this technique to compare the voltage output of two series-arrays, each containing 2,076 Josephson junctions and biased to $V \approx 1$ volt, to a precision of $\Delta V/V \approx 2x10^{-17}$. In all the above measurements both Josephson devices were biased by the same rf source to check for deviations from $i_S = 0$ when both Josephson devices were biased to the same step (n). Other experiments[14] have been performed which used two different rf sources of known detuning ($\Delta\omega$) to bias the junctions and thus to generate an increasing i_S consistent with Eq. 1 and the prediction that $\Delta V = \Delta V_0 = nh\Delta\omega/4\pi e$, where h is Planck's constant, and e is the magnitude of the charge of the electron. This method of measurement may provide a powerful new method of detecting very tiny frequency differences generated in gravity wave detectors and Sagnac-effect gyroscopes [6].

Referring to Eq. 1, it is clear that an arbitrarily precise measurement of ΔV may be made by either decreasing L, increasing the length of the measurement interval, or both. The value of the self-inductance (L) of the MQC need not be limited by the SQUID's input inductance [13]. Due to the interaction of the two Josephson devices via the superconductive loop, the measured ΔV is not simply equal to $\Delta V_0 = nh\Delta\omega/4\pi e$. Rather there exists a correction which depends on the value of L, and which becomes large as L approaches the Josephson inductance $L_J = h/2eI_c$, where I_c is the junction's critical current. These corrections, which have been calculated within the framework of the Steward-McCumber model [4] in the absence of external noise, are detailed in the proceedings of ISEC '91 [5].

These JAVS may be used as an ultra-stable voltage source provided that the load is very quiet and that the compliance current never exceeds the voltage state's critical current (which is on the order of 100 μA). In such applications (which may include ultra-accurate drift tube measurements at cryogenic temperatures) the voltage resolution and stability is directly determined by the linewidth and center-frequency stability of the biasing oscillator. Conventional microwave locker/counters can servo the center frequency of Gunn diode oscillators to within a few Hertz out of 75 - 95 GHz. Superconductive cavity oscillators have been developed which have $Q > 10^6$ over their full tuning range [15]. This cavity oscillator may be tuned with a proposed ultra-stable superconductive positioner [16] in order to obtain about a factor of one thousand improvement over conventional center-frequency stability.

Acknowledgement

This work is supported by the U.S.A. Department of Energy under Contract Number DE-AC04-76DP00789.

References

1. F. L. Lloyd, C. A. Hamilton, J. A. Beall, D. Go, R. H. Ono, and R. E. Harris, IEEE Elec. Dev. Lett. **8**, 449 (1987); R. Popel, J. Niemeyer, R. Fromknecht, W. Meier, L. Grimm, and F. W. Dunschede, IEEE Trans. Instrum. Meas. **40**, 298 (1991); C. Hamilton, R. Kautz, M. Stieg, K. Chieh, W. Avrin, and M. Simmonds, IEEE Trans. Instrum. Meas. **40**, 301 (1991); Y. Sakamoto, H. Yoshida, T. Sakuraba, A. Odawara, Y. Murayama, and T. Endo, IEEE Trans. Instrum. Meas. **40**, 312 (1991).
2. R.L. Kautz and F.L. Lloyd, Appl. Phys. Lett. **51**, 2043 (1987).
3. In collaboration with D. H. Dunlap of the University of New Mexico.
4. D. E. McCumber, J. Appl. Phys. **39**, 3113 (1968); W. C. Stewart, Appl. Phys. Lett. **12**, 277 (1968).
5. D. H. Dunlap and R. V. Duncan, Procedings of ISEC '91, Glasgow, (1991).
6. L. Z. Wang and R. V. Duncan, in preparation.
7. R. L. Steiner and B. F. Field, IEEE Trans. Instrum. Meas. **38**, 296 (1989).
8. R. V. Duncan, Bull. Am. Phys. Soc. **34**, 1535 (1989).
9. R. V. Duncan, IEEE Trans. Instrum. Meas. **40**, 326 (1991).
10. P. J. Spreadbury, IEEE Trans. Instrum. Meas. **40**, 343 (1991).
11. R.V. Duncan, Physica B **165**, 101 (1990).
12. J. Clarke, Phys. Rev. Lett. **21**, 1566 (1968).
13. A.K. Jain, J.E. Lukens, and J.S. Tsai, Phys. Rev. Lett. **58**, 1165 (1987).
14. T. Dan Bracken and W. O. Hamilton, Phys. Rev. **B6**, 2603 (1972).
15. B. Hughey, T. Gentile, and D. Kleppner, Rev. Sci. Instrum. **61**, 1940 (1990).
16. R. V. Duncan, in preparation.

Use of a Resistive SQUID for Noise Thermometry

R.J. Soulen, Jr., W.E. Fogle, and J.H. Colwell*

National Institute of Standards and Technology,
Gaithersburg, MD 20899, USA
*Present address: Naval Research Laboratory, Code 6344,
Washington, DC 20375-5000, USA

Abstract. The recently-developed international temperature scale, ITS 90, does not extend below 0.65K. However, a considerable body of research, particularly that devoted to defining the physical properties of liquid ^{3}He, has been conducted below this temperature without benefit of a common temperature scale. Accordingly, we have been developing a temperature scale in this region. The primary thermometer used in this effort is a special form of noise thermometer consisting of a resistively-shunted Josephson junction, or an R-SQUID. The R-SQUID offers advantages over conventional analogue noise thermometry in that the influence of the measuring circuits on the measured noise temperature is considerably reduced. A review is given of the decade of experience gained at NIST in the operation of this thermometer and in modelling its systematic errors. The reproducibility of the temperatures measured using the R-SQUID noise thermometer will also be presented, as well as comparisons of the noise temperature scale with scales defined by other thermometers.

1. Introduction

The latest international temperature scale, ITS-90, forms the basis for all present temperature measurements and calibrations. It does not extend below 0.65K, however, where many low temperature experiments are conducted. To assist those engaged in such research, we have been developing a temperature scale below 1 K. We have used several types of thermometers in this effort, but the ultimate accuracy of this scale rests principally upon a single absolute thermometer, a special form of noise thermometer. This is an R-SQUID which is a circuit comprised of a resistor R at temperature T connected in parallel with a Josephson junction. Thermal noise generated in R is rigorously proportional to T as defined by Nyquist's formula. The purpose of the Josephson junction is to convert the noise voltage fluctuations into frequency fluctuations which then are measured with a frequency counter [1]. A computer calculates the variance from the measured frequencies, from which a temperature may be calculated. Connected to the R-SQUID is a measurement circuit which unfortunately introduces additional noise. Therefore the influence of the measurement circuit on the measured properties of the R-SQUID must be taken into account if temperatures with the accuracy desired for the thermometry scale (~0.1%) are to be attained. The purpose of this paper is to report on a study of these systematic effects.

2. Principles of Operation of the R-SQUID

A very schematic diagram of the R-SQUID and measurement circuit are shown in the upper part of Fig.1. The R-SQUID consists of a resistor R (roughly 12.9 $\mu\Omega$) connected in parallel to a Josephson junction, J, which is an adjustable point-contact made between two Nb screws. The Josephson junction is adjusted so that its critical current, I_c, is roughly 1 μA. A dc current I_o (roughly 13 μA) passes through the R-SQUID, biasing the resistor at voltage, V_o which causes the Josephson junction to oscillate at an audio frequency, f_o (approximately 83 kHz). Thermal noise voltages, V_n, generally much smaller than V_o, cause the Josephson frequency to vary about f_o; that is, Johnson noise adds frequency modulation noise to the carrier at f_o. Coupled to the R-SQUID is a circuit which consists of several amplifiers and other components. Only the two most critical ones, amplifiers A_1 and A_2 are shown in Fig. 1. Amplifier A_1 , since it provides the first stage of amplification, generates the dominant amplitude noise of the system, which referred to its input is V_{A1}. The bandwidth of the system is determined by the amplifier with the narrowest bandwidth, which is amplifier A_2. By construction, its transfer function $|H_2(f)|^2$ approximates a single-pole Butterworth filter response. Following A_2 is a frequency counter, F, and a computer. A digital multimeter, DMM, is connected in parallel to F and is used to measure the signal and noise at its input. The input to A_2 may be switched to the R-SQUID (upper circuit in Fig. 1) or to another circuit (lower) whose purpose will be explained later.

A representation of what the signal looks like at the input to F is shown in Fig. 2. We point out two important features of this signal. Consider first the time axis, where we see that the signal oscillates at a Josephson frequency, f_o. Superimposed upon this frequency is an additional *frequency* modulation due to Johnson noise, although it is small and hard to see here. Secondly we note that superimposed upon the amplitude of the Josephson signal is an additional *amplitude* modulation noise (often referred to as "additive" noise) due to amplitude noise at the first stage of amplification, i.e., due to V_{A1}.

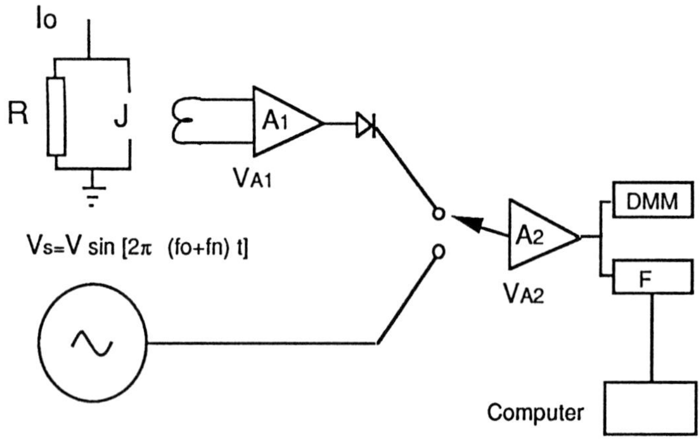

Fig.1. Schematic representation of the R-SQUID circuit (upper) and the room temperature simulator (lower).

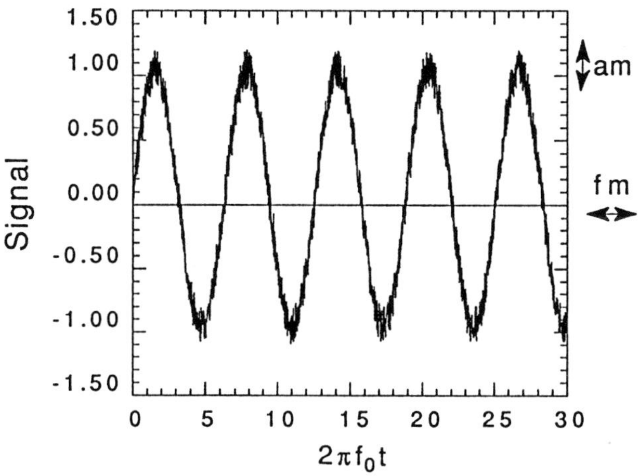

Fig. 2. The Josephson signal oscillating at frequency f_0 upon which is superimposed frequency modulation as well as amplitude modulation.

Both effects shown in Fig. 2 will cause the measured frequency to fluctuate. A measure of the magnitude of these noise effects is the variance σ^2 of these frequency variations, which is calculated by the computer. The Allan variance [2], one of the many different variances which may be calculated from the data, is best suited for these measurements. It is given below in two forms: the first defines it in terms of the algorithm used by the computer to calculate it, while the second defines it by a theoretical expression involving the spectral power density [2] which generates the frequency fluctuations

$$\sigma^2(2, r, \tau) = \frac{1}{2} \sum_{i=1}^{N} \frac{(f_{i+r} - f_i)^2}{N} \tag{1}$$

$$\sigma^2(2, r, \tau) = 2 \int_0^{\infty} S(f)\, |H_2(f)|^2 \, \frac{\sin^2(\pi f \tau)}{(\pi f \tau)^2} \sin^2(r \pi f \tau)\, df\,, \tag{2}$$

where f_i is an individual frequency counter reading, τ is the measurement time (nearly equal to the "gate time") of the frequency counter, r is a measure of the time between the start of two successive frequency measurements, and S(f) is the spectral power density generating the frequency fluctuations. Since the variance is a statistical quantity, its value will also fluctuate; its standard deviation, $\delta\sigma^2$, is given by

$$\frac{\delta\sigma^2}{\sigma^2} = \sqrt{\frac{2}{N}}\,. \tag{3}$$

To evaluate this integral we need to specify S(f) and the transfer function of the system which we take to be equal to that of A_2. For S(f) we use the expressions for white noise (S independent of f) and additive noise (S proportional to the square of f)

$$S(f) = \left(\frac{4k_B R T}{\phi_0^2} + \frac{1}{\omega_2} \left(\frac{V_{A1}^2}{V_S^2} \right) f^2 \right) \qquad (4)$$

where the amplitude of the signal is V_S and the noise at the input to A_1 is V_{A1}. Careful measurements showed that the transfer function of A_2 was fit reasonably well by the Butterworth expression

$$| H_2 (f) |^2 = 1 / [1 + ((f - f_0) / f_2)^2] \qquad (5)$$

where f_2 is one-half of the full width at half maximum. The value of f_2 found from fitting Eqn. 5 to the data was 2650.9Hz. The value of $\omega_2=2\pi f_2$ was thus 16656 s^{-1}. Putting these results into Eqn. 2 and carrying out the integration, we obtain the result[3,4]

$$\sigma^2 (2, r, \tau) = \frac{\sigma_T^2}{\tau} \left[1 - \frac{1}{\omega_2 \tau} (1 - e^{-\omega_2 \tau}) \right] + \frac{C\, \omega_2}{\tau^2} \left[(1 - e^{-\omega_2 \tau}) - (m_4) \right] \qquad (6)$$

$$\sigma_T^2 \equiv \frac{2 k_B R T}{\phi_0^2} \qquad (7)$$

$$C = \frac{1}{4\pi^2} \frac{\langle V_A^2 \rangle}{V_S^2} \equiv \frac{1}{4\pi^2} \left(\frac{N}{S} \right)^2 \qquad (8)$$

where a small departure of $|H_2(f)|^2$ from Eqn. 5 is completely taken into account in the variance by the inclusion of the extra parameter, m_4~0.0095, in Eqn. 6.

Thus we have shown that the variance given in Eqn. 6 is the sum of two terms. The first is due to Johnson noise generated by the resistor in the R-SQUID which is converted into frequency fluctuations and is thus proportional to T. The second is due to amplitude noise at the first stage of amplification which causes false triggering of the frequency counter and thus appears as an additional variance. The second term is thus proportional to the amplifier noise-to-signal ratio, (N/S), and it is independent of T. The function of the digital multimeter DMM is to measure the quantity (N/S).

3. Room-Temperature Simulation of the Variance

It would be desirable if we could measure each effect independently and compare such data with the appropriate theoretical expression given above. Unfortunately, it is not possible to separate them under experimental conditions because the R-SQUID cannot be cooled to the limit T=0 where the first term in the variance would vanish, while the amplifier noise can never be reduced to zero so that the second term never vanishes either. Thus the variance measured during the cryogenic experiments is always the sum of both terms.

Fortunately we found a way to simulate each effect independently using room temperature circuitry. Referring once again to Fig. 1, the circuit containing A_2, F

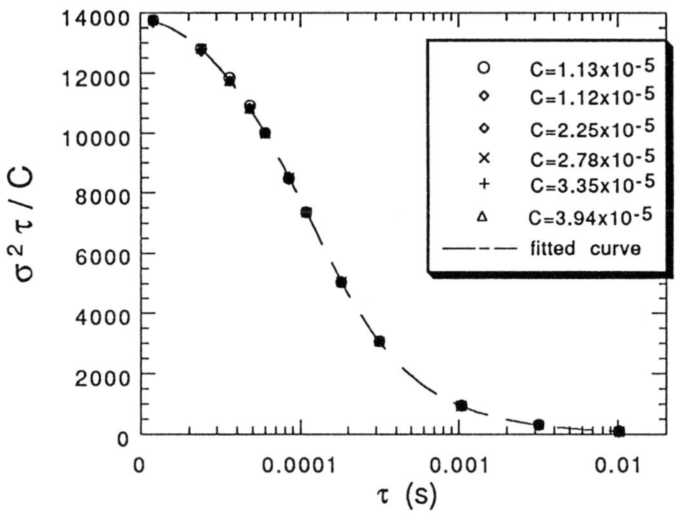

Fig. 3. Simulation of the effect of additive noise on the variance.

and the computer is switched to the lower circuit which is simply a high quality audio frequency oscillator. The oscillator was operated in one of two modes thereby simulating either additive noise or white noise. In the former circumstance, the oscillator, by virtue of its construction, had virtually no frequency modulation, and it thus simulated the R-SQUID at T=0. Thus additive noise was determined in this case by adjusting the amplitude of the oscillator so that the ratio of the noise from amplifier A_2 to V_s would simulate those values of (N/S) encountered in the noise thermometer experiments. The variance was then measured with a statistical imprecision of 0.1% as a function of τ for several values of the oscillator voltage, and thus for several values of (N/S). In the latter case, to simulate white noise, we chose signal amplitudes large enough that N was negligible, and we connected an external white noise source to the external frequency modulation port of the oscillator. The variance was measured (also with an imprecision of 0.1%) as a function of τ for several values of the fm amplitude which in this way simulated the white noise from the R-SQUID without additive noise.

We first show the results for the simulation of the additive noise. In Fig. 3 we plot a series of 6 sets of variance data, each measured for 12 values of τ, for different values of (S/N). The values of (S/N) were chosen to span those encountered in the low temperature experiments. The second term in Eqn. 6 was fitted to the data to obtain values for ω_2, C and m_4. The average value (1σ uncertainty) of ω_2=16597±(438) s^{-1} was within 1% of the value obtained from measurements of $|H_2(f)|^2$. The fitted values of C agreed to within 8% of those values calculated from Eqn. 8 and the measured values of (S/N). We also found the average value of m_4 to be 0.0095. To make the presentation easier on a common vertical axis, each curve was first divided by the fitted value of C and multiplied by τ before plotting. According to Eqn. 6, the data so treated should define a universal curve. We see in Fig. 3 that all the variance data fall on a common curve (usually within the 0.1% imprecision of the data) and strongly confirm the applicability of the model for the variance.

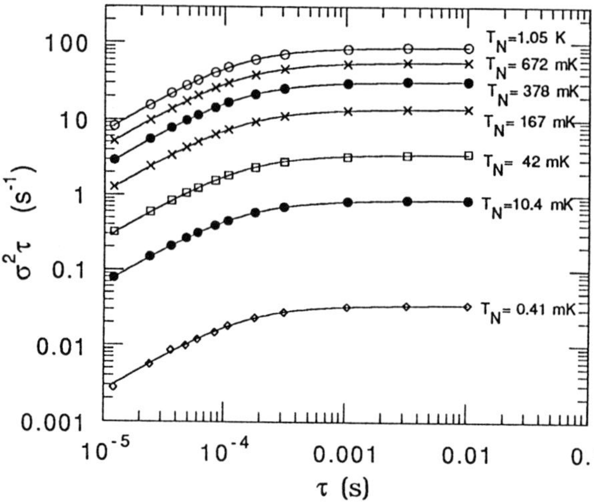

Fig. 4. Simulation of the effect of white noise on the variance.

We next turn to the simulation of the Johnson noise. Here the results are plotted in Fig. 4 for several values of the fm amplitude- the square of which is proportional to the "noise temperature", "T_N". The fm amplitude was calibrated by a technique described in Ref. 4, where we found the *relative* imprecision to be 0.1% for "T_N" >20 mK, but which rapidly degraded to 10% at "T_N" =0.4mK. The absolute value of "T_N" was established by dividing the value of the fitted value $\sigma_T{}^2$ for the largest fm amplitude by the square of that quantity. In this way we found that these curves simulate the variance for equivalent white noise temperatures varying from 1049 mK to 0.413 mK- a range exceeding that realized in the cryogenic experiments. Again the fit of the first part of Eqn. 6 to the data is very nearly equal to the 0.1% statistical imprecision of the variance measurements for all values of τ. The average value found for ω_2 was 15927 s^{-1} with a 1σ deviation of 424 s^{-1}, the first being 4% lower than the value obtained from $|H_2(f)|^2$.

4. Low Temperature Data

Having verified the model for each term in the variance, we now turn to the actual noise thermometer data taken at cryogenic temperatures. Over 400 data sets were taken with this noise thermometer over the past decade. Each set was studied by other statistical methods to demonstrate that only white noise and additive noise were present and that the imprecision of the data was in accordance with Eqn. 3. A sample of 7 of these sets is shown in Fig. 5. We fit Eqn. 6 to these data and found that the deviations of the fits from the variance data are generally within the statistical imprecision (0.1%) of the latter. The noise temperature given to the right of each curve is the one calculated from Eqn. 7 using the fitted value of $\sigma_T{}^2$ and measured values of R . We found that ω_2=16779± (356) s^{-1} for this collection of data sets.

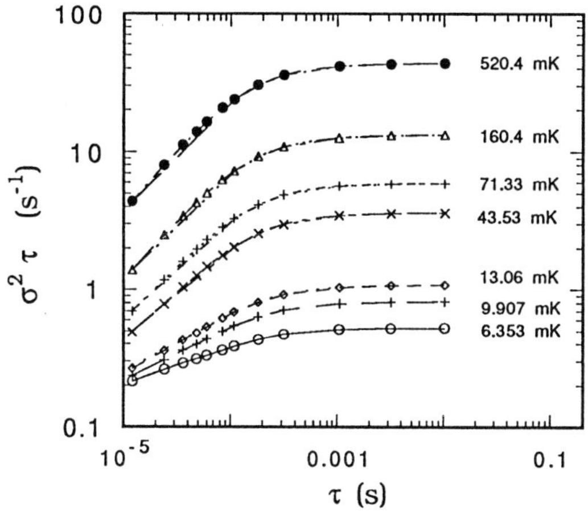

Fig. 5. Variance measured versus τ for several temperatures.

5. Reproducibility of the Noise Thermometer

The reproducibility of the temperatures calculated from the R-SQUID were checked by comparing them with two SRM superconductive fixed point devices [5] and with the melting curve of ^{3}He. The results of the first are summarized in Table I below (the other results will be reported elsewhere [4]).

Table I

Comparison of R-SQUID Noise Thermometer with SRM 767 and 768
Jan 1982-Jan 1991

Material	Number Comp	$<T_c>$ (mK)	δT_c (mK)	(%)
W	55	15.514	0.048	0.31
Be	47	22.711	0.101	0.45
Ir	29	99.378	0.271	0.27
$AuAl_2$	35	159.88	0.467	0.30
$AuIn_2$	47	204.48	0.319	0.15
Cd	35	521.06	0.786	0.15

Here the average value of the measured superconductive transition T_c is denoted by $<T_c>$, and the average 1σ deviation by δT_c. Comparisons were made over a 10-year period and we see that the percentage standard deviation of the measurements over this period for the lower T_c materials were somewhat larger than the expected value of 0.1%. These results will be discussed elsewhere [4].

6. Comparison of the Noise Thermometer with other Temperature Scales

The noise thermometer was also directly compared with a paramagnetic salt (lanthanum-diluted cerium magnesium nitrate) at 21 temperatures between 6.3 mK to 204 mK. Fitting a Curie-Weiss law to these data leads us to conclude that the temperature scale defined by the R-SQUID is linear to 0.1% over this range. Finally, the noise thermometer was directly compared with a second absolute thermometer (a nuclear orientation thermometer) [6] at the low temperature range of the former. The results of that comparison are shown in Fig. 6, where we plot the difference between the temperature defined by the nuclear orientation thermometer (T_γ) and that of the noise thermometer (T_N) versus temperature. The combined statistical uncertainty of both thermometers at each temperature is shown by the error flags, and we see that the agreement is within those error bars (0.5%) for all temperatures.

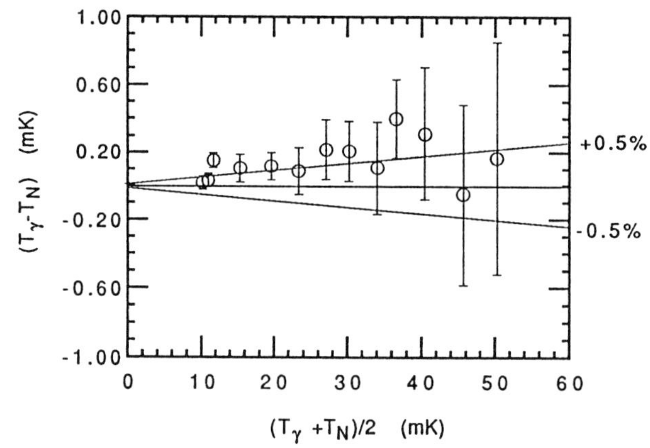

Fig. 6. Comparison of noise thermometer with nuclear orientation thermometer.

7. Conclusions

We have simulated the two types of noise (white and additive) present in an R-SQUID noise thermometer with room temperature circuitry and have shown that they are in excellent agreement with the standard frequency modulation model. We then studied the noise from the R-SQUID and showed that it was also consistent with the same model. From the latter data we developed a temperature scale which we show to be reproducible (<0.5%), and accurate (<0.5%).

References

[1] R. A. Kamper and J. E. Zimmerman, *J. Appl. Phys.* **42**, 132,1971.

[2] J. A. Barnes, A. R. Chi, L. S. Cutler, D. J. Healey, D. B. Leeson, T. E. McGunigal, J. A. Mullen, W. L. Smith, R. Snydor, R. F. C. Vessot, and G. M. Winkler, *IEEE Trans Instru. Meas.* **IM-20**(2), 105,1971.

[3] R. J. Soulen, Jr., Deborah Van Vechten, and H. Seppa, *Rev. Sci. Instru.* **53**, 1355,1982.
[4] R. J. Soulen, Jr, W. E. Fogle, and J. H. Colwell, to be published.
[5] R. J. Soulen, Jr. and R. B. Dove, NBS Special Publication 260-44, US Govt. Printing Office; 003-003-02047-8, 1979.
[6] R. J. Soulen, Jr. and H. Marshak, *Cryogenics* **7**,408 (1980).

Analysis of a Josephson Junction Noise Thermometer with a DC-SQUID Preamplifier

H. Seppä

Metrology Research Institute, VTT, Technical Research Centre of Finland, Otakaari 7 B, SF-02150 Espoo 15, Finland

Abstract. A Josephson junction noise thermometer directly coupled to a low noise preamplifier is analyzed. The absolute determination of temperature is disturbed by several effects: mixed down noise, additive noise, post-detection filter, etc. The analysis presented here takes these error sources into account, and the accuracy of the thermometer is estimated. The analysis shows that, exactly like an rf-biased R-SQUID noise thermometer, a directly coupled noise thermometer is ultimately limited by the noise temperature of the preamplifier. Since the noise temperature of a state-of-the-art dc SQUID amplifier is excellent, sufficiently low uncertainty (less than 1%) can be obtained, even below 1 mK.

1. Introduction

Since a Josephson junction converts voltage into frequency it can be used as an ordinary voltmeter to detect small voltages. High thermal noise caused by the small shunt resistor, however, restricts its use in practice. For applications where the shunt resistor is necessary, the Josephson junction can be profitably utilized. The most consequential application is to establish an accurate temperature scale by measuring voltage fluctuations across the shunt resistor, as proposed by Kamper and Zimmerman [1], with further development by Soulen and his co-workers [2,3] and by Hoffmann and Buchholz [4]. Traditionally, the Josephson junction and the shunt resistor are incorporated into a superconducting loop, forming a resistively-shunted rf SQUID. The SQUID loop is excited by an rf signal and the nonlinearity of the Josephson junction converts the low-frequency oscillations into rf sidebands in the tank circuit. The sidebands are amplified and detected by a demodulation circuit, a mixer or a diode detector, and the audio signal is fed into the frequency counter through a bandpass filter.

The rf electronics markedly impedes the use of the rf biased R-SQUID as a cryogenic temperature standard. The preamplifier should be very carefully matched to the R-SQUID in order to minimize the mixed down noise and the additive noise. In addition, the noise temperature of the first transistor stage is usually reduced by immersing it in liquid nitrogen or in liquid helium.

The most obvious method to determine the cryogenic temperature is to monitor current fluctuations in a dc SQUID input coil shunted by a small resistor. This method requires calibrations, and therefore, the Josephson junction noise thermometer (JJNT) is, in principle, advantageous for maintaining the

Springer Proceedings in Physics, Vol. 64 Superconducting Devices and Their Applications
Editors: H. Koch, H. Lübbig

cryogenic temperature scale. We will show, however, that the analysis and the calibration of the readout electronics is also essential for accurate temperature measurements carried out with the JJNT.

2. Josephson Junction Noise Thermometer

A block diagram of a Josephson junction noise thermometer directly coupled to a preamplifier is shown in Fig. 1a. The current through the circuit maintains a voltage drop across the junction and causes it to oscillate. If the junctions are disturbed by thermal noise, the motion of the quantum phase difference φ obeys the following equation [5]:

$$d\varphi/dt = \omega_0(1 - i\sin\varphi + i_n)\,, \tag{1}$$

where $i = I_c/I_0$ and $i_n = I_n/I_0$ denote the normalized critical current and the normalized noise current, respectively. The angular frequency of the oscillation becomes $\omega_0 = 2\pi R I_0/\Phi_0$ for high bias currents. If $i << 1$ and if the current fluctuations originate only from the shunt resistor, *i.e.*, $\langle {i_n}^2\rangle = 4k_bT/R$, the frequency variance for the gate time τ is ${\sigma_t}^2 = 2k_bRT/(\Phi_0\tau)$[6]. Following the calculations given in Ref. [5] and the equivalent circuit diagram in Fig. 1b, the variance σ_v^2 can be written as

$$\frac{\sigma_v^2}{\sigma_t^2} = \frac{1}{1-i^2}\left[1 + \frac{1}{2}(1 + \frac{1}{2}r\frac{T_a}{T})i^2\right] \approx 1 + \frac{3}{2}i^2 + \frac{1}{4}r\frac{T_a}{T}i^2\,, \tag{2}$$

The term $(1-i^2)^{-1}$, which describes the normalized dynamic resistance arises from the nonlinearity of the Josephson junction. The term proportional to $(1+\frac{1}{2}rT_a/T)$, where T_a is the minimum noise temperature of the preamplifier, results from the mixing between the Josephson oscillations and the noise cur-

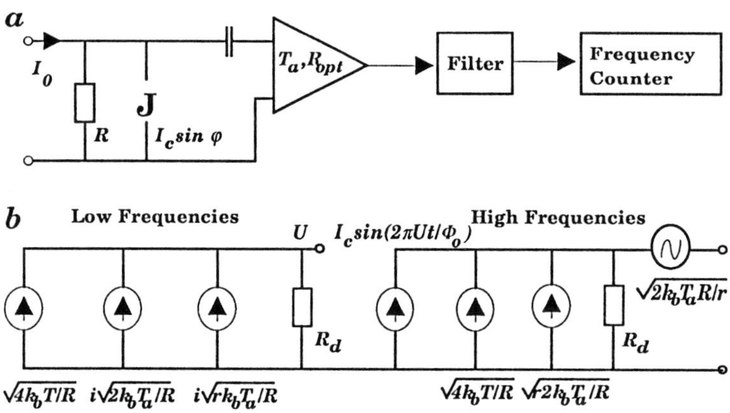

Fig. 1. a) Block diagram of a Josephson junction noise thermometer directly coupled to a low-noise preamplifier, and *b*) the equivalent circuit.

rents flowing through the junction. Both the thermal noise of the shunt resistor and the current noise of the preamplifier are of importance in this context. In Fig. 1b the equivalent noise current source and voltage source are described by the optimal source resistance, R_{opt}, and by T_a. The noise mismatch between the junction and the amplifier is denoted by the normalized optimal source resistance $r = R/R_{opt}$. The equivalent voltage noise source e_n and the current noise source i_n define T_a as $T_a = \sqrt{\langle i_n^2\rangle\langle e_n^2\rangle}/(2k_b\Delta f)$. Figure 1b shows the equivalent circuit for a Josepshon junction disturbed by thermal noise. The circuit is divided into the high and low frequency parts: The current sources on the left-hand side influence the frequency fluctuations while the right-hand side sources on the additive noise. The Josephson oscillations are described by an equivalent current source $i_s \approx I_c \sin(2\pi Ut/\Phi_0)$, where U is the low frequency voltage across the Josephson junction. The dynamic resistance of the junction $R_d = R/(1 - i^2)$.

3. Sources of Error

If the noise temperature of the preamplifier, T_a, cannot be accurately determined, the mixed down noise causes the error $E_{mix} = ri^2T_a/4T \approx ri^2\omega_0 t_a/4T = ri\omega_c t_a/4T$, where $t_a = T_a/\omega$ and $\omega_c = 2\pi RI_c/\Phi_0$. The noise temperature of the dc SQUID as well as many other amplifiers is linearly proportional to the frequency and thus it is convenient to use the frequency independent noise temperature t_a instead of T_a. If the critical current can be measured within an uncertainty $e_c = \Delta I_c/I_c$, the error caused by the dynamic resistance and the mixed down noise is reduced to $E_{dyn} = 3e_c i^2$.

Since additive noise increases the detected frequency fluctuations, the signal is usually fed through a narrow post-detection filter. Unfortunately, the filter also cuts the linewidth of the noise-broadened Josephson oscillation. In the model presented here, the additive noise results from the shunt resistor and the preamplifier. Evidently, the following amplifier stages can also add some noise, but their contributions can be incorporated into the noise temperature of the first amplifier. Following the calculations given in Ref. [6], the systematic fractional error from the additive noise can be given in the form:

$$\begin{aligned} E_{add} &= \frac{\omega^2{}_{fil}}{\omega_c{}^2}\left[1 + \frac{1}{2}\frac{T_a}{T}(r + \frac{1}{r})\right] \\ &= \frac{\omega^2{}_{fil}}{\omega_c{}^2} + \frac{1}{2}\frac{T_a}{T}(r + \frac{1}{r})\frac{\omega^2{}_{fil}}{\omega_c{}^2} = E^1_{add} + E^2_{add}\,, \end{aligned} \tag{3}$$

where $\omega^2_{fil} = \omega_1/\tau$. Here the bandpass filter is characterized by its half bandwidth ω_1. The first term of the systematic error can be partly eliminated taking into account the calibrated value for the critical current. This leads to the expression $E^1_{add} = 2e_c\omega_{fil}/\omega_c{}^2$. The last term, E^2_{add}, proportional to T_a remains unchanged since the noise temperature is very difficult to determine accurately.

If the signal is fed through a single-pole Butterworth filter, the frequency variance will be decreased [6] by $1/(\omega_1 \tau) = 1/(\omega_{fil}\tau)^2$. The half bandwidth of the filter can be determined accurately; here we assume that $\Delta\omega_1/\omega_1 = e_{fil}$ and thus $E_{fil} = e_{fil}/(\omega_{fil}\tau)^2$. The variance is a statistical quantity and the finite measurement time produces a random uncertainty. If N is the number of all frequency counts, the total measurement time Υ is related to N as $\Upsilon = \tau N$. If the Allan variance is used, the random error is $\sqrt{2/N}$ and thus $E_{freq} = \sqrt{2\tau/\Upsilon}$. In addition to the error sources discussed above, the bias current heats up the shunt resistor, impeding the determination of the path temperature. This temperature rise is simply $\Delta T = R_t R I_0{}^2$ and thus $E_{heat} = \Delta T/T = R_t R I_0{}^2/T$, where R_t is the thermal resistance between the shunt resistance and the heat path.

4. Optimization of the Thermometer

Table 1 summarizes the error sources discussed in Sections 2 and 3. The noise of the preamplifier systematically increases the frequency variance via the mixing effect and via the additive noise effect. We start our optimization by minimizing the contribution of the preamplifier, *i.e.*, $E_{amp} = E_{mix} + E_{add}$, by varying $\omega_c ir$. The minimum $E_{amp} = 2^{-1/2}\omega_{fil} t_a (1+r)^{1/2}/T$ occurs when $(\omega_c ir)^2 = 2\omega_{fil}^2(1+r)$. Setting $r \approx 1$ the maximum systematic error caused by the preamplifier becomes $E_{amp} = \omega_{fil} t_a/T_a$ and $i\omega_c = 2\omega_{fil}$. Substituting $2\omega_{fil}/i$ for ω_c in the expression for $E^1{}_{add}$, we find $E^1_{add} = \frac{1}{2}e_c i^2$. Since $E^1_{add} << E_{dyn}$, E^1_{add} can be neglected. The column entitled "step 2" summarizes the error sources after minimizing the contribution of the preamplifier.

The contributions of E_{amp}, E_{fil}, and E_{freq} can be minimized by varying τ and ω_{fil} leading to

Table 1. Error sources in the temperature measurements are given in column "step 1". Column "step 2" presents the errors after the contribution of the preamplifier is minimized. The results in column "step 3" are found by varying the gate time and the half bandwidth of the post-detection filter. The last row in the last column shows the error source found by varying the bias current.

Source of Error	Step 1	Step 2	Step 3	Step 4
E_{mix}	$\frac{1}{4} ri\omega_c t_a/T$	$\omega_{fil} t_a/T$	$2e_{fil}^{1/7}(t_a/T\Upsilon)^{2/7}$	
$E^1{}_{add}$	$2e_c\omega^2{}_{fil}/\omega_c{}^2$			
$E^2{}_{add}$	$\frac{1}{2}\omega_{fil}^2(t_a/T)(1+r^2)/(\omega_c ir)$			
E_{fil}	$e_{fil}(\tau\omega_{fil})^{-2}$	$e_{fil}(\tau\omega_{fil})^{-2}$		
E_{fre}	$(2\tau/\Upsilon)^{1/2}$	$(2\tau/\Upsilon)^{1/2}$		$2e_{fil}^{1/7}(t_a/T\Upsilon)^{2/7}$
E_{heat}	$R_t R I_0{}^2/T$	$R_t R I_0{}^2/T$	$R_t R I_0{}^2/T$	
E_{dyn}	$3e_c i^2$	$3e_c i^2$	$3e_c i^2$	$5(\frac{R_t}{TR})^{1/3}(e_c\frac{\Phi_0\omega_{fil}}{2\pi})^{2/3}$

$$\tau_{opt} = e_{fil}^{2/7} \Upsilon (t_a/\Upsilon T)^{4/7} \approx 0.3 \Upsilon (t_a/\Upsilon T)^{4/7} \,, \tag{4}$$

$$\omega_{1opt} = e_{fil}^{4/7} \Upsilon^{-1} (t_a/\Upsilon T)^{-6/7} \approx 0.07 \Upsilon^{-1} (t_a/\Upsilon T)^{-6/7} \,. \tag{5}$$

The approximations on the right-hand sides assume $e_{fil} = 0.01$. In principle, the half bandwidth can be determined very accurately, but because any real filter does not exactly obey the Butterworth response, the precise value is useless in this context. The final error can be written in the form $E_{tot} \approx (t_a/\Upsilon T)^{2/7}$, which is exactly the same result derived for the rf-biased R-SQUID [7]. The error sources caused by self-heating and the dynamic resistance can be separately mimimized by varying the bias current. The result is seen in Table 1 in column "step 4". This error can be minimized by increasing the shunt resisistor. However, both the normalized temperature $\Gamma = 2\pi k_b T/(I_c \Phi_0)$ and $\beta_c = 2\pi I_c R^2 C/\Phi_0$ should remain much below unity.

5. DC SQUID as a Preamplifier

The dc SQUID is a very suitable preamplifier for the Josepshon junction noise thermometer because of its low optimal source resistance and low noise temperature. In addition the long signal coil effectively prevents microwave signals from entering the input coil. It should be emphasized that it is extremely important to eliminate the mixing between the SQUID oscillations and the noise from the shunt resistor. Following the calculations given in Ref. [8], the noise temperature of the dc SQUID can be written in the form: $t_a = T_a/\omega \approx T_{sq}/((R_j/2)\partial I/\partial \Phi)$, where R_j is the shunt resistance of the Josephson junction and T_{sq} is the temperature of the SQUID. The flux-to-current conversion gain $\partial I/\partial \Phi$ is near independent of the point of operation [9]. If $R_j \approx 3\ \Omega$ and $\partial I/\partial \Phi = 80\ \mu A/\Phi_0$, $t_a \approx 1$ nKs. For $\Upsilon = 1$ h and $T = 1$ mK, the analysis predicts that the total uncertainty cannot be lower than 0.2%.

6. Conclusion

The total systematic error of the optimized Josephson junction noise thermometer depends on the normalized noise temperature of the preamplifier used to detect the Josephson oscillation directly or indirectly by the aid of parametric upconversion. This also means that the rf-biased R-SQUID provides no power gain, and serves only as noise matching circuit. Since noise matching can also be made via high$-Q$ transformers, the directly coupled electronics should be much more convenient. The dc SQUID is a very suitable preamplifier for this purpose because its optimal source resistance is sufficiently low. In addition, the low noise temperature of state-of-the-art dc SQUIDs should enable accurate temperature determination even below 1 mK.

References

[1] R.A. Kamper and J.E. Zimmerman, *J. Appl. Phys.* **132** (1971).
[2] R.J. Soulen, Jr., R. Rusby, and D. Van Vechten, *J. Low Temp. Phys.* **40**, 553 (1980).
[3] R.J. Soulen, Jr. and H. Marshak, *Cryogenics* **7**, 408 (1980).
[4] A. Hoffmann and B. Buchholz, *J. Phys. E.: Sci. Instrum.* **17**, 1035 (1984).
[5] H. Seppä, *J. Appl. Phys.* **55**, 1578 (1984).
[6] R.J. Soulen, Jr., D. Van Vechten, and H. Seppä, *Rev. Sci. Instrum.* **53**, 1355 (1982).
[7] H. Seppä, *J. Low Temp. Phys.* **62**, 329 (1986).
[8] K.R. Carroll, *IEEE Trans. Mag.* **MAG-27**, 3017 (1991).
[9] T. Ryhänen, H. Seppä, R. Cantor, D. Drung, H. Koch, and D. Veldhuis, this volume.

Review on Superconducting Tunnel Junctions as Ionizing-Radiation Detectors

M. Kurakado

Advanced Materials & Technology Research Laboratories,
Nippon Steel Corporation, 1618 Ida, Nakahara-ku, Kawasaki-shi,
Kanagawa 211, Japan

Abstract. Superconducting tunnel junctions (STJs) have a large possibility of being utilized as radiation detectors of ultra-high energy-resolution if excess quasiparticles excited by a radiation efficiently contribute to the signal. Researches of STJ detectors including our recent studies on Nb/Al-AlOx/Nb junctions with a single crystal Nb film and on series-connected STJ detectors are briefly reviewed.

1. Introduction

The energy resolution of a radiation detector is essentially limited by the statistical fluctuation (ΔN) of the number (N) of electrons excited by a radiation of energy (E). The statistical limit of energy resolution ΔE (FWHM) is given as $\Delta E/E = \Delta N/\langle N\rangle = 2.355\ (\varepsilon F/E)^{1/2}$, where $\langle N\rangle$ is the mean value of N, F is the Fano factor, and ε ($\equiv E/\langle N\rangle$) is the mean energy required to excite one electron in the detector. Thus, a smaller ε results in a higher energy-resolution.

According to a numerical simulation of the cascade excitation process of quasiparticles and phonons in a bulk Sn at 0 K, ε is 1.7Δ ($\approx$ 1 meV) which is about three orders smaller than the case of semiconductors. Therefore, ultra-high resolution detectors become feasible with STJs if the excess quasiparticles efficiently pass through the tunnel barrier and contribute to a signal charge. In fact, Sn/SnOx/Sn junctions as x-ray detectors have already shown about three times higher energy resolution than Si semiconductor detectors. Recently, in addition to Sn junctions, studies of new-type STJ detectors are in progress.

2. Review of Earlier Studies

In 1969, G.H.Wood and B.L.White reported that α particles were detected with an STJ [1,2]. They used a crossed-film-type Sn/SnOx/Sn junction cooled to 1.2 K. The DC Josephson current was suppressed by applying a magnetic field parallel to the junction. The STJ was obliquely irradiated with 5.1-MeV α particles, of which energy loss in the junction was 140 ~ 500 keV. Regarding the mean energy ε required to excite one quasiparticle in the detector, they estimated $\varepsilon \leq 8.2$ meV by taking into account the quasiparticles which might have recombined before passing through the tunnel barrier. The energy ε_{eff} per quasiparticle which actu-

Springer Proceedings in Physics, Vol. 64 Superconducting Devices and Their Applications
Editors: H. Koch, H. Lübbig

ally tunneled was estimated as $\varepsilon_{eff} \leq 0.145$ eV. They suggested that the STJs can be utilized as high energy resolution detectors because ε of STJ was much smaller than that of semiconductor detectors.

No clear peak corresponding to the deposited energy could be observed in the pulse-height spectrum of the signals. They concluded that the actual limit of energy resolution was probably determined by the stochastic variations in insulator thickness across the junction overlap area.

The mechanism of signal appearance was considered as follows. A superconductor is brought into a nonequilibrium state immediately after the energy deposition, but the superconductor recovers to an equilibrium state within a period of ~ 10^{-9} s. The temperature of the equilibrium state is higher than that before the incidence of radiation and it gradually decreases. Consequently the change of tunnel current following the temperature rise and fall causes the signal.

After the study of Wood and White the next step in research on the STJ detectors was performed at Kyoto University from 1977 to 1982 [3-10]. The Sn/SnOx /Sn junctions were also adopted as detectors.

First, a junction was perpendicularly irradiated with 5.3-MeV α particles. The temperature dependence and the bias current dependence of the height of the signals were investigated [3]. Secondly, by irradiating an STJ fabricated on a glass substrate with pulsed lights of 300-ns duration-time, decay-time constants of the signals at about 1.4 K were studied under different pressures of heat-exchange He gas. When the pressure was low, the signal had two decay-time constants (~350 ns, ~3500 ns) [4]. These experiments made it clear that the excess quasiparticles, i.e., electron-like ones and hole-like ones, in a nonequilibrium state are essential to the signal appearance from the STJ detectors.

A small gap energy E_g of a detector does not always result in a small value of ε. Using theories of nonequilibrium superconductivity , ε and F of superconducting Sn were estimated [6,8]. Quasiparticles excited by an incident radiation mostly have energies greater than several eV. The dominant interaction of the quasiparticles having energy greater than $\hbar\bar{\omega}$ [$\propto(\hbar\omega_D E_F)^{1/2}$, on the order of 100 meV] is the Cooper pair breaking, i.e. the excitation of other quasiparticles, where E_F is the Fermi energy and $\hbar\omega_D$ (on the order of 10 meV) is the Debye energy. Quasiparticles with energy (E) of $\Delta \leq E < \hbar\bar{\omega}$ expend their energy mainly in phonon excitation, where 2Δ (E_g) is the gap energy. Therefore, if the phonons do not excite quasiparticles, as is the case with semiconductor detectors, ε may be nearly equal to $\hbar\bar{\omega}$ even when the gap energy E_g is much smaller than $\hbar\bar{\omega}$.

Since the gap energy of a superconductor is usually much smaller than $\hbar\omega_D$, the phonons of energy $\Omega \geq 2\Delta$, which are emitted by higher energy quasiparticles, can produce other quasiparticles by breaking Cooper pairs. The resulting quasiparticles can also emit phonons. By a Monte Carlo simulation of the cascade excitation process between phonons and quasiparticles, the ε and F obtained for a bulk Sn at 0 K were

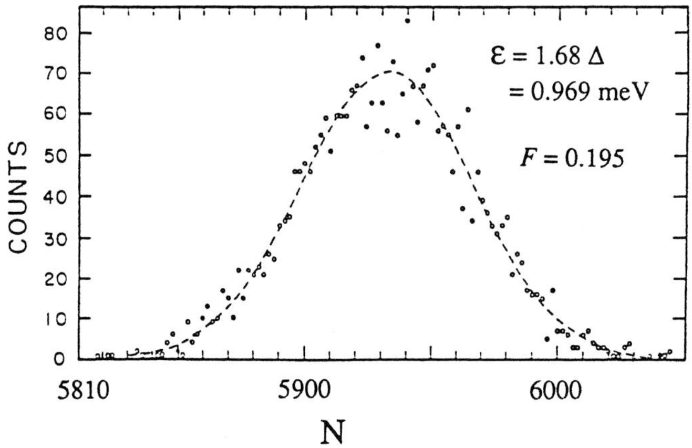

Fig. 1. A spectrum of N for 3000 radiations of energy 5.75 eV (=10000 Δ) [8].

$$\varepsilon \equiv E / \langle N \rangle = 1.68\,\Delta = 0.969 \text{ meV},$$

$$F \equiv \langle (N - \langle N \rangle)^2 \rangle / \langle N \rangle \approx 0.2,$$

where N is the number of quasiparticles excited by a radiation of energy E. The result of the Monte Carlo simulation is shown in Fig. 1. These ε and F mean that the limitation of energy resolution due to the statistical fluctuation of N is one or two orders smaller than those of semiconductor detectors, e.g., FWHM = 2.5 eV for 5.9 keV radiations in the STJ [8].

The ε value obtained by the computer simulation (nonequilibrium model) is completely different from that expected from an equilibrium model. If quasiparticles are excited by the temperature increase due to the incidence of a radiation, the energy E of the radiation will be shared between the electron system and the phonon system according to their specific heats C_e and C_p. Hence ε largely depends on temperature and becomes infinitely large at very low temperatures ($T / T_c \approx 0$) because $C_e / C_p \approx 0$ at such low temperatures. On the contrary, the equation $\varepsilon = 1.68\,\Delta$ means that about 60 % of the incident energy is assigned to the electron system immediately after the cascade excitation process of quasiparticles, in spite of its small C_e at 0 K..

Figure 2 shows the ε value numerically obtained as a function of the injection energy of phonons. The phonons of energy $\Omega \geq 2\,\Delta$ can produce quasiparticles as efficiently as a higher energy radiation. From this result, it was pointed out that the STJ detectors may be highly sensitive not only to ionizing events but also to non-ionizing events [8,9].

If the stochastic variation in the thickness of the tunnel barrier had been the main origin of the broadness of the energy spectra obtained in the previous works [1,3], it might be extremely difficult to fabricate a high energy-resolution detector

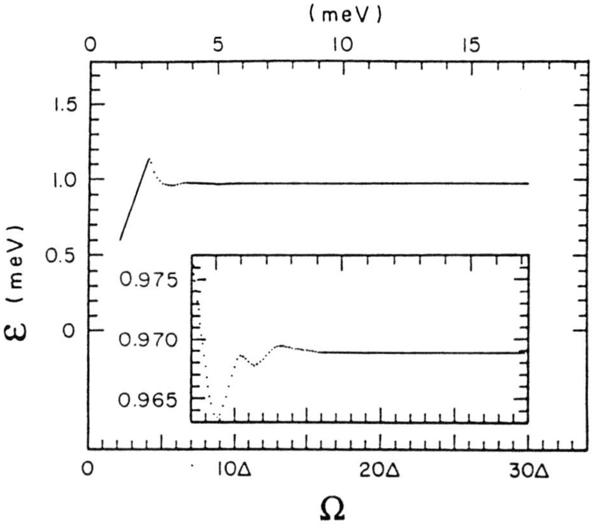

Fig. 2. The ε values ($\varepsilon \equiv \Omega/\langle N\rangle$) as a function of phonon injection energy Ω [8].

with an STJ. On the basis of the studies mentioned above [3-8], an experimental study to achieve a high energy-resolution with STJs was carried out [9,10].

STJs were cooled down to 0.32 K using a ^{3}He cryostat, instead of ^{4}He cryostats, in order to decrease the density of thermally excited quasiparticles. Although the current through junctions hardly decreased with decreasing temperature below about 0.7 K probably because of leakage current, the signal-to-noise ratio was much improved. A peak in a spectrum obtained without collimator was broad. When a hole of a collimator was set between the center and the edge of a junction, a sharp peak having a tail on the lower energy side was obtained [9]. On the contrary, a sharp pulse-height spectrum without tail was obtained by irradiating α particles only to the central part of a junction, making use of a collimator. The width of the peak was almost determined only by the broadening due to noise. This study clarified that the main origin of the broadness of the pulse-height spectra obtained before was not the stochastic variation in the thickness of the tunnel barrier and also clarified that the energy resolution of STJ detectors can be greatly improved if the leakage current of the junction and the diffusion of excess quasiparticles away from the junction to its leads are sufficiently suppressed [10].

Independently, Barone et al. demonstrated that an island type Nb/NbOx/Pb junction with narrow leads can achieve a higher energy-resolution than a conventional crossed-film-type one [11].

3. Review of Recent Studies

After the researches mentioned above[1-11], many sophisticated studies have been carried out [12]. Using Sn/SnOx/Sn junctions with small leakage current, a group of Technische Universität München (TUM) [13,14] and a group of Paul Scherrer Institute (PSI; formerly SIN) [15,16] succeeded in detection of Mn-K_α and K_β x-rays from ^{55}Fe. The junctions were operated at temperatures around 0.3 K. The group of TUM achieved a high energy-resolution of 88 eV for 5.9 keV x-rays with an Sn junction of which electric leads were made of Pb. The energy gap of Pb is larger than that of the junction material Sn. Thus, the difference of the energy gaps suppresses the diffusion of quasiparticles from the junction to the electric leads.[14]. The PSI group achieved an energy resolution of 48 eV at 5.9 keV, which is about three times higher energy-resolution than that of high energy-resolution semiconductor detectors (~150 eV). The resolution was obtained with an Sn junction having 4 μm-wide electric leads [16].

The thickness of the tunnel barrier of STJs is only a few nm, and therefore the resulting large electric capacitance (typically 10 $\mu F/cm^2$) makes it difficult to utilize large-area STJs as high resolution detectors. Since the sensitive area and volume of the STJs that showed high resolution were small, i.e., of the order of $100 \times 100\ \mu m^2 \times 100$ nm, their detection efficiencies were too low for many practical uses. Several kinds of STJ detectors were proposed and have been studied in order to overcome the drawback of the usual STJ detectors: for example,

(i) Quasiparticle-trapping-detectors, which consist of a large superconductor with a energy gap Δ_1 and a small STJ with a smaller energy gap Δ_2 than Δ_1 [17, 18],

(ii) Acoustic detectors composed of an insulating single crystal as a radiation absorber and STJs as phonon sensors [19, 20], and

(iii) Series-connected STJ detectors [21,22].

Good resolutions have been achieved with Sn/SnOx/Sn junctions [10,13-16] probably because of their small leakage current and high quality of Sn layers. However, it is well known that the Sn/SnOx/Sn junctions are easily destroyed by thermal cycles. Since Nb/Al-AlOx/Nb junctions are more resistant to thermal cycles than Sn junctions, extensive studies of Nb/Al-AlOx/Nb junctions as an x-ray detector have recently been carried out [23-27].

The leakage current of the STJs with a bottom superconductor of single crystal had been larger than those of usual STJs with a polycrystalline bottom layer. A large leakage current is fatal to the STJ as a radiation detector. We developed a new type Nb/Al-AlOx/Nb junctions making use of a single crystal Nb film as the bottom layer [27]. The upper surface of the single crystal Nb was coated with a thin polycrystalline Nb film. The leakage current of the STJs was almost the same as or smaller than that of usual polycrystalline junctions. The STJs were applied to the detection of x-rays. Properties and ε_{eff} of the samples are listed in Table I. The energy resolution were not good, i.e., around 1 keV, probably because of the diffusion of excess quasiparticles into the leads. However, ε_{eff} of an STJ (e) was about 6.6 meV (4.3 Δ), which is the smallest value so far obtained with Nb/Al-

Table I. Properties and ε_{eff} of each sample [27].

Sample	(a)	(b)	(c)	(d)	(e)
Bottom Nb	poly	single	single	single	single
Area(μm^2)	20^2	50^2	50^2	80^2	20^2
Width of leads(μm)	5	5	5	20	5
$R_N(\Omega)$[a]	5.0	0.13	0.11	0.044	0.91
$dV/dI(\Omega)$[b]	2000	600	500	460	2600
RRR[c]	8	13	21	32	33
ε_{eff} (meV)	550	100	28	21	~ 6.6

[a] Normal state resistance. [b] Dynamic resistance at the bias point where X-rays were measured. [c] RRR ≡ R(300 K)/R(10 K).

AlOx/Nb STJs and is about 1/500 that of Si detectors. A high signal-to-noise ratio is essential for a high energy-resolution detector. The broadness of the spectrum due to noise was only 30 eV (FWHM) for the STJ. This is the lowest noise level attained with Nb/Al-AlOx/Nb STJs to date, and it is about 3 times smaller than those of high resolution Si detectors. The capability of high rate detection is important in many practical uses of radiation detectors. The rise time, that is the charge collection time, of the STJ was about 200 ns, which is about one order of magnitude shorter than those obtained with Sn junctions.

More recently, we obtained an energy resolution of 170 eV for 5.9-keV x rays with an 178×178-μm^2 STJ with lower leakage current ($(dV/dI)/R_N \approx 140000$) [28].

The voltage height of a signal from any detector is desired to be as large as possible. From a simple model of series-connected detector, we found that the effective capacitance C_{eff} (≡ signal charge / signal voltage) is not given by $C/n + C'$ but given by $C+nC'$, where n is the number of STJs connected in series, C is the electric capacitance of an STJ, and C' is the capacitance parallel to the detector, which usually comes from the preamplifier and signal cables [21]. The C_{eff} is minimized when n is equal to $(SC_0/C')^{1/2}$, and then $C_{eff} = 2\,(SC_0C')^{1/2} \propto S^{1/2}$, where S is the total area of the STJs and C_0 is the specific capacitance of the junctions [29].

Series-connected STJ detectors (8000 STJs of 20×20 μm^2 or 960 STJs of 100 ×100 μm^2) were applied to detection of nonthermal phonons ($\Omega \geq 2\Delta$) induced by α particles in sapphire substrates. The pulse-height spectra showed sharp peaks corresponding to the energy of the α particles [21].

Series-connected STJ detectors with optimized geometry and small leakage current will have excellent properties [29]. For example, with the values of C' =

20 pF and $C_0 = 6$ μF/cm^2, effective capacitances C_{eff} of 0.022 μF for S = 1 cm^2 and 0.22 μF for S = 100 cm^2 are possible.

References

[1] G. H. Wood and B. L. White, Appl. Phys. Lett. **15**, 237 (1969).
[2] G. H. Wood and B. L. White, Can. J. Phys. **51**, 2032 (1973).
[3] M. Kurakado and H. Mazaki, Phys. Rev. **B22**, 168 (1980).
[4] M. Kurakado, S. Tachi, R. Katano and H. Mazaki, Bull. Inst. Chem. Res., Kyoto Univ. **59**, 106 (1981).
[5] M. Kurakado and H. Mazaki, Bull. Inst. Chem. Res., Kyoto Univ. **60**, 43, (1982).
[6] M. Kurakado and H. Mazaki, Nucl. Inst. and Meth. **185**, 141 (1981).
[7] M. Kurakado and H. Mazaki, Nucl. Inst. and Meth. **185**, 149 (1981).
[8] M. Kurakado, Nucl. Inst. and Meth. **196**, 275 (1982); thesis, Kyoto Univ. (1981), unpublished.
[9] M. Kurakado, Oyo Buturi, **53**, 532 (1984) [in Japanese].
[10] M. Kurakado, J. Appl. Phys. **55**, 3185 (1984).
[11] A. Barone, G. Darbo, S. De Stefano, G. Gallinaro, A. Siri, R. Vaglio and S. Vitale, Nucl. Inst. and Meth. **A234**, 61 (1985).
[12] See, for example, "*Superconductive Particle Detectors*" ed. by A. Barone (World Scientific, Singapore, 1988), and references therein.
[13] H. Kraus, Th. Peterreins, F. Pröbst, F. v. Feilitzsch, R. L. Mössbauer, V. Zacek and E. Umlauf, Europhys. Lett. **1**, 161 (1986)
[14] F. v. Feilitzsch, T. Hertrich, H. Kraus, Th. Peterreins, F. Pröbst and W. Seidel, Nucl. Inst. and Meth. **A271**, 332 (1988).
[15] D. Twerenbold, Europhys. Lett. **1**, 209 (1986).
[16] W. Rothmund and A. Zehnder, p.52 in ref. [12].
[17] N. E. Booth, Appl. Phys. Lett. **50**, 293 (1987).
[18] H. Kraus, F. v. Feilitzsch, J. Jochum, R. L. Mössbauer, Th. Peterreins and F. Pröbst, Phys. Lett. **B231**, 195 (1989).
[19] B. Neuhauser, B. Cabrera, C. J. Matoff and B. A. Young, IEEE Trans. Magn. **MAG-25**, 469 (1987).
[20] Th. Peterreins, F. Pröbst, F. v. Feilitzsch, R. L. Mössbauer and H. Kraus, Phys. Lett. **B202**, 161 (1988).
[21] M.Kurakado, A. Matsumura, T. Takahashi, S, Ito, R. Katano and Y. Isozumi, Rev. Sci. Inst. **62**, 156 (1991); Proceedings of the Workshop on *Superconducting Tunnel Junctions for X-ray Detection*, Dec. 1990, to be published, ed. by A. Barone (World Scientific, 1991).
[22] D. J. Goldie, Proceedings of the Workshop on *Superconducting Tunnel Junctions for X-ray Detection*, Dec. 1990, to be published, ed. by A. Barone (World Scientific, 1991).
[23] M. Kurakado and A. Matsumura, Jpn. J. Appl. Phys. **28**, L459 (1989).

[24] P. Gare, R. Engelhardt, A. Peacock, D. Twerenbold, J. Lumley and R. E. Somekh, IEEE Trans. Magn. **MAG-25**, 1351 (1989).
[25] K. Ishibashi, K. Takeno, Y. Oae, T. Sakae, Y. Matsumoto, A. Katase, S. Takada, H. Akoh and H. Nakagawa, IEEE Trans. Magn. **Mag-25**, 1354 (1989).
[26] M. Kurakado and A. Matsumura, Sensors and Actuators, **A21-23**, 33 (1990).
[27] M. Kurakado, T. Takahashi and A. Matsumura, Appl. Phys. Lett. **57**, 1933 (1990).
[28] A. Matsumura, T. Takahashi and M.Kurakado, Nucl. Inst. and Meth. in press.
[29] M. Kurakado, to be published.

Particle Detection with Superconducting Tunnel Junctions–Modelling the Non-Equilibrium State Generated by Particle Interactions

D.J. Goldie, N.E. Booth, R.J. Gaitskell, and G. Salmon

Department of Physics, University of Oxford,
Nuclear Physics Laboratory, Keble Road, Oxford OX1 3RH, UK

Abstract Superconducting tunnel junctions (STJ's) have attracted much interest as high resolution detectors for nuclear particles. Detection schemes have been implemented where the STJ provides the read-out for particle interactions in the superconducting films which comprise the junction. Alternatively the STJ may provide read-out for particle absorption in a dielectric or superconducting absorber crystal. We give here an overview of current work directed towards understanding the details of the phonon and quasiparticle yields in superconductors, and thus establishing a theoretical basis for the ultimate energy resolution and how it varies from one material to another.

1. Introduction

A number of groups are investigating superconducting tunnel junctions (STJ's) for particle detection[1]. Detection schemes may be subdivided into two distinct groups. In the first the superconducting films which comprise the STJ also provide the absorbing medium for the incident nuclear radiation. Since these films are of necessity thin (less than 1 μm thick), these devices are applicable only for low energy X-ray detection where the absorption length of the photon is not much greater than the film thickness. Larger volume superconducting detectors can be achieved by using quasiparticle trapping[2].

In the second class, the STJ detects the phonons above the superconducting pair-breaking threshold which are generated in an insulating, semi-conducting or superconducting crystal. More recently, series arrays of STJ's (SASTJ's) have been developed to give much improved area coverage and sensitivity[3].

This paper considers the mechanism by which an incident X- or γ-ray leads to the production of detectable phonons or quasiparticles (broken Cooper pairs). In the case of a dielectric crystal the detection is phonon mediated and the interaction between optical and acoustic phonon branches is of prime concern. In semiconductors the role of electron-hole pairs must also be included. These complications will not be discussed here. For superconductors the concern lies in the energy partition between quasiparticles and phonons.

Many experiments have shown that excess quasiparticles can be generated both in thin film and bulk single crystal superconductors by pulsed irradiation with optical photons. Work by our group has confirmed this for single crystals of Pb, In and Nb. However, we have also found that γ-rays do *not* produce detectable quasiparticles in In or Pb, whereas they do in Nb. In fact, for Nb, densely ionizing α-particles give results

Springer Proceedings in Physics, Vol. 64 Superconducting Devices and Their Applications
Editors: H. Koch, H. Lübbig

similar to those obtained from a diffuse source of optical photons[4]. It is important to understand why these materials behave differently, and what are the factors which determine the ultimate energy resolution.

The major part of this paper will concentrate on attempts to understand the mechanisms of quasiparticle-phonon interactions in thin film and large single crystal superconductors from the shortest times after a particle interaction.

2. Particle Interactions in Superconductors

Much current work has studied particle interactions in thin film STJ's and demonstrated good energy resolution for low energy X-rays. Tin tunnel junctions operated at about 360 mK have detected 5.89 keV X-rays from ^{55}Fe with a full-width-at-half-maximum (FWHM) of 40 eV[5]. Aluminium tunnel junctions at 100 mK have provided read-out of slightly larger area Sn films (active area of order 0.05 mm^2) with comparable energy resolution[6].

The energy required to break a Cooper pair in the STJ detector is equal to twice the superconducting energy gap Δ. In practice, more energy than this is required since some of the initial energy deposition resides in undetectable sub-gap phonons of energy $\Omega < 2\Delta$. It is common to introduce an effective energy gap $\epsilon \geq \Delta$ such that the statistical energy resolution (δE) is given by

$$(\delta E)_{\mathrm{FWHM}} = 2.35\sqrt{FE\epsilon} \qquad (1).$$

Here F is the Fano factor and E is the deposited energy. Many factors influence the observed value of ϵ. Any mechanism which leads to a loss of quasiparticles from the STJ, such as recombination loss at trapped flux points or diffusion loss from the sensitive junction area, contributes to an effective increase in its value. For Sn, values of ϵ of order 5 to 20 times Δ have been reported[7] and for Nb, $\epsilon \simeq 2\Delta$[8]. In Oxford, experiments on particle interactions in large single crystals of In, Pb and Nb indicate that for In and Pb very few quasiparticles are produced whereas for Nb the quasiparticle yield is much greater.

Most models of the STJ detector make a number of assumptions about the initial distribution of quasiparticles and phonons in the superconductor and use as a starting point the Rothwarf-Taylor (R-T) equations[9]. In the spirit of these equations all of the quasiparticles are assumed to have energies close to $E = \Delta$ and all the phonons have energy $\Omega = 2\Delta$. The quasiparticles and phonons form a coupled system and obey the rate equations [10]

$$\begin{aligned}\frac{dn_{qp}}{dt} &= 2\Gamma_{pb} - 2\Gamma_R n_{qp} - Rn_{qp}^2 - \Gamma_{tun} n_{qp},\\ \frac{dn_\phi}{dt} &= -\Gamma_{pb} n_\phi + \Gamma_R n_{qp} + Rn_{qp}^2 - \Gamma_\gamma n_\phi\end{aligned} \qquad (2),$$

where n_{qp}, n_ϕ are excess quasiparticle and phonon densities, Γ_{pb} is the rate at which a phonon breaks a Cooper pair producing two quasiparticles, $\Gamma_R = 2Rn_{th}$ is the quasiparticle recombination rate at the thermal density n_{th} and R is the recombination coefficient. The temperature and energy dependence of these quantities can be calculated using the equations of Kaplan *et al*[11]. The quantity Γ_{tun} is the tunneling rate while Γ_{γ} is the rate of phonon loss usually to the thin film substrate. Additional loss terms may also be included.

These equations assume spatial uniformity of the quasiparticles and phonons, although inclusion of diffusion terms is possible and vital for the description at short times after an interaction or for the case of large volume single crystals. Even with these refinements a problem remains. Is the basic assumption of low energy quasiparticles valid? How is the energy of an X-ray transformed into these long-lived excitations and what can be understood about relative quasiparticle-phonon yields?

Equations (2) do not include the effects of quasiparticle scattering with phonon emission (rate Γ_s) or phonon scattering by quasiparticles (rate $\Gamma_{\phi\, s}$) which must be important in the initial stages following the particle interaction where $E,\ \Omega \gg \Delta$. We include these by introducing the total scattering rates for the quasiparticles and phonons $\Gamma_{qp\, I}$, $\Gamma_{\phi\, I}$ such that

$$\begin{aligned}\Gamma_{qp\, I} &= \Gamma_R + \Gamma_s,\\ \Gamma_{\phi\, I} &= \Gamma_{pb} + \Gamma_{\phi\, s}\end{aligned} \qquad (3).$$

A numerical solution of the coupled rate equations for a superconductor far from equilibrium, including both the energy dependence of the interaction rates and the effects of quasiparticle and phonon diffusion, is not trivial. However an insight into the problem can be obtained in a straight-forward way.

3. The Hotspot

We want to understand how a single quantum of energy of perhaps 10 or 100 keV is transformed into many low-energy excitations. Gray[12] and more recently Van Vechten *et al*[13] have noted that at very short times (perhaps on a scale of order 100 ps) a small volume of the superconducting absorber is in a state far from equilibrium. It is important to understand how and on what scales in time and space this highly excited state relaxes to the non-equilibrium superconducting state which can be described in terms similar to equations (2).

The interaction of a high energy X-ray or photon in a material proceeds by a chain of events. The initial interaction with an atom produces a high energy electron and possibly lower energy X-rays. Photo-electrons have energies nearly equal to that of the the incident photon and a mean range given by[14]

$$R[\mu\text{m}] = \frac{0.033E[\text{keV}]^{1.70}}{\rho[\text{g cm}^{-3}]} \qquad (4).$$

Table 1 gives typical photo-electron ranges in a number of materials. The integrated path length for electrons of these energies is about twice this range.

Measurements of laser pulse relaxation times in thin gold films indicate that on a time-scale of order a few ps the initial very high energy electrons relax producing

Table 1 Electron ranges in selected superconductors at two energies.

	R in μm 10 keV	R in μm 1 keV
Nb	0.19	0.004
Sn	0.23	0.005
Al	0.61	0.012

Table 2 Materials parameters and enthalpies for selected superconductors.

	$\gamma \times 10^3$ $\mathrm{Jmol^{-1}K^{-1}}$	T_c K	Θ_D K	$\int C_{es}\mathrm{d}T$ keV/μm^3	$\int C_L\mathrm{d}T$ keV/μm^3
Nb	7.8	9.25	276	209	97
Pb	3.1	7.19	96	30	500
Sn	1.78	3.75	195	5.2	5.0
In	1.67	3.41	109	4.2	20.1
Al	1.35	1.21	420	0.7	0.01

a distribution of lower energy electrons and high energy phonons[15]. Thus on this time-scale the energy of an initial 10 keV photon becomes distributed over a volume of material of order $(0.1\,\mu\mathrm{m})^3$ giving energy densities of order 1 MeV μm^{-3} and typical electron and phonon energies of order Ω_D, the Debye energy.

This energy density can be compared with the total enthalpy H_c of the superconducting state between the bath temperature ($\simeq 0$) and the superconducting transition temperature T_c. We have

$$H_c = \int_0^{T_c} C_{es}\,\mathrm{d}T + \int_0^{T_c} C_L\,\mathrm{d}T \tag{5}$$

where C_{es}, C_L are the (superconducting) electronic and lattice heat capacities respectively. The integrals can be performed and lead to

$$\begin{aligned} \int_0^{T_c} C_{es}\,\mathrm{d}T &= \frac{1.1}{2}\gamma T_c^2\left(\frac{\rho}{A}\right) \\ \int_0^{T_c} C_L\,\mathrm{d}T &= \frac{1944T_c^4}{4\,\Theta_D^3}\left(\frac{\rho}{A}\right) \end{aligned} \tag{6}$$

where A is the atomic mass, ρ is the density and γ is the normal state electronic specific heat capacity. Suitable units are (keV/μm^3).

Table 2 gives the calculated enthalpies and illustrates that, for most common superconductors, X-rays of quite low energy lead to energy densities which imply a superconducting to normal transition. The first question is: does such a transition occur in some small volume around the particle interaction? The answer is— initially no.

The self-consistent BCS energy gap Δ is given by

$$\ln\left(\frac{\Delta(0)}{\Delta}\right) = \int_0^{\infty} \mathrm{d}E\,\frac{\rho(E)f(E)}{E} \tag{7}$$

where $\Delta(0)$ is the zero temperature energy gap, $f(E)$ is the Fermi distribution and $\rho(E) = \Re(E/(E^2 - \Delta^2)^{1/2})$ is the quasiparticle density of states. Quasiparticles with energies greater than a few times Δ have little effect on the energy gap. Thus any gap reduction will become significant only when electrons and phonons have relaxed to close to these energies. The problem now is to determine the subsequent evolution of the electrons and phonons from energies of order Ω_D to energies comparable with the superconducting energy gap and ultimately into detectable quasiparticles.

The cascade process can be examined utilizing the quasiparticle and phonon total scattering rates $\Gamma_{qp\,I}$ and $\Gamma_{\phi\,I}$ computed from the equations and characteristic times given by Kaplan *et al*[11]. Caution is needed however since these equations apply to a near-equilibrium situation which is not the case in the hotspot. Gray[12] has argued that the average energy of the phonon emitted by a quasiparticle of energy E is $\Omega = 3E/4$ and that the average electron energy created by a phonon of energy Ω is $E = \Omega/2$. Thus each cycle of the quasiparticle-phonon system reduces the average energy to $E_{i+1} = 3\,E_i/8$. For Al ($\Omega_D = 36$ meV, $\Delta(0) = 175$ μeV) some five cycles are necessary. For Nb or In only three are required.

At high energies $\Gamma_{qp\,I}$ exceeds $\Gamma_{\phi\,I}$ so that the cascade at any given instant consists mainly of phonons. This is illustrated in Fig. 1 which shows the dependence of $\Gamma_{qp\,I}$ (solid curves) and $\Gamma_{\phi\,I}$ (dotted curves) calculated as a function of energy in units of $\Delta(0)$ for Al, Nb and In.

For Al the rates become equal at energies of order $9\Delta(0)$. Consequently, rapid quasiparticle diffusion sets in before significant gap reduction occurs. The time-scale for $E \simeq \Delta(0)$ is now determined by the last quasiparticle scattering event at $E \simeq 4\Delta(0)$ with a rate of about 10 ns^{-1}. The quasiparticles have now expanded to a radius of order $\sqrt{6D/\Gamma_{qp\,I}} \simeq 10$ μm. The quasiparticle diffusion coefficient D is given by $D = \frac{1}{3}v_{qp}^2/\Gamma_i$ where v_{qp} is the quasiparticle velocity and Γ_i is the scattering rate from impurities and defects. Typical values of D range from some 0.2 m^2s^{-1} for thin films to 10 m^2s^{-1} for high purity single crystals. This quasiparticle radius exceeds the low temperature characteristic phonon length $L_\phi = c_t/\Gamma_{pb}$, where c_t is the transverse sound velocity, even in disordered thin films. The time evolution of the hotspot thus changes dramatically when $\Gamma_{qp\,I} \leq \Gamma_{\phi\,I}$. The rapid dilution of the energy density by quasiparticle diffusion leads to the effective end of the hotspot.

For Nb, $\Gamma_{qp\,I}$ exceeds $\Gamma_{\phi\,I}$ at all energies above $2\Delta(0)$ when the rate is about 5 ps^{-1}. Thus some 10 ps after the hotspot formation, quasiparticle propagation dominates and rapid cooling occurs.

For In, $\Gamma_{qp\,I}$ exceeds $\Gamma_{\phi\,I}$ for all energies at $T = T_c$. Thus In looks like a material where significant hotspot effects may be observed. After a time of order 0.5 ns the average energy is about $\Delta(0)$ but the hotspot has only expanded to a radius of order $c_t/\Gamma_{\phi,I} = 0.5$ μm. Quasiparticle-phonon decoupling has not yet occurred. The energy gap in this region will be reduced so that recombination phonons emitted into the bulk will have insufficient energy to break pairs thus losing energy from the quasiparticle system. Similarly the quasiparticles have energies below the bulk energy gap $\Delta(0)$ and cannot expand.

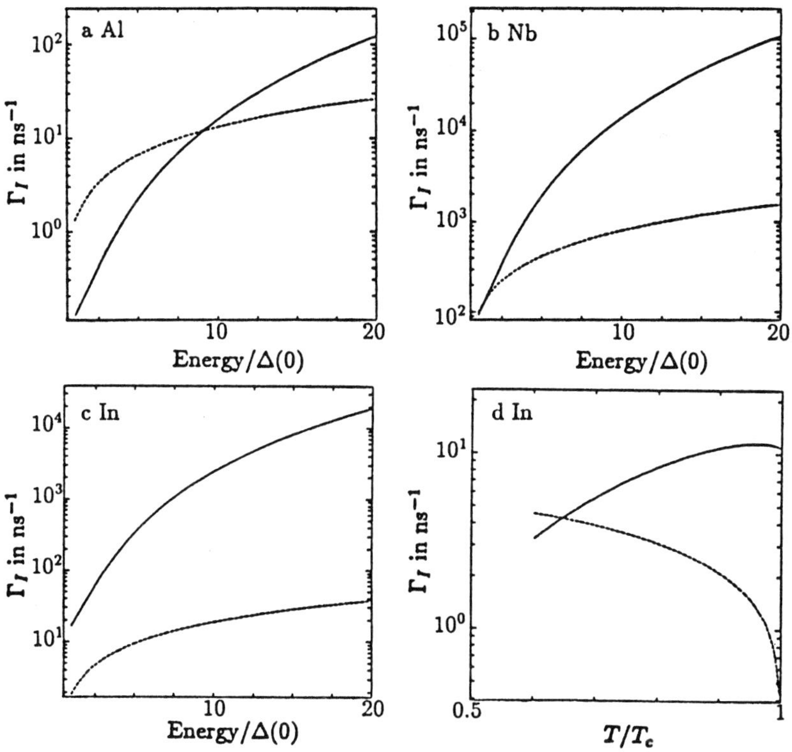

Fig. 1 Quasiparticle and phonon (dotted lines) scattering rates in (a) Al, (b) Nb and (c) In as a function of energy $T = T_C$. (d) The rates for In as a function of T/T_c.

At this stage of the complex relaxation process the coupled quasiparticle-phonon system evolves as shown in Fig. 2. If the local energy gap in the vicinity of the hotspot is not reduced, then quasiparticles will diffuse away as the effective temperature $T^*(x,t)$ decreases and $\Gamma_{qp\,I}$ drops below $\Gamma_{\phi\,I}$. The quasiparticles and phonons decouple and we expect to have a low value of ϵ, i.e. a large number of quasiparticles. However, if $\Gamma_{qp\,I}$ remains above $\Gamma_{\phi\,I}$ as the system relaxes, we expect the local value of Δ to be reduced and energy will be lost from the quasiparticle system via the emission of recombination phonons with $\Omega < 2\Delta(0)$ into the bulk superconductor, as indicated in Fig. 2 by the loss term $\Gamma_\gamma(x,t)$. In the extreme, the result will be a large ϵ and most of the deposited energy will appear in the form of sub-gap phonons.

This picture is qualitative and very approximate. However, it correlates with our results for In, Pb and Nb, and with the relatively large value of ϵ found for Sn as compared with Nb. Moreover it has predictive powers: materials with low Ω_D such as Pb, In, Hg, and Tl should have high values of ϵ, while materials such as Ta, Zn, Al and Mo, with high Ω_D should give small ϵ and large quasiparticle yields. A figure of merit is roughly given by Ω_D^3/T_c^2[4]. For intermediate materials such as Sn (and to some extent Nb) the observed ϵ will depend on the quasiparticle diffusion coefficient which is strongly affected by impurities and microstructure.

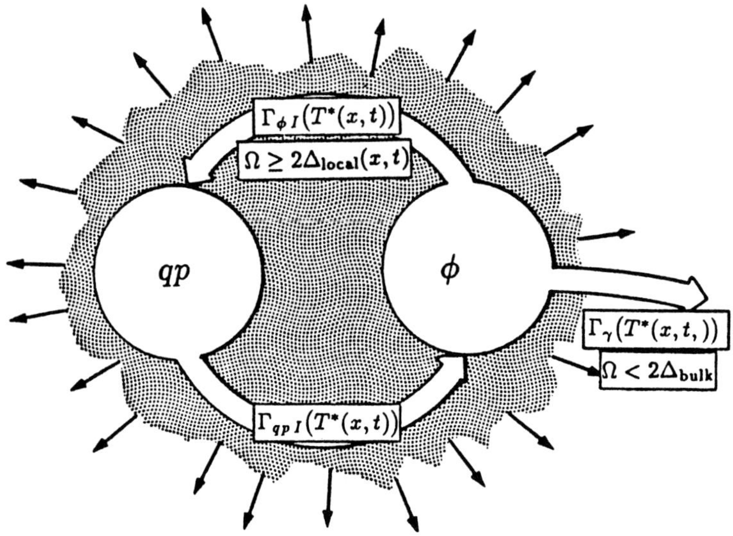

Fig. 2 Schematic of the coupled quasiparticle-phonon system.

4. Conclusions

A clear understanding of the time-evolution of the far from equilibrium state created by a particle interaction in a superconductor is necessary. Experiments indicate significant material variations in the numbers of quasiparticles and phonons created by particle interactions and this can be correlated by a simple model with material properties such as Ω_D, T_c and the electron-phonon interaction strength.

References

[1a] *Low Temperature Detectors for Neutrinos and Dark Matter II*, eds. L.Gonzalez-Mestres and D.Perret-Gallix, (Editions Frontieres,1988).

[1b] *Superconductive Particle Detectors* , ed. A.Barone, (World Scientific, 1988).

[1c] *Low Temperature Detectors for Neutrinos and Dark Matter III*, eds. L.Brogiato, D.V.Camin and E.Fiorini, (Editions Frontieres, 1990).

[1d] *Proceedings of the Workshop on Tunnel Junction Detectors for X-rays*, eds. A.Barone, R.Cristiano and S.Pagano, (World Scientific, 1991).

[2] D.J.Goldie, N.E.Booth, C.Patel and G.L.Salmon, Phys. Rev. Lett. **64**, 954 (1990).

[3] D.J.Goldie in Ref. [1d].

[4] N.E.Booth, R.J.Gaitskell, D.J.Goldie, C.Patel and G.L.Salmon in Ref [1d].

[5] W.Rothmund and A.Zehnder in Ref. [1b].

[6] H.Kraus, F.V.Feilitzsch, J.Jochum, R.L.Mossbauer Th.Peterreins and F.Probst, Phys. Lett. B. 231, 195 (1989).

[7] H.Kraus, Th.Peterreins, F.Probst F.v.Feilitzsch, R.L Mossbauer and V.Zacek, Euro. Phys. Lett. **1**, 161 (1986).

[8] P.Garé, R.Engelhardt, A.Peacock, D.Twerenbold, J.Lumley and R.E.Somekh, IEEE Trans. Mag. **MAG-25** 1351 (1989).

[9] A.Rothwarf and B.N .Taylor, Phys. Rev. Lett. **19** 27 (1967).

[10] W.Rothmund and A.Zehnder in Ref. [1a].

[11] S.B.Kaplan, C.C.Chi, D.N.Langenberg, J.J.Chang, S.Jafarey and D.J.Scalapino, Phys. Rev. B **14**, 4854 (1976).

[12] K.E.Gray in Ref. [1b].

[13] D.Van Vechten and K.S.Wood, submitted to Phys. Rev. B (1990).

[14] L.Katz and A.S.Penfold, Rev. Mod. Phys. **24** 28 (1952).

[15] R.W.Schoenlein, W.Z.Lin, J.G.Fujimoto and G.L.Eesley, Phys. Rev. Lett. **58**, 1680 (1987).

A Calorimetric Particle Detector Using a SQUID Monitored Transition Edge Thermometer

D. Dummer

Max-Planck-Institut für Physik, Föhringer Ring 6,
W-8000 München 40, Fed. Rep. of Germany

Abstract. At the Technische Universität München and MPI München, we are studying calorimetric particle detectors utilizing absorber crystals and superconducting transition edge thermometry. High resolution results from the low absorber heat capacity and the strong temperature dependance of the resistance of the thermometer at the superconducting – normal phase transition. The thermometer signals are monitored with a SQUID, and subsequent pulse height analysis is performed. Detector performance is discussed relative to SQUID characteristics, and SQUID system slew rate limitations are specifically addressed. Prototype detector results are presented.

1 Introduction

This work is being performed in collaboration with S. Cooper, M. Frank, J. Igalson, F. Pröbst, A. Rulofs, and W. Seidel of the Max-Planck-Institut für Physik and F. von Feilitzsch, G. Forster, and E. Kellner of the Technische Universität München. Previous results are described in W. Seidel, *et al.*, Phys. Lett. **B 236** (1990) 483.

We are studying prototype calorimeters for use in the detection of rare events in particle physics. Possible uses for such detectors are the detection of low energy neutrinos, searches for dark matter candidates, or rare decay events such as double-beta decay. Such a detector should have a relatively high mass (high event rate) without sacrificing too much in energy resolution.

An idealized calorimeter consists of an absorber material in weak thermal contact with a heat bath. The absorber is initially at a steady state temperature which is determined by the heat bath temperature. A scattering event deposits an energy δE into the absorber. The temperature of the absorber rises by an amount

$$\delta T_o = \frac{\delta E}{C} , \tag{1}$$

and then decays back to the steady state temperature with the characteristic time

$$\tau = \frac{C}{g} , \tag{2}$$

where C is the heat capacity of the absorber and g is the thermal conductance between the absorber and the heat bath. We take advantage of the low spe-

Springer Proceedings in Physics, Vol. 64 Superconducting Devices and Their Applications
Editors: H. Koch, H. Lübbig

cific heat of dielectric crystals at low temperatures for the absorbers in these calorimeters.

To measure δT_o with high resolution we utilize the dramatic change in resistance of a superconducting strip at the superconducting – normal phase transition. Because of the low impedance ($< 1\,\Omega$) of the superconducting strip, a SQUID is better suited than an FET to monitor the small changes in the strip resistance.

2 The Si Absorber – Ir Strip Calorimeter

The structure of the calorimeter currently under study at MPI is shown in Fig. 1. It consists primarily of a 0.66 mole Si single crystal with an Ir strip thermometer (vertical in the figure) evaporated on one surface. The Ir strip dimensions are 2 mm x 1 mm x 200 nm, and the overlayed contact pads (horizontal in the figure) are of Al. The Cu sample holder serves as the heat bath. The assembly shown in Fig. 1 (except for the SQUID) is suspended from the mixing chamber of a dilution refrigerator. The calorimeter and mixing chamber are housed in magnetic shielding.

The electrical circuit consists of the reference resistor, $R_s = 49\,\mathrm{m}\Omega$, and SQUID inductor, $L_s = 2\,\mu\mathrm{H}$, in parallel with the Ir thermometer R_{Th}. (For some of the data the effective L_s was increased to 270 μH.) This particular Ir thermometer has a transition temperature of 230 mK, a transition width of 6 mK, and a normal state resistance of 250 mΩ. This network is driven with constant current from a battery powered current supply.

When an energy absorption event occurs in the absorber, the thermometer temperature and resistance increase. This causes the current through the SQUID branch of the circuit (containing R_s and L_s) to increase an amount $\delta I(t)$. This is the quantity measured by the SQUID. A typical pulse is shown

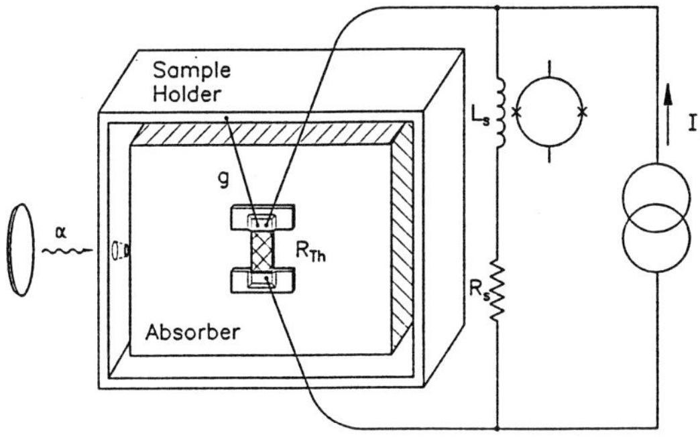

Figure 1. The detector schematic. The Ir strip thermometer R_{Th} is deposited directly on the Si absorber crystal. The whole calorimeter is encased in the Cu sample holder. A small hole in the sample holder admits radiation from an α-particle source. The thermal conductance g is made by a set of Au wires bonded to the Al contact pads of the thermometer and to the sample holder.

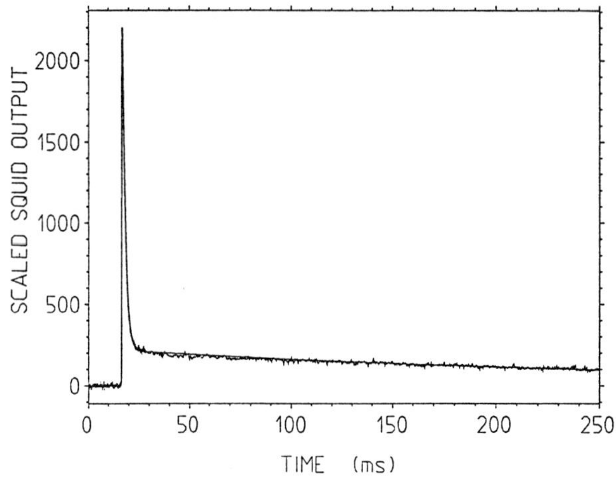

Figure 2. A typical thermometric pulse measured by the SQUID resulting from the absorption of a 5.8 MeV α-particle by the calorimeter. For this measurement the effective $L_s = 270\ \mu$H. The pulse can be characterized by a fast rise time, $\tau_{rise} \approx 0.2$ ms, and two decay times. These decay times are a short decay time, $\tau_s \approx 1.3$ ms, and a long decay time, $\tau_l \approx 220$ ms. The pulse height corresponds to a temperature rise of 90 μK.

in Fig. 2. The time dependance is not as simple as described in the introduction. The observed signals can be approximately described by

$$\delta I(t) = -(A_s + A_l)e^{-t/\tau_{rise}} + A_s e^{-t/\tau_s} + A_l e^{-t/\tau_l} , \tag{3}$$

where $\tau_{rise} < \tau_s \ll \tau_l$ and $A_l \ll A_s$. The amplitude A_l associated with the long time constant τ_l corresponds to the expected temperature rise δT_o. The amplitude A_s associated with the short decay time τ_s is an overshoot which can exceed ten times A_l. This overshoot is still being studied. A typical pulse height spectrum is shown in Fig. 3.

Higher signals result from a sharper phase transition (larger $\frac{\partial R}{\partial T}$) and higher drive current I_o. However, the SQUID system slew rate (100 mA/s through L_s) limits the size of the signals which can be tracked because of their fast rise time ($\tau_{rise} = 20\,\mu$s for $L_s = 2\,\mu$H). The rise time can be increased to allow stable operation at higher drive current by inclusion of an additional inductor in series with L_s. The rise time then becomes determined by the L-R time constant, $(L + L_s)/(R_{Th} + R_s)$. This, however, decreases the pulse height when the value of the L-R time constant approaches that of the short decay time. With only the intrinsic SQUID inductor in place, the Si-Ir calorimeter is limited by SQUID slew rate to operation with drive current less than 8 μA. The pulse height obtained in this configuration due to the absorption of a 5.8 MeV α-particle is 250 nA through L_s.

At present the SQUID noise is less than that from other sources, such as Johnson noise from the resistors. However theses other contributions should

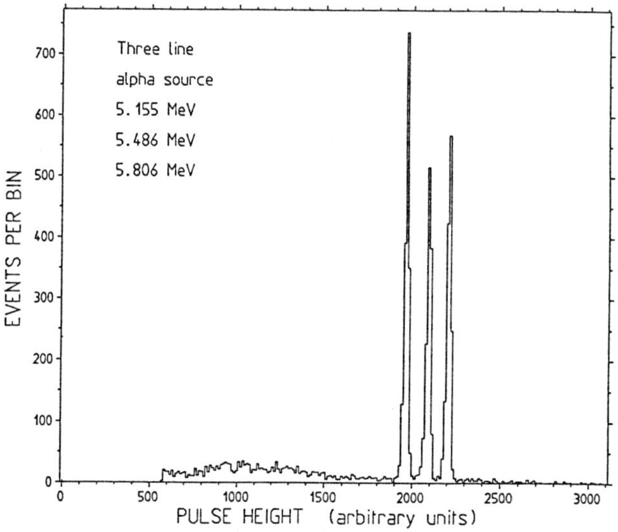

Figure 3. A typical pulse height spectrum measured by the calorimeter. The three peaks result from the absorption by the calorimeter of α-particles emitted by the three line source. The zero of the x-axis corresponds to zero energy events. A trigger threshhold was set at a pulse height of 570 units. For this measurement the effective $L_s = 270\ \mu$H. The energy resolution of the calorimeter is 50 keV.

decrease significantly when we go to lower operating temperatures (e.g. 15 mK with a W thermometer), so that the intrinsic noise of our SQUID may become dominant.

3 Desired SQUID System Improvements

The following items are desired in SQUID systems to be used with future calorimeters. In order of decreasing importance to us these desired features are:

- **High slew rate.** We are currently slew rate limited to operating the described calorimeter at drive currents ten times below the maximum value allowed by stability constraints imposed by thermometer self-heating.

- **Low noise.** Future calorimeters will operate at lower temperatures thus bringing the Johnson noise and the absorber fluctuation noise contributions below the present intrinsic SQUID noise.

- **Multiple sensor element capacity.** Multiple sensor capacity is desirable for detector arrays and for calorimeters with multiple thermometers per absorber.

- **Custom feedback option.** The ability to alter the feedback loop would enable the operation of the thermometer strip at constant voltage. Whether this is significantly advantageous has yet to be tested.

Multichannel Instrumentation for Biomagnetism

V. Foglietti

IESS-CNR, Via Cineto Romano 42, Rome, Italy

Abstract. A review of recent developments of multichannel instrumentation for Biomagnetism is presented. The main factors affecting the design, with different source configuration, is examined. Problems related to the SQUID sensors, the detection coils and the cryogenic aspects are examined. The existing large array multichannel systems and of those one that will be ready in the near future are described.

1. Introduction

The last few years have been characterized by the development of a new generation of biomagnetic systems, mostly devoted to neuromagnetic and cardiomagnetic measurements, allowing simultaneous detection of magnetic fields at several sites of the body surface. The systems developed between 1984 and 1988 featured four to seven adjacent magnetic sensors [1,2,3,4,5]. The major effect of those systems was only that to speed up the recording time, not an insignificant advantage when dealing with clinical studies. Nevertheless, both the number of adjacent measuring sites, and the overall extension of the resulting measuring grid, were clearly insufficient for a proper investigation of bioelectric phenomena in the heart and in the brain. Therefore, sequential positioning of the systems over different body surface areas was still needed, with possible introduction of positioning errors, and consequent worsening of the localization accuracy. As a consequence, larger systems have been developed during 1989 [6,7,8,4] and others will be ready in the near future [9,10,11]. This further development in multichannel instrumentation has been made possible thanks to the progress in dc-SQUID fabrication technology which allows to rely on robust and low-noise magnetic sensors with very low 1/f corner frequency. The performance of dc-SQUID devices enables the most complex experiments making biomagnetism a highly sophisticated non-invasive diagnostic tool that offers the possibility to localize functional activity with an excellent spatial and temporal resolution, on a millimeter and millisecond scale, respectively [12,13]. The basic concepts in the design of sensor arrays will be discussed in the second section of this paper. The third section will deal with the SQUID sensor, the electronic system suitable for biomagnetic applications and the cryogenic and measuring environment. The existing biomagnetic multichannel systems having a large number of sensors ≥ 19 and those that will be ready in the next future, will be described in the last section.

2. Considerations on design of detector arrays

There are many factors affecting the design of a detector array: the optimum size of the recording grid, the sensor size and type. The shape of the detector array is also another factor of concern, depending on the type of measurement and on the type of model that is used to solve the inverse problem. In designing a multichannel system for brain research, an appropriate model is the current dipole in a uniform sphere and this leads to the best grid configuration having the sensors displaced over a spherical surface to roughly follow the curvature of the head. On the contrary the current dipole in the infinite half-space, a more appropriate model for the heart signal, leads to a flat array of sensors. A recording grid size close to the spatial distribution of a magnetic field pattern generated by the source of interest, allows a " one-shot " measurement without the need of repositioning the sensor, this is the most interesting feature of the new generation of multichannel systems. The optimization of the size for the recording grid has obviously not a general rule. We could tentatively try to identify a minimum size for the grid in order to get a "one-shot" measurement by considering the extension of the spatial distribution of the magnetic field, say, over the scalp, in consequence of a specific celebral event, such as an evoked cortical response or an abnormal epileptic (focal) activity. Due to the relatively more "concentrated" activity at a cortical level, in response to a simple sensory stimulation, the resulting scalp distribution of the magnetic field may be confined to a scalp area of about 200 cm^2, particularly if the active neuronal pool lies at a depth of 2-3 cm beneath the scalp. The scalp area containing the fundamental characters essential for source localization of a spike-and-wave complex in case of temporal lobe epilepsy is wider, say, of the order of 300 cm^2. These considerations permits to infer that a good value of the inner radius of the dewar tail containing the detectors is about 9 cm. Similar considerations for MCG would lead to an optimum surface of $\simeq 500\ cm^2$ with the detectors arranged on a flat surface.
How many detection coils should we fit in these areas? An excessive spatial sampling will only increase the complexity of the system without giving any real improvement in the localization accuracy . This has been recently demonstrated in a very elegant way by Kemppainen and Ilmoniemi applying Shannon's theory of communication [14]. However,

just taking into account the inevitable spacing between the inner and outer surface of the dewar (approximately 15 mm), we evince that we do not need to localize sources closer than 25-30 mm to the surface containing the coils. This consequently sets an optimum number of coils to be fitted in the recording areas. This number may be identified with $\simeq$ 50 sensors inside the circular surface of $\simeq$ 300 cm^2, and, considering the deeper sources in heart measurements, to 40 detectors arranged on the flat surface specified above.

The locating accuracy of the dipole current source is the key parameter in the optimization of a detector array. The geometry of the detection coils strongly affects the ability of the system to locate and separate different sources in the celebral cortex. Different types of detection coils behave in a different way, as far as the localization capability is concerned. This is mainly due to a different field or gradient sensitivity that is characteristic of a coil configuration.

Recently Drung et al.[15,16] have developed an integrated dc-SQUID magnetometer where the pick-up coil constitutes the inductance of the dc-SQUID itself. A low inductance is obtained with a "fractional-turn" loop configuration [17]. This solution leads to an improvement in the field sensitivity with respect to the conventional scheme, where the pick-up coil is "transformer coupled" to the SQUID by means of the input coil. The magnetometer configuration has been succesfully used in heavily shielded room [6] but it can be extended to moderately shielded environment by utilizing electronic balancing techniques.

In the axial gradiometer configuration, where the axial derivative $\partial B_z/\partial z$ is utilized, the baselenght is usually long compared with the distance of the active coil from the source, the far away field is cancelled by the compensation coils. In this case we can still define a field sensitivity, that is degraded with respect to the magnetometer configuration, due to the extra inductance of the compensation coils[18].

Instead of the axial derivative, the tangential derivatives $\partial B_z/\partial x$ and $\partial B_z/\partial y$ can be utilized as well. These values can be measured with a configuration of coils laying on a surface. The main advantage of such a gradiometer is the possibility to fabricate them with thin-film techniques and eventually integrated with the SQUID sensor itself. All the coils act as pick-up and compensating coils at the same time, with a baselenght usually small compared with the distance of the coils from the source. The latter property does not allow to define a field sensitivity for this kind of gradiometer.

The planar gradiometer configuration is affected by a reduction of the net field signal sensed by the device, with respect to axial gradiometer configurations. This signal is equal $(\partial B/\partial x)\cdot l$, where l is the baseline. A larger value of the baseline l in the planar gradiometer would increase

the signal but this choice is obviously not compatible with a fixed area of detector arrays.
A useful value of the localization accuracy of the array, that decrease monotonically as the signal to noise ratio of a single sensor increases, must be of the oder of a few millimeters. The planar gradiometers need SQUID sensors with a three or four times better energy sensitivity with respect to SQUIDs coupled to axial gradiometers in order to not degrade the localization accuracy.

3. SQUID sensor, electronic read-out and measuring environment

Over the last few years the microfabricated low-noise dc-SQUIDs have demonstrated to accomplish the reliability and sensitivity requirements for a multichannel system and they have almost completely replaced the rf-SQUID in multichannel biomagnetic instrumentation[16,19]. The reliability of the device has been greatly improved due to the progress in thin film multilevel fabrication technology [20,21,22].
Dc-SQUIDs are nowadays mostly fabricated as a washer type SQUID ring with a planar input coil on top[23]. The dynamics and noise performance of these devices are well understood, typical energy sensitivities ranging from $5 \cdot 10^{-32}$ to 10^{-31} J/Hz are normally achieved in the white noise region [24,25]. A typical problem in practical devices is the presence of parasitic resonances due to the tightly coupled input coil, which give rise to irregularities in the $V - \Phi$ characteristic of the device. These features are usually accompanied by excess noise that degrades the device performance. The problem is more critical in flux locked loop operation [26], where the SQUID operating point is switched between two or more values, making this technique much more sensitive to excess noise and irregularities in the $V - \Phi$ curve. Many solution to eliminate the parasitic effects has been successfully proposed [27,28,29,30].
The double washer dc-SQUID proposed by Ketchen [31] with the SQUID inductance resistively shunted, as suggested by Enpuku [27], has proved to be succesfull in removing most of the parasitic resonances in the $V - \Phi$ curves [29,32]. The design used is a variation of the usual design [23], two washers are connected in parallel, thus reducing by half the inductance of the SQUID, but the input inductance is wound on each hole separately thus enabling a good coupling efficiency to the SQUID itself. With a proper wire configuration the double washer SQUID can act as a first order planar gradiometer with a fairly good intrinsic balance. Such device is particularly attractive when integrated with planar gradiometers pick-up coil designs[31], leading to useful gradient sensitivities with intrinisic gradiometer balance $\simeq 10^{-4}$ against uniform field changes [33].

The intrinsic gradiometer configuration of the double washer design makes it particularly attractive to reduce the problem of the stray inductive coupling between devices. This allow the integration of many active devices on the same chip, a very important step toward the reduction of complexity of multichannel systems. However, if the integration density is not very high, the stray coupling is not a severe limitation and the problem can be overcomed using different solutions [7].
Many efforts have also been devoted to overcome the complexity of the room temperature electronics necessary to increase the poor dynamic range of the device [15,34,35], in order to improve the overall reliability of the multichannel system. The approach used by Drung et al. [15,16] has proved to be the most successful one, the feedback electronics does not require any flux modulation technique or impedance matching between the SQUID and the room temperature amplifier. The Direct Offset Integration Technique circuit, DOIT[16], includes a novel "Additional Positive Feedback " which steepens the $V-\Phi$ characteristics at one working point. This very ingenious idea does not modify the signal to noise ratio of the device itself, but the increased gain of the SQUID will reduce the noise degradation of the room temperature preamplifier.
A further advantage of the DOIT scheme is the complete elimination of the crosstalk at the modulation frequencies of the single devices. In multichannel systems using the conventional flux locked loop, where the modulation frequency range from 100 KHz to 500 KHz[36], this annoying problem can be handled using different modulation frequencies. A difference, say, $\simeq$ 10 KHz, between contiguous channels avoids that mixing frequencies fall in the measurement frequency range.
A second source of crosstalk is generated by the the shielding current in the superconducting flux transformers, leading to a distortion of the original magnetic field. If the inductances of the coils and their mutual inductance are known, the correction can be calculated [19]. An elegant alternative solution has been presented by ter Brake et al. [37]. The external feedback is applied in the input transformer circuit. In this way the current in the coil is used as the error signal in the feedback loop, thus eliminating crosstalk between the detection coils.
The neuromagnetic measurements require good resolution at frequencies down to 0.1 Hz or less, and the level of 1/f noise becomes a very important factor of concern. There are two separate sources of 1/f noise[38]. The first arises from 1/f fluctuation in the critical current of the Josephson junction, and the mechanism for this process is the trapping and detrapping of electrons in the defect of the barrier. Using some approximation[39] it is possible to write the 1/f component of the flux noise spectral density:

$$S_\Phi(1/f) \simeq \frac{1}{2}\frac{\eta}{A}(LI_c)^2 , \qquad (1)$$

where η is function of the barrier type and quality [40], A and I_c is the area and the critical current of the junctions, L is the inductance of the SQUID. The equation 1 has been verified experimentally[29] for SQUID fabricated at IBM [22], the value of $\eta \simeq 8{\cdot}10^{-12}$ μm^2 leads to an extremely low value of the 1/f noise in those devices [29], making them particularly attractive for biomagnetic applications [9]. This source of noise can be eventually reduced using particular electronic systems[41,39,42] but the level of the noise obtainable with the available technologies satisfy the requirements of biomagnetism, making this smart modulation schemes not very useful. A second source of 1/f noise may be related to the motion of flux lines trapped in the SQUID washer, the mechanism leads to an apparent flux, like if an external flux noise was applied to the SQUID. This noise is less well understood, can be ascribed to the structural properties and contaminations of the Nb film, it cannot be cancelled by modulation techniques and when it is present [38,36,32], it affects the low frequency performance of the device very seriously.

One major factor of concern as far as the sensitivity of high resolution biomagnetic measurement is concerned, is the thermal noise arising from the radiation shields of the dewar. The Johnson noise in normal metals, in the form of local current fluctuation, generates magnetic field fluctuation. The noise due to conducting plates has been considered both theoretically[43] and experimentally[44]. The result clearly shows that the noise arising from the currently used radiation shield layers is of the order of a few $fT/\sqrt{Hz}$ [43] and this is confirmed by many experimental results[6,45]. In the planar gradiometer configuration the power noise increases by a factor 2, with respect to axial gradiometers. This is mainly due to the close position of all the coils to the dewar bottom which has the more significant effect on the thermal noise than the shields positioned at the side of the dewar. A reduction in the noise level may be possible with a proper shape of the radiation shields[6], it is clear however that the future challenge to furtherly increase the resolution of biomagnetic systems will have to deal with the dewar noise rather than with the intrinsic sensitivity of the dc-SQUID sensors.

Most of the existing multichannel systems operate in shielded rooms. The magnetometer array is seldomly used [6] and only in heavily shielded rooms. The most used array configuration inside moderately shielded rooms, i.e. with a shielding factor $> 10^4$ above 10 Hz, is the first order gradiometer array. The second order gradiometer array can afford the unshielded environmental noise, with the help of comb filtering and noise cancellation techniques[46], without degrading consistently the system sensivity. The balance of gradiometers in a multichannel instrumentation cannot be simply improved by placing small superconducting tabs at appropriate locations. The presence of these tabs will affect the balance of adjacent gradiometers and the overall balance will need a tedious iterative balancing procedure. This procedure has been successfully used by Nowak et al.[47] for a 5 channel system and a similar procedure will be used for an 11 channel, dedicated to cardiac studies outside a shielded room[45]. The technique cannot be obviously extrapolated to multichannel systems having more than 20 channels. A further improvement in the noise cancellation techniques or active shielding systems [48] is necessary in order support the use of large multichannel systems in very unfavorable clinical environment.

4. Overview of multichannel systems

This section will describe the new generation of magnetometers, having more than 19 channels, that are currently in use in many laboratories, and those systems that will be ready in the near future. A detailed review of the first generation of multichannel systems, with a number of channels ≤ 10, has been given by Ilmoniemi et al. [19].

4.1. The 24 channel system in the Helsinki University of Technology

The 24-channel gradiometer in Helsinki is the first successful application of planar gradiometers[12] specifically designed to perform neuromagnetic measurements. The pick-up coils consist of 12 sensor units arranged on a spherical surface. Each unit comprises a machinable-glass former, two wire wound orthogonal figure-eight shaped planar coils, and two dc-SQUIDs made by IBM[49]. The sensor units, shown in fig. 1, are placed on the spherical end cap of the dewar, manifactured by CTF Inc.[50]. The

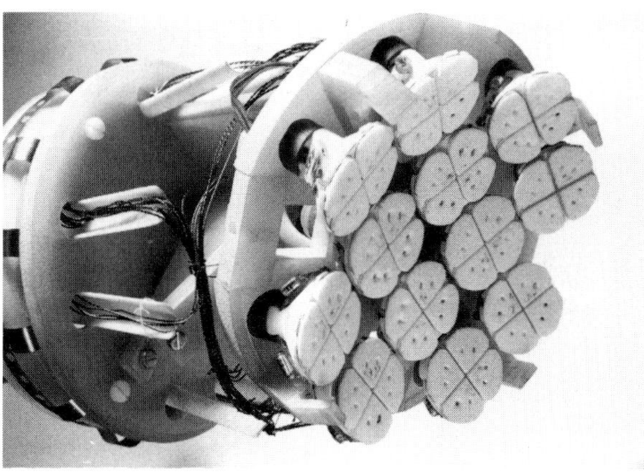

Fig.1. Arrangement of the 24-planar gradiometer coils of the Helsinki multichannel system.

radius of curvature of 125 mm and the coils cover an area with a diameter of 125 mm. The pick-up coils are sensitive to both the tangential derivatives $\partial B_z/\partial x$ and $\partial B_z/\partial y$ of the radial field component B_z. This solution permits to overcome the problem of the reduced sensitivity in planar gradiometers due to the reduced baseline with respect to axial gradiometer. The area of each gradiometer loop is 3.7 cm^2 and the effective baseline is 1.3 cm The gradient noise level is about $3 - 5fT/(cm \cdot \sqrt{Hz})$. The dc-SQUIDs use conventional read-out electronics[5], digital outputs ports allow the control of the SQUID electronics, low pass filters and stimulus generation. A real time measurement software has been developed, for evoked-response and for spontaneous activity.

A 122 channel system is under construction in the same laboratory. This system covers the whole head, the detectors are arranged in 61 planar sensors resembling those used in the 24 channel. The sensor structure is modified to incorporate thin-film pick-up coils fabricated on a silicon substrate, IBM will supply the SQUIDs.

4.2. The 37 channel system developed by BTI

The most recent multichannel system presented by BTI[4] is a 37 channel system called " MAGNES". The detectors are first order gradiometers arranged on a spherical surface of 14 cm diamter. There are also additional 8 channels used for noise compensation. The field noise is $\simeq 7\ fT/\sqrt{Hz}$. The system with its own shielded room has been already

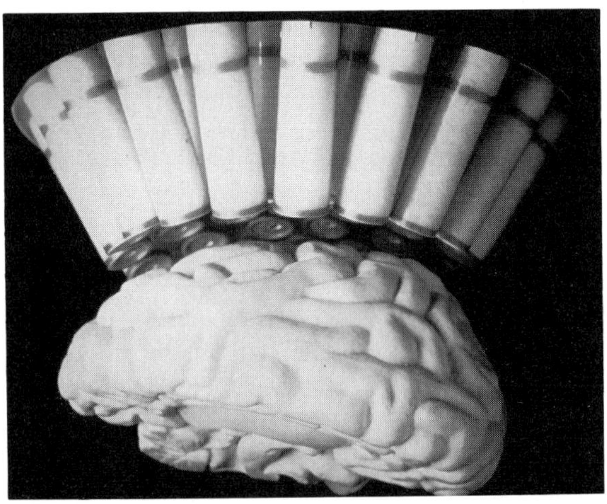

Fig. 2. The array of thde bare 37 channel detection coils of the BTI system positioned over model of brain. The actual coil-to-brain surface distance is $\simeq$ 3 cm.

installed in Hospital environments. A computer controlled head position indicator permits to combine the magnetic source localization with the magnetic resonance imaging.

4.3. The 37 channel system developed by SIEMENS

SIEMENS[7,51] has developed a multichannel system for brain and heart measurements. The sensor array is made by axial first order gradiometers ingeniously manifactured as a flexible superconducting printed circuit. In order to form gradiometer loops together with stripline interconnects and feedback transformers a two dimensional gradiometer foil is folded to a three dimensional structure and then bonded onto a carrier made from glassfiber enforced epoxy. The $KRENIKON^R$ system consists of 37 axial gradiometers arranged to a circular flat array of 19 cm diameter, additionally 3 orthogonal magnetometers are provided. The single turn pick-up loops have an area of 6 cm^2, with a baseline of 70 mm. The SQUIDs with integrated thin film input coils are arranged tenfold, the magnetic decoupling is performed by using superconducting ring structures. The sensor array is placed in a dewar with flat bottom, at about 20 mm from the body surface, the typical field noise level is less than 10 $fT/\sqrt{Hz}$. The complete system is housed in a moderately shielded room. Due to the planar arrangement of the arrays, the system is ideal for cardiac studies. However the excellent results obtained with brain studies have strengthened the confidence in the neuromagnetic capabilities of this system[52]. The system is equipped with several tools designed to combine the localization result with diagnostic images, like magnetic resonance or computerized tomography.

4.4. The 37 channel magnetometer system at the PTB-Berlin.

A 37 channel SQUID magnetometer system is operative inside an heavily shielded room at the PTB laboratories in Berlin, it is the only large area magnetometer system operating in the world. The sensors are arranged on concentric circles on a total circular surface of 21 cm diameter. As already described in the third section, the SQUID loop serves as the active sensing element, with a sensitivity of 0.4 nT/Φ_o. The magnetometer chip, shown in fig. 3, covers an area of 81 mm^2, 8 loops in parallel cover the sensing coil area. The sensors are located 3 cm above the outer surface of a flat dewar bottom. The " DOIT" scheme is used as read-out electronic system[16]. The system has a sensitivity of $\simeq 5$ $fT/\sqrt{Hz}$ in the white noise region. A novel noise compensation technique permits to reduce the field noise to 10 $fT/\sqrt{Hz}$ at 1Hz[53]. These excellent results are obtained with a careful optimization of the complete system, with particular care

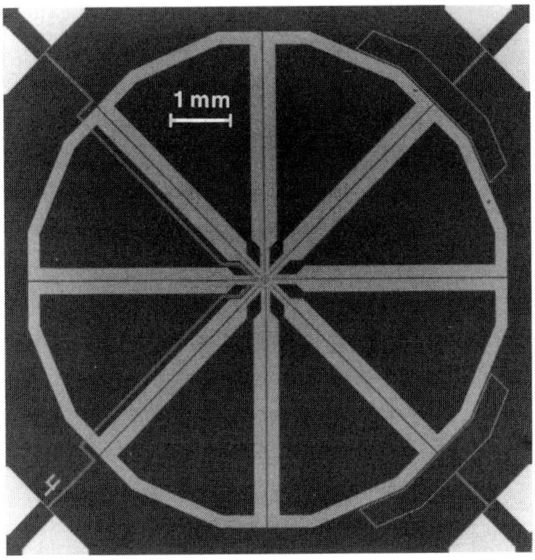

Fig. 3. Micrograph of the integrated SQUID magnetometer made at PTB, Berlin.

on the home-made dewar[6]. One of the most important features of this system is the modular design. Every sensor is autonomous and can be easily replaced. The warm-up of single sensors in case of trapped flux or the replacement of a failing single unit can be easily achieved. The system has given a relevant contribution to the magnetocardiography.

4.5. The 19 channel system developed by PHILIPS

Philips has developed a flexible modular system consisting of 38 channels in two dewars, with possible reduction to 19 channels or a variable number $\leq$ 19, in one dewar. Every single channel consists of a bottom part including the first order axial gradiometer and the SQUID housing and a top part with a sealing plug to achieve good thermal insulation. Each array covers a circular area of 12 cm diameter. In order to measure the radial field components with straight removable measurement rods, a tilted gradiometer configuration has been adopted[55]. The SQUIDs are based on the materials NbN-MgO-NbN and made using thin film technology. They are very robust and withstand thermal cycling without any changes[54]. The noise level is below 15 $fT/\sqrt{Hz}$ above 0.3 Hz [55].

4.6. The 28 channel system developed by DORNIER

DORNIER [57] has developed a 28-channel system for MEG for Ulm University. The system is in operation since 1990. There are 28 rf-SQUIDs connected to pick-up coil array, the configuration of this array can be chosen at room temperature, depending on the measuring problem. The sensors are contained in two dewars for better adaption of the shape of the patient's head. Their tips are spherical and fit together to form part of a sphere of 50 mm diameter. The sensitivity is $\simeq 20\ fT\sqrt{Hz}$ in the magnetometer configuration.

4.7. The 28 channel system at IESS-CNR in Roma

A 28 channel system will be operative in the next future at the Istituto di Elettronica dello Stato Solido in Roma[9], supported by " Progetto Finalizzato Tecnologie Superconduttive e Criogeniche". The sensor array is a novel "hybrid" design consisting of 16 first order axial gradiometers and 12 first order planar gradiometers. This configuration permits to optimize the available cycindrical volume of the dewar tail, manufactured by CTF[50]. The detecting array covers a spherical surface of 16 cm diameter, and permits simultaneous magnetic measuremnts over a circular scalp region of 200 cm^2. The detection coils are arranged on concentric circumferences: the 16 inner coils are axial gradiometers, with 1.8 coil diameter and 8.5 cm baseline, the external row is formed by 12 planar gradimeters, with 1cm coil diameter and and 1.5 cm baseline. The SQUID used are made by IBM with a double washer configuration[31]. The devices have an extremely low 1/f noise. The field sensitivity mesured with 1 channel is $\simeq 3\ fT/\sqrt{Hz}$ measured above 10Hz[56].

4.8. The 19 channel system of Twente University

A 19 channel neuromagnetometer will be operative in the next future at the Twente University[10]. The sensors are first order gradiometrs distributed in an hexagonal configuration over a spherical surface with radius of curvature of 125 mm. The external feedback to the flux transformer circuit is applied, in order to eliminate the crosstalk between channels, as already mentioned in the previous section[37]. The predicted noise is $5\ fT\sqrt{Hz}$.

5. Conclusions

In the state of art biomagnetic systems the recent progress in the fabrication of reliable and low noise dc-SQUID has played an important role. New SQUID systems developed very recently have improved the performance of the device and simplified the read-out electronics, reducing the complexity of multichannel systems. These new ideas will determine a large impact of Biomagnetic instrumentation outside research laboratories in the near future. The author wish to thank V. Pizzella and G.L. Romani for the help given in the preparation of the manuscript.

References

[1] R.J. Ilmoniemi and J. Knuutila in : Proc. 10th International Cryogenic Engineering Conf. ICEC 10, edited by H. Collan, P. Berglund and M. Krusius (Butterwoths, London) p.457 (1984).

[2] S.J. Williamson, M.Pellizzone, Y.Okada, L.Kaufman, D.B. Crum and J.R. Marsden in: Biomagnetism: Application and Theory, eds. H. Weiberg, G. Stroink and T. Katila (Pergamon Press New York) p.46 (1985).

[3] G.L. Romani, R. Leoni and C. Salustri in : SQUID 85, Superconducting Quantum Interference Devices and their applications, edited by H.D. Hahlbohm and H. Lubbig (De Gruyter, Berlin) p.919 (1985).

[4] Biomagnetic Technologies Inc., 9727 Pacific Heights Boulevard, San Diego (CA) 92121, USA.

[5] J. Knuutila, S. Ahlfors, A.I. Ahonen, J. Hallstrom, M.J. Kajola, O.V. Lounasmaa, V. Vilkman and C.D. Tesche, Rev. Scient. Instr. 58, 2145(1987).

[6] H.Koch, R. Cantor, D. Drung, S.N. Ernè, K.P. Matthies, M.Peters, T. Ryhanen, H.J. Scheer and H.D. Halbohm, IEEE Trans on Magn. MAG-27, 2793(1991).

[7] H.E. Hoenig, G.M. Daalmans, L.Bar, F. Bommel, A. Paulus, D. Uhl, H.J. Weisse, S. Schneider, H.Seifert, H.Reichenberger, K. Abraham-Fuchs, IEEE Trans. on Magn. MAG-27, 2777(1991).

[8] M. Kajola, S. Ahlfors, G.J. Ehnholm, J Hallstrom, M.S. Hamalainen, R.J. Ilmoniemi, M. Kiviranta, J. Knuutila, O.V. Lounasmaa, C.D. Tesche and V. Vilkman in : Advances in Biomagnetism, edited by S.J. Williamson, M. Hoke, G. Stroink and M. Kotani (Plenum Press, New York) p.693 (1990).

[9] V. Foglietti, C. Del Gratta, A. Pasquarelli, V. Pizzella, G.Torrioli, G.L. Romani, W.J. Gallagher, M.B. Ketchen, A.W. Kleinsasser, R.L. Sandstrom, IEEE Trans. on Magn. MAG-27,2959(1991).

[10] H.J.M. ter Brake, J. Flokstra, W. Jaszczuk, R. Stammis, G.K. van Ancum, A. Martinez and H. Rogalla, Clin. Phys. and Physiological Meas., in press.

[11] M. Kajola, A. Ahonen, M.S. Hamalainen, J. Knuutila, O.V. Lounasmaa, J.Simola and V. Vilkman , Clin. Phys. and Physiological Meas., in press.

[12] A.I. Ahonen, M.S. Hamalainen, M.J. Kajola, J.E.T. Knuutila, O.V. Lounasmaa, J.T. Simola, C.D. Tesche and V.A. Vilkman, IEEE Trans. on Magn. MAG-27, 2786(1991).

[13] G.L. Romani in: Advances in Biomagnetism, edited by S.J. Williamson, M. Hoke, G. Stroink and M. Kotani (Plenum Press, New York) p.33 (1990).

[14] P.K. Kemppainen and R.J. Ilmoniemi in : Advances in Biomagnetism, edited by S.J. Williamson, M. Hoke, G. Stroink and M. Kotani (Plenum Press, New York) p.693 (1990).

[15] D.Drung, R. Cantor, M. Peters, H.J. Scheer and H.Koch Appl. Phys. Lett. 57,406(1990).

[16] D. Drung, to appear in proceedings of ISEC 91 (Glasgow, June 1991).

[17] J.E. Zimmerman J.Appl.Phys. 42,4483(1971).

[18] G.L. Romani, S.J. Williamson, L. Kaufman, Rev.Sci.Instr. 53, 1815 (1982).

[19] R.Ilmoniemi, J.Knuutila, T.Ryhanen, H.Seppa, Prog. Low Temp. Phys. 12, 271(1989).

[20] R. Cantor, D. Drung, M. Peters and H. Koch, J. Appl. Phys. 67, 3038 (1990).

[21] M.B. Ketchen, M. Bhushan, S. B. Kaplan and W.J. Gallagher, IEEE Trans on Magn MAG-27 ,3005 (1991).

[22] R.L. Sandstrom, A.W. Kleinsasser, W.J. Gallagher and S.I. Raider, IEEE Trans.Magn. MAG-23, 1984 (1987).

[23] J.M. Jaycox and M.B. Ketchen, IEEE Trans. on Magn. MAG-17, 400 (1981).

[24] for a review : T. Ryhanen, H. Seppa, R. Ilmoniemi, J. Knuutila, J. Low Temp. Phys. 76, 287 (1989)

[25] J. Clarke in *Superconducting Electronics*, edited by H. Weinstock and M. Nisenoff, Springer New York 1989, p.87.

[26] J. Clarke, W. M. Goubau and M.B. Ketchen, J. Low Temp. Phys. 25, 99(1976).

[27] K.Enpuku, K. Yoshida and S. Kohjiro, J. Appl. Phys. 60, 4218 (1986).

[28] J. Knuutila, M. Kajola, H. Seppa, R. Mutikainen and J. Salmi, J. Low Temp. Phys. 71, 369(1988).

[29] V. Foglietti, W.J. Gallagher, M. B. Ketchen, A.W. Kleinsasser, R.H. Koch and R.L. Sandstrom, Appl. Phys. Lett. 55, 1451(1989).

[30] R. Cantor, T. Ryhanen, D. Drung and H. Koch, IEEE Trans. on Magn. MAG-27, 2927(1991).

[31] M.B. Ketchen, J. Appl. Phys. 58, 4322(1985).

[32] V. Foglietti, M.E. Giannini and G. Petrocco, IEEE Trans. on Magn. MAG-27, 2989 (1991).

[33] R.H. Koch, unpublished.

[34] H. Matz, D. Drung, E. Crocoll, R. Herwig, G. Kramer, M. Neuhaus and W. Jutzi, IEEE Trans. on Magn. MAG-27, 2979 (1991).

[35] M. Muck, IEEE Trans. on Magn. MAG-27, 2986(1991).

[36] F. Wellstood, C. Heiden and J. Clarke, Rev. Sci. Instr. 55, 952(1984).

[37] H.J.M. ter Brake, F.H. Fleuren, J.A. Ulfman and J. Flokstra, Cryogenics 26, 667 (1986).

[38] R.H. Koch, J.Clarke, W.M. Goubau, J.M. Martinis, C.M. Pegrum and D.J. Van Harlingen, J. Low Temp. Phys. 51, 207 (1983).

[39] V.Foglietti, W.J. Gallagher, M.B. Ketchen, A.W. Kleinsasser, R.H. Koch, S.I. Raider and R.L. Sandstrom, Appl, Phys. Lett. 49, 1393 (1986).

[40] C.T. Rogers and R.A. Buhrman, IEEE Trans. on Magn. MAG-21, 126(1985).

[41] M.B. Simmonds and R.P. Giffard, U.S. Patent No. 4 389 612(1983).

[42] O.Dossel, B.David, M. Fuchs, W.H. Kuhlman, K.M. Ludeke, IEEE Trans. on Magn. MAG-27, 2797(1991).

[43] J. Nenonen, T. Katila and J. Montonen, in: Advances in Biomagnetism, edited by S.J. Williamson, M. Hoke, G. Stroink and M. Kotani (Plenum Press, New York) p.729 (1990).

[44] J. Nenonen and T. Katila in : Proc. of the 6th Intern. Conf. on Biomagn. (Denki Univ. Press, Tokyo) p.426(1988).

[45] A.Pasquarelli, S. Casciardi, M.E. Giannini, V. Foglietti, V. Pizzella, G. Torrioli and G.L. Romani, these proceedings.

[46] J. Vrba, J. Mc Cubbin, S. Lee, A.A. Fife and M.B. Burbank in: Advances in Biomagnetism, edited by S.J. Williamson, M. Hoke, G. Stroink and M. Kotani (Plenum Press, New York) p.733 (1990).

[47] H. Nowak and F. Giessler, Clin. Phys. and Physiological Meas., to be published.

[48] H. Seifert, A. Wirth and S. Schneider, these proceedings.

[49] C.D. Tesche, K.H. Brown, A.C. Callegari, M.M. Chen, J.H. Greiner, H.C. Jones, M.B. Ketchen, K.K. Kim, A.W. Kleinsasser, H.A. Notarys, G.Proto, R.H. Wang and T.Yogi, IEEE Trans. on Magn. MAG-21, 1032(1985).

[50] CTF Systems Inc. 15-1750 McLean Avenue, Port Coquitlam, British Columbia, Canada V3C 1M9.

[51] S. Schneider, E. Hoenig and D. Erlangen, these proceedings.

[52] S. Schneider and K. Abraham-Fuchs Clin. Phys. and Physiological Meas., to be published.

[53] S.N. Ernè, patent pending.

[54] B. David, O. Dossel, G. Kursten, G. Rabe, to appear in proceedings of ISEC 91 (Glasgow, June 1991).

[55] O.Dossel, B. David, M. Fuchs, J. Kruger, W.H. Kullman, K.M. Ludeke, these proceedings.

[56] G. Torrioli, S. Casciardi, C. Del Gratta, V. Foglietti, W.J. Gallagher, M.B. Ketchen, A.W. Kleinsasser, A. Pasquarelli, V. Pizzella, G.L. Romani, R.L. Sandstrom, these proceedings.

[57] Dornier GmbH FOEP, D-7990 Friedrichshafen 1, Germany.

Design and Operation of a Biomagnetic Multichannel System

S. Schneider[1], *H. Seifert*[1], *H.E. Hoenig*[2], *and H. Reichenberger*[1]

[1]Siemens Medical Engineering Group, GG 31, Henkestr. 127, W-8520 Erlangen, Fed. Rep. of Germany
[2]Siemens Corporate Research, ZFE ME TPH 1, Paul-Gossen Str. 100, W-8520 Erlangen, Fed. Rep. of Germany

Abstract. This paper describes the design and optimization of the KRENIKON® system for the measurement of magnetic fields of the brain and the heart. Aspects for the configuration of the sensor array are discussed. Special emphasis is given to the suppression of interferences by a combination of magnetic shielding with first-order gradiometers and to the construction of the gradiometer array in foil technology.

1. Introduction

During the last decade multichannel systems have been considered as an essential tool to investigate the clinical relevance of biomagnetism. An overview of the present state of multichannel instrumentation is given by Foglietti [1].

Since 1980 the technology for producing SQUIDs has been developed at Siemens Corporate Research. This was the prerequisite for the development of a multichannel system for recording biomagnetic fields. Design of the multichannel sensor is based on chips integrating ten DC-SQUIDs [2]. The first inset constructed with this ten-channel SQUID-chip and first-order gradiometers was tested during 1984/85 in the magnetically shielded room of the PTB at Berlin.

At the Siemens Medical Engineering Group clinical studies with a ten-channel system started in August 1987. An engineering prototype with four SQUID-modules went into operation in February 1988, first with 31 channels and later upgraded to 37 channels [3]. This system was the first operating multichannel system worldwide. During 1988 and 1989 experience was gained both with technical tests and extended clinical measurements in cooperation with our clinical partners [4].

Two further systems modified and improved on the basis of this experience have been put into operation, one at the Biomagnetic Center of the Erlangen University Clinic in July 1990 and the other one in our laboratories.

2. System Conception and Design

Using magnetic fields to localize electrical activity in the human body requires the acquisition of complete field distributions. The number of channels necessary

Springer Proceedings in Physics, Vol. 64 Superconducting Devices and Their Applications
Editors: H. Koch, H. Lübbig

follows from the area to be investigated and the spacing of the pickup coils, which is determined by the sampling theorem. The localization accuracy of current dipoles is then determined only by the signal to noise ratio.

The shortest distance between current source and gradiometer determines the minimum separation between the two poles of the magnetic field distribution which again establishes the sensor spacing. This gap is the sum of the distance between the source inside the body and the surface of the body, the thickness of electrodes (taking EEG or ECG in parallel) plus a few millimeters spacing to avoid patient contact with the cryostat and the distance between the outside of the cryostat and the pickup coils inside. This shortest distance is between three and four cm. For a dipolar source this corresponds to a minimal pickup coil spacing of 2 - 2.4 cm [5]. This enables localizing superficial dipolar sources in the cortex.

The KRENIKON® was designed as a multipurpose device for clinical research of the brain and the heart. Simulations were performed for a current dipole located in a sphere for the former and an infinite half space for the latter. The localization accuracy was calculated for a flat and a curved array. The localization error is 20 % less for a curved array exactly matched to the curvature of a sphere than for a flat array positioned on this sphere. But for a flat object the error goes up to 200 % when a curved array instead of a flat array is used [6]. Therefore, a flat array is a good compromise for a multipurpose device.

With 19 cm the diameter of the sensor array is large enough to cover one hemisphere of the head or the area of the heart. A cryostat with an inner diameter of 20 cm is technically still feasible while keeping the bottom sufficiently thin and liquid He-consumption low.

The sensor array plane with 19 cm diameter is covered by 37 pickup coils with a spacing of 2.7 cm, slightly larger than required by the sampling theorem for superficial sources in order to increase signal to noise for deeper sources.

Although with a SQUID-system the theoretically achievable noise limit is given by the uncertainty principle, in the real world magnetic interference is the main problem since this interference can be several orders of magnitude higher than the signal of interest from the body. In principle there are two ways to solve this problem: passive or active shielding, or compensation by gradiometers. For passive shielding of magnetic fields, four multilayer cabins were in use worldwide at the time when the decision for the concept of the KRENIKON® had to be made. These multilayer cabins provide high shielding but are too expensive to be used in a clinical every day system.

Many research groups perform biomagnetic measurements with axial gradiometers of higher order without magnetic shielding. Yet, especially with respect to multichannel systems there arise several problems:

- The signals of interest are reduced significantly by subtracting those of compensating coils. This is demonstrated in Fig. 1, which shows the input current of a SQUID connected via a flux transformer to different sensor configurations. A current dipole 40 mm horizontally offset from the center of a sphere at different depth is assumed to be the signal source. The flux transformer is optimized in each case in order to maximize the flux in the SQUID. For a magnetometer two turns are optimal, for a first-order

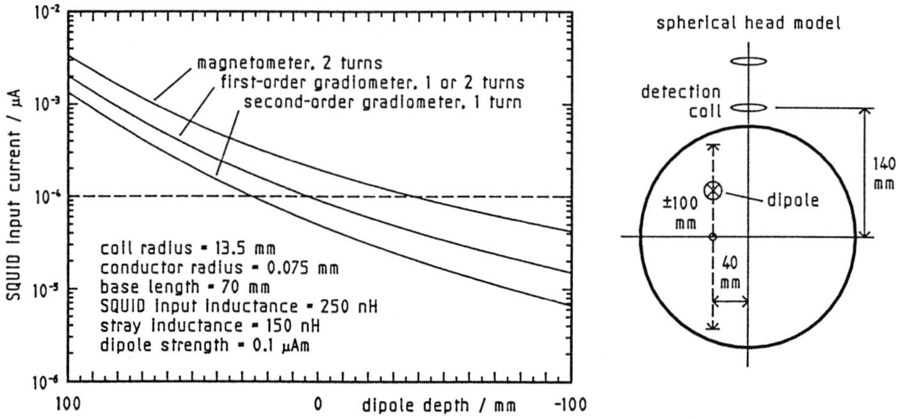

Fig. 1: SQUID input current calculated for different types of flux transformers connected to the KRENIKON® SQUID. The horizontal line gives a typical value of the equivalent RMS SQUID input noise current (e.g. 5 pA/√Hz SQUID input noise current and 400 Hz bandwidth).

gradiometer both one or two turns are feasible, for a second-order gradiometer one turn is optimal while 50 turns are used at the SQUID side. From these calculations follows that a magnetometer gives the highest SQUID input current, and a gradiometer of second-order the lowest. The horizontal line in Fig. 1 gives a typical value of the equivalent RMS SQUID input noise current. Using a second-order gradiometer, to achieve the same signal to SQUID-noise ratio as with a gradiometer of first-order the source must be about 2 cm closer to the surface.

- Gradiometers of higher order have to be constructed by connecting a number of spatially separated coils in series. Large spacing between these coils is required in order not to degrade the penetration depth too much.
- Higher-order gradiometers have to be balanced carefully, whereby balancing has to be repeated for each position and direction in space. For large scale systems up to now there is no practical solution to this problem, since balancing cannot be performed in an easy mechanical way due to mutual interaction between channels.
- A feature of magnetometers is that they are more sensitive than gradiometers, not only for signals from the region of interest but also for signals from other sources within the body or from the environment.

As a consequence of these various aspects, the KRENIKON® was designed with first-order gradiometers in conjunction with a three-layer shielded room.

3. Shielded Room

The shielded room consists of two layers of Mumetal® (Nickel-Iron alloy having a high permeability; Vacuumschmelze, Hanau) which absorbs the environmental field until the material is saturated. Between these two layers a highly conductive layer made from aluminum is placed to give eddy current shielding. Eddy current shielding sets on already at frequencies above 0.1 Hz. The Mumetal® of the inner layer of the shielded room screens thermal magnetic noise originating from the eddy current shield.

The shielded room of our engineering prototype system was the first of this type. In the meantime it has become a standard for several systems. Since 1985 35 shielded rooms of this type have been built by Vacuumschmelze [7]. Most of them are used for biomagnetic systems.

For the KRENIKON® only welded aluminum shields are used. This avoids junctions of different materials (e.g. aluminum/copper), that with temperature variations could give rise to thermo-currents in the layer causing changes of the DC magnetic field inside the shielded room up to the order of nT/K.

Another source of interference are signals originating from vibrations of the detector coils or the layers of the shielded room, since inside there is a residual magnetic field that is constant in time but inhomogeneous in space. At frequencies much higher than the resonance frequency mechanical vibrations can be damped with shock absorbers. For biomagnetic measurements, however, frequencies down to DC are of interest. Vibration amplitudes are kept low

- by mounting the magnetically shielded room on a specially designed concrete foundation that is separated from the surroundings and
- by using a rigid construction for shielded room and cryostat support anchored directly to the foundation.

For the improved shielded room of the KRENIKON® the space between the Mumetal® panels and the aluminum wall is filled with rigid expanded plastic against which the panels are pressed. Thus spectral vibration could be reduced from $s = 80\ nm/\sqrt{Hz}$ at 5 Hz for our first prototype to $25\ nm/\sqrt{Hz}$ now.

4. Gradiometer Array

The array of 37 axial first-order gradiometers is manufactured in a photolithographic process. This allows the design of arbitrary shapes of the coils with minimum space in between so that the flux sensitive area is maximized. In this way the advantage of axial gradiometers - their high sensitivity - is combined with the advantage of easy simultaneous and reproducible manufacturing of the whole gradiometer array.

The design is based on a special process, which allows deposition of a superconducting circuit on both sides of a flexible polyimide foil with superconducting feed-throughs between the layers. The superconductor is a lead-tin alloy, so that simple solder joints are possible. After fabrication, the two-dimensional polyimide foil is folded and glued to a rigid glassfiber epoxy carrier so

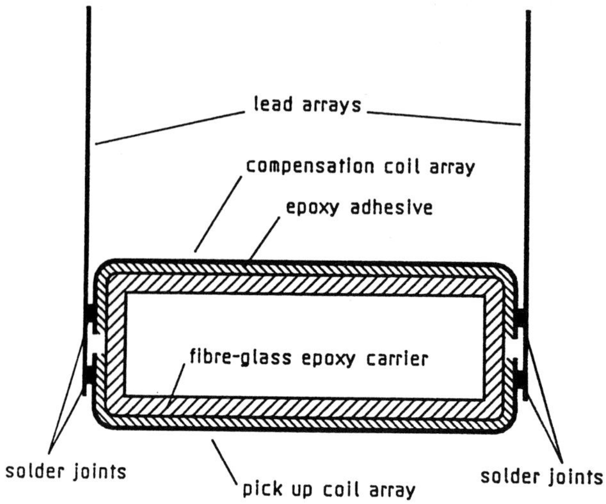

Fig. 2: Principle of the mechanical setup of the gradiometer array.

that a set of axial gradiometers results. By this procedure the gradiometer array is fabricated as a block with no need to handle single coils.

Figure 2 shows a scheme of the gradiometer arrangement. Pickup coils and compensation coils are connected by four foils with superconducting striplines, that also establish the connections for the flux transport to the SQUID chips.

The inductance and magnetic field sensitivity of flux transformer leads is usually reduced by twisting or shielding with superconducting tubes. These techniques cannot be used with our foil concept. Here the structure is open and the reduction of lead inductance and magnetic field sensitivity of the conducting strips has to be established by compensation methods. The simplest form of compensation is achieved by a dual layer stripline where flow and counterflow of current have minimum spacing.

However, as the spacing between layers is limited by the fabrication process the magnetic flux still threading through the stripline has to be compensated by additional coils or other striplines. Parallel striplines allow a local compensation even if the magnetic field along the line is varying.

In the case described here two similar methods for compensation are used:

1. For the flux transformer two parallel striplines are used which are connected in series with opposite polarity.
2. Within the two symmetrical coil arrays the stripline is formed by parallel strips on the same side of the foil which are covered by a broad pad on the other side acting as a superconducting groundplane. Then the current distribution is very similar to that of a double stripline because a current in the strips causes a mirror current in the groundplane of approximately equal magnitude.

Figure 3 shows the complete circuit diagram of one channel of the gradiometer unit. An additional coil in the flux transformer near the SQUID output

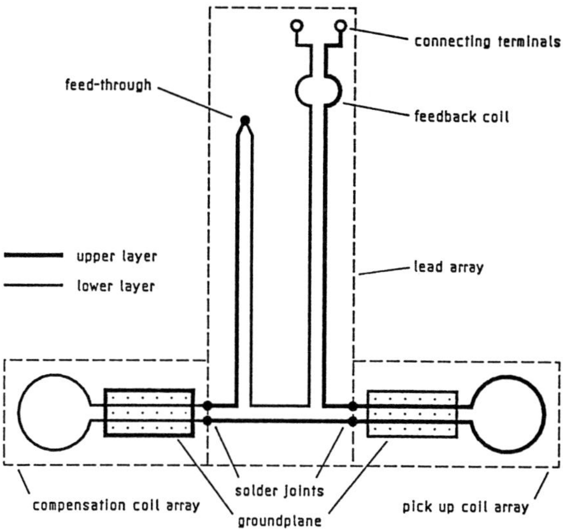

Fig. 3: Equivalent circuit diagram of one gradiometer channel.

serves as secondary winding of a transformer in the feedback loop of the DC-SQUID. The primary winding is placed in the SQUID module. In this way, the feedback flux is applied directly to the flux transformer, so that the current of the sensor coils and the flux transformer is kept constant, which reduces crosstalk between neighboring gradiometers [8].

Figure 4 shows the block of the gradiometer array with 37 coils in a honeycomb arrangement. Three magnetometer coils near the connection terminals are used for recording the external field.

The signals of the SQUID modules are fed to preamplifiers at room temperature with resistive stripline leads of low thermal conductivity on flexible foils (Fig. 5).

Due to foil technology He-consumption of the probe head is kept below 1.5 l/day; total He-consumption of probe head and cryostat is less than 6 l/day. For the first prototype a cryostat with a bottom thickness of 23.5 mm and a helium capacity of 25 l made by HOXAN Corp. was used. Then we switched to cryostats of our own design with a helium capacity of 30 l and a thickness of the bottom of 15.5 mm and 13.5 mm, for two types of dewars respectively.

Fig. 4

Fig. 5

Fig. 4: Flat array of 37 hexagonal shaped first-order gradiometers and 3 magnetometers made in foil technology. The centers of neighboring coils have a distance of 27 mm. The area of each coil is 6 cm .

Fig.: 5 KRENIKON® probe head containing gradiometer array, SQUID modules and preamplifiers.

5. Signal Evaluation

The DC-SQUIDs are operated in a flux locked loop with a lock-in frequency of 100 kHz. Signals then are processed with notch filters to prevent interference from the line frequency and with high-pass and antialiasing filters to improve the signal to noise ratio. Data are digitized with up to 6.000 Hz and stored. Data evaluation and system operation is performed with a workstation.

Additional digital signal processing allows the separation of background activity, removal of periodic signals e.g. interference from heart activity recorded together with the magnetic fields from the brain [9], and averaging of sporadic events, e.g. from epileptic activity or from cardiac arrhythmias [10].

Evaluation of signals for diagnostic purposes is performed essentially at three levels [4]:

- Calculation and display of the distribution of the biomagnetic flux density at the site of the pickup coils.
- Identification of locations and strength of sources that generate the field distribution. At present with clinical data this is mainly performed by an

iterative procedure using simple source and object models. More sophisticated evaluation schemes are under evaluation [11].

- Fusion of source locations and sequences of locations with 3D-anatomical images derived from MR or CT.

Thus the extremely weak magnetic signals recorded with the pickup coils finally are transformed into functional information about spatial and temporal distribution of electric activity within the body.

6. Prospect

The combination of first-order gradiometers with a three-layer shielded room has proven to be a robust solution which has been adapted by several groups while systems using second-order gradiometers without a shielded room have not prevailed.

Successful operation of the KRENIKON® has been demonstrated for more than three years. Up to now more than 240 patients and 70 volunteers have been investigated. An example is given in Fig. 6, showing the 37 channel recording of α-waves from the occipital region of the brain. Applications up to now extend to WPW syndrome, heart arrhythmias, epilepsy, brain infarction and event related fields [12-18]. Although very promissing results have been achieved clinical

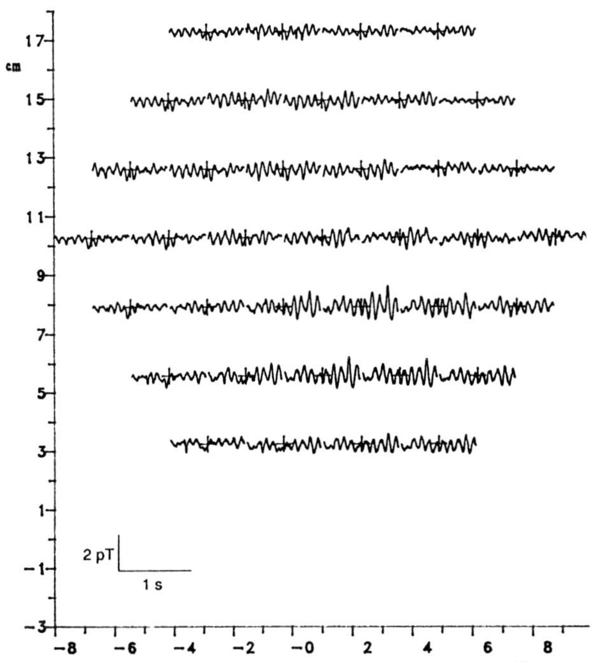

Fig. 6: MEG recording of α-waves from the occipital region of the brain.

relevance of biomagnetism has now to be confirmed by investigating a representative number of patients. It is expected that further applications will be found during this process. The KRENIKON® is ready for systematic clinical evaluation to explore the full potential of biomagnetism.

References

[1] V. Foglietti, this volume

[2] G. Daalmans, H. Seifert, F. Bömmel, H.E. Hoenig, Proc. 2nd Workshop High Temp. Supercond. Electr., Hokkaido, Japan, June 7-9, 219-222 (1989)

[3] H.E. Hoenig, G. Daalmans, W. Folberth, H. Reichenberger, S. Schneider, H. Seifert, Cryogen., **29**, 809-813 (1989)

[4] S. Schneider, H.E. Hoenig, H. Reichenberger, K. Abraham-Fuchs, W. Moshage, A. Oppelt, H. Stefan, A. Weikl, A. Wirth, Radiology, **176**, 825-830 (1990)

[5] R. Graumann, S. Schneider, H. Seifert, A. Oppelt, to be published in Physics in Medicine and Biology (1991)

[6] A. Oppelt, F. Bömmel, R. Graumann, J. Uebler, Symposium for Opening of the Biomagnetic Center Erlangen, Oct. 5 (1990), to be published by Springer, Berlin

[7] K.J. Best, Vacuumschmelze, Hanau, Germany, private communication

[8] H.J.M. ter Brake, F.H. Fleuren, J.A. Ulfman and J. Flokstra, Cryogenics, **26**, 667 (1986)

[9] K. Abraham-Fuchs, W. Härer, S. Schneider, H. Reichenberger, A. Oppelt, SPIE 1351, Dig. Im. Synth. Inv. Opt. 427-437 (1990)

[10] K. Abraham-Fuchs, W. Härer, S. Schneider, Medical & Biological Engineering & Computing, **28**, 398-406 (1990)

[11] R. Graumann, K. Abraham-Fuchs, F. Bömmel, S. Schneider, accepted for publ. in Clin. Phys. and Phys. Meas. (1991)

[12] W. Moshage, A. Weikl, S. Achenbach, K. Abraham-Fuchs, S. Schneider, Frontiers in European Radiology **8**, 1-19 (1991) Springer, Berlin

[13] P. Weismüller, K. Abraham-Fuchs, S. Schneider, P. Richter, W. Härer, M. Kochs, J. Edrich, V. Hombach, Conference Digest of 8th Int. Conf. on Biomag. 463-464 (1991) Münster, Germany

[14] H. Stefan, S. Schneider, K. Abraham-Fuchs, G. Pawlik, H. Feistel, J. Bauer, U. Neubauer, W.J. Huk, V. Holthoff, EEG and Clin. Neurophys., **79**, 1-10 (1991)

[15] J. Vieth, P. Grummich, G. Sack, H. Kober, S. Schneider, K. Abraham-Fuchs, U. Kerber, O. Ganslandt, T. Schmidt, Biomed. Technik **25**, 238-239 (1990)

[16] E. Hellstrand, K. Abraham-Fuchs, B. Jernberg, L. Kihlström, E. Knutsson, C. Lindqvist, S. Schneider, A. Wirth, to be publ. in Clin. Phys. and Phys. Meas. (1991)

[17] T. Elbert, Chr. Braun, B. Rockstroh, S. Schneider, Conference Digest of 8th Int. Conf. on Biomag., 257-258 (1991) Münster, Germany

[18] P.B.C. Fenwick, J. Vieth, P. Grummich, J. Lumsden, H. Kober, G.W. Fenton, Proceedings of New Developments in Event-Related Potentials, Hannover, Germany, May 15 (1991)

Biomagnetic Multi-Channel System Consisting of Several Self-Contained Autonomous Small-Size Units

A. Matlashov, A. Bakharev, Y. Zhuravlev, and V. Slobodchikov

Institute of Radio Engineering and Electronics, Academy of Sciences, Marx ave. 18, GSP-3, 103907 Moscow, Russia

Abstract. A new approach is suggested to the design of multi-channel systems for biomagnetic applications, based on a combination of several autonomous small-size measuring modules within one system. Small-diameter second-order gradiometers have been developed for practical realization of the new design; the gradiometers are based on integrated dc-SQUIDs and have an intrinsic noise level of $5 \cdot 10^{-15} T/\sqrt{Hz}$. Small-size 1.3 liter fiberglass helium dewars have been designed; the operating period of such dewars is more than two days. Tests have been carried out on a 3-channel version of the modular system. Ways are suggested for future development of modular multi-channel measuring system.

1. Introduction

The ultimate aim of biomagnetic investigations is to obtain information about the structure of bioelectric processes by solving the inverse problem. Since the accuracy of the inverse problem solution is higher the more initial experimental data are available, many investigators try to increase the number of signal channels. Designers of modern measuring systems, too, devote their main attention to this task. In the late 1980s there appeared models of biomagnetic multi-channel systems with more than 30 signal channels [1], which provide the means for obtaining and analyzing "magnetic images" of bioelectric processes.

The use of magnetically shielded rooms allowed the design of multi-channel systems to be greatly simplified. These measuring complexes can be used for carrying out unique biomagnetic investigations. However, because of their bulkiness and high cost they are not accessible to many investigators and this considerably hinders wide-scale development of biomagnetic techniques. Unfortunately alternative trends in the development of biomagnetic apparatus do not receive adequate attention in the world.

A serious drawback of multi-channel systems using large dewars is that the channels cannot be moved with respect to each other. Despite the large number of channels some of them may prove to be informational ineffective. The magnetic field area being measured is limited by the size of the dewar and by the fact that the signal channels cannot be arranged in accordance with the real shape of the head.

A multi-channel system can be built by combining several autonomous small-size measuring modules. Each module may contain one or several signal channels. Such makeup of biomagnetic multi-channel systems is highly attractive for investigators who have practical experience in neuromagnetic measurements. Small-size measuring modules can be mounted at the most preferable points in each experiment, for instance along one arc. The modular multi-channel system can be easily combined with simultaneous recording of EEG.

A number of difficulties arise in practical realization of modular multi-channel systems. The most difficult problems are encountered in the the development of small-diameter gradiometers with a high magnetic field resolution and small size fiberglass dewars with a long operating period.

Below we describe the first practical steps in the realization of this idea which seem to be quite successful.

2. Description of Modular Multi-channel System

Fig. 1 shows a modular system consisting of seven identical small-size autonomous modules. Each measuring module is designed as a small-size fiberglass helium dewar containing a cryogenic probe. All dewars are equipped with independent flexible holders. A system version made up of three single-channel measuring modules has been tested experimentally. The weight of one single-channel measuring module prepared for measurements is about 4 kg.

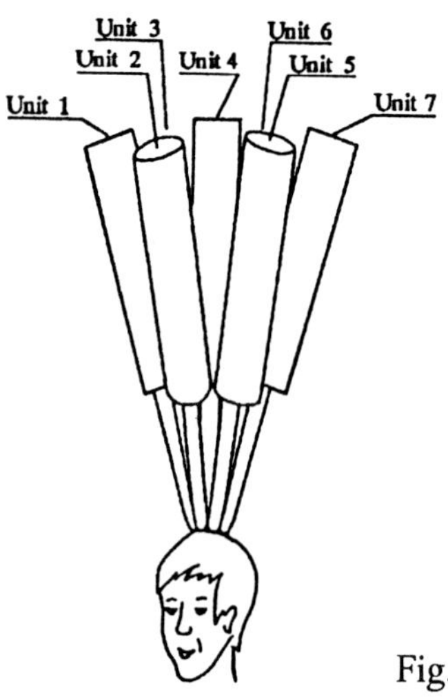

Fig.1 The neuromagnetic seven-channel system

2.1 Fiberglass Helium Dewar

Fig 2. shows the construction of a fiberglass dewar. The size of the dewar has been selected such that when several dewars are simultaneously arranged above a human head close to each other their axes are directed radially and the distance between the tail centers is 32 mm. The inner diameter of the neck and tail is 17 mm. The distance from the external to the internal surface of the tail at the bottom in the cooled state does not exceed 6 mm.

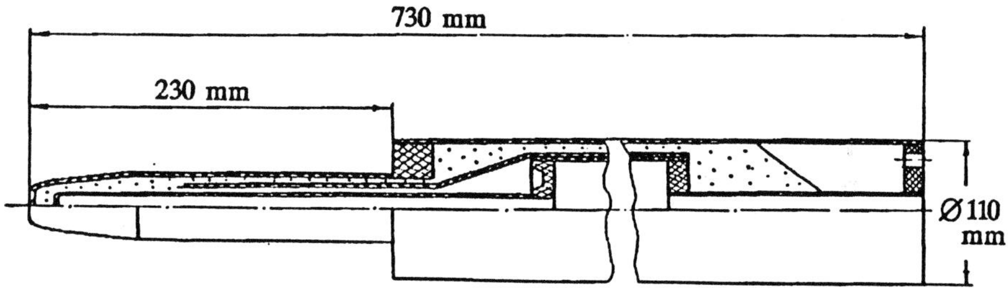

Fig.2 The miniature fiberglass dewar

The volume of liquid helium put into the dewar is 1.3 liter; the operating time of the dewar without the measuring probe is more than two days. Thanks to a small volume of liquid helium the filling procedure is very simple and takes several minutes. The small helium volume of the dewar and the provision of a safety valve makes the dewar safe when the vacuum insulation fails.

2.2. Measuring Cryogenic Probe

Small-size dewars are used with a permanently balanced single-channel measuring probe and with a probe having provisions for fine mechanical balancing, specially designed for the purpose. The flux transformer is an axial second-order gradiometer with a diameter of 15 mm and a base of 50 mm. The intrinsic balance level of second-order gradiometer is about 10^3. Mechanical balancing allows this level to be improved by a factor of more than 10^2

The system uses integrated niobium dc-SQUIDs with shunted tunnel junctions [2]. The SQUID and the matching transformer are enclosed in a superconducting screen. The intrinsic noise of the measuring channel is $5 \cdot 10^{-15} T/\sqrt{Hz}$. When measurements are carried out without shielded room in the center of an industrial city the total noise level is within the limits of 10 to 20 $fT/\sqrt{Hz}$, depending on the time of the day.

Apart from a second-order gradiometer the permanently balanced probe contains three reference channels in the form of a vector magnetometer. In

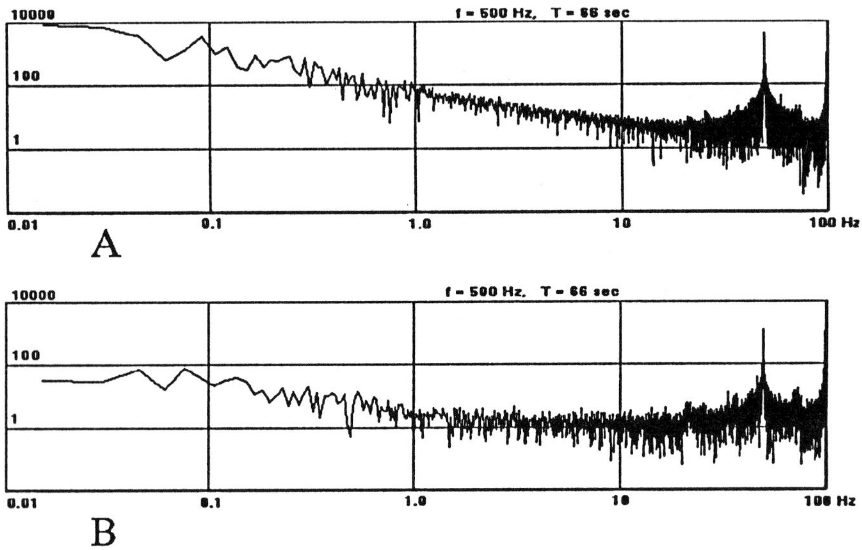

Fig.3 The spectra of an output signal for the second-order gradiometer with balans a level 10^3 : without /A/ and with /B/ electronic noise suppression system

the reference channels the vector components of the magnetic field are measured by means of three mutually orthogonal dc-SQUIDs without flux transformers. The vector magnetometer makes it possible to obtain a simple but effective noise suppression system [3]. Fig. 3 shows the spectra of an output signal from one of the signal channels, which demonstrate the operating efficiency of the system. The vector magnetometer can also be used for coordinate measurements and orientation of the measuring gradiometer.

2.3. Electronic Units

The electronic circuitry of multi-channel modular systems provides for autonomy and independence of all channels. The electronic section of the signal channel and the electronic section of three reference channels are designed as small-size units mounted on top of the dewar. The reference channels do not use input signal modu- lation and do not have input matching circuits.

The small-size electronic units are tuned and adjusted for various operating modes by remote control with the help of a sequential digital code. The channel electronic units are connected to the control unit which simultaneously controls 16 channels in the manual or automatic mode. In the latter case it must be connected to a personal computer compatible with the IBM PC AT.

3. Prospects for the Development of Modular Multi-Channel Systems

The number of signal channels in one autonomous measuring module may be increased to three. When three reference channels are used the total number of the channels will be six. Fig. 4 shows the tangential second-order gradiometer. A combination of two such gradiometers plus an axial second-order gradiometer arranged coaxial forms a vector gradiometer. In this geometry the signal channels measure, at the same point in space, three vector components of the magnetic field set up by a closely located source. When seven modules of this type are used simultaneously the total number of signal channels will be 21, which is commensurate with the number of channels in modern systems.

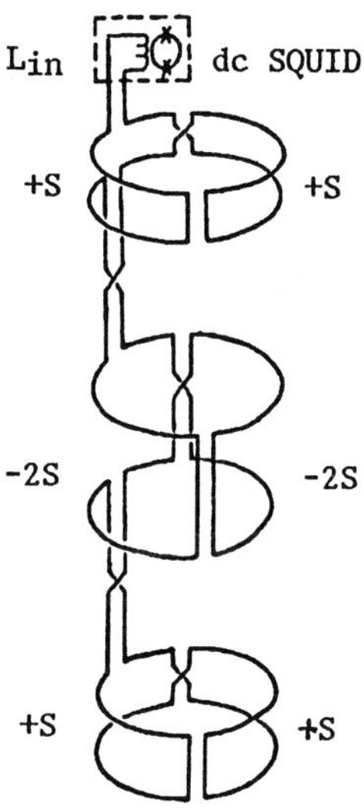

Fig.4 The tangential second-order gradiometer

Modular multi-channel systems can be used for development of magnetometers for biomagnetic investigations, based on high-temperature superconducting materials. The use of liquid nitrogen for cooling will make it possible to develop very small and reliable measuring modules. An important advantage of such systems will always be their structural flexibility.

Acknowledgements

We want to thank Dr. S. Schneider and Dr. H. Seifert for the assistance in studying of a single-channel measuring module.

References

[1] V. Foglietti, Multichannel Instrumentation for Biomagnetism, this volume.

[2] V. Koshelets, A. Matlashov, I. Serpuchenko, L. Fillipenko, Y. Zhuravlev, DC-SQUID Preamplifier for DC-SQUID Magnetometer, IEEE Trans. Magn. MAG-25 (2), 1182-1185 (1989)

[3] A. Matlashov, Y. Zhuravlev, A. Lipovich, A. Alexandrov, E. Mazaev, V. Slobodchikov, O. Wazhievski, Electronic Noise Suppression in Multichannel Neuromagneic System, in: S. Williamson, M. Hoke, G. Stroink, M. Kotani (Eds.), Advances in Biomagnetism, Plenum Press: New York and London 1989, pp.725-728.

A Modular 19-Channel SQUID System for Biomagnetic Measurements

O. Dössel, B. David, M. Fuchs, J. Krüger, W.H. Kullmann, and K.-M. Lüdeke

Philips GmbH, Forschungslaboratorium Hamburg, Postfach 540840, W-2000 Hamburg 54, Fed. Rep. of Germany

Abstract. A low noise 19-channel SQUID system was developed. It is part of a complete system for functional diagnosis of human brain or heart.

The thin film DC-SQUIDs with integrated coupling coils are based on the refractive materials NbN-MgO-NbN. First order wire wound gradiometers with 20mm diameter are used. The cryogenic system is modular in a way that every channel can be optimized or replaced individually without warming up of the dewar. The electronics are based on the flux locked loop scheme with a new bias modulation technique. The system is installed inside a magnetically shielded chamber. The noise level of all 19 channels measured simultaneously is below $12fT/\sqrt{Hz}$.

1. Introduction

Multichannel SQUID systems for functional diagnosis of the brain and the heart have been developed using different approaches [1-5]. The use of thin film DC-SQUIDs, first order gradiometers and magnetically shielded chambers is well established. Differences can be observed concerning the SQUID materials ($Nb/Al_2O_3/Nb$ versus NbN/MgO/NbN), the gradiometer design (thin films versus wires), the cryogenic design (modular systems versus compact systems) and the electronics (simplified versions versus complex modulation techniques). The choices made for this 19 channel system are described and discussed and the results are presented.

2. SQUIDs and Gradiometers

The properties of the NbN/MgO/NbN-SQUIDs used in this system are reported in detail elsewhere [6]. The SQUIDs show a low white noise level better than $10\mu\Phi_o/\sqrt{Hz}$ and a relatively large 1/f noise, which can be completely suppressed by flux and bias modulation techniques. The advantage of using these materials is the robustness of the SQUIDs. They withstand thermal cycling without any changes, are not degraded by humidity and survive moderate electrostatic discharges.

The gradiometers are made of superconducting wire that is precisely laid into grooves in a ceramic holder. The diameter is 20mm, the baseline is 50mm, the number of windings in the pick up coil is 3 resulting in an inductance of 820nH. Compared to other fabrication techniques this method has some important advantages: It is very flexible so that first order or second order gradiometers with various baselines can easily be produced. Compared to foil gradiometers [3] the balancing "as made" seems to be significantly better, in our case it is better than 0.1%.

3. The Cryogenic System

The modular approach chosen in this system (Figure 1) has the clear advantage, that every single channel can be optimized or replaced individually without warming up the dewar. Compared to compact block systems additional measures have to be taken against vibrations of the gradiometers. A clamping technique was developed consisting of two supporting plates that are fixed in the tail of the cryostat. Special springs made of fiber reinforced plastic hold the gradiometers tight. So no vibration frequencies show up in the noise spectra (Figure 2).

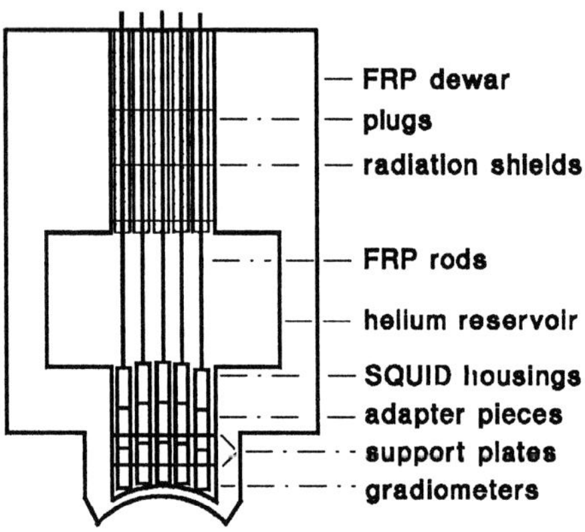

Figure 1: Modular cryogenic system

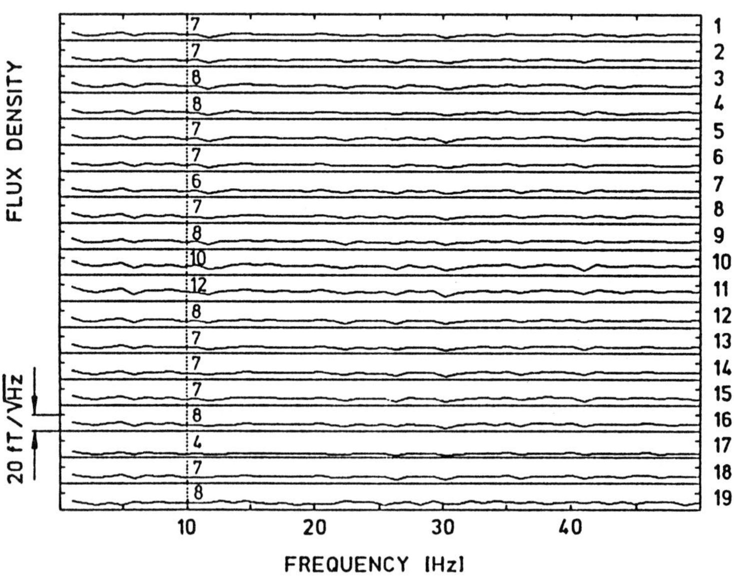

Figure 2: Noise spectra of all 19 channels measured simultaneously (at cursor position: noise of each channel in fT/√Hz)

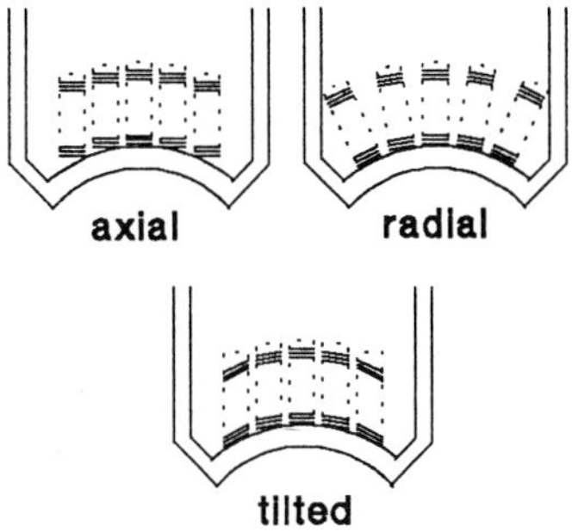

Figure 3: Different gradiometer arrangements

Concerning the thermal losses special rods made of fiber reinforced plastic with thin walls and low thermal conductivity are used for the connection between the liquid helium end and the room temperature end. Inside these rods are the shielded electrical leads which give the main contribution to the thermal losses. The boil off rate of the dewar [7] without any inset is 3.5 l/day and the boil off rate of the complete 19 channel system is 5.5 l/day, so the boil off rate per channel is only 0.1 l/day.

The system presented here is equipped with conventional axial gradiometers (Figure 3). The measurement of radial field components is also possible using tilted gradiometers. Computer simulations show, that the information about the field distribution is nearly identical to the information gained with radial gradiometers. Also volume current contributions can significantly be reduced compared to flat gradiometer arrangements.

4. Electronics

The electronics are based on the conventional flux locked loop scheme improved by a new bias modulation technique described elsewhere [8]. For flux and for bias modulation identical frequencies are used with a phase shift of 90° and the signal is detected at double frequency. Simplified electronics described e.g. in [5] cannot be used because of the large 1/f noise of the NbN/MgO/NbN SQUIDs. The complete SQUID electronics fit onto a single board of 130mm*180mm size. These electronics modules are contained in cassettes that can be plugged together to a multichannel system. The complete 19 channel electronics is located inside the magnetically shielded chamber [9] on top of the gantry. The noise spectra (Figure 2) prove, that there is no interference between electronics and gradiometers.

5. Conclusions

NbN/MgO/NbN SQUIDs are adequate for multichannel SQUID systems provided that flux and bias modulation schemes are used. Wirewound gradiometers are still the most flexible approach with a good balancing "as made". A modular cryogenic system can be constructed if adequate measures against vibrations and thermal losses are taken. And the complete SQUID electronics including flux and bias modulation schemes can be realized on a single board that can be located inside the magnetically shielded chamber relatively near to the gradiometer array.

Acknowledgments

Substantial contributions from M. Dargatz, W. Edeler, W. Hoppe, R. Kobs, J. Köhler, G. Kürsten, H. Laudan, M. Lemke, G. Rabe, G. Weidinger and L.Wittorf are gratefully acknowledged.

References

[1] W.C. Black, Contribution to 1990 Applied Superconductivity Conference (1990)

[2] P. Carelli, C. Del Gratta, V. Foglietti, I. Modena, A. Pasquarelli, V. Pizzella, M. Pullano, G.L. Romani, G. Torrioli, in: S.J. Williamson, M. Hoke, G. Stroink, M. Kotani (Eds.), Advances in Biomagnetism, Plenum Press, New York, London, 665-668 (1989)

[3] H.E. Hoenig, G.M. Daalmans, L. Bär, F. Bömmel, A. Paulus, D. Uhl, H.J. Weisse, S. Schneider, H. Seifert, H. Reichenberger, K. Abraham-Fuchs, IEEE Trans. Magn. MAG-27, 2777-2785 (1990)

[4] A.I. Ahonen, M.S. Hämäläinen, M.J. Kajola, J.E.T. Knuutila, O.V. Lounasmaa, J.T. Simola, C.D. Tesche, V.A. Vilkman, IEEE Trans. Magn. MAG-27, 2786 (1990)

[5] H. Koch, R. Cantor, D. Drung, S.N. Erne, K.P. Matthies, M. Peters, T. Ryhänen, H.J. Scheer, H.D. Hahlbohm, IEEE Trans. Magn. MAG-27, 2793-2796 (1990)

[6] O. Dössel, B. David, R. Kobs, K.M. Lüdeke, D. Grundler, this issue (1991)

[7] SST-160/25, CTF Systems Inc., Pt. Copquitlam, B.C., Canada

[8] O. Dössel, B. David, M. Fuchs, W.H. Kullmann, K.M. Lüdeke, IEEE Trans. Magn. MAG-27, 2797-2800 (1991)

[9] AK3a, Vakuumschmelze, Hanau, Germany

On the SQUID-Modules for the UT Multichannel Neuromagnetometer

H.J.M. ter Brake, J. Flokstra, E.P. Houwman, D. Veldhuis, W. Jaszczuk, R. Stammis, G.K. van Ancum, and H. Rogalla

University of Twente, Faculty of Applied Physics, P.O. Box 217, NL-7500 AE Enschede, The Netherlands

Abstract. A 19-channel dc-SQUID based neuromagnetometer is constructed. Except for the cryostat, all elements of the system were developed at the UT. In the paper we focus on the SQUID modules: their shielding performance and their effect on the gradiometer balance. Further superconducting bond connections are considered.

1. Introduction

At the University of Twente (UT) a 19-channel dc-SQUID based neuromagnetometer is constructed. The neuromagnetometer is equipped with 19 wire-wound first-order gradiometers. They are arranged side by side in a hexagonal configuration over the bottom of a CTF-SST 140 cryostat. The gradiometers have a diameter of 20 mm and a baseline length of 40 mm (2 sections of 3 turns). Crosstalk between the channels is eliminated by means of external feedback [1]. Including noise references the magnetometer comprises 22 to 24 SQUID systems. Each gradiometer is connected to a module consisting of a planar dc-SQUID placed in a superconducting enclosure. The sensitivity of the electronics [2] is set to 1 V at the output for 1 ϕ_0 in the SQUID. The dynamic range of 20 V corresponds to an effective field of about 10 nT in the gradiometer. In the paper the SQUID modules are considered in detail. Attention is paid to the superconducting body and shield: their shielding performance and their effect on the balance of the gradiometers. Further, the specifications of the SQUIDs and the bond connections between the gradiometers and the SQUID input coils are considered.

2. Superconducting body and shield

Each SQUID is placed in a solid niobium body that fits into a cylindrical lead shield (figures 1 and 2). The body is divided into compartments in order to maximize the shielding of the different sections. As can be seen in the figures the gradiometer wires and the external feedback transformer are connected to the SQUID input coil via niobium screw terminals.

The shielding performance of the superconducting body and the shield can be estimated by means of a model presented by Rigby et al.[3]. Magnetic

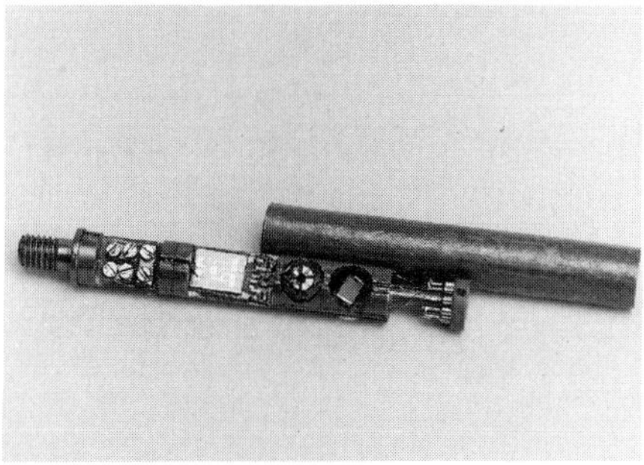

Fig.1: dc-SQUID module

Fig.2: detail of the module, showing screw terminals, feedback transformer, SQUID and bonds

fields in the axial direction are far better shielded than fields in the radial direction. Within the module the field decays as exp(-0.41z) with z the distance to the module end piece (in mm). Because of the relatively small z, the largest sensitivity for external fields arises from the loop that is inevitably formed at the screw connections in the input circuit. The calculated worst-case imbalance contribution of the loop in our prototype module was 2.10^{-4}. An improved module in which the distance between the screw blocks and the module end piece is increased from 1 mm to 8 mm, yields an imbalance contribution of 2.10^{-5}, which is acceptable.

We determined the reducing effect of the superconducting body and the shield on the flux that is coupled to the SQUID by an applied external field. For this purpose we made a small loop under the screws and connected it to the SQUID input coil. In the calculations we assume that the magnetic field is symmetrically distorted by the niobium blocks and that the wire is directly on the surface of the blocks. The measured flux in the prototype module was a factor of 160 smaller than the flux that would have been detected without any shielding. With the improved module a factor of 3100 was measured. The model of Rigby gave reduction factors of 130 and 2600, respectively. We therefore conclude that (1) the Rigby-model can be used for worst-case estimates, (2) the improved module has at least one order of magnitude better shielding than the prototype, and (3) with respect to shielding lead is an appropriate material for the cylindrical shield.

To evaluate the effect of the modules on balance of the gradiometers, consider a single module with radius a, which is aligned parallel to a gradiometer with baseline b. The smallest axial distance is z and the lateral displacement r. In general, the magnetic field near superconducting material is analogous to the velocity field of a fluid near a rigid obstacle [4]. Therefore, in the case of an axial field the module can be considered as a circular disk with radius a. The contribution of this disk to the imbalance, repesented by C_b, is evaluated with ref.[4] and is given by

$$C_b = \alpha^3 \cdot 2/\pi \cdot [1-(1+\rho^2)/3] \cdot (1+\rho^2)^{-5/2} \cdot [1-(1+\beta^2)^{-3}]$$

with $\alpha = a/z$ (<0.1); $\rho = r/z$ and $\beta = b/z$. We calculated the imbalance contribution for our case (a = 5.1 mm, b = 40 mm, 24 modules positioned at a circle with a diameter of 122 mm) for different positions of a gradiometer. In the 19-channel neuromagnetometer the imbalance should be below 10^{-4}, which can be realized with z larger than 16 cm. In the neuromagnetometer we have z = 17 cm ($C_b < 9.10^{-5}$).

3. SQUID section

The UT dc-SQUIDs are fabricated based on the Nb/Al technology [5]. In the neuromagnetometer we apply standard Ketchen-type SQUIDs and resistively shunted SQUIDs (inductances about 0.3 nH, forward transfers resp. 70 $\mu V/\phi_0$ and 110 $\mu V/\phi_0$). Both SQUID configurations have an input inductance of 0.1 μH and an input coupling factor 0.8. The white noise level is below 10^{-5} $\phi_0/\sqrt{Hz}$ corresponding to an effective field noise of 5 $fT/\sqrt{Hz}$.

Each SQUID is glued on to a printed circuit board that has bonding pads for bias, modulation and output-voltage connections. To connect the gradiometer to the SQUID input coil we developed a method for bonding niobium wire to a niobium thin-film pad. Before bonding, the wire is annealed in a vacuum chamber at roughly 2200 ^{0}C for about 5 minutes. The Vickers hardness of the wire is thus reduced from typically 180 to 80. The resistance ratio (R(300 K) to R(10 K)) increases from typically 8 to 80. The wire is bonded to a niobium film covered with a thin layer of palladium (4 nm). Superconducting bonds have been realized and tested to a sensitivity level of 6.10^{-15} Ω. Details on the bonding method will be presented elsewhere [6]. The critical current of the bonds is typically 1 mA at 4.2 K. For two extreme cases (0.2 mA and 2 mA) we measured the critical current as a function of the temperature. In figure 3a the temperature dependence of the

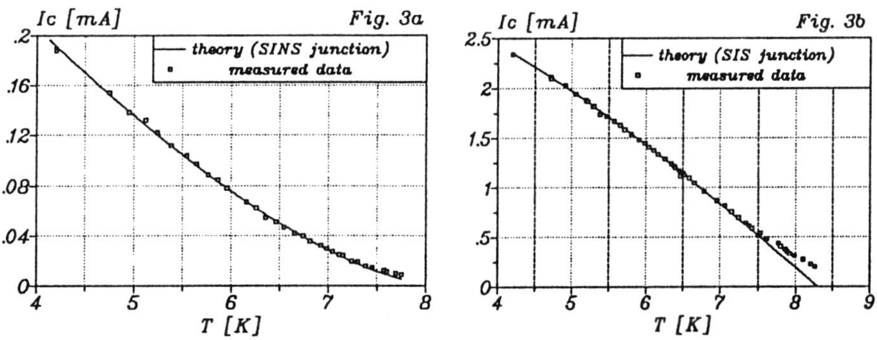

Fig.3: Two measurements of the critical current of two bonds in series as a function of the temperature. Fig. 3a: low-I_c case; Fig. 3b: high-I_c case

critical current for a low-I_c case is depicted and in figure 3b that for a high-I_c case. Apparently, a low I_c corresponds to an SINS-junction and a high I_c with an SIS-junction [7].

4. Conclusions

The SQUID module that is described in this paper can be used in the UT 19-channel neuromagnetometer. The imbalance contribution due to imperfect shielding is limited to 2.10^{-5}. The effect of all the modules on the balance of the gradiometers is calculated to be smaller than 10^{-4}. The bonding of a niobium wire to a niobium thin film appeared to be successful. Critical currents of the bonds of typically 1 mA were measured and the temperature dependence of the critical current appeared to correspond to that of an SINS-junction for a low I_c (ca. 0.2 mA at 4.2 K) and to that of an SIS-junction for a high I_c (ca. 2 mA at 4.2 K).

References

1. H.J.M. ter Brake, F.M. Fleuren, J.A. Ulfman and J. Flokstra: Cryogenics **26**, 667 (1986)
2. J. Flokstra, H.J.M. ter Brake, E.P. Houwman, D. Veldhuis, W. Jaszczuk, H. Rogalla, A. Martinez and C.Rillo: Sensors and Actuators **27**, 781 (1991)
3. K.W. Rigby, D. Marek and T.C.P. Chui: Rev. Sci. Instrum. **61**, 834 (1990)
4. J.A. Overweg and M.J. Walter-Peters: Cryogenics **18**, 529 (1978)
5. E.P. Houwman, D. Veldhuis, J. Flokstra, H.J.M. ter Brake, W. Jaszczuk, A. Martinez and H. Rogalla: IEEE Trans. Mag. **27**, 2955 (1991)
6. W. Jaszczuk, H.J.M. ter Brake, J. Flokstra, D. Veldhuis, R. Stammis and H. Rogalla: "Bonding of a niobium wire to a niobium thin film", submitted to Measurement Science and Technology
7. A. Barone and G. Paternò: *Physics and Applications of the Josephson Effect* (John Wiley & Sons, New York, 1982)

11-Channel Multipurpose Biomagnetic System for Operation in Unshielded Environment

A. Pasquarelli[1], *S. Casciardi*[2], *M. Giannini*[1], *V. Foglietti*[1], *V. Pizzella*[1], *G. Torrioli*[1], *and G.L. Romani*[2]

[1]Istituto di Elettronica dello Stato Solido, C.N.R., Via Cineto Romano, 42, I-00156 Rome, Italy

[2]Istituto di Fisica Medica, Università "G. D'annunzio", Via dei Vestini, I-66100 Chieti, Italy

Abstract. Progress toward the realization of a medium size multipurpose biomagnetic system is described. Eleven second-order gradiometers are coupled with as many dc-SQUIDs manifactured in our laboratory. The geometry of the detecting coils consists of seven sensors arranged in a straight line and four sensors placed around the center. By means of this configurationit is possible to scan the chest or the abdomen with the seven aligned sensors, to measure the head with the seven central sensors whereas the whole system can provide significant information for "single shot" cardiomagnetic measurements in clinical studies.

1. Introduction

In the last few years many researcher groups and industries have devoted their efforts to develop multisensor systems for biomagnetic measurement. Most of these systems are designed to work in a shielded environment and show an increasing number of channels and feature different geometries for different investigating areas of human body.

In this report we will present the features of a medium size system which has been developed to work in an unshielded environment at the Institute of Medical Physics at the University of Chieti.

2. Instrument Description

Seven sensors are arranged in a straight line and spaced by 35 mm. In this way, by simple scanning, it is possible to cover a large area of the chest or of the abdomen to perform both cardiomagnetic and gastro-intestinal measurements. Additionally, four more sensors are placed around the center, in order to cover, together with the three central sensors laying in the straight line, a 92 mm diameter circle. By means of this configuration, the seven "central" channels can be used for neuromagnetic measurements whereas the whole system can provide significant informations for "single shot" cardiomagnetic measurements in clinical studies. A schematic of the gradiometer array is shown in fig 1a.

Springer Proceedings in Physics, Vol. 64 **Superconducting Devices and Their Applications**
Editors: H. Koch, H. Lübbig

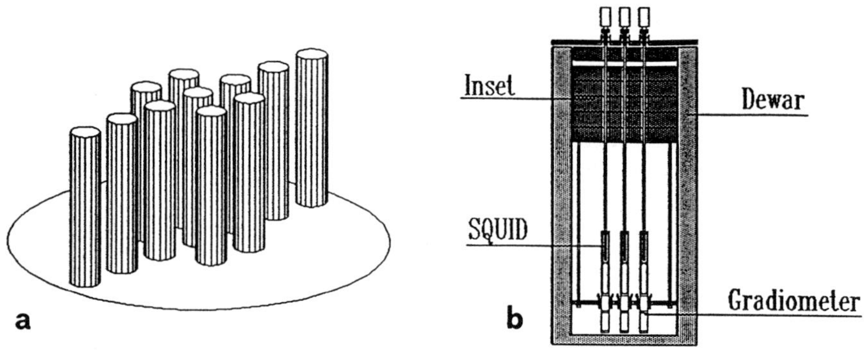

Fig 1: a) Schematic illustration of the gradiometer array. b) Schematic cross-section of the system

Fig 2: View of the inset of the dewar with one cryogenic probe mounted

The dewar consists of two main parts, a double walled neckless container with a reservoir of 30 liters of liquid helium and an inset. In fig 1b is shown a schematic cross-section of the instrument.

An important feature of this system is that each channel is constituted by a single cryogenic probe which is mechanically independent with respect to the others. This means that each one can be easily removed from and inserted into the dewar for maintenance even in the operating condition, when the system is immersed in a liquid helium bath.

Each probe consist of a second order gradiometer coupled to a dc-SQUID sensor by means of a flux-transformer. The room temperature electronics is connected to the corresponding SQUID by means of connections enclosed in a stainless steel pipe. All the probes are kept together with a collective holder which constitute the inset of the dewar as well. A view of the dewar inset with one cryogenic probe mounted is shown in fig 2.

The gradiometers are made of 80 μm niobium wire wound around a MACOR cylinder with a 22 mm diameter and 80 mm baseline. The dc-SQUIDs, microfabricated in our institute, have a double washer geometry [1] which reduce their inductance without decreasing the coupling to the input coil and allow the SQUIDs to behave like a first-order planar gradiometer with respect to background magnetic field.

The input coil, patterned over each washer, has a inductance of 0.6 μH and a mutual inductance with SQUID loop of 7 nH and a coefficient coupling of 0.9. The measured flux transforming ratio between the SQUID loops and the gradiometers is 6.8 10^{-3}, in a good agreement with the computed value.

Each SQUID has been set on a plexiglas holder which contains two coils used to send the external modulation signal to the SQUID and the feedback signal to the input circuit which allows to cancel the cross-talk between gradiometers [2]. Each holder has been enclosed in a superconducting shield made of a copper cylinder galvanically coated with lead in all its surfaces. Each SQUID is connected to a single electronic system which is substantially the same developed in our laboratory for the 28-channel hybrid system [3].

3. Preliminary results

The performance of one probe, inserted in the dewar, was tested inside the shielded room and the white noise level was found to be about 2.9 $fT/(Hz)^{1/2}$, as shown in fig 3. In order to test the probe in an unshielded environmental we shunted the input coil with a 10 Ω resistor and a 330 nF capacitor in series and we added a supplementary rf shield. In these conditions the white noise level increased up to about 6 $fT/(Hz)^{1/2}$. The liquid helium evaporation rate of the dewar was 4 liters per day. This result lead us to estimate an evaporation rate, with all probes mounted, lower than 5 liters per day.

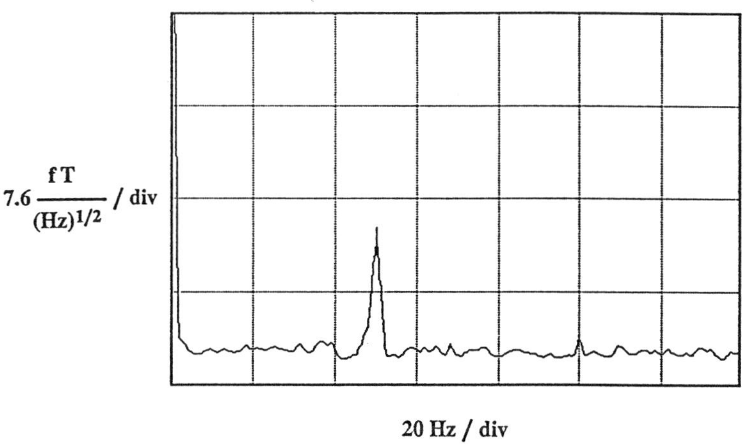

Fig 3: Field noise spectral density of tested probe recorded inside the shielded room

Acknowledgments

The authors thank S.N. Erné for fruitful discussions and the Physikalisch-Technische Bundesanstalt, Berlin, for providing the dewar. This project has been partially supported by Progetto Finalizzato Superconductive and Cryogenic Technologies, C.N.R.

References

[1] V. Foglietti, M. E. Giannini, G. Petrocco. A double dc-SQUID device for flux locked loop operation. IEEE Trans. Mag. **27**, 2989 (1990).

[2] V. Foglietti, C. Del Gratta, A. Pasquarelli, V. Pizzella, G. Torrioli, G. L. Romani, W. J. Gallagher, M. B. Ketchen, A. W. Kleinsasser, R. L. Sandstrom. 28-channel hybrid system for neuromagnetic measurements. IEEE Trans. Mag. **27**, 2959 (1990).

[3] H. J. M. ter Brake, F. H. Fleuren, J. A. Ulfman, J. Flokstra. Elimination of flux-transformer crosstalk in multichannel SQUID magnetometers. Cryogenics 26, 667 (1986).

A New Method of Adjusting Bias Currents of Neuromagnetometers

M. Hotta[1], *H. Kado*[2], *K. Makie*[1], *and K. Okajima*[1]

[1]Central Research Laboratory, Hitachi, Ltd., Kobubunji, Tokyo 185, Japan
[2]Electrotechnical Laboratory, Tsukuba 305, Japan

Abstract. In this paper, a new method of adjusting the bias current of a SQUID is proposed. The bias current can be adjusted to the optimum value which maximizes the SNR by measuring the second harmonic of the output signal of the SQUID.

1. Introduction

This paper proposes a novel method of adjusting the bias current of DC-SQUIDs. In recent years, multichannel biomagnetometer systems have been developed for localizing the neural current sources accurately. Biomagnetometers with over 30 channels have been reported[1]. In such multichannel systems, the adjustment of the bias current is one of the key techniques for achieving high performance, because the SNR of the system depends on the bias current. Furthermore, it is troublesome to adjust the bias current of each SQUID, because the adjustment is conventionally done by measuring the voltage-flux characteristics. The method proposed in this paper makes it possible to easily adjust the bias current to its optimum value, for a magnetometer with the FLL (Flux-Locked Loop) configuration.

2. Analytic Methods

Figure 1 shows the FLL configuration for the DC SQUID magnetometer. In this figure, ϕ_{in} is the input signal flux, K is the transfer function of the SQUID, A is the total gain of amplifiers, and β is the transfer function of feedback circuits. The simplified FLL configuration is illustrated in Figure 2, where x is the equivalent input noise of the amplifier. The output voltage of the FLL is given by Eq.(1), and the SNR is calculated from Eq.(2).

$$V_o = \frac{KA\phi_{in}}{1 - \beta KA} + \frac{Ax}{1 - \beta KA} \tag{1}$$

$$SNR = \frac{K\phi_{in}}{x} \tag{2}$$

From Eq.(2), increasing K increases the SNR. Here, K corresponds to $dV/d\Phi$ in the V-Φ characteristic and depends on the bias current[2]. However, Eq.(1) indicates that the optimization of K by measuring V_o is difficult because the closed-loop gain of the FLL is constant at $1/\beta$, when KA is much greater than 1. Therefore, a new method for optimizing K is required.

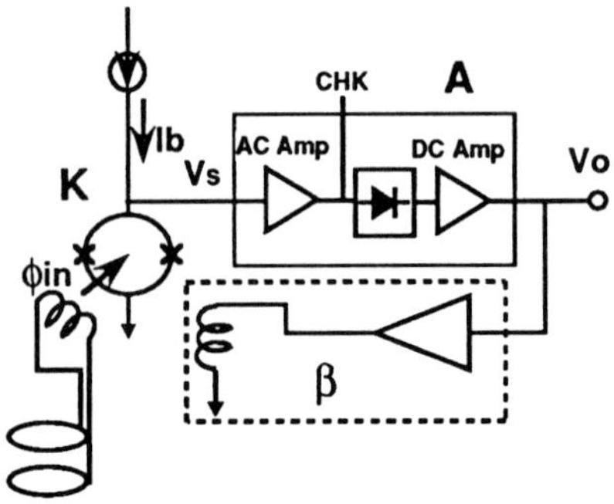

Fig.1 Flux-Locked Loop (FLL) configuration
for DC SQUID magnetometer.

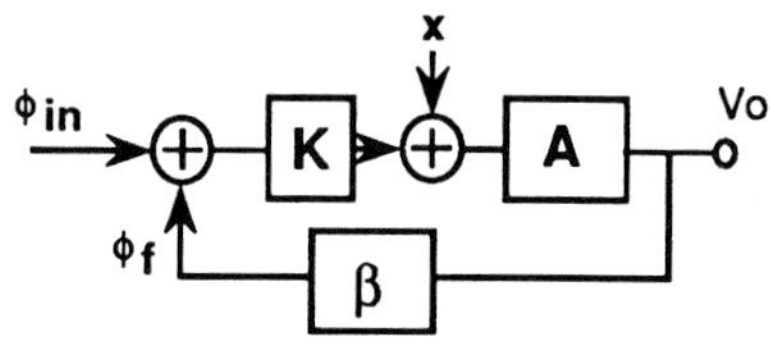

Fig.2 Simplified block diagram of the FLL.

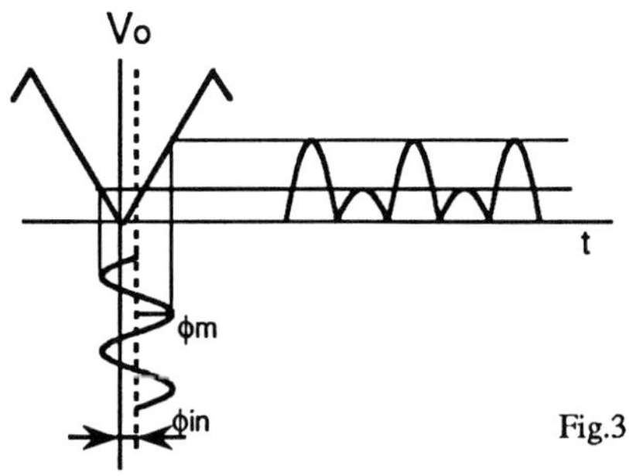

Fig.3 Flux moduration sheme showing voltage across SQUID for an input signal flux ϕin.

To simplify the analysis, the voltage-flux characteristic (V-Φ characteristic) is approximated by the folded line shown in Figure 3. When a modulation flux with amplitude ϕ_m and input signal flux ϕ_{in} are fed into the SQUID, the output waveform is similar to a rectified sine wave, as shown in the figure. This output voltage $V_{s(t)}$ is given approximately by Eq.(3), using Fourier series.

$$V_{s(t)} = \frac{2K\phi_m}{\pi} + K\phi_{in}\sin\omega t - \frac{4K\phi_m}{\pi}\sum_{n=1}^{\infty}\frac{\cos 2n\omega t}{4n^2-1} \tag{3}$$

From Eq.(3), the second harmonic component $V_{s2(t)}$ is given as

$$V_{s2(t)} = -\frac{4K\phi_m}{3\pi}\cos 2\omega t \tag{4}$$

From Eq.(4), it is clear that the second harmonic component is proportional to K, the transfer function of flux-to-voltage. Because the amplitude of the modulation flux is chosen as a constant $\phi_m = \Phi_o/4$, the bias current can be adjusted to the optimum value which maximizes K, by measuring V_{s2}.

3. Experiment

Figure 4 shows the V-Φ characteristics of a DC-SQUID with a 40-turn input coil, at different bias currents. The maximum range of output voltage is obtained at a bias current of 11 μA. The signal and noise voltage at the output terminal of the FLL are plotted as a function of bias current I_b in Figure 5. The SNRs calculated from Figure 5 are shown in Figure 6. The bias current which maximizes the SNR is around 10 μA, coinciding with the value for the maximum V-Φ characteristic. The harmonics of the SQUID output voltage, amplified by an AC amplifier, are plotted in Figure 7, where V_{s1},V_{s2}, and V_{s3} are the fundamental, second and third components. This figure shows that the second harmonic component V_{s2} is much greater than the fundamental and third harmonics, and has its maximum at a bias current I_b of about 10 μA. This is the same as the value of I_b which maximizes the SNR. Therefore, the adjustment of the bias current can be accomplished by measuring the second harmonic of the output signal of the SQUID, even when the FLL is working.

An evoked auditory response is shown in Figure 8. This response was measured by a neuromagnetometer with its bias current adjusted to the optimum value by the method proposed above.

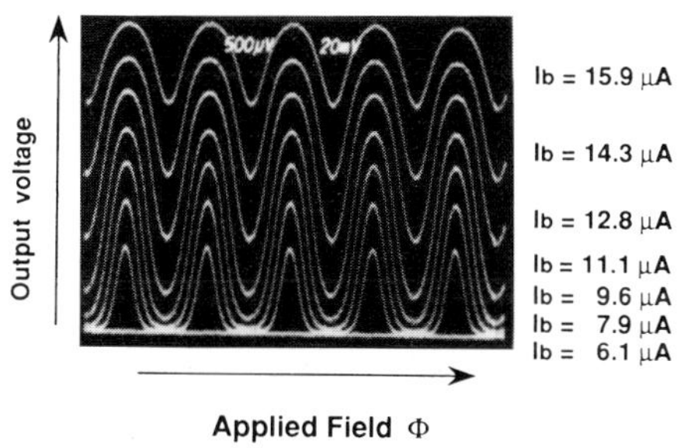

Fig.4 V-Φ characteristic of DC SQUID at differential bias curent.

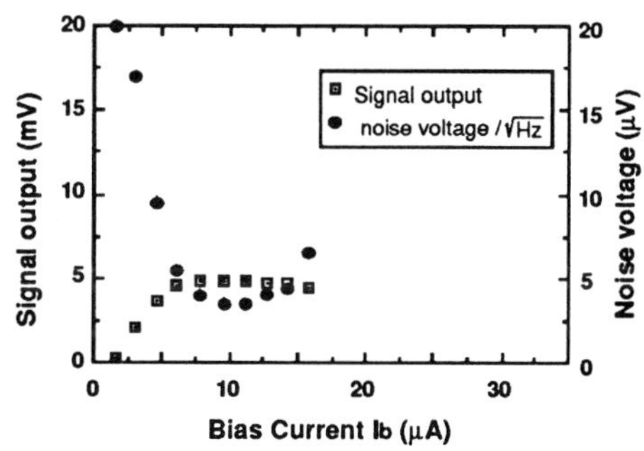

Fig.5 Signal and noise voltage of FLL as a function of bias current Ib.

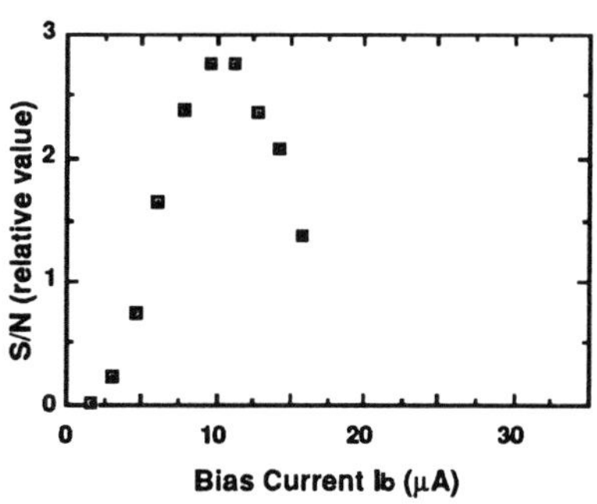

Fig.6 Rerationship between SNR and bias current for FLL.

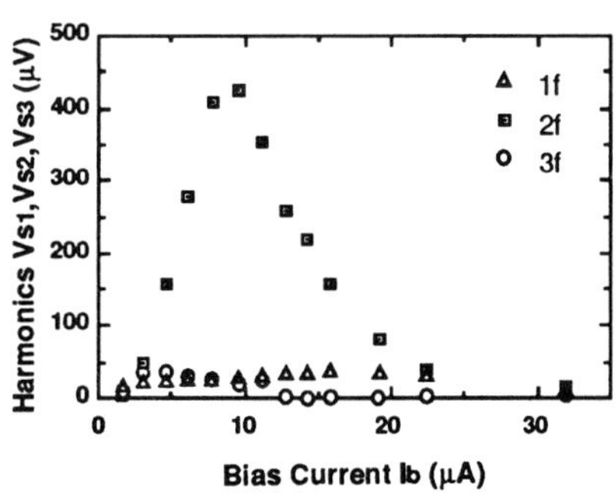

Fig.7 Harmonics of SQUID output voltage amplified by an AC amplifier.

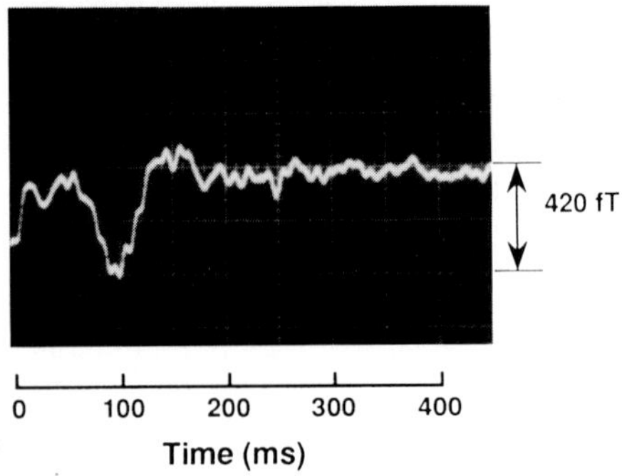

Fig.8 An auditory evoked magnetic field.

4. Conclusion

A new method of adjusting the bias current of a SQUID to maximize the SNR has been is proposed. The key point in achieving the adjustment in FLL operation without an external calibration signal is to measure the second harmonic of the output voltage of SQUID. The effectiveness of the proposed method was shown proved by experiments. Moreover, this method can be used to monitor abnormal states in SQUIDs such as trapped flux.

REFERENCES

[1] G. Daalmans, H. Seifert, W. Folberth, F. Bommel, H.E. Hoenig, S. Schneider ; 31-Channel DC SQUID Gradiometer Array for Biomagnetic Diagnosis, Extended Abstracts of ISEC'89, SS-1,(1989)

[2] M. B. Ketchen ; DC SQUID 1980: The State of Art, IEEE Trans. Magn., MAG-17(1981), pp387 - 394

Active Shield with SQUID for Biomagnetic Measurement

K. Fujioka, K. Matsumoto, Y. Yamagishi, T. Noda, and Y. Kuraoka

Central Research Laboratory, Hoxan Corporation,
16-2, Ninomiya 3-chome, Tsukuba, Ibaraki 305, Japan

Abstract. A new method for the cancellation of the environmental magnetic noise (EMN) is presented in this paper : a SQUID based active shield (SAS). The principle of the SAS is based on a flux-locked-loop operation of a SQUID system. The SAS has many advantages for the multichannel biomagnetic measurement. Features of the SAS have been described by results of calculations. The preliminary experiment also has been performed successfully.

1. Introduction

A new simple and effective method for the cancellation of the environmental magnetic noise (EMN) is presented in this paper. This cancellation system of the new method is an active shield with a SQUID. The principle of the method is based on a flux-locked -loop operation of a SQUID system.

This SQUID based active shield (SAS) has several special merits for the biomagnetic measurement. The first merit is that the EMN is practically canceled by the SAS in a certain area which is required by the multichannel pickup coils for the whole head or chest. The second advantage is that a high cancellation effect of the EMN is obtained in the low frequency region. This fact is very important for the biomagnetic measurement, because the EMN increases in almost inverse proportion to its frequency, [1]. One more remarkable feature is that the SAS does not need the orthogonal relation among its coils for canceling the EMN. It becomes very easy to contract and use the SAS by this feature.

2. Principle

The principle of the SAS is based on a flux-locked-loop operation of a SQUID system in which a SQUID ring is operated usually as a null detector. In the SAS circuit, the feedback current for the flux-locked-loop is supplied directly to a large-sized noise cancellation coil instead of the usual feedback circuit. The noise cancellation coil is located in a room temperature area. The pickup coil of the SQUID is placed near the center of the noise cancellation coil and its operated as a null detector instead of a SQUID ring. The magnetic fluctuation of the EMN in the pickup coil is compensated by a magnetic field which is generated by the noise cancelation coil. Then, automatically, the EMN around the center is almost canceled. The SAS circuit is essentially equal to a usual SQUID circuit. Therefore the SAS is effective basically in the same frequency range as a SQUID. For the system noise, the SAS is also fundamentally of the same order of a SQUID.

Springer Proceedings in Physics, Vol. 64 Superconducting Devices and Their Applications
Editors: H. Koch, H. Lübbig

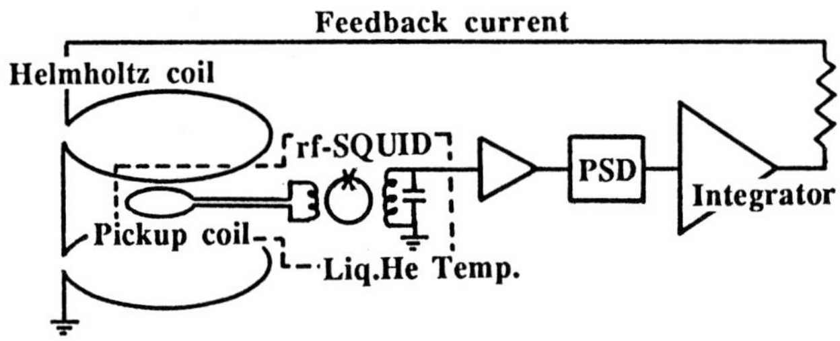

Fig.1 Basic arrangement of the SQUID based active shield

Fig. 1 shows the basic arrangement of the SAS. This basic SAS consists of a SQUID magnetometer and a Helmholtz coil as a noise canceling coil. A Helmholtz coil can generate a uniform field around the center of the coil, so that the basic SAS can cancel the uniform component of the EMN.

Using one set of the basic SAS, A single direction of the EMN along the axis of a pickup coil can be canceled. In order to cancel the EMN of all directions, the three dimensional SAS (3-D SAS) has to be used, which consists of three sets of the SAS. The remarkable fact has been found out that the orthogonal relation among all of coils of the 3-D SAS is not required to cancel the EMN of all directions. In order to simplify a description, we will prove this fact on the two dimensional SAS (2-D SAS) which consists of two sets of the SAS, 1 and 2. Let us start from the condition that respective magnetic fluctuations in pickup coils of the SAS 1 and 2 are individually to be zero. In the non-orthogonal 2-D SAS, magnetic flux in a pickup coil is obtained as the sum of magnetic flux generated by two cancellation coils and a contribution of the EMN. Thus the conditions are given by

$$B_1\cos\phi_1 - B_2\cos(\phi-\phi_1) + B_x\cos\phi_1 + B_y\sin\phi_1 = 0, \quad (1)$$

$$-B_1\cos(\phi+\phi_2) + B_2\cos\phi_2 - B_x\cos(\phi+\phi_2) + B_y\sin(\phi+\phi_2) = 0, \quad (2)$$

where B_1 and B_2 are magnetic fields generated by the cancellation coils of SAS 1 and 2, respectively, B_x and B_y are two components of the EMN, ϕ_1 and ϕ_2 are angles between the cancellation coil and the pickup coil in the SAS 1 and 2, respectively ($-\pi/2<\phi_1,\phi_2<\pi/2$), and ϕ is an angle between two cancellation coils ($0<\phi<\pi$). B_x is given the same direction as B_1. Rearranging them, we obtain two simple equations :

$$B_x + B_1 - B_2\cos\phi = 0, \quad (3)$$

$$B_y + B_2\sin\phi = 0. \quad (4)$$

These equations hold for all values of ϕ and never include ϕ_1 and ϕ_2. It has become clear that the EMN can be canceled independent of coil directions, so that the orthogonal relation of the SAS is not essential. This remarkable feature makes a great advantage on practical application. The SAS can give full play to its ability without complicated adjustment for its positioning.

3. Performance of the SAS

Performance of the SAS has been demonstrated using some calculations. The calculation has been performed on the basic SAS and the advanced SAS which can cancel the EMN up to first-order gradient field. The advanced SAS consist of the basic SAS for a uniform component of the EMN and the SAS for a component of first-order gradient of it. This type of the SAS has a first-order gradiometer and a cancellation coil which generates a first-order gradient field. This cancellation coil is a pair of coils which keeps a distance to be $\sqrt{3}$ times of its radius and is wound in opposite directions. This coil can generate a uniform gradient field around its center. This SAS and the basic SAS are located co-axially and have the same center.

Fig.2(a) shows size of a shielded space by the basic SAS and the advanced SAS using shielding factors as a parameter. The vertical axis indicates the zone length from the center on the axis of the SAS. The horizontal axis gives the distance of a magnetic dipole as a noise source on the axis of the SAS. These values are normalized by the radius of the cancellation coil. It is seen that the advanced SAS has much higher performance than the basic SAS and can make a large space of a high shielding factor against a noise source in the middle distance. Fig.2(b) shows the field profile in the SAS when the distance of a noise source is 10 (relative units). The vertical axis indicates the central axis of the SAS. The low noise area spreads also to the radius direction, as shown in Fig.2(b). For example, in the case that radius of the cancellation coils is 1 m, the area of 23 cm in height and 60 cm in diameter has more than 1000 of a shielding factor. These results indicate that the advance SAS is practical and useful for the multichannel biomagnetic measurement.

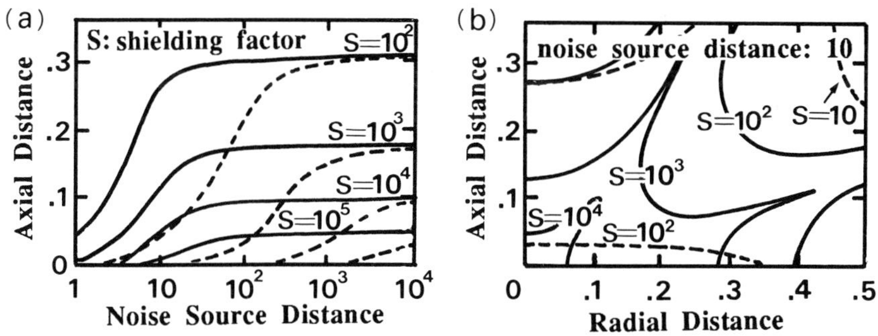

Fig.2 Shielding performance of the advanced SAS and the basic SAS
(a) Noise source distance vs. Shielding space, (b) Field profile in the SAS
Solid line ; the advanced SAS, Dashed line ; the basic SAS

4. Experiment

The preliminary experiment on the basic SAS has been performed, [2]. The experimental system consists of an rf-SQUID magnetometer and a Helmholtz coil with a diameter of 1.59 m. The magnetic noise has been measured by another rf-SQUID magnetometer located at a distance of 10 cm from the pickup coil of the SAS. Fig.3(a) and (b) show magnetic noise when the SAS is dead and alive, respectively. The large noise peak on the chart of Fig.3 (a) is generated by an

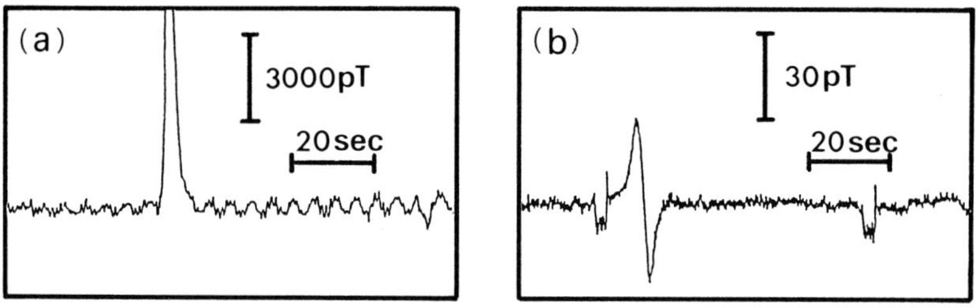

Fig.3 Magnetic field noise measured by a SQUID magnetometer
(a) SAS OFF, (b) SAS ON

automobile on a nearby road. As shown Fig. 3 (b), this peak noise and baseline noise have been reduced to less than two hundredths by the SAS. The effectiveness of the SAS has been actually proved by this experiment.

5. Conclusion

The SQUID based active shield has remarkable advantages for the multichannel biomagnetic measurement. It is the most important point that the SAS actually cancels the EMN in a large area without adjustment. This point is different from another method of electronic noise cancellation, [3]. Another important point is that the SAS is effective in the low frequency region. And the system noise of the SAS is comparable with that of a SQUID. Therefore, the SAS can be used in a magnetically shielded room to enhance the shielding performance. One more important point is that the SAS does not need the orthogonal relation of its coils so that the SAS is easy to construct and use.

References.

1. Sullivan, G.W. and Flynn, E.R., Proc. Biomagnetism'87,486 (1987)
2. Matsumoto, K., Yamagishi, Y., Wakusawa, A., Noda, T., Fujioka, K. and Kuraoka, Y., to be contributed to Biomagnetism'91
3. Robinson, S.E., Advances in Biomagnetism, Edited by Williamson, S.J., Plenum Press, New York, 721, (1989)

Active Magnetic Shielding Support for Biomagnetic Instruments

H. Seifert, A. Wirth, and S. Schneider

Siemens Medical Engineering Group, GG31, Henkestr. 127,
W-8520 Erlangen, Fed. Rep. of Germany

Abstract. An active magnetic shield is described which improves the efficiency of passively shielded rooms at low frequencies. The function of the active shield is limited to homogeneous vertical interference fields which turned out to be most severe. A large area inductance magnetometer surrounding the whole measuring site serves as interference sensor and minimizes interferences with local fields.

1. Problem

In order to measure biomagnetic signals from brain or heart a suppression of ambient magnetic noise of about five orders of magnitude is necessary. At present, the favored solution to reach this value consists of a moderately shielded room with two layers of mu-metal and one layer of aluminum in combination with first-order gradiometers as detection coil system [1]. In the presence of slowly varying magnetic interference fields with frequencies below 1 Hz the total shielding of this configuration is not quite sufficient because eddy currents in the aluminum shell vanish and, in addition, room-made gradients are detected if the measuring position is asymmetric within the room.

Several methods can be used to improve the low frequency shielding, e.g., bigger efforts in room shielding [2], gradiometers of higher order, software gradiometers within fixed arrays or active shielding [3]. Great expenses, deterioration of signal to noise ratio, losses in information or limited applicability imply that only a combination of different methods will improve the solution of the problem.

In this paper an approach with active shielding is described with the intent to reduce interferences which arise from the town mains supply network connecting private consumers along the streets. For safety reasons the neutral conductor forms a network with nodes at the crossings of the streets, while the phases are branch-like installed. So, the electric current flowing through the phase-branch from one direction to a consumer is distributed in two directions when flowing back through the neutral conductor network. Flow and counter-flow of current do not compensate each other resulting in a total current along the whole street on both sides of the consumer. The magnetic field of the uncompensated part of current linearly decreases with distance, hence having a very long range. At a distance of 100 m a current of 1 A produces a nearly perfect homogeneous field of 2 nT in vertical direction.

Low frequency interference results because many electric equipments use half-wave rectifiers in order to reduce power consumption. When such a device switches on and off, slowly varying magnetic fields are produced that are recorded by the biomagnetic system due to reduced total shielding in this frequency range.

In the described case the thermostatic temperature-control of a photocopier 100 m away from the measuring site generates an interference which looks in front of the shielded room like square-wave pulses of 2 nT amplitude and 1 second length following each other either immediately or in a 10 to 20 seconds rhythm. Within the shielded room the slopes of these pulses are rounded due to the low-pass behavior of the shielding factor (Fig. 2).

2. Method

In the simplest case an active magnetic shield consists of field coil and field sensor where the latter detects the superposition of the coil field and the interference fields. An electronic circuit driven by the sensor controls the current through the coil in a way that the total field at the place of the sensor becomes zero. Unfortunately such a system only suppresses that component of the field which is detected by the sensor, e.g., if a magnetometer is used the field in one direction of space is reduced while perpendicular directions and gradients remain unaffected. So, this type of active shielding is most advantageous only if a special type of interference has to be suppressed as in the described case.

In order to handle the vertical interference mentioned above, two adjacend 50-turn coils have been installed horizontally around the shielded room, approximately 1 m distant from the outer mu-metal wall and 2.2 m above the floor. One coil is used as inductance magnetometer to detect the interferences, the other generates the compensation field. With a wire diameter of 0.4 mm and a standard low noise preamplifier (OP27) the magnetometer reaches a sensitivity of 1 pT in unity bandwidth at 1 Hz, sufficient for an application outside the shielded room. The large area of the detection coil increases the sensitivity to homogeneous fields and lowers it to local interference sources which cannot be compensated by the arrangement. Both coils have an inductance of 85 mH, a mutual inductance of 64 mH, a self-resonance of 3 kHz and a resistance of 172 Ohm.

Figure 1 shows the block diagram of the complete set-up. The control amplifier provides the loop gain and suppresses the self-resonance of the coils by zero input and output impedance. The feedback-path through the linearization network compensates for the frequency response of the coil system and generates an integral characteristic of the open-loop gain. Due to the zero impedance loading of the coils, the detection coil behaves like a high-pass, the field coil like a low-pass filter, both together giving a bandpass response with center frequency around 500 Hz. The maximum output voltage of the amplifier is limited to 10 V which feeds a current of 60 mA through the coil. This corresponds to a central field of approximately 600 nT which is sufficient for most interferences and cannot saturate the mu-metal shells of the room.

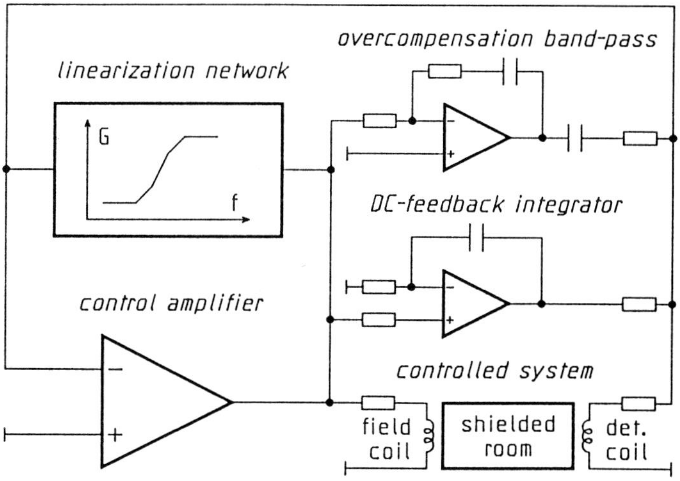

Fig. 1 Block diagram of the active shield

The DC-feedback improves the stability of the feedback loop. At low frequencies the coil assembly has a differential frequency response with vanishing transfer coefficient toward zero frequency. Therefore, in order to reach a low frequency cut-off of the active shield below 0.01 Hz a very high gain of the control amplifier is needed. With lacking feedback through the coil system offsets and low-frequency fluctuations of the amplifier generate magnetic interferences in the field coil. The additional electronic DC-path in parallel to the coil system restores the feedback. By integral path-gain there are no DC-currents in the field coil and the whole arrangement becomes insensitive to mechanical vibrations.

The magnetic field generated by the coil is inhomogeneous. In order to compensate for a homogeneous interference field at the place of the shielded room the total compensation flux through the coil has to be higher than the interference flux. This overcompensation is also achieved by an additional feedback-path in parallel to the coil assembly. It consists of a band-pass filter modelling the frequency response of the coils with inverted polarity. Thus the field coil is forced to generate a field higher than necessary to achieve zero signal from the detection coil.

The described active shield has been installed around a shielded room in the Siemens biomagnetism laboratories in Erlangen. The room was a preliminary version of the standard shielded room available from Vacuumschmelze now. Since 1989 the room is equipped with the 37-channel KRENIKON® SQUID-System serving for MCG and MEG patient measurements.

3. Results

Figure 2 shows the output of 1 magnetometer and 31 gradiometer channels of the biomagnetic system arranged in a vertical but asymmetric position in the shielded

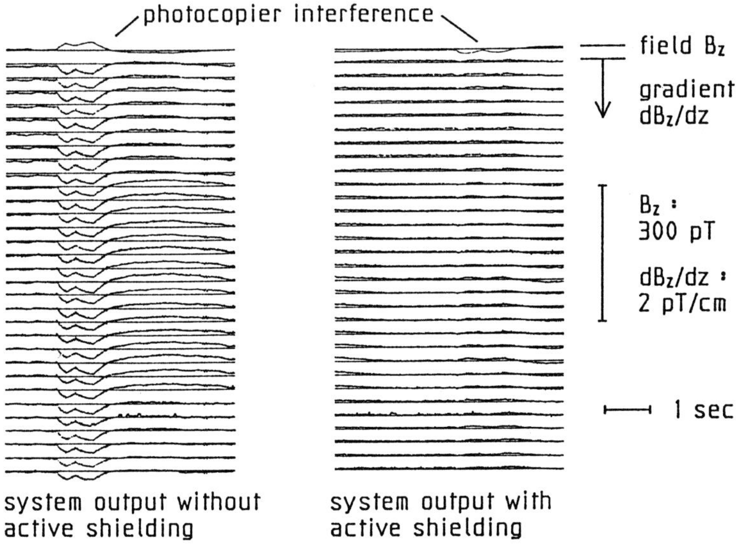

Fig. 2 Biomagnetic system output with interference

room with and without active shielding. The interference is a double pulse of the copier. With active shield the interference in the gradiometer channels nearly vanishes while the signal of the magnetometer is inverted. This is due to the inhomogeneous field generated by the compensation coil which does not allow to suppress fields and gradients at the same time. Also the remaining signals in the gradiometer channels differ slightly due to second order gradients. The noise spectrum of the gradiometer channels shows a reduction of low frequency noise by a factor of 5 when the active shield is switched on in presence of the copier interference.

It turned out that not only the interference originating from the photocopier but also other low frequency interferences of unknown origin are reduced implying that the interference mechanism mentioned above is rather common. The effect of nearby interference sources, such as moving chairs with ferromagnetic parts, is intensified. This behavior is a disadvantage of all active shields of the described type. However, by the use of a large area inductance magnetometer the effect is largely reduced.

References

[1] H.E. Hoenig, G.M. Daalmans, L. Bär, F. Bömmel, A. Paulus, D. Uhl, H.J. Weisse, S. Schneider, H. Seifert, H. Reichenberger, and K. Abraham-Fuchs IEEE Trans. Magn. **MAG-27**, 2777 (1991)

[2] S.N. Erné, H.D. Hahlbohm, H. Scheer, and Z. Trontelj
in: Biomagnetism, eds.: S.N. Erné, H.D. Hahlbohm, and
H. Lübbig, de Gruyter, Berlin, 79 (1981)
[3] V.O. Kelhä, J.M. Pukki, R.S. Peltonen, A.J. Pentinen,
R.J. Ilmoniemi, and J.J. Heino
IEEE Trans. Magn. **MAG-18**, 260 (1982)

Performance of an Electronic Gradiometer in Noisy Environments

D. Drung

Physikalisch-Technische Bundesanstalt (PTB), Abbestr. 10–12, W-1000 Berlin 10, Fed. Rep. of Germany

Abstract. An electronic first-order gradiometer consisting of two fully integrated dc SQUID magnetometers has been operated successfully in different noisy environments: at the PTB Berlin inside and outside the *heavily* shielded room, at a Berlin clinic without a shielded room, and at Siemens Erlangen in a *moderately* shielded room. In the Siemens room, a maximum feedback range of ±3 nT for each magnetometer (10 % of the standard magnetometer feedback range) has been sufficient for safe operation, and noise levels of 14 $fT/\sqrt{Hz}$ at 1 Hz and 540 $fT/\sqrt{Hz}$ at 0.1 Hz have been achieved with a baseline of 20 cm. Based on the noise measurements presented, a novel modular 55-channel system is proposed.

1. Introduction

Present biomagnetic multi-channel systems usually consist of axial first-order gradiometers which are operated inside a *moderately* shielded room [1]. Recently, a fully integrated dc SQUID magnetometer has been developed for the PTB 37-channel system which is designed for operation inside our *heavily* shielded room (BMSR) [2]. The low frequency magnetic noise inside the BMSR can be suppressed strongly by electronically forming gradiometers using additional magnetometers as references, or by using the average signal of some or all sensor magnetometers as a reference [3]. In contrast to standard axial gradiometer systems, magnetometer based systems allow the use of reliable fully integrated sensors which considerably simplify the cryogenic construction and increase the system flexibility [4]. In this paper, it is shown that magnetometer systems can also be operated in a moderately shielded room in noisy clinical environments.

2. Noise Measurements

The tests have been performed using an electronic axial first-order gradiometer consisting of two independent integrated magnetometers [5] with a distance (baseline) of 20 cm from each other and with a balance of about 0.3 %. The noise level of the magnetometers (measured inside a superconducting shield) was about 2 $fT/\sqrt{Hz}$ in the white noise region, 5 $fT/\sqrt{Hz}$ at 1 Hz and 15 $fT/\sqrt{Hz}$ at 0.1 Hz. For operation in shielded rooms, the SQUID system was inserted into a biomagnetic

Springer Proceedings in Physics, Vol. 64 Superconducting Devices and Their Applications
Editors: H. Koch, H. Lübbig

dewar (BMD-15, S.H.E. Corporation) which increased the white noise level of the magnetometers to about 5 $fT/\sqrt{Hz}$. The measurements outside shielded rooms were done in an Al liquid helium storage container which reduced the 50 Hz signal by about a factor of 10 and acted as an efficient rf shield, but did not influence the low frequency magnetic fields below 10 Hz. For unshielded operation, the 4.7 kΩ feedback resistors (R_6 in Ref. [5]) have been replaced by 1 kΩ resistors in order to increase the dynamic range of the magnetometers from the standard ±30 nT to ±140 nT.

The output signal of the lower magnetometer (MAG signal) and the difference signal between both magnetometers (GRAD signal) have been recorded simultaneously and analyzed in the time and frequency domain using a FFT signal analyzer (HP 3562A). For each noise spectrum, at least 90 single FFT spectra have been averaged in order to get accurate results. For the low frequency range (0.1 - 40 Hz), this leads to measurement times of at least half an hour.

At the PTB Berlin, the noise has been measured outside and inside the BMSR (Fig. 1a). Outside the BMSR, the low frequency (0.1 Hz) noise reduction factor (i.e. MAG signal noise level divided by GRAD signal noise level) is about 100, whereas inside the BMSR it is only about 20. The smaller noise reduction factor inside the shielded room may be explained by the deformation of uniform noise fields due to the room walls [6]. This is demonstrated in Fig. 1b where a partial correlation between the MAG signal and the GRAD signal is clearly visible. The low frequency fluctuations inside the BMSR can be minimized (i.e., reduced by about a factor of two) if 3 % of the MAG signal are subtracted from the GRAD signal (curve GRAD - 0.03×MAG in Fig. 1b). Inside the BMSR, the MAG noise is typically 35 $fT/\sqrt{Hz}$ at 1 Hz and 3000 $fT/\sqrt{Hz}$ at 0.1 Hz. It is in good agreement with the noise level extrapolated from the performance of the unshielded system (dashed line in Fig. 1a). The GRAD noise is typically 14 $fT/\sqrt{Hz}$ at 1 Hz and 150 $fT/\sqrt{Hz}$ at 0.1 Hz. The extrapolated low frequency GRAD noise is about a factor of two lower than the measured noise if the above-mentioned deformation of the uniform noise fields is taken into account (dotted line in Fig. 1a).

In a clinical environment (Klinikum Steglitz) in Berlin, the noise of the unshielded system was slightly lower than at the PTB (Fig. 1c). The 0.1 Hz noise reduction factor was again about 100. The low frequency MAG and GRAD signal fluctuations are almost uncorrelated (Fig. 1d). In this environment, the GRAD noise inside a moderately shieded room (dashed and dotted lines in Fig. 1c) is expected to be comparable with the MAG noise inside the BMSR.

The SQUID system has also been operated at Siemens Erlangen inside a moderately shielded room [1]. A maximum feedback range of ±3 nT for each magnetometer (10 % of the standard magnetometer feedback range) has been sufficient for safe operation. The 0.1 Hz noise reduction factor was about 25, and GRAD noise levels of 14 $fT/\sqrt{Hz}$ at 1 Hz and 540 $fT/\sqrt{Hz}$ at 0.1 Hz have been achieved (Fig. 2). The GRAD noise is sufficiently low above about 1 Hz, i.e., for most heart measurements. For sensitive brain studies the noise level below 1 Hz should be further improved, e.g., by using second-order electronic gradiometers.

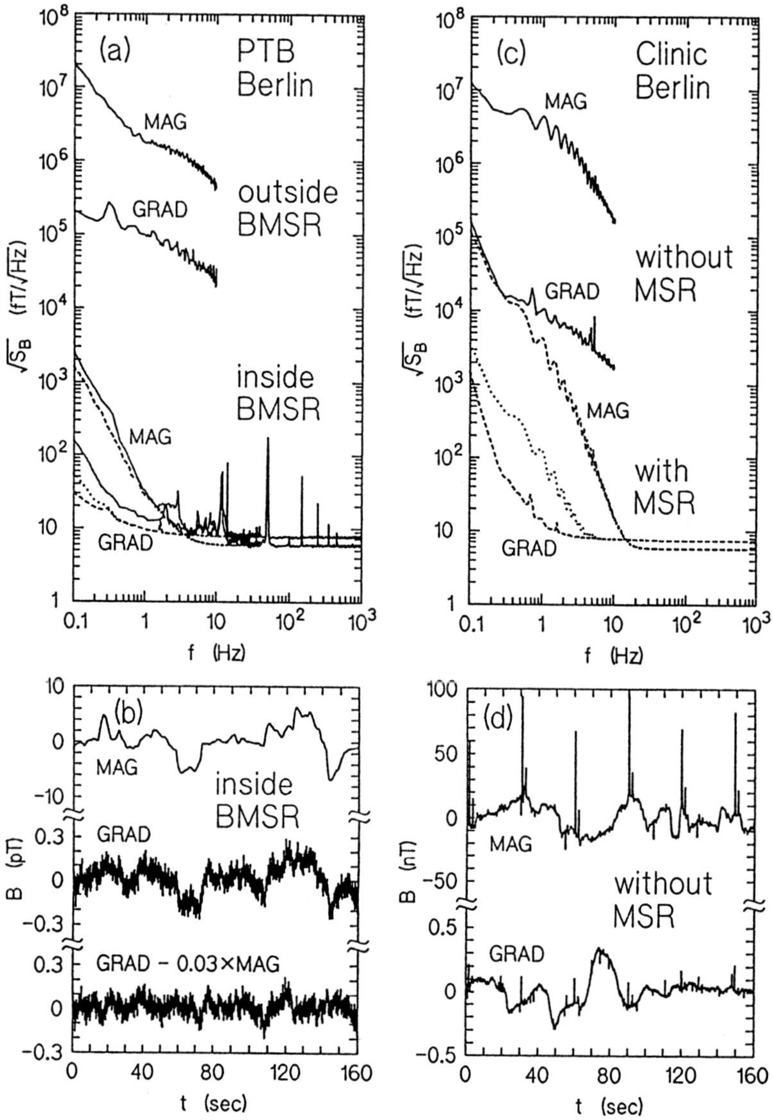

Fig. 1. MAG and GRAD noise (a),(b) at the PTB Berlin and (c),(d) at a clinic (Klinikum Steglitz) in Berlin. Solid lines are measured, dashed lines correspond to the measured noise of the unshielded system divided by the respective shielding factor of (a) the BMSR and (c) a moderately shielded room MSR [1]. The dotted lines are calculated from the dashed lines by adding 3 % of the MAG noise to the GRAD noise. In (b) and (d), the bandwidth is 0.016 - 10 Hz. The curve labelled GRAD - 0.03×MAG is calculated using the digitized MAG and GRAD signals.

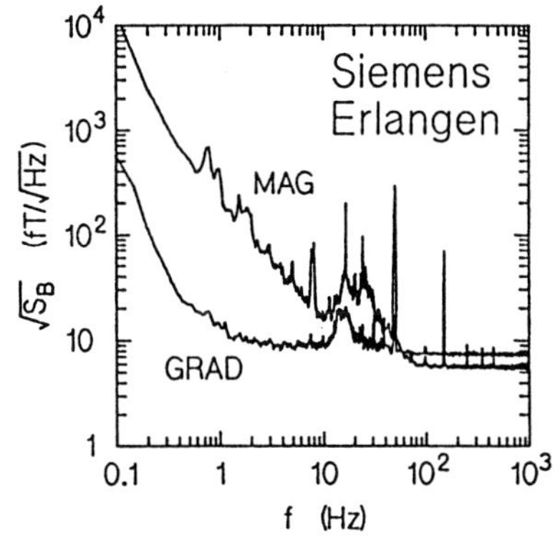

Fig. 2. MAG and GRAD noise spectra measured at Siemens Erlangen inside a moderately shielded room [1]. Very strong magnetic disturbances (caused, e.g., by cars or trucks which sometimes passed by the shielded room at a distance of ≈10 meters) have been suppressed by pausing the average process of the FFT analyzer during the disturbance.

3. Conclusions

It has been demonstrated experimentally that electronic gradiometers can be operated in clinical environments if a *moderately* shielded room is used. This makes it possible to use fully integrated sensors for practical multi-channel systems in clinics. For heart measurements, electronic first-order gradiometers will be sufficient in most cases, whereas for sensitive brain measurements second-order gradiometers are recommended.

Based on these results we propose a modular 55-channel system for clinical shielded rooms (Fig. 3). It consists of seven 7-channel insets with common reference and of six 1-channel insets. The sensors are arranged in a flat hexagonal array with a grid constant of 30 mm, covering an almost circular area with an

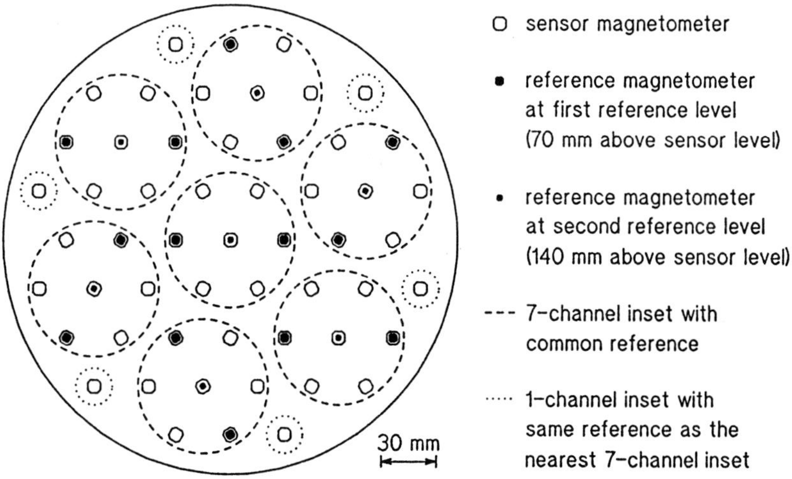

Fig. 3. Modular 55-channel magnetometer. The sensors are arranged in a flat hexagonal array with a grid constant of 30 mm, covering an almost circular area with an average diameter of 205 mm.

average diameter of 205 mm. Each 7-channel unit consists of seven sensor magnetometers, two reference magnetometers at the first reference level (70 mm above the sensor level), and one reference magnetometer at the second reference level (140 mm above the sensor level). The reference signal of each 7-channel unit may be formed using either only reference magnetometers belonging to that unit, or the average signal of all reference magnetometers at the required reference level(s). In the latter case, the white noise contribution of the reference magnetometers is reduced to a negligible value at the expense of a somewhat worse low frequency noise suppression. The proposed system allows the realization of first-order gradiometers with short and long baselines as well as second-order gradiometers, depending on the actual measurement requirements. It will be a very flexible and efficient tool for biomagnetic applications.

Acknowledgements. I would like to thank the Siemens Medical Engineering Group for offering the opportunity to carry out noise measurements in their magnetically shielded room in Erlangen, and Dr. H. Seifert for his assistance during these measurements. This work has been partially supported by the Senator für Wirtschaft und Arbeit, Berlin, under ERP-contract #2597.

References

[1] H.E. Hoenig, G.M. Daalmans, L. Bär, F. Bömmel, A. Paulus, D. Uhl, H.J. Weisse, S. Schneider, H. Seifert, H. Reichenberger, and K. Abraham-Fuchs, IEEE Trans. Magn. **MAG-27**, 2777 (1991).

[2] H. Koch, R. Cantor, D. Drung, S.N. Erné, K.P. Matthies, M. Peters, T. Ryhänen, H.J. Scheer, and H.D. Hahlbohm, IEEE Trans. Magn. **MAG-27**, 2793 (1991).

[3] D. Drung, R. Zimmermann, R. Cantor, S.N. Erné, H. Koch, K.P. Matthies, M. Peters, H.J. Scheer, and D. Stollfuß, Clin. Phys. Physiol. Meas. (in press).

[4] D. Drung, in *Extended Abstracts of Third Int. Superconductive Electronics Conf.* (Glasgow, 1991), p. 252 and Supercond. Sci. Technol. (to be published).

[5] D. Drung, these Proceedings.

[6] H.J.M. ter Brake, H.J. Wieringa, and H. Rogalla, Meas. Sci. Technol. **2**, 596 (1991).

Gradiometer Response Analysis for Current Multipole Sources

W. Haberkorn

Physikalisch-Technische Bundesanstalt, Institut Berlin,
Abbestr. 2–12, W-1000 Berlin 10, Fed. Rep. of Germany

Abstract. The paper presents an analytical study on the first order gradiometer effects for current multipoles as biomagnetic field sources. A tensor representation of the field expansion up to the current octupole is given and applied for the multipole description of an elementary source model. The sensitivity of first order gradiometers to the higher order poles is analysed and evaluated.

1. Introduction

The main purpose of the measurement of biomagnetic fields and their derivatives is to obtain information about the field sources. For a simple response analysis the current dipole can be used as source model. More complex source structures can be described by means of the current multipole expansion. In this paper a tensor formulation of the field expansion performed up to the current octupole term is presented. In order to determine the gradiometer response the first derivatives of the field expansion are calculated. The general expressions are applied for a detailed study of the cylindrical source model. For this elementary source type the information content of the multipole moments is investigated. Furthermore, the sensitivity of the first order gradiometer response to the higher order poles with respect to the dipole term is analysed. The results indicate that this sensitivity can be larger than that of the magnetometer.

2. Tensor analysis

The current multipole expansion of the source magnetic field due to primary current distributions can be written as

$$B_a = B_a{}^d + B_a{}^q + B_a{}^o + \cdots .$$

The first term represents the well-known current dipole field

$$B_a{}^d = -(\mu_o/4\pi r^3) g_{abc} x_b P_c , \qquad P_c = \int j_c(r') dv' .$$

Here, g_{abc} equals $+1(-1)$ for abc an even (odd) permutation of 123 and zero otherwise; $x_1 = x$, $x_2 = y$, and $x_3 = z$ are the Cartesian coordinates. P_c denotes the current dipole components, and j_c represents the components of the primary current density. The integration is performed over the volume of the current distribution. Furthermore, the summation convention is applied.

The second contribution to the magnetic field describes the current quadrupole and is given by

$$B_a^q = -(\mu_o/4\pi r^5)g_{abc}(3x_ix_b - r^2\delta_{ib})Q_{ic}\ , \qquad Q_{ic} = \int x_i' j_c(\mathbf{r}')dv'\ ,$$

where δ_{ib} is the Kronecker delta, and Q_{ic} denotes the current quadrupole tensor [1].

The third term of the field expansion represents the field of the current octupole which can be shown to be

$$B_a^o = -(\mu_o/8\pi r^7)g_{abc}(5x_ix_jx_b - 2x_jr^2\delta_{ib})O_{ijc}\ ,$$
$$O_{ijc} = \int(3x_i'x_j' - r'^2\delta_{ij})j_c(\mathbf{r}')dv'\ .$$

O_{ijc} denotes the current octupole tensor introduced here. This third-rank tensor satisfies

$$O_{ijc} = O_{jic} \qquad \text{and} \qquad O_{iic} = 0\ .$$

From these features it follows that there are at most 15 independent current octupole tensor components.

The current dipole components P_c may be shown to be independent of the origin chosen for the multipole expansion, while the higher order moments depend on the origin. A displacement of the origin to a point $\mathbf{r}^*$ yields for the new current quadrupole tensor Q_{ic}^* the equation

$$Q_{ic}^* = Q_{ic} - x_i^*P_c$$

and for the new current octupole tensor O_{ijc}^* the relation

$$O_{ijc}^* = O_{ijc} - 3x_i^*Q_{jc} - 3x_j^*Q_{ic} + 2\delta_{ij}x_n^*Q_{nc} + (3x_i^*x_j^* - r^{*2}\delta_{ij})P_c\ .$$

These shift equations can be used to relate two current multipole expansions computed for the same source but choosing different origins.

Note that a multipole representation can also be obtained using spherical coordinates or other types of orthogonal coordinate systems. The relationship between the tensor formulation given here and the spherical harmonic multipole expansion may be found on the basis of [2].

The analysis of the first order gradiometer effects demands the calculation of the field derivatives $\partial B_a/\partial x_k = B_{a,k}$ which represent the components of a second-rank traceless tensor. The corresponding contributions of the current dipole, the current quadrupole, and the current octupole are found to be

$$B_{a,k}^d = (\mu_o/4\pi r^5)g_{abc}(3x_kx_b - r^2\delta_{kb})P_c\ ,$$
$$B_{a,k}^q = (3\mu_o/4\pi r^7)g_{abc}(5x_ix_kx_b - r^2f)Q_{ic}\ ,$$

$$f = x_i\delta_{kb} + x_k\delta_{ib} + x_b\delta_{ik} \ ,$$

$$B_{a,k}{}^{o} = (\mu_o/8\pi r^9) g_{abc}(35x_ix_jx_kx_b - 5r^2h + 2r^4\delta_{ib}\delta_{jk})O_{ijc} \ ,$$

$$h = x_ix_j\delta_{kb} + x_ix_b\delta_{jk} + x_jx_b\delta_{ik} + 2x_jx_k\delta_{ib} \ .$$

The current multipole is assumed to be positioned in a homogeneous volume conductor with a surface parallel to the x-y plane. For the response analysis we treat the field component normal to the surface, i. e. the z-component B_3, which is not influenced by the half-space model.

3. Gradiometer effects

The response ΔB_3 of a $B_{3,k}$ gradiometer is defined by

$$\Delta B_3/\Delta x_k = - B_{3,k} \ ,$$

where Δx_k is the gradiometer baseline. To evaluate the gradiometer effects we compare the sensitivity of ΔB_3 to the higher order poles with that of the magnetometer response B_3. In the source model considered here the primary current density component $j_1(x)$ is assumed to be confined to a cylinder of radius R and of end surfaces at $x = \pm L$. This elementary source model can be used to describe a variety of existing biomagnetic field sources, e. g. brain sources, heart sources, and nerves. For the current multipole moments we obtain

$$P_1 = F\int j_1(x')dx' \ , \qquad Q_{11} = F\int x' j_1(x')dx' \ ,$$

$$O_{111} = 2F\int x'^2 j_1(x')dx' - R^2P_1/2 \ , \qquad O_{221} = O_{331} = - O_{111}/2 \ .$$

$F = \pi R^2$ is the cross-sectional area of the cylinder. The integration is carried out along the cylinder axis. The different multipole moments contain detailed information about the shape and the constitution of the current source. For constant current densities Q_{11} vanishes and the other moments are given by

$$P_1 = 2LFj_1 \ , \qquad O_{111} = (4L^2 - 3R^2)P_1/6 \ .$$

Considering for the $B_{3,3}$ gradiometer the ratio of the gradiometer response due to the quadrupole and the dipole one obtains

$$\Delta B_3{}^q/\Delta B_3{}^d = (5/3)\, B_3{}^q/B_3{}^d \ ,$$

where $B_3{}^q = 0$ for $x = 0$. This result shows that the sensitivity of the gradiometer to the current quadrupole is larger than that of the magnetometer.

Furthermore, the ratio of the octupole term to the dipole is found to be

$$\Delta B_3{}^o/\Delta B_3{}^d = (\alpha 5/3)\, B_3{}^o/B_3{}^d \ ,$$

where $\alpha = (7x^2 - r^2)/(5x^2 - r^2)$. Considering the special case of $x = 0$ we have $\alpha = 1$. For $4x^2 = y^2 + z^2$ the ratio $\Delta B_3{}^o/\Delta B_3{}^d$ is given by

$$\Delta B_3^{o}/\Delta B_3^{d} = 2O_{111}/5(y^2 + z^2)P_1 \ ,$$

whereas the ratio B_3^{o}/B_3^{d} equals zero.

Note that a similar analysis for $B_{3,1}$ and $B_{3,2}$ gradiometers leads to more complicated relations.

In conclusion, the response of first order gradiometers to current multipoles as field sources has been analysed. The main result shown here is that the sensitivity to higher order poles with respect to the dipole term may be larger for gradiometers compared with magnetometers, which may be of particular interest for biomagnetic applications.

Acknowledgment

The author wishes to thank H. Lübbig for helpful comments and valuable suggestions.

References

[1] T. Katila: Il Nuovo Cimento **2D**, 660 (1983)

[2] G. Albrecht, M. Burghoff, F. Gießler, W. Haberkorn, H. Nowak: In *Biomagnetism*, ed. by S. J. Williamson, M. Hoke, G. Stroink, M. Kotani (Plenum Press, New York 1989) pp. 349-352

Analyzing Multichannel Magnetometer Data

C.D. Tesche

IBM Research Division, T.J. Watson Research Center,
P.O. Box 218, Yorktown Heights, NY 10598, USA

Abstract A method based on lead field analysis is described for generating a unique set of current dipole sources which reproduce exactly the magnetic flux detected by an array of SQUID magnetometers. The dipole sources are located on a grid of points which have been selected considering all available information about the source. This information may include anatomical or structural data and/or current source distributions obtained from conventional inversion routines for the magnetometer data. The number of current dipole sources scales with the number of magnetometers in the array without involving a lengthy iteration scheme. The technique is particularly appropriate for complex current sources which cannot be reconstructed by a small number of current dipoles. In addition, a figure of merit is defined for the reconstruction of a given current dipole distribution. This figure of merit is used to evaluate the performance of a variety of pick-up coil arrays.

1. Introduction

Arrays of SQUID magnetometers are now used with other imaging modalities to measure current distribution in biological and non-biological samples. However, analysis of magnetometer data is complicated by the fact that there is no unique inversion of magnetic field data which has been measured at points external to a source volume into a current distribution defined within the source volume. A common inversion technique is to approximate the current distribution by a finite number of current dipoles. [1] The strength and location of the dipoles are taken as variables. This technique involves a lengthy iteration of the source locations for even a small number of sources. In addition, the reconstructed sources may be inconsistent with anatomical or structural features of the sample. The inversion problem in SQUID magnetometer arrays is further complicated by the fact that not all components of an arbitrary current distribution link a net flux into an array of SQUID pick-up coils.

2. Lead Field Analysis

The problem is to find current distributions which reproduce the measured flux values exactly, and then eliminate those which are inconsistent with other data

Springer Proceedings in Physics, Vol. 64 Superconducting Devices and Their Applications
Editors: H. Koch, H. Lübbig

sets and/or cannot be detected with the coil array. Lead field analysis is useful because it automatically restricts the set of current reconstructions to those which produce responses in the SQUID coils. [3,4]

The coil lead field, $\vec{L}_i(\vec{r})$, which links a current source, $\vec{J}(\vec{r})$, to the flux, ϕ_i, coupled into the ith pick-up coil in an array of N magnetometers satisfies [2]

$$\phi_i = \int \vec{L}_i(\vec{r}) \cdot \vec{J}(\vec{r}) dV. \tag{1}$$

The coil lead field is determined at all points in the source volume by the geometry of the pick-up coil and the conductivity of the source volume. For simplicity, we restrict this discussion to geometries in which volume currents are insignificant. The analysis can be extended to apply to any linear conductor model. The coil lead field for the ith magnetometer is

$$\vec{L}_i(\vec{r}) = (\mu_0/4\pi)\oint d\vec{l}_s/|\vec{r} - \vec{r}_s|, \tag{2}$$

where $\vec{r}_s$ is the location of the line element $d\vec{l}_s$ integrated around the pick-up coil.

A current source $\vec{J}_c(\vec{r})$ which reproduces exactly the measured flux values can be reconstructed as a vector sum over the lead fields [3,4]

$$\vec{J}_c(\vec{r}) = \sum_{i=1}^{N} \alpha_i \vec{L}_i(\vec{r}), \tag{3}$$

The reconstruction coefficients α_i are determined by the inner product of the lead fields,

$$< L_i, L_j > = \int \vec{L}_i(\vec{r}) \cdot \vec{L}_j(\vec{r}) dV. \tag{4}$$

and the flux signals ϕ_i. It is convenient to define a signal vector $\Phi = (\phi_1, \ldots, \phi_N)^T$ and a vector of reconstruction coefficients $A = (\alpha_1, \ldots, \alpha_N)^T$. The reconstruction coefficients then satisfy $\Phi = \Gamma A$, where $\Gamma_{ij} = < L_i, L_j >$ is the inner product of the lead fields. For orthogonal lead fields, $\Gamma_{ij} = \delta_{ij}$, and $\alpha_i = \phi_i$. [4]

The lead fields in Eqn. 3 can be computed at any point $\vec{r}$. However, this generates a reconstructed current source which is dominated by the structure of the lead fields. Thus, although this reconstruction is consistent with the information about the source which has been detected by the coils, it clearly contains arbitrary components which may be misleading. However, if the number of points over which the lead field reconstruction is determined is limited to N/2 for a planar sample and N/3 for a three dimensional source, then Eqn. 3 defines a unique current dipole at each point in the grid of reconstruction points. This reconstructed current distribution is the smallest number of current dipoles defined on this specific grid of points which is consistent with the pick-up coil structure and measured flux values. As a result, if the reconstruction grid points are chosen using all available information as a guide, a well defined current dipole distribution is determined which is consistent with information available both from the magnetometer data and from all other data sets.

If a finer mesh of reconstruction points is desired, either the number of magnetometer channels must be increased or a series of measurements at shifted locations of the coils with respect to the sample must be made. Interpolation between the grid of points does not, in general, yield dipole sources which are meaningful in the sense of generating the observed magnetometer signals. In addition, the reconstructed current dipole sources reflect the presence of additive noise in the SQUID magnetometer system. As a result, the reconstructed dipole sources for a fixed grid of points depend on the signal to noise ratio for each magnetometer channel, and thus on the match between the coil lead field and the actual current distribution in the source volume.

3. Examples

The elements of a multichannel array are most sensitive to nearby sources. To illustrate the algorithm, a subset of coils are used to reconstruct current sources on a nearby grid of points. Reconstructed current distributions are computed for three arrays of pick-up coils: vertical, planar and axial (Fig. 1). In the vertical array, four pairs of square coils are mounted vertically around the perimeter of the pick-up area. Each coil is connected in series with a coplanar balance coil at a height Z_v. The planar array consists of eight first order gradiometer coils. In the axial array, nine square coplanar pick-up coils are each connected in series with a balance coil located at a distance Z_a above the plane of the pick-up coils. The area spanned by all the coil sets is the same. For MEG applications the area is taken to be 4 cm x 4 cm, with Z_v,Z_a = 10 cm. The reconstruction grid is located in a plane parallel to the pick-up coils at a distance D (Fig. 1). The planar and vertical arrays are centered above the grid. The axial set is most effective when offset by (-0.4 cm, 0.7 cm).

The reconstruction of a current dipole source is illustrated in Figs. 2-3. In Fig. 2, the source is located on a reconstruction grid point. The dipole strength

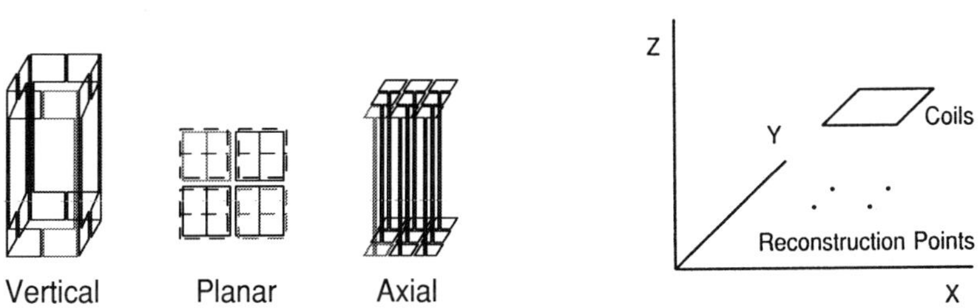

Figure 1. Configuration of the vertical, planar and axial pick-up coil arrays. The relative position of the coils and source plane are shown on the right.

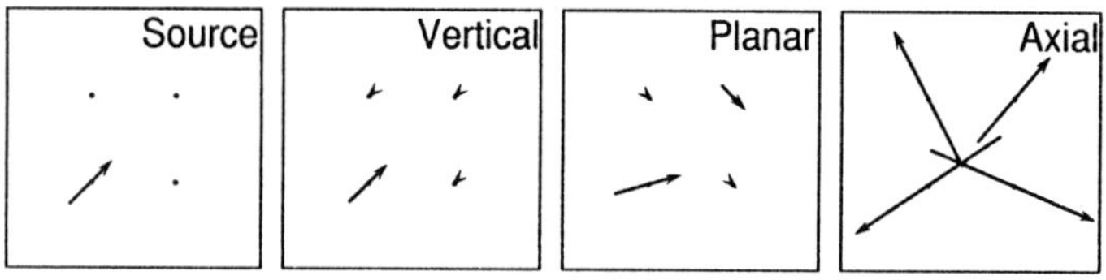

Figure 2. Reconstruction of a 20 pAm source located on a reconstruction grid point at a depth of 2 cm by the various arrays. The solutions for the axial set are reduced by a factor of 20.

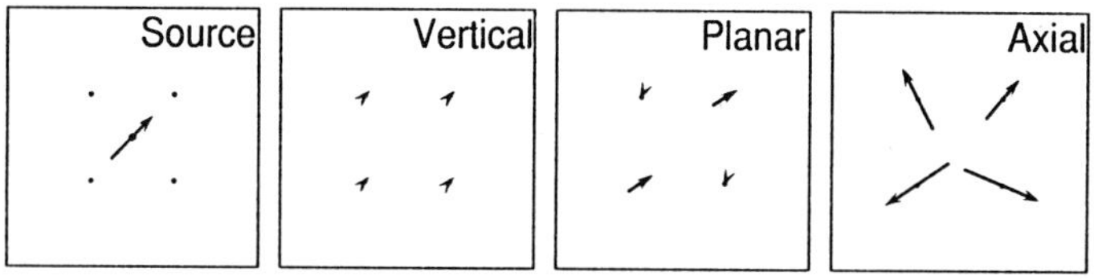

Figure 3. Reconstruction of a 0.1 nAm source located in the center of the reconstruction grid at a depth of 2 cm by the various arrays. The solutions for the axial set are reduced by a factor of 20.

is 20 pAm and the SQUID noise is 7 x $10^{-7}\phi_0/\sqrt{Hz}$ for 100 averages of the data in a 1 KHz bandwidth. If the signal strength scales with the SQUID noise, the reconstruction remains constant. As can be seen from the figure, the vertical coil array produces an excellent reconstruction of the data at this signal to noise level. The commonly used axial array gives a relatively poor reconstruction. All three coil arrays reproduce the current source exactly in the absence of noise.

Although the coils and grid may be selected to give the best possible match to known features of the sources, it is likely that the actual current sources may lie off the grid points. The reconstruction of a current source located at the center of the four grid points is illustrated in Fig. 3.

Reasonable reconstruction of the signal is obtained at a source level of 1 nAm with both the vertical and planar coils, although the reconstruction by the planar coils is somewhat more sensitive to the SQUID noise. The axial set fail to give an adequate reconstruction of the source even at this enhanced signal strength, even though the flux values in the coils are reproduced exactly.

4. Figure of Merit for Reconstruction of a Source

A useful figure of merit for the ability of an array of SQUID magnetometers to reconstruct a specific current source over a fixed grid of reconstruction points is the average over the grid of the norm of the difference between the reconstructed current and the source current. It is convenient to normalize the difference by the average of the norm of the source current. Thus, the Differ-

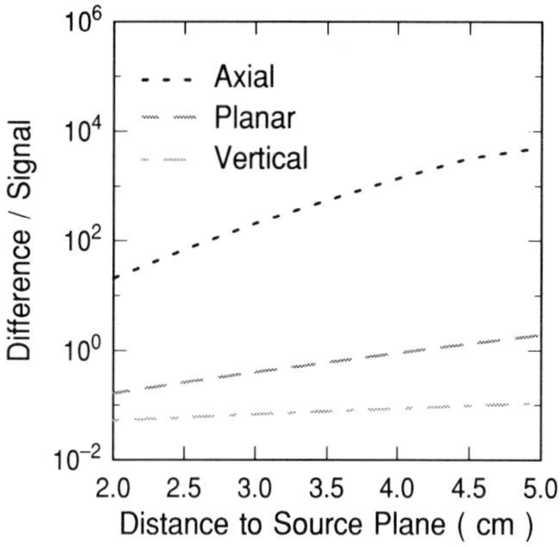

Figure 4. Dependence of the figure of merit for the reconstruction of the current source, Difference / Signal, as a function of the distance D to the source plane for various coil arrays.

ence / Signal, D/S , is

$$D/S = \frac{\sum_{i=1}^{N} \|\vec{J}(\vec{r}_i) - \vec{J}_c(\vec{r}_i)\|}{\sum_{i=1}^{N} \|\vec{J}(\vec{r}_i)\|} . \tag{5}$$

The figure of merit is plotted in Figure 4 for the various coil configurations as a function of the distance D to the source plane. The dipole strength is 0.1 nAm. The vertical coil array is clearly superior over the range D = 2 - 5 cm. This coil array size, depth and signal-to-noise ratio are typical of MEG arrays (both SQUID noise and signal are increased by x 10).

5. Conclusions

Lead field analysis has been used to reconstruct the smallest set of current dipoles defined over a grid of points which reproduces the measured flux values exactly. The reconstruction grid is chosen utilizing all available data which may include conventional reconstructions of magnetometer data and/or data from other modalities. This provides a simple method for integrating the data sets. The technique is particularly appropriate for complex sources which are appropriately modeled by more than one current dipole. In addition, calculations of a figure of merit for reconstruction indicates that vertical pick-up coils are superior to axial and planar gradiometer coils when used with this algorithm.

References

1. G. L. Romani and R. Leoni, " Localization of cerebral sources with neuromagnetic measurements", Biomagnetism: Applications and Theory, Pergammon Press, (1985) New York-Toronto, pp. 205-220.

2. S. J. Williamson and L. Kaufman, "Magnetic Fields of the cerebral cortex", Biomagnetism, Walter de Gruyter, (1981) Berlin, pp. 353-402.

3. M. S. Hamalainen and R. J. Ilmoniemi, "Interpreting measured magnetic fields of the brain: Estimates of current distributions," Tech. Rep. TKK-F-A559, 1984) Helsinki University of Technology.

4. M. S. Hamalainen and R. J. Ilmoniemi, "Interpreting magnetic fields of the brain", submitted to IEEE Transactions on Biomedical Engineering.

Measurements on Magnetic Samples with a DC-SQUID Working at mK Temperatures in Magnetic Fields up to 6 mT

M. Bühler[1], *E. Umlauf*[1], *D. Drung*[2], *and H. Koch*[2]

[1]Walther-Meissner-Institut, Walther-Meissner-Str. 8, W-8046 Garching, Fed. Rep. of Germany

[2]Physikalisch-Technisches Bundesanstalt, Institut Berlin, Abbestr. 2–12, W-1000 Berlin 10, Fed. Rep. of Germany

A dc SQUID magnetometer, integrated on a 7.2 x 7.2 mm^2 chip with 8 pick-up loops in parallel has been applied at millikelvin temperatures with a direct coupling to the magnetic samples. The SQUID was mounted in a superconducting hollow cylinder with frozen-in fields up to 6 mT, and only the Josephson junctions were shielded by attaching a foil of Nb_3Sn (diameter 1mm) on the chip. The experimental set-up is used as a magnetic calorimeter for particle detection with high energy resolution.

1. Introduction

An integrated dc SQUID magnetometer /1/ is applied in connection with a magnetic calorimeter. It is well known that paramagnetic samples can be used for thermometry. If such a sample is suspended semiadiabatically, the increase of temperature ΔT on inserting a heat-pulse changes the magnetic susceptibility by

$$\Delta\chi = - C\,(T + \Theta)^{-2}\,\Delta T \qquad (1)$$

with C being the Curie-constant and Θ the Curie-Weiß temperature of the magnetic material respectively. This gives a step-like or pulse-shaped signal, depending on the thermal link of the sample to the bath. Heat pulses can be applied by the absorption of light pulses or single α-particles or γ-quanta.

When the magnetic thermometer is in thermal contact with another sample, which is the absorber, the increase of the temperature ΔT on energy absorption ΔE depends on the heat capacities C_A and C_T of the absorber and the thermometer, respectively:

$$\Delta E = (C_A + C_T)\,\Delta T \qquad (2)$$

If C_T has been determined by performing a measurement with the thermometer alone, then additional heat capacities $C_A(T)$ can be measured. This is the way to investigate the specific heat of materials. On the other hand, we can perform a calibration of the compound calorimeter with well defined heat pulses and then measure unknown absorbed energies of particles, quanta or light pulses. This is our application of the calorimeter. The aim of this project is to detect neutrinos through the recoil energy transferred to a nucleus by coherent elastic scattering /2/. In such an experiment, absorbers, as big as possible (about 1 kg), should be applied. In order to keep the heat capacity C_A small enough, very pure insulating materials with high Debye-temperatures must be used, e. g. single crystals of sapphire or silicon, and very low temperatures. For a given compound detector the energy resolution is given by the temperature resolution (equ. 2) and that is determined by the smallest detectable value of Dc (equ. 1) with the SQUID magnetometer. For detailed considerations in respect to the energy sensitivity of the magnetic calorimeter see refs. 3 and 4.

In earlier measurements we applied the conventional experimental set-up with a commercial SQUID at 4.2 K in connection with a flux transformer and the sample at lower temperature in a magnetic field. In this way we could not fully utilize the sensitivity of the SQUID, due to background signals, arising mainly from vibrations. Therefore, we tried to mount an integrated SQUID magnetometer on the mixing chamber of a home-made refrigerator, in order to obtain a direct coupling of the sample to the pick-up coil of the SQUID. The SQUID magnetometer had a planar design and was sputtered on a silicon chip.

2. The SQUID characteristics

We investigated two chips (SQ1 and SQ2) with the same type of SQUID magnetometer, because SQ1 had been damaged mechanically after some time. Both magnetometers exhibited qualitatively the same behavior. Nevertheless we indicate, on the data given below, which SQUID had been investigated. In zero field (earth field) both SQUIDs worked without problems at lower temperatures, down to 30 mK, exhibiting a nearly unchanged SQUID critical current I_c. The noise was not reduced with decreasing temperature, because it was determined by the electronics, designed similarly to that described in ref. 1.

The main problem was, as expected, the application of magnetic fields of 3-6 mT, necessary for our measurements. On applying 1 mT we first found on SQ2 a reduction of the critical current I_c by a factor of 2 and a reduction of V_F = dV/dF at the optimum bias current. With that, the noise increased considerably and no sensitive measurement could be performed. In order to shield the Josephson junctions which are in the center of eight pick-up loops in parallel, we covered this region with a foil of Nb_3Sn (diameter 1 mm), glued on the chip. Of course, this foil could not shield the magnetic field completely, and consequently we still found a reduction of I_c and V_F, but the influence of the magnetic field was now much lower. As an example we show in Fig. 1 the I-V characteristics of SQ2 with zero field and 3 mT. The reduction of I_c is only 25 percent. At zero field the optimum bias current (indicated by the arrow) is in

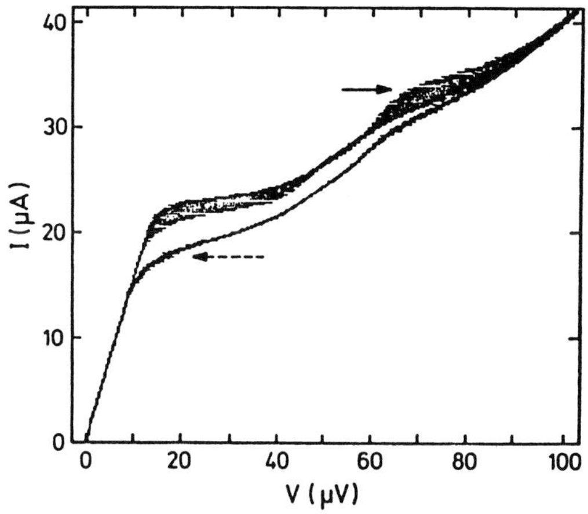

Fig. 1. The I-V characteristic of the SQUID SQ2, in zero field (upper plot) and with 4 mT.The arrows indicate the applied bias current for optimum dV/dΦ.

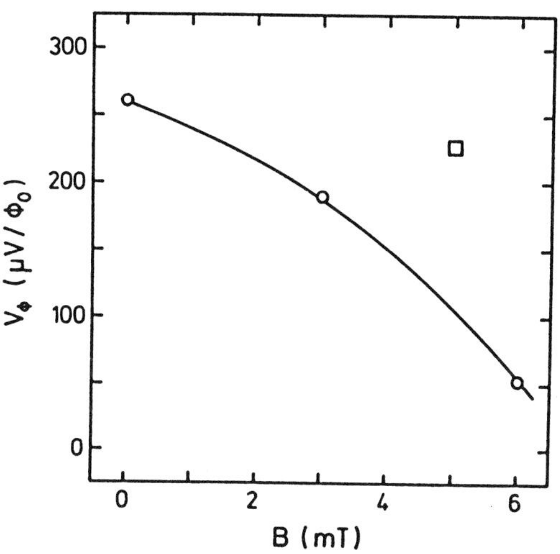

Fig. 2. V_Φ vs. the applied magnetic field B.Bias currents at the first resonance (circles) and near the critical current (square), cf. Fig. 1.

the range of the resonance-step, because the first step near I_C is hysteretic. With the applied field this hysteresis vanishes and the optimum working point was found with a bias current adjusted to the step at I_C (dashed arrow). The I-V characteristic at zero field has been measured with the superimposed 50 Hz field of the surroundings. On applying 3 mT with a solenoid the 50 Hz field was enhanced. Therefore the displayed modulation of the I-V characteristic is reduced due to the filter characteristic of the preamplifier. The initial slope is determined by the resistance of the current leads. These measurements were performed at 4.2 K, above the transition temperature of the superconducting hollow cylinder around the SQUID. This hollow cylinder made from tin is used for shielding external magnetic fields as well as for freezing-in magnetic fields. In Fig. 2 V_F at the optimum bias current is plotted vs. the applied field. The values decrease monotonically if the bias current remains adjusted to the resonance step. However, if the bias current can be changed to the step at I_C, because the hysteresis vanishes in a magnetic field, then V_F is enhanced. The corresponding value at 5 mT (square point) shows, that in this way with a magnetic field V_F can take approximately the same value as in zero field.

The rms noise spectrum of the SQUID (SQ1) measured at zero field in the flux locked mode has been found to be very close to that shown in ref. 1. The white noise level was found to be 3fT/ÖHz. On applying magnetic fields, the noise level increased only according to the reduction of V_F, as given Fig. 2, but no remarkable enhancement of the noise due to vibrations has been found.

3. Experimental problems

Applying the SQUID in magnetic fields some peripheral problems had to be solved, arising especially at very low temperatures. The SQUID-chip was glued on a mounting plate of glass fiber which contains magnetic impurities and gives therefore a signal for any change of temperature. Due to this effect, the temperature dependent magnetic susceptibility of the samples could not be measured exactly. In addition, below about 0.3 K, the SQUID-chip and the mounting plate were slightly heated up by the feedback current, presumably due to a resistance at the bond pads. This effect resulted in disturbed detector signals. We substituted the mounting plate by another one of pure silicon and found the disturbing signals to be considerably lower. We assume that the effects still remaining result from the palladium, sputtered on the contact pads and for resistive elements in the integrated circiut. Finally, we substituted the feedback loop by another one to avoid the heating and in this way we were able to measure undisturbed detector signals. This feedback loop was simply made with a superconducting wire glued to the backside of the mounting plate.

The superconducting hollow cylinder for freezing in the magnetic field had an inner diameter of 25 mm corresponding to the mounting plate of the SQUID. The length of the cylinder was 120 mm and the SQUID was mounted at the midpoint. We measured the shielding of the SQUID with respect to small external magnetic fields and found a shielding factor of 1.7×10^5. In addition we mounted a superconducting wire lattice on each hole of the cylinder; one of these lattices however had a hole of 8 mm diameter for the sample holder. In this way the shielding factor could be enhanced to 1.4×10^6. Furthermore, it should be mentioned, that the earth field had not been shielded when cooling down the SQUID.

The energy coupling coefficient of a disc-shaped sample (diameter 7 mm, thickness 2 mm) was found to be only about 30 %. This small value is partly due to the special geometry of the pick-up coil. With a simple circular pick-up coil one could expect a value of about 60 % and a correspondingly higher energy resolution. A new design of the SQUID is in preparation, where the 8 loops in parallel are maintained, but nevertheless the coupling coefficient of 60 % should be reached approximately.

Finally we want to emphasize that measurements on the level of the SQUID noise are possible only if the experimental set-up is made very carefully, in order to avoid disturbances arising mainly from vibrations. In particular, superconducting leads must be fixed, even far outside the shielding hollow cylinder. In order to avoid the transmission of vibration to the cryostat we constructed a single-shot dilution refrigerator, which works with absorption pumps only and which has no liquid nitrogen bath. Furthermore, the cryostat can be suspended with a single nylon rope from the ceiling.

4. Summary

The integrated SQUID magnetometer can be applied in magnetic fields up to at least 6 mT without an essential loss in sensitivity. In this way, measurements on samples in corresponding fields can be performed at all temperatures below 4.2 K, without using a flux transformer. Measurements at still higher fields with improved shielding of the Josephson junctions might be possible. Applying this SQUID magnetometer we were able to improve the energy sensitivity of the calorimetric detector by more than an order of magnitude, compared with the earlier measurements, because we were able to utilize fully the sensitivity of the integrated SQUID magnetometer.

This work was supported by the Bundesministerium für Forschung und Technologie.

References

[1] D. Drung, R. Cantor, M. Peters, T. Ryhänen and H. Koch, IEEE Trans. Magn. **MAG-27**, 3001 (1991)

[2] A. Drukier and L. Stodolsky, Phys. Rev. **D 30**, 2295 (1984)

[3] M. Bühler and E. Umlauf, Europhysics Lett. **5**, 297 (1988)

[4] E. Umlauf and M. Bühler, in Superconducting and Low Temperature Particle Detectors (Edited by G. Waysand and G. Chardin) Elsevier Science Publ., Amsterdam (1989)

A Miniature DC-SQUID Magnetometer with Current Injection Feedback

G.J. Sloggett, L. Wieczorek, K.E. Leslie, R.A. Binks, and A. Bendavid

CSIRO Division of Applied Physics, P.O. Box 218,
Lindfield, NSW 2070, Australia

Abstract. A miniature niobium d.c. SQUID magnetometer into which feedback flux is coupled by direct current injection is described. Its sensitivity in flux-locked loop operation is shown to be indistinguishable from that obtained using an external coil for feedback flux coupling. The field sensitivity is below 21 $pT/\sqrt{Hz}$ for frequencies greater than 1 Hz and the energy sensitivity at 1 kHz is 210 times Planck's constant, h.

The SQUID magnetometer described here was developed primarily as a field sensor for use in electronic balancing of SQUID gradiometers, along the lines first suggested by Williamson et al [1]. For this application the field sensitivity required is modest. A symmetrically designed thin-film gradiometer should have an effective unbalance area of 10^{-4} cm^2 or less provided it is not affected by nearby superconducting and conducting materials. If the gradiometer is tightly coupled via an n-turn input coil to a SQUID, then, making reasonable assumptions for a state-of-the-art d.c. SQUID (n=25, flux noise density = 2 $\mu\Phi_o/\sqrt{Hz}$), the gradiometer field noise density will be no less than 20 $pT/\sqrt{Hz}$. Electronic balancing will not degrade the gradiometer sensitivity provided the field noise density of the magnetometer does not exceed this figure. A field noise of 20 $pT/\sqrt{Hz}$ can be achieved by a bare 4.2 K d.c. SQUID magnetometer, for example one with a flux capture area greater than 1000 μm^2 and a flux noise below 10 $\mu\Phi_o/\sqrt{Hz}$.

We propose a design which requires neither a separate pickup coil nor a flux feedback coil for flux-locked loop operation of the device, direct current injection being used to provide flux feedback. Current injection into SQUIDs has been used in the past, usually as a means of coupling input signals into the device [2-4]. Here its principal benefits are to provide compact efficient coupling of feedback flux into a small SQUID and to allow simple device fabrication. In this mode of operation field and current distributions within the SQUID differ from those existing with mutual inductance feedback from an external coil. Thus the possibility exists of different levels of flux noise in the two modes of operation. We report below measurements aimed at determining the extent of such differences.

The SQUID magnetometer (figure 1) has a square loop with inner and outer dimensions of 20 μm and 40 μm respectively. The effective area of the loop is greater than the theoretical value of about 800 μm^2 [5] due to flux focussing

Springer Proceedings in Physics, Vol. 64 Superconducting Devices and Their Applications
Editors: H. Koch, H. Lübbig

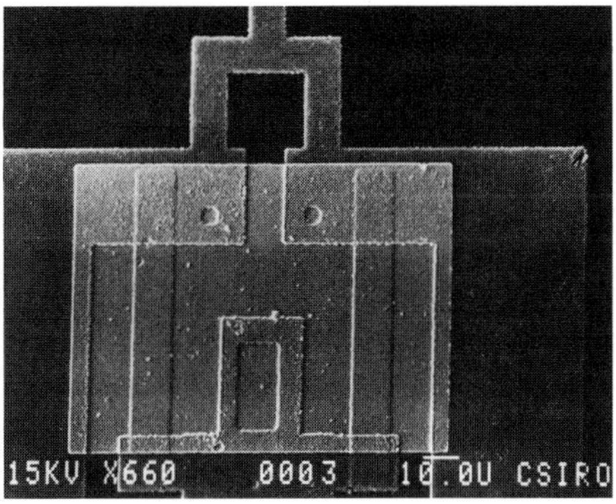

Figure 1. Scanning electron micrograph of d.c. SQUID magnetometer.

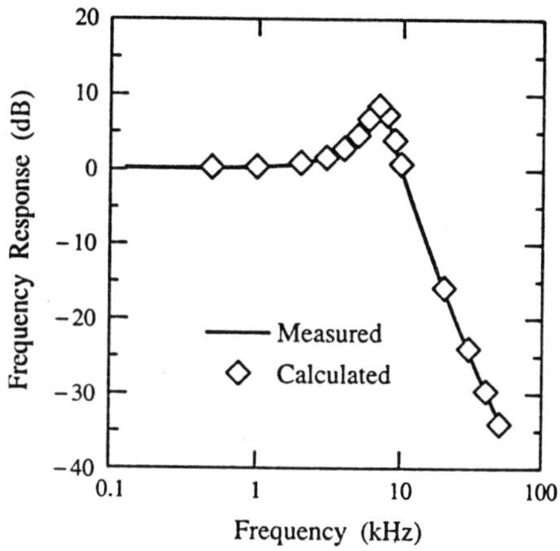

Figure 2. FLL frequency response.

by the counterelectrode. The theoretical inductance is about 40 pH. Feedback flux is injected into the SQUID loop at one of two current injection tracks (left and right, figure 1) and leaves either via the SQUID common terminal (top, figure 1) or via the second current injection track. These allow single-ended or push-pull operation, respectively. The devices reported here have Nb/Al/AlO_x/Nb junctions defined by windows in the SiO insulation layer, the fabrication process resembling that of Savo et al [6]. A Nb/Al/AlO_x/Nb trilayer fabrication process with either lead-indium (as in figure 1) or niobium wiring is under development. Details of these processes will be published elsewhere.

The flux-locked loop (FLL) electronics is of the carrier frequency type first described by Clarke et al [7]. A sinusoidal modulation flux of frequency 455 kHz is used, its amplitude being adjusted so as to maximise the SQUID output voltage. The SQUID is impedance-matched to a low-noise FET (2SK147) preamplifier by a cooled L-C resonator. The system is intended for use in environments where the dominant noise source is expected to be mains frequency (50 Hz) interference. We have specified an open-loop gain of at least 1000 at this frequency in order to ensure that such signals are accurately tracked without undue distortion of wanted signals of potentially much smaller amplitude. The electronics includes two integrators giving a rapid rolloff of open-loop frequency response, and a zero to compensate for the phase shift of the tuned circuit. Figure 2 shows the measured closed-loop frequency response for a SQUID for which the tuned circuit Q is 57 ($R_{dyn} = 2.3\ \Omega$) and the open-loop gain is 1860 at 50 Hz. The closed-loop response is flat over the frequency range of interest (0-1 kHz) with a resonant peak of 8 dB at about 7 kHz. The measured frequency response is in good agreement with calculations from open-loop characteristics.

Table 1. SQUID magnetometer parameters.

Junction parameters:	
area	25 μm^2
critical current	50 μA
shunt resistance	1.0 Ω
Inductance	40 pH
β	1.9
β_c †	0.23
Feedback inductance	
(single-ended)	20 pH
Flux capture area	1950 μm^2
Flux noise density	
(1 Hz)	20 $\mu\Phi_0/\sqrt{Hz}$
(1 kHz)	1.6 $\mu\Phi_0/\sqrt{Hz}$
Field noise density	
(1 Hz)	21 $pT/\sqrt{Hz}$
(1 kHz)	1.7 $pT/\sqrt{Hz}$
Energy sensitivity	
(1 kHz)	210 h

†Calculation assumes specific junction capacitance of 0.06 $pF/\mu m^2$.

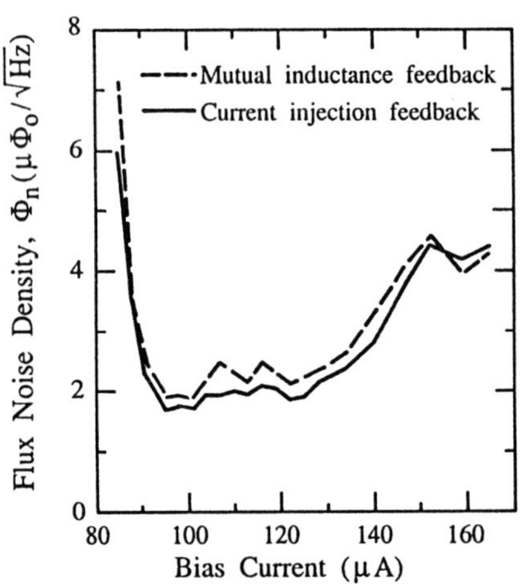

Figure 3. Flux noise density at 1 kHz as a function of bias.

Table 1 summarises measured and calculated parameters for a prototype device. Flux-locked loop noise measurements were carried out on the magnetically shielded SQUID using both single-ended current injection flux feedback and mutual inductance feedback. For the latter measurements a 3.5 mm diameter 44-turn coil with a measured mutual inductance to the SQUID of 3.1 pH was fixed to the underside of the device header. Figures 3 and 4 show

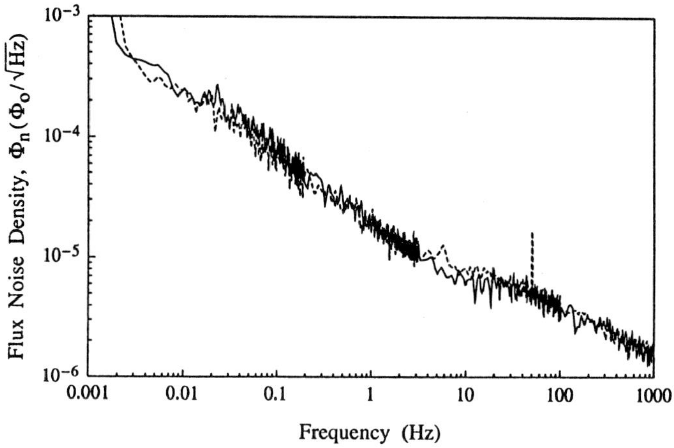

Figure 4. Flux noise density at optimum bias current for SQUID magnetometer in FLL with current injection (full line) and mutual inductance (broken line) flux feedback.

closed-loop flux noise density Φ_n as a function of bias current and frequency, respectively. These measurements were quite repeatable and the device showed little susceptibility to flux trapping over a number of cool-down cycles. At 1 kHz and optimum bias current the flux noise density is 1.6 $\mu\Phi_0/\sqrt{\text{Hz}}$, corresponding to an energy sensitivity of 210 h, where h=6.63×10^{-34} J/Hz is Planck's constant. The noise level decreases with increasing frequency throughout the six decade range, indicating the existence of excess noise processes other than Johnson noise. For frequencies below about 2 Hz, Φ_n varies as $f^{-0.59}$. For frequencies between 50 Hz and 1 kHz, the dependence is $f^{-0.41}$, suggesting that a different noise mechanism may be dominating in this range.

The SQUID has been operated successfully without magnetic shields. By rotating the device in the horizontal component of the Earth's magnetic field (21 μT) we have determined the flux capture area and field noise densities given in Table 1. The magnetometer meets the target field sensitivity of 20 pT/$\sqrt{\text{Hz}}$ for frequencies greater than 1.1 Hz.

There is considerable scope for further improvement in sensitivity. Most obviously, the device could be redesigned to provide a larger flux capture area, still without recourse to a flux transformer. However, even within the constraints of the present design, improvements are possible. Better optimisation of device parameters, particularly β_c, should provide a small improvement in energy sensitivity, at least so far as white noise is concerned. More importantly, the source of the excess noise in the device should be determined and corrected. Devices fabricated using a conventional trilayer process will be used to investigate the relationship of the excess noise to the junction technology.

We have found current injection to be an entirely satisfactory means of SQUID flux-locked loop operation. Measurements down to 1 mHz reveal no adverse effects on device sensitivity. This observation remains true whether the device is locked onto a positive-going or a negative-going peak of the V-Φ characteristics. Only minor differences in device noise between the mutual coupling and current injection feedback modes are seen. We attribute these to a slight difference in the level of flux modulation in the two modes of operation. Current injection allows compact efficient coupling to very small SQUIDs and reduced lead count. The absence of both a feedback coupling coil and an input coil in the magnetometer described here allows simple device fabrication and minimises problems associated with L-C resonances in the SQUID characteristics. Current injection feedback may be a useful mode of operation for high-T_c SQUIDs with, for example, grain boundary or step-edge junctions, which can be fabricated with current injection in a single superconducting layer.

Acknowledgement. We thank P.R. Dencher and Z.X. Yang for contributions to electronics design and fabrication.

References

1. S.J. Williamson, M. Pellizone, Y. Okada, L. Kaufman, D.B. Crum and J.R. Marsden, in Biomagnetism: Theory and Applications, edited by H. Weinberg, G. Stroink and T. Katila (Pergamon, New York, 1985) 46.
2. J. Clarke, Phil. Mag. **13**, 115 (1966).
3. M.B. Ketchen, W.M. Goubau, J. Clarke and G.B. Donaldson, J. Appl. Phys. **49**, 4111 (1978).
4. T.R. Gheewala, IBM J. Res. Develop. **24**, 130 (1980).
5. M.B. Ketchen, W.J. Gallagher, A.W. Kleinsasser, S. Murphy and J.R. Clem, in SQUID'85, edited by H.D. Hahlbohm and H. Lubbig (Walter de Gruyter, Berlin, New York, 1985) 865.
6. B. Savo, F.C. Wellstood and J. Clarke, Appl. Phys. Lett. **50**, 1757 (1987).
7. J. Clarke, W.M. Goubau and M.B. Ketchen, J. Low. Temp. Phys. **25**, 99 (1976).

The DC-SQUID Amplifier for the NAUTILUS Gravitational Wave Detector at CERN: Preliminary Measurements

C. Cosmelli[1,2], *M.G. Castellano*[1,3], *and P. Carelli*[1,4]

[1]INFN, sez. di Roma, P.A. Moro 2, I-00185 Rome, Italy
[2]Dipartimento di Fisica, Università di Salerno, I-84100 Salerno, Italy
[3]Istituto di Elettronica dello Stato Solido, C.N.R., Via Cineto Romano, 42, I-00156 Rome, Italy
[4]Dipartimento di Energetica, Monteluco di Roio, L'Aquila, Italy

Abstract. NAUTILUS, a third generation Cryogenic Gravitational Radiation Detector, now being tested at CERN by the Rome group, uses a dc SQUID as low noise amplifier. For such a system we have developed a superconducting transformer to realize the noise matching between the SQUID and the gravitational detector, and an OFHC holder for the SQUID to be cooled at $T \simeq 100\ mK$. Some preliminary measurements has been made on the SQUID connected to a small dilution refrigerator obtaining a flux noise of $\phi_n = 7 \cdot 10^{-7} \phi_o/\sqrt{Hz}$ at $0.2\ K$. Further measurements show that this noise is not due to the SQUID but to a noise source associated to the SQUID holder.

1. Introduction

The goal of the experiment for searching gravitational radiation is to detect a pulse of energy $\Delta E \approx 10^{-29} \div 10^{-28}\ J$ at a frequency of $1\ kHz$ or, in terms of the so called effective noise temperature, to reach a $T_{eff} = \Delta E / k_B$ of a few μK.

It can be demonstrated [1] that the lower limit for the effective noise temperature is $2T_n$, T_n being the noise temperature of the amplifier connected to the apparatus. The use of a dc SQUID is then mandatory to reach the desired sensitivity.

NAUTILUS[2] is a new resonant bar detector of the Rome group operating with an antenna of 2500 kg at a temperature of about 100 mK, reached by means of a special ^{3}He–^{4}He dilution refrigerator.

2. Experimental Apparatus

As in the other detectors of the Rome group, the transducer which converts the bar vibrations into electrical signals is a capacitive resonant transducer[3], a mushroom shaped oscillator, much lighter than the antenna and with a resonant frequency tuned to that of the antenna, connected to the end face of the cylinder. The mechanical coupling between the two oscillators results into two normal modes, in which the bar displacement is amplified. The transducer is faced to a plate and makes up a capacitor, which is biased by a high electrostatic field. The transducer motion modulates the capacitance gap and is converted into a voltage signal. By means of a proper matching circuit made of a large superconducting transformer, the voltage signal is in turn converted into current, which is fed to the input coil of a tunnel junction dc SQUID that forms the amplification stage.

This scheme has been successfully used in the detectors Explorer at CERN and ALTAIR in Frascati, both working at a temperature between 2 and 5 K. The matching circuit between the SQUID and the transducer is composed by a superconducting transformer, air wound, having an inductance of 7 H on the primary and 1 μH on the secondary connected to the input coil of the SQUID; its quality factor at 1 kHz is of the order of $6 \cdot 10^4$. The dc SQUID is a multiloop tunnel junction made of Nb-NbOx-PbAuIn [4], that proved to be very stable and reliable. The superconducting transformer surrounded by a Lead shield, and the SQUID holder are inserted into an Aluminum block covered by a thin layer of Sn-Pb.

The SQUID holder is made of OFHC copper to get good thermal contact, covered with a layer of Sn-Pb, 0.1 mm thick. In its body three separate cavities are milled, one for the SQUID chip, one for the tank circuit and the third for the output filtering stage .

The SQUID chip is mounted on a fiberglass base, realized as a printed circuit board. The copper printed circuit is bonded on one side to the SQUID pads, on the other to the wires leading to the tank circuit and to the current bias.

The matching of the SQUID with the room temperature low noise amplifier[5] is realised with a conventional tank circuit. The cables connecting the SQUID to the room temperature electronics are made with twisted pairs of high resistance wire except for the SQUID voltage output wires, which are made of Niobium wire (150μm diameter) for the low temperature part(3 meters), and copper wire from the 4.2 K stage to room temperature(2 meters). All the cables (a continuous length from room temperature to low temperature, with thermal anchoring along several stages of the system) are covered with aluminum foil for electromagnetic shielding, with the shield grounded all along the insert body.

In order to investigate the noise temperature of the dc SQUID amplifier we have performed measurements on the flux noise at low temperature in a small dilution refrigerator, reproducing the ambient condition that the device will find in the final system. The SQUID holder has been bolted to an OFHC copper bracket, which in turn is bolted to the mixing chamber of the dilution refrigerator. The SQUID temperature has been estimated by means of a Ge thermometer put into a slot, inside a fiberglass sandwich of proper thickness, such as to reproduce the SQUID thermal contact with the holder body. Hot electron effect is not supposed to be relevant at our operating temperature ($T > 100\ mK$), so the SQUID temperature should be the same of the macroscopic chip temperature. No μ-metal shield was placed around the experimental apparatus and all the measurements were done during a standard working day with all the instrumentation of our and a near laboratory in operation. NAUTILUS, in fact is a cryogenic dewar 6 meters long by 4 meters high and cannot be shielded nor disconnected from the pumping system and from the dilution refrigerator.

3. Noise Measurements

The noise measurements have been done with the SQUID in flux locked loop, with a band of a few hundreds Hertz, measuring the noise spectral density in the flat region at about $300 \div 400 Hz$.

The noise measurements have been taken at three different stable temperatures of the SQUID: 4.2 K, obtained by putting He gas into the vacuum can to keep the SQUID in thermal contact with the Helium bath; 1.4 K, obtained by using the 1 K stage of the refrigerator; finally 200 mK, with the refrigerator in operation, limited by the thermal inputs and by the thermal contact.

The measurements have been taken with the dilution refrigerator in operation. Turning off the refrigerator or detaching the thermometer cables did not affect the measured noise level.

Fig. 1 shows the behaviour of ϕ_n^2 versus T. It can be noted that while the dependence on T is linear, a constant term remains at 0 K; the data are fitted by the curve

$$\phi_n^2 = a^2 + bT \ . \tag{1}$$

In other words, it appears that a further term, constant with the temperature, is added to the expected noise. From fig.1, it is found $a = 6.7 \cdot 10^{-7}\ \phi_o/\sqrt{Hz}$.

Looking for a possible cause of this noise, we investigated the flux noise limit set by the amplifier. On the amplifier alone, we measured $e_n \simeq 0.6 nV/\sqrt{Hz}$ and $i_n \simeq 40 fA/\sqrt{Hz}$, which agree with the planned values.

With a SQUID mean responsivity $V_\phi = 40 \div 60 \mu V/\phi_o$ and a tank quality factor $Q_{tank} = 60 \div 90$, the amplifier limit (calculated in terms of

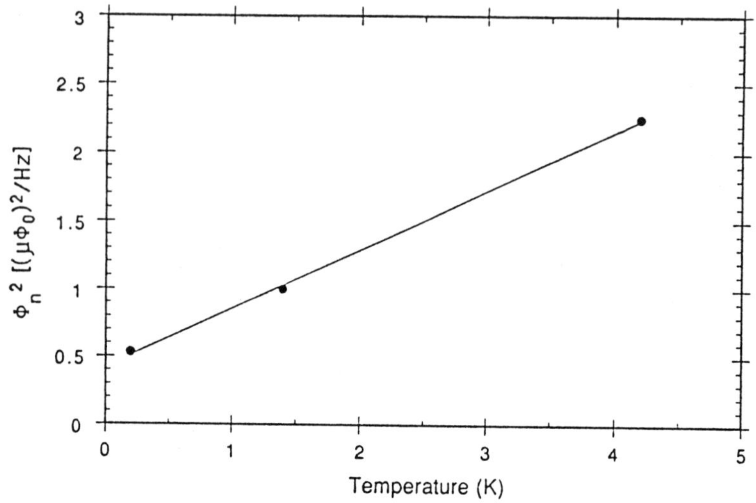

Fig.1 - Measured SQUID flux noise versus temperature.

equivalent flux noise at the SQUID input) is $\phi_n^{ampl} \simeq 2 \div 3 \cdot 10^{-7} \phi_o/\sqrt{Hz}$ lower than the additional noise we are looking for. We then measured the noise from the tank circuit. The SQUID was unbiased by setting $I_b = 0$ and it was checked that there was no SQUID modulation. In this condition we measured a current noise $i'_n \simeq 250\ fA/\sqrt{Hz}$, much higher than the contribution from the amplifier and sufficient to explain the additional flux noise measured on the SQUID.

A set of mesurements has been made on all the parts of the apparatus: Amplifier alone, loaded with the low temperature cables, loaded with only the tank circuit etc... In no one of these measurements we measured the additional current noise over the predicted one. In particular the noise measured with the tank circuit shorted, i.e. replacing the SQUID with a short gave the noise expected from the contribution of the amplifier and from the Johnson noise of the tank circuit losses:

$$V_{peak}^2 = \left[e_n^2 + (i_n^2 + \frac{4KT}{RQ_{tank}^2})(RQ_{tank}^2)^2 \right] A_{diff}^2 \quad V^2/Hz \, . \qquad (2)$$

We concluded that this extra noise source is related to the SQUID chip and/or to the fiberglass mounting base and printed circuit pads.

A new test has been made on the system having removed a small fiberglass sheet glued onto the pads going to the SQUID. A preliminary measurement showed no extra current noise with the SQUID off ($I_b = 0$) but connected. Work is in progress to cool as soon as possible the SQUID with the dilution refrigerator to see if we have indeed eliminated the extra noise limiting the intrinsic flux noise of the SQUID.

References

[1.] R.P.Giffard,Phys. Rev. D **14**,2487(1976).
[2.] M. Bassan, E. Coccia, I. Modena, G. Pizzella, P. Rapagnani, F. Ricci, Proc. of the Fifth Marcel Grossmann Meeting on General Relativity (World Scientific), 1723 (1988)
[3.] P. Rapagnani, Il Nuovo Cimento **5 C**, 385 (1982)
[4.] P. Carelli, V. Foglietti, R. Leoni, M. Pullano, IEEE Trans. Mag. **MAG-25**, 1026 (1989)
[5.] G.V. Pallottino, T. Lupi, Rev. Sci. Instr. **61**, 2436 (1990)

An Improved System for the Nondestructive Evaluation of Steel

H. Weinstock[1], *R.B. Mignogna*[2], *R.S. Schechter*[2], *and K.E. Simmonds*[2]

[1]Air Force Office of Scientific Research, Washington, DC 20332-6448, USA
[2]Naval Research Laboratory, Washington, DC 20375-6000, USA

Abstract. A non-magnetic horizontal load frame has been constructed allowing for x-y motion and rotation about the specimen axis in order to investigate the state of stress in steel components with a SQUID gradiometer. Tensile and cyclic loading can be performed under load or strain control. Data are taken by a Masscomp data acquisition system. Initial tests of this system have been made using a vertically-mounted SQUID gradiometer.

1. Introduction

A study by Weinstock and Nisenoff [1] on the passive magnetic field response to stress applied to a vertical steel specimen showed that even when the second-derivative gradiometer SQUID sensing coil was far from the specimen under test, the response was quite large and revealed qualitatively different behavior under different stress levels. The most dramatic aspect of this behavior was the reversal of sign in the magnetoelastic coefficient when the stress exceeded about 60% of that required to reach the elastic limit. That is, at the initial application of stress, there is an increase of the magnetic field strength in the region near the specimen. However, at about 60% of the elastic limit stress level, the ambient field started to decrease as the stress level was increased. This negative slope of field versus stress was found to be approximately constant at still higher stress levels until the region of the elastic limit was reached. Upon exceeding the elastic limit there was a transition to a smaller value of negative slope. Subsequent studies by Mignogna and Chaskelis [2] showed that this same behavior could be observed by placing a flux-gate magnetometer quite close to a steel bar, whereas in the SQUID-based measurements [1] the tip of the dewar was at least 20 cm away from the bar and at a less favorable angle to it. Despite this disparity in proximity, the SQUID gradiometer proved far more sensitive to the stress-related magnetic response.

There were two major impediments to both studies cited above. In order to make a detailed mapping of the magnetic field in order to determine the region of maximum stress, one must have the freedom to move the magnetic sensor or the test object at will. This was not very practical with a vertically-mounted specimen in a standard tensile-testing system. Such a system is stationary, and it is generally unwise to construct a mobile, orientable SQUID-dewar support system, particularly because a large angle to the vertical would inhibit the useable capacity of the liquid helium reservoir, while the geometry of the tensile-testing unit precludes the

Springer Proceedings in Physics, Vol. 64 Superconducting Devices and Their Applications
Editors: H. Koch, H. Lübbig

opportunity to position the dewar close to the unit with its axis vertical and inhibits complete azimuthal access. The second major impediment is that a standard tensile-testing system is comprised of structural steel components. This creates a rather hostile magnetic environment, as was found in the earlier cited work. The problem would be aggravated for the case of an entire field mapping and could lead to false conclusions about the field due only to the specimen under test.

2. A Horizontal, Movable, Nonmagnetic Test Bed

Given the impediments enumerated above, it was decided to construct a test bed comprised entirely of nonmagnetic materials, with specimen grips and load cell in a horizontal (x-y) plane, and with two orthogonal drive motors for controlled movement in this plane. Furthermore, in the unit constructed, the frame can be rotated so that the specimen under test may be subjected to a maximum angular displacement (about its own axis) of 135 degrees in either direction. It is expected that specimen symmetry will be sufficient to make it unnecessary to perform measurements for larger rotations, but the grips also have the freedom to rotate up to 90 degrees independent of the frame's rotation. Thus, it is possible to access all azimuthal angles.

The drive motors are controlled by an IBM model AT personal computer and have a maximum travel of 50 cm in both the x and y directions. They are programmed so that incremental motion is on a line parallel to the specimen axis. After each increment of travel, the movement is stopped until data have been taken, then the frame moves on to the next stopping point. After the end point in a given line has been reached, the frame retraces its path back to the other extremum, moves an incremental distance parallel to the specimen axis, and then proceeds to take data along this new path, i.e., the data are always taken sequentially in the same direction. However, a serpertine scan can be programmed as well.

There is no automated drive for the vertical (z) direction, but the dewar or its wooden frame may be adjusted manually so that the tip of the dewar can travel from just above the surface of the test specimen to a vertical displacement of up to about 20 cm.

The entire load frame was constructed of nonmagnetic material, mostly aluminum, with some titanium where greater strength was required. Only the commercial load cell contains ferromagnetic components. This cell is placed along the specimen axis at a distance of approximately one meter from the specimen and appears to produce no magnetic field anomaly that can be detected when a field survey is made with a nonmagnetic test specimen.

Specimens may be accommodated with a gauge length of up to 25 cm and a total length of up to 50 cm. Flat specimens are held at each end by a brass grip which is secured by a titanium locking pin fitted tightly through the specimen and the grip on either side of it. Cylindrical specimens are threaded into a different set of grips at each end.

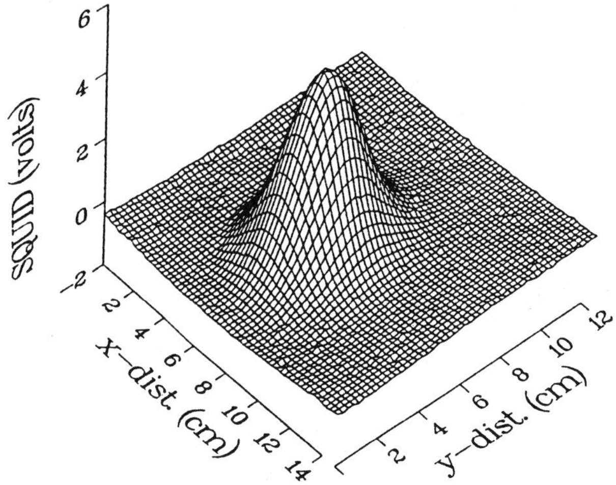

FIG. 1. Mapping of the z component of magnetic field over an x-y plane, about 8.5 cm above the top of an 8-turn coil, 1 cm in diameter, carrying 210 mA.

Figure 1 shows a plot of the vertical component of the second derivative of the magnetic field obtained by a SQUID sensing coil with a 2.3-cm diameter and situated about 8.5 cm above the top of a test coil comprised of 8 turns of copper on a 1 cm mandrel and carrying 210 mA. The z axis in this figure shows the relative output voltage of the SQUID electronics unit. Each intersection in the mesh represents a point at which data were taken. Although not all points are shown, over a period of about 30 minutes, a total of 3,600 were taken, 60 points at 2.5 mm intervals along each of 60 tracks displaced by 2.5 mm.

The data were recorded digitally on a Masscomp 5600 data acquisition sytem which has a sampling rate of 1 MHz, 64 input channels and 2 Mb of memory. The geometrical interval between data points is adjustable, so that initial scans can be made in a shorter time period, while regions in which the magnetic field varies more dramatically can be scanned with a finer mesh.

Some preliminary plots also have been obtained with a (1018 cold-rolled) steel bar with a gauge length of 14 cm and a cross section of 25 mm by 1.3 mm. Although measuring the magnetic response of the bar to stress showed qualitatively the same behavior as that seen earlier [1], the field in the z direction showed no outstanding anomaly at constant stress, even when a transverse slot was made in the bar. The inability to detect a field anomaly is because the pick-up coil was about 20 cm from the slot. This large separation was used to avoid output voltage resets, but future measurements will be made at closer range. Subsequent measurements (at this same separation) with a large-area steel plate containing a 2.5 cm hole, did exhibit a characteristic vertical field anomaly related to this hole. The problems of sensitivity, instrument resets and separation are addressed below.

3. Future Activity

Currently, the versatility of this system is limited by the large-diameter (2.3-cm) pick-up coil (in comparison to the size of the defect structure). This situation will be alleviated soon when a new SQUID system is installed. The new unit contains a 3-axis first-derivative gradiometer with 1-cm-diameter pick-up coils, a reduction in coil area of 5. Ultimately, it will be necessary to use even smaller pick-up coils to obtain sufficient resolution to find the region of maximum stress in a steel bar.

Generally, sensitivity is limited by the larger of (1) the standoff distance between the field source and the pickup coil, and (2) the diameter of the coil itself. For the case of the steel specimens used, a relatively small standoff distance produced one or more resets of the system's output voltage. At the present time, a protocol for adding these resets to the recorded voltage is being adapted. In the longer run, circuitry with wider dynamic range will be employed, although a digital SQUID system would be most desirable and versatile.

References

1. H. Weinstock and M. Nisenoff, in *SQUID '85*, edited by H. D. Hahlbohm and H. Lubbig (de Gruyter, Berlin, 1985), pp. 853-858.

2. R. B. Mignogna and H. H. Chaskelis, in *Review of Progress in Quantitative Nondestructive Evaluation*, edited by D. O. Thompson and D. E. Chimenti (Plenum, New York, 1989), pp. 551-558.

Improved Techniques for Structural NDT Using SQUIDs

A. Cochran and G.B. Donaldson

University of Strathclyde, Department of Physics and Applied Physics,
John Anderson Building, 107 Rottenrow, Glasgow G4 0NG, UK

Although the SQUID has been identified as having possible applications in structural nondestructive testing (NDT), the 10 to 20 mm minimum lift-off (or separation between pick-up coils and the test subject) imposed by a typical conventional cryostat makes high resolution SQUID inspection impossible. Here, we report on the development of a 4.2K cryostat which allows the lift-off to be reduced to 2 mm. A complementary planar gradiometric pick-up coil assembly with coil diameters of 2 and 4 mm is also outlined. The increased inspection resolution should significantly widen the scope of laboratory assessment of structural NDE using conventional SQUIDs. Preliminary results are presented to support this conclusion, and their relevance to SQUIDs operating at LN_2 temperatures is noted.

1. Introduction

In the past, the high sensitivity and uniform field rejection performance of SQUIDs compared with other electromagnetic detection techniques has motivated some NDT research [1][2]. The practical attraction of demonstration systems has been an ability to detect gross defects at lift-offs of several centimetres. However, their reliance on LHe forms an unsurmounted barrier to their adoption in the field. With the advent of LN_2 temperature superconductors, it is likely that SQUID systems will be developed with sufficient sensitivity and spatial resolution advantages over other techniques to make them commercially attractive. To study some properties of such systems while HTS devices are still unavailable, an existing LHe SQUID NDT system has been refined and used to investigate a number of electromagnetic NDT techniques.

2. System development

The main drawback of the existing system was the 20 mm minimum lift-off imposed by the design of the tail of the metal cryostat. Since this in turn imposes a limit on the spatial resolution of the system [2], it was disadvantageous to configure the first-order planar gradiometric pick-up coil to have an inner coil

Springer Proceedings in Physics, Vol. 64 Superconducting Devices and Their Applications
Editors: H. Koch, H. Lübbig

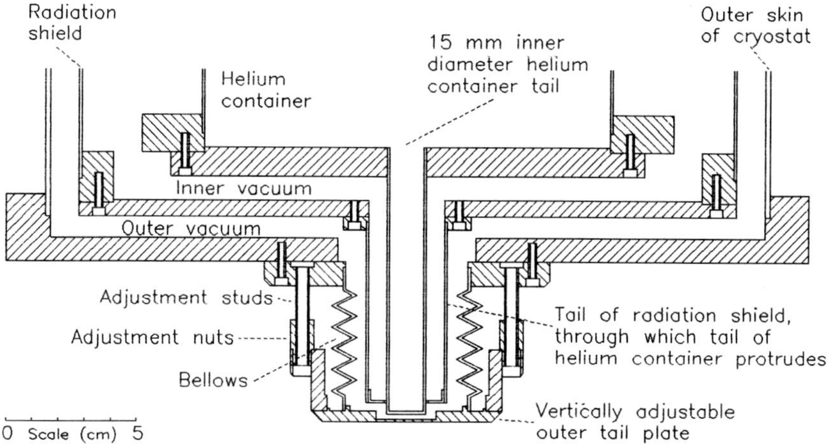

Figure 1 The cryostat tail design allowing a minimum lift-off of 2 mm.

diameter of less than 20 mm. To overcome this limitation, the existing tail was replaced by the arrangement shown in Figure 1.

When the cryostat is warm, the adjustment nuts stretch the bellows to provide clearance between the outer tail plate and the LHe container tail. When the cryostat is cooled, thermal contraction moves this tail a few millimetres upwards. The resulting increased separation between the LHe and the external atmosphere is then reduced by readjusting the bellows until the outer tail plate approaches the LHe container tail. It has been found that a 0.5 mm vacuum gap does not increase the LHe evaporation rate unduly. With this gap, and tail and outer tail plate thicknesses of 0.75 mm, the distance between LHe and the external atmosphere is 2 mm.

In association with the new cryostat tail, a probe has been developed which supports a planar pick-up coil comprising a 20 turn, 2 mm diameter inner coil and a counterwound 5 turn, 4 mm diameter outer coil. These coils detect first-order magnetic field gradients perpendicular to their axis and are connected to a conventional RF SQUID. To overcome the inevitable imbalance of hand-wound coils, access is provided for manual vertical adjustment of a superconducting core which disproportionately excludes flux from the inner coil. The new cryostat and probe are part of experimental apparatus which allows 2D scanning of large test subjects in a normal, unscreened laboratory. Cryostat height adjustment and scanning accuracies of less than 1 mm can be achieved. Data are recorded by a personal computer, with the results transferred to a DEC VAX for processing and presentation.

3. Results

Various techniques have been used for detection of features of metal test subjects using the SQUID system. They may be defined by the method of generation of the magnetic field to be measured by the SQUID, the electromagnetic properties of the test subject, and the mechanism of distortion of the field by features of the test subject.

During previous work [2], a superconducting solenoid applied a static magnetic field perpendicular to ferromagnetic test subjects. The pick-up coils were positioned within this field, and simulated flaws in the test subject were detected by their geometrical distortion of the magnetic field. This technique has been applied in the revised system. The typical result shown in Figure 2 was obtained by scanning a steel plate containing 20, 40 and 80 mm long slots of 6 x 6 mm cross-section. A combination of the reduced pick-up coil dimensions and a 12 mm lift-off has both improved resolution and reduced background interference. However, because field distortion is proportional to the volume or cross-sectional area of each feature in the test subject in this applied field arrangement, the sensitivity of the technique is limited for real flaws of relatively small dimensions, such as fatigue cracks.

Flaws can also be detected by their disturbance of electrical current flow in a test subject, in turn distorting an external magnetic field. The result of a preliminary experiment to detect a 40 mm long slot of 6 x 6 mm cross section in aluminium by injecting a 125 Hz current parallel to its length (parallel to the lateral axis) is shown in Figure 3. Qualitatively, it is in agreement with expectations. While the slot dimensions are large, the experiment is encouraging because the 1 A p-p current is, in NDE terms, small and the 3.5 mm lift-off is

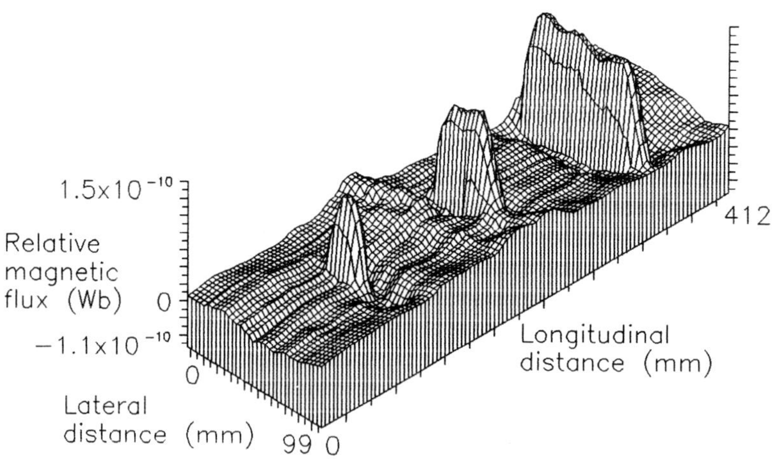

Figure 2 The field distortion detected above a steel plate containing machined slots.

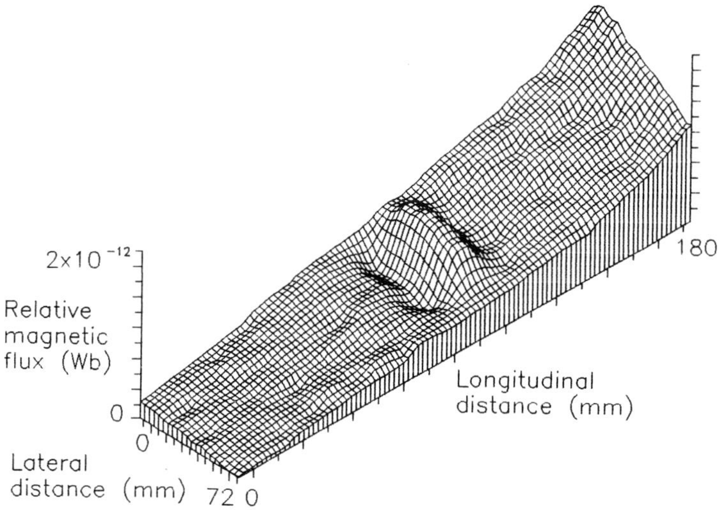

Figure 3 The magnetic field due to a current flowing parallel to a slot in an aluminium plate.

quite large. Other points of interest are a dependence of measured field on the relative orientation of the current and slot, indicating a potential for further processing to improve feature characterisation [3], and the fact that sensitivity should be increased by stronger skin effects if the technique is applied to, for example, a steel test subject.

Although measurement of directly injected currents is possible, it requires (undesirable) contact with the test subject. As an alternative, induced currents can be used. By applying an AC signal to an additional, non-contacting coil, eddy currents can be generated in the test subject. Distortion of the external magnetic field as these currents are disturbed by features of the test subject can again be detected by a SQUID system. To study this, an aluminium test subject containing full-width slots of different depths was scanned unidimensionally with a 6.3 mm lift-off. The result shown in Figure 4 demonstrates that the slots can be detected, and that the magnitude of the field perturbation is dependent on their dimensions.

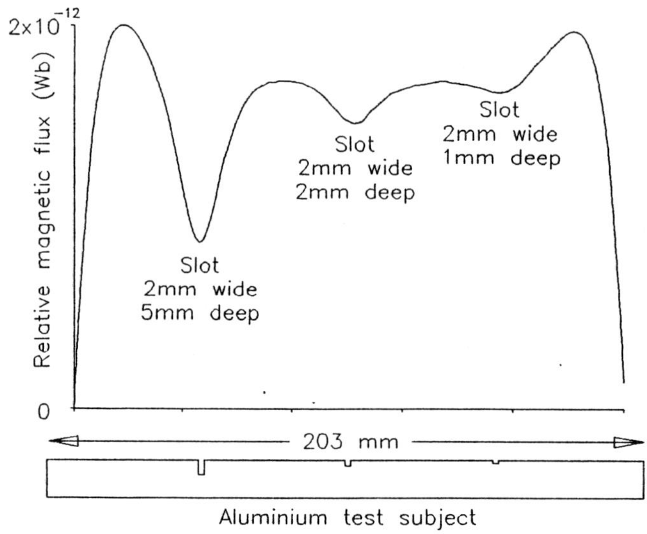

Figure 4 Magnetic field variation due to eddy current distortion by slots in aluminium.

4. Conclusions

Refinement of an existing 4.2K SQUID NDT system has improved spatial resolution and discrimination between signals and interference and allowed exploration of a number of electromagnetic NDT techniques. Since the system operates in an unscreened environment, higher intrinsic noise will not be a major disadvantage in the use of the HTS detection systems just becoming available, and the increased convenience of LN_2 compared with LHe should be very significant.

References

[1] H.Weinstock, IEEE Trans. Mag. 27 (2), Mar. 1991, p.3231-3236

[2] S.Evanson, Ph.D. Thesis, University of Strathclyde, Scotland, 1988.

[3] B.J.Roth, N.G.Sepulveda & J.P.Wikswo,Jr., J.Appl.Phys. 65 (1) Jan. 1989, p.361-372

Application of DC-SQUID Magnetometers for Nondestructive Testing of Multilayer Electronic Cards

Yu.E. Zhuravlev, A.A. Bakharev, A.N. Matlashov, V.Yu. Slobodchikov, I.D. Velt, S.L. Nikulin, and R.V. Kalashnikov

Institute of Radio Engineering and Electronics, Academy of Sciences, Marx ave. 18, 103907 Moscow, Russia

Abstract. We have studied the possibility to utilize second order dc-SQUID gradiometers to localize interlayer point defects of multilayer electronic cards. First a source model in the form of flat arbitrarily shaped current loop was tested. After that we have measured the magnetic field from the card. We used a regulization reconstruction technique to obtain the current density pattern. A defect localization accuracy of about 2 mm was obtained.

1. Introduction

Development of shortcircuiting interlayer point defects during the production of multilayer electronic cards makes their further application impossible. Localization of point defects is important for their removal. Ordinary thermovision or high frequency electromagnetic detectors are not sensitive enough to detect small defects [1].

The proposed method of localizing interlayer point defects is based on measurements of the magnetic field induced by current between shortcircuited layers. To avoid destruction of the layers by testing high currents cannot be used. For this reason it is necessary to use a highly sensitive magnetometer. Furthermore the magnetometer allows more precise localization.

2. Method

For this purpose a superconducting second order gradiometer with thin film dc-SQUID was used. The gradiometer diameter is 15 mm, its base is 50 mm. The system sensitivity was $20\,fT/\sqrt{Hz}$. A miniature dewar with a liquid helium volume of 1.3 liters and 3 days recycling period was used. The minimum distance between the pick-up coil and testing card is about 6 mm. To verify the localization ability of the system the source model in the form of an arbitrarily shaped current loop (Fig.1a) was tested.

The multilayer electronic card with shortcircuited layers to be tested is placed horizontally on a movable two-coordinate fiberglass table. The dewar with a dc-SQUID magnetometer is fixed vertically above the card. The applied current is 1 mA amplitude and 40 Hz frequency. The vertical component of the magnetic field is measured at 10x10 points of a 1 cm spaced grid. After that nonlinear interpolation is used to obtain a fine grid of 64x64 points. The inverse solution is based on a

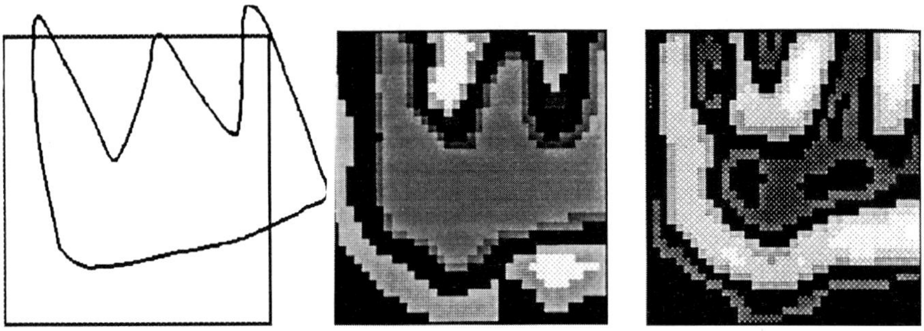

A B C

Fig.1 (a) Source model in the form of flat arbitrarily shaped current loop, (b) magnetic field distribution in the same region, (c) reconstructed current path.

regularization technique [2] and makes it possible to reconstruct the current distribution from the magnetic field.

3. Results and Discussion

Fig.1 illustrates the magnetic field (Fig.1b) and current distribution (Fig.1c) of the model current loop. Exact current position cannot be obtained directly from the Fig.1c because of the poor image resolution, but we used for this an amplitude value of each pixel of the interpolated map. The system can detect the complex current flowpath with a few millimeters accuracy. A part of the current loop was not in the measured region but nevertheless it is not significant for the current flowpath reconstruction. The results of testing the multilayer electronic card are shown in Fig.2.

The magnetic field pattern (Fig.2b) was generated by the leakage current through the shortcircuiting defect. The reconstructed current distribution (Fig.2c) allows the

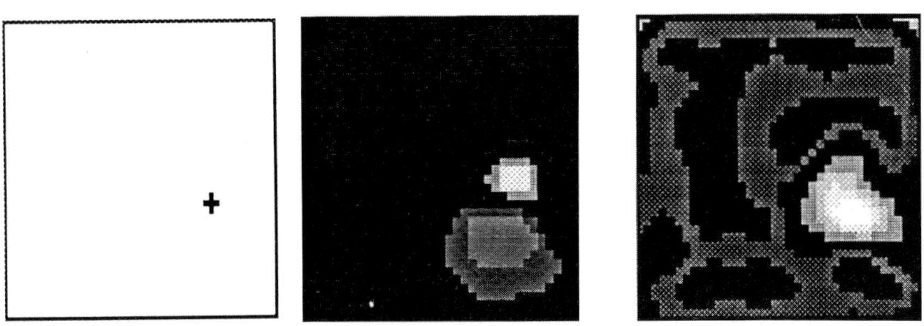

A B C

Fig.2 (a) Multilayer electronic card region with shortcircuiting interlayer point defect marked by the cross, (b) measured magnetic field, (c) reconstructed current density.

defect to be localized. An accuracy of about 2 mm was obtained and verified by making the hole in the defect region that is shown by a cross on Fig 2a.

4. Conclusions

We have demonstrated the potentialities of utilizing SQUID magnetometers for non-destructive testing of electronic cards. The localization accuracy for current paths could be increased by the use of liquid nitrogen HiTc SQUIDs which are already available.

References

[1] N.I.Yakovlev. Contactless measuring electronics for testing of electronic equipment. Leningrad.,Energoatomizdat,1990

[2] S.L.Nikulin, Yu.E.Zhuravlev, A.N. Matlashov, A.Ya Lipovich. Tichonov regularization approach to dynamic biomagnetic imaging COMAC-Biomagnetism workshop,Swansea, UK,1990

Characterization of Normal/Superconducting Interfaces: A Novel Technique

J.H. Claassen[1], *J.E. Evetts*[2], *R.E. Somekh*[2], *and Z.H. Barker*[2]

[1]Naval Research Laboratory, Code 6344, Washington, DC 20375, USA
[2]University of Cambridge, Department of Materials Science and Metallurgy, Pembroke Street, Cambridge CB2 3QZ, UK

Abstract. A new technique for demonstrating the presence of a proximity effect in a thin superconductor/normal metal bilayer has been demonstrated. It is based on a measurement of the kinetic inductance of the film, which shows a characteristic deviation from the temperature dependence of a single superconducting film at moderately low temperatures. The effect has been observed in Nb/Cu and NbN/Al bilayers, and should be applicable to high temperature superconductors as well. The results are very sensitive to cleanliness of the interface, as is expected for the proximity effect.

1. Introduction

Many proposed devices using the high T_c superconductors involve proximity coupling to a normal metal (usually Ag or Au). In other applications it is important to achieve a very low contact resistance to a normal metal. It is well known that the amplitude of the induced pair potential in the normal metal depends critically on the nature of the interface with the superconductor. For the oxide superconductors the interface properties are presumably a sensitive function of processing procedures, crystallographic orientation, etc. It would thus be desirable to be able to study the degree to which superconductivity is induced in a normal metal in a systematic way, without the necessity of fabricating a complex structure (such as an SNS microbridge). We have investigated a novel technique to characterize the proximity effect in a superconductor/normal metal bilayer that requires no processing of the sample. To date it has been used to observe the proximity effect in Nb/Cu and NbN/Al bilayers.

2. Experimental Results

We use a two-coil mutual inductance system as described by Fiory et. al.[1] to measure the kinetic inductance $L_k=\mu_0\lambda^2/d_s$. This expression applies when the thickness d_s of the superconducting film is less than its penetration depth λ. Below T_c the mutual inductance is proportional to L_k. The data for a single layer Nb or NbN film can be fit very well to a two-fluid temperature dependence $\lambda^{-2}\sim\{1-(T/T_c)^4\}$. Here the proportionality between mutual inductance and λ^2/d, which is not known with much accuracy in the experiment, is treated as an adjustable parameter. A fit to the two-fluid expression with an adjustable offset is consistent with a BCS temperature dependence for λ [2].

Springer Proceedings in Physics, Vol. 64 Superconducting Devices and Their Applications
Editors: H. Koch, H. Lübbig

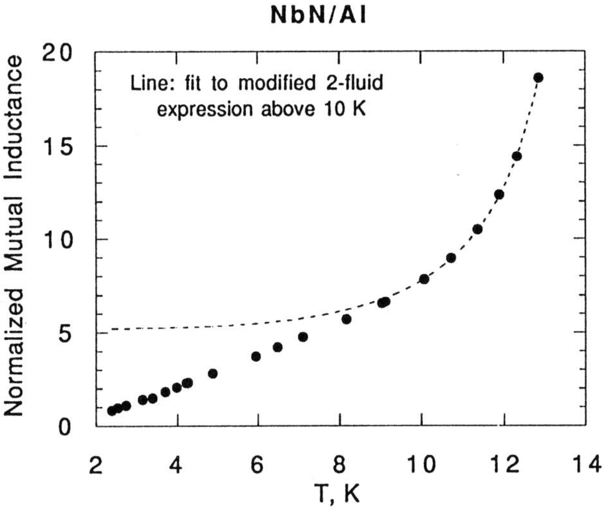

Fig. 1 Plot of normalized mutual inductance for a NbN/Al film. The dashed line is a fit to the two-fluid model for the data near T_c and represents the contribution from the superconducting film alone. NbN thickness, ~500 Å; Al thickness, 1100 Å.

Fig. 1 shows a plot of mutual inductance as a function of temperature for a NbN/Al bilayer, where the Al layer was deposited without breaking vacuum in an UHV sputtering system. Note that near T_c the mutual inductance can still be fit to a two-fluid expression, but at low temperatures it deviates significantly. To verify that this low temperature depression is due to the proximity effect, a second NbN/Al sample was prepared with a thin AlO_x barrier between the layers. The barrier layer was produced by a process that is used to fabricate Josephson junctions, so is very thin (≈nm). The mutual inductance data for this bilayer was indistinguishable from a single layer NbN film, indicating that a negligible proximity effect was present.

3. Discussion

It can be shown[3] that for thin enough films the kinetic inductance is generally inversely proportional to the areal density of superconducting electrons: $L_k^{-1} \sim \int n_s dz$, where n_s is the density of superconducting electrons and z is normal to the film surface. In the case of a single superconducting film, n_s is constant through the film. Since $\lambda^2 \sim 1/n_s$ one recovers the expression $L_k \sim \lambda^2/d_s$. In the case of a bilayer, the situation is more complicated: n_s in the superconductor, which is proportional to the square of the order parameter, is reduced at the interface due to a changed boundary condition[4]. This is most pronounced near the transition and results in the well-known reduction in the transition temperature due to the proximity effect. In the normal metal n_s is proportional to the condensation amplitude, which has a finite value at the interface and decays as $\exp(-z/\xi_N)$ into the normal region. The normal metal coherence length ξ_N has a temperature dependence $\sim 1/\sqrt{T}$ in the (usually applicable) dirty limit. Thus there is a contribution to L_k from the superelectrons in the normal metal which can potentially grow quite large at low enough temperatures. Our data shows that at ≈ 2 K the effect due to the induced superconductivity in the normal layer can be many times larger than that due to the superconductor alone.

4. Comparison to theory

The capsule outline of the proximity theory given above strictly applies to the region near the transition temperature, where the rather intractable general formalism may be simplified somewhat. Unfortunately, the region of interest for our data is at low temperatures where no simple theory applies. A Ginzburg-Landau formalism may be used near T_c which models the effect of the interface by a boundary condition on $d(\ln\Psi)/dz$, where Ψ is the order parameter. Assuming that the order parameter is continuous across the interface, we have numerically calculated $\int n_s dz$ near T_c. Over a broad range of parameters this has virtually the same temperature dependence as a superconducting film alone. Thus the presence of a proximity effect cannot be inferred from the behavior near T_c.

The Ginzburg-Landau calculation can be formally extended to low temperatures, and it is found that it reproduces the general behavior of our data. An attempt to fit our results in detail is of dubious value, given that the model has no sound theoretical basis at low temperature.

There are two added complications in attempting to fit the low temperature data from a proximity bilayer. The first is that at some point one will begin to have exclusion of flux from the normal metal, corresponding to the condition $d_s > \lambda$ in the case of a superconductor alone. When this occurs the mutual inductance will drop faster than the inverse of the areal density of superconducting electrons.

A second complication occurs when the thickness of the normal film becomes comparable with ξ_N. Presumably its contribution to the kinetic inductance then saturates. The noble metals all have $\xi_N \approx 1500$ Å $\{(RR-1)/T\}^{1/2}$, where RR is the resistivity ratio. Using RR=10, which is typical for thin films, we find that saturation will occur at ~1K for a film thickness of ~4500 Å.

5. Application to oxide superconductors

There is speculation that the order parameter is strongly reduced at the interface between a "good" metal and a high T_c superconductor due to a mismatch of material parameters [5] or an intrinsic barrier [6]. The method described here is expected to be readily applicable to the oxide superconductors, and should be quite helpful in resolving this question. The requirement that d_s be less than the penetration depth is easily met, due to the relatively large values of λ in these materials. Films can now be grown with either the c-axis or the a-axis normal to the surface, so the orientation dependence of the proximity coupling can be studied. Moreover, for temperatures of interest (<5 K) there should be no residual temperature dependence of the kinetic inductance from the superconductor contribution. Thus any drop in L_k at low temperatures is due to proximity coupling.

References

1. A. T. Fiory, A. F. Hebard, P. M. Mankiewich, and R. E. Howard, Appl. Phys Lett. **52**, 2165-2167 (1988).
2. A. F. Hebard, A. T. Fiory, and D. R. Harshman, Phys. Rev. Lett. **62**, 2885 (1989).
3. J. H. Claassen, unpublished.
4. "Proximity Effects", G. Deutscher and P. G. de Gennes, in *Superconductivity*, ed. R. D. Parks, Marcel Dekker, Inc., New York, 1969.
5. G. Deutscher and R. W. Simon, to be published in Appl.. Phys. Lett.
6. Mark Lee and M. R. Beasley, preprint.

Transistor Action in Three-Terminal Structures Having Nb Electrodes on an Insulating $SrTiO_3$ Substrate

H. Tamura, A. Yoshida, H. Takauchi, and S. Hasuo

Fujitsu Limited, 10-1 Morinosato-Wakamiya, Atsugi 243-01, Japan

Abstract. Transistor action with both voltage and current gain exceeding 1 was observed at 4.2 K in structures consisting of Nb electrodes, insulating barriers, and high-dielectric-constant insulating $SrTiO_3$ substrates. Essentially no current was required to control the collector current, which ranged from a few microamperes to a few milliamperes.

1. Introduction

Much effort has gone into work on three-terminal superconducting devices, but those proposed thus far [1-7] have not been very successful in terms of gain and input-output isolation. To overcome such difficulties, we proposed a tunneling transistor that uses a high-permittivity dielectric-base layer [8]. In such a transistor, current through the dielectric-base is controlled by changing the base electric potential. The base dielectric is used as a medium for transporting the controlling electrode potential change. The potential controlling mechanism can be used to construct transistors where carriers move in the base layer via nontunneling mechanisms such as diffusion and drift. This paper describes the operating principles and results of experiments with a dielectric-base transistor having nontunneling carrier transport in the base.

2. Principles

We proposed a transistor that uses a high-permittivity dielectric base layer as a medium where the voltage change in the controlling electrode is transported [8]. In such a transistor, or dielectric-base transistor, the dielectric base layer is separated from the emitter and collector by intervening low-dielectric-constant barriers (Figs. 1(a) and (b)).

Carriers go through dielectric materials by transport mechanisms such as diffusion, drift, and tunneling. The base layer dielectric onstant must be very high to transport the electric potential change in the controlling electrode, placed on the base layer, into the base dielectric layer at a high coupling efficiency. Transistor action occurs because the current injected from the emitter into the base region varies with the voltage change across the emitter-base barrier.

Springer Proceedings in Physics, Vol. 64 Superconducting Devices and Their Applications
Editors: H. Koch, H. Lübbig

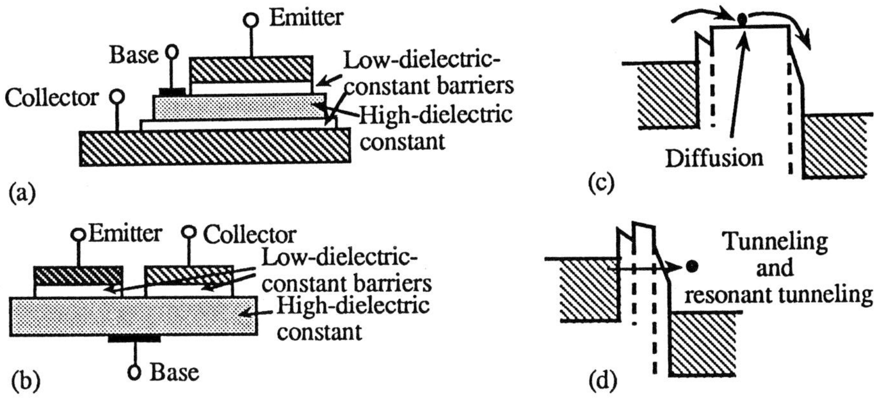

Fig. 1 Proposed transistor: (a) and (b) possible structures and (c) and (d) band diagrams.

Figure 1 (c) is a band diagram of the dielectric base transistor where carriers travel in the conduction band of the high-dielectric-constant base by diffusion. If the dielectric-base is very thin, there may be a tunneling current between the emitter and collector (Fig. 1(d)).

3. Dielectric-Base Transistor Fabrication

We made transistors on insulating (100) single-crystal $SrTiO_3$ substrates. The low-frequency (10-kHz) substrate dielectric constant was measured to be 250 at room temperature, and rose to $2x10^4$ below 10 K.

We fabricated two types of transistors having a coplanar configuration (Fig. 2). In one, the collector was placed directly, i.e., without an intervening barrier, on the $SrTiO_3$, whereas the Nb emitter was separated from the $SrTiO_3$ by a low-dielectric-constant barrier. We made both Si and SiO_2 barrier devices for this structure. The other structure had a symmetrical emitter and collector electrode, where electrodes of the same shape were separated from the $SrTiO_3$ by Si barriers. In asymmetrical emitter-collector devices or devices where the collector was placed directly on the $SrTiO_3$, the emitter was 30 μm in diameter and circular, with a minimum spacing 15 μm between the emitter and collector (Fig. 2 (a)-(b)). Symmetrical emitter-collector devices had 20-by-40-μm rectangular emitters and collectors, with an emitter-collector spacing or channel length of 2 μm and channels 40 μm wide (Fig. 2 (c)-(d)). Base leads were made with silver paste on the back of the $SrTiO_3$ wafers in both configurations.

Using RF magnetron sputtering, we deposited 6-nm-thick Si and 260-nm-thick Nb films over the entire wafer. We also used a 2-nm-thick SiO_2 barrier layer instead of Si for the asymmetrical device structure.

After the Nb/barrier layer was deposited, the layer was patterned by photolithography and reactive ion etching (RIE) using CF_4+5%O_2 gas to

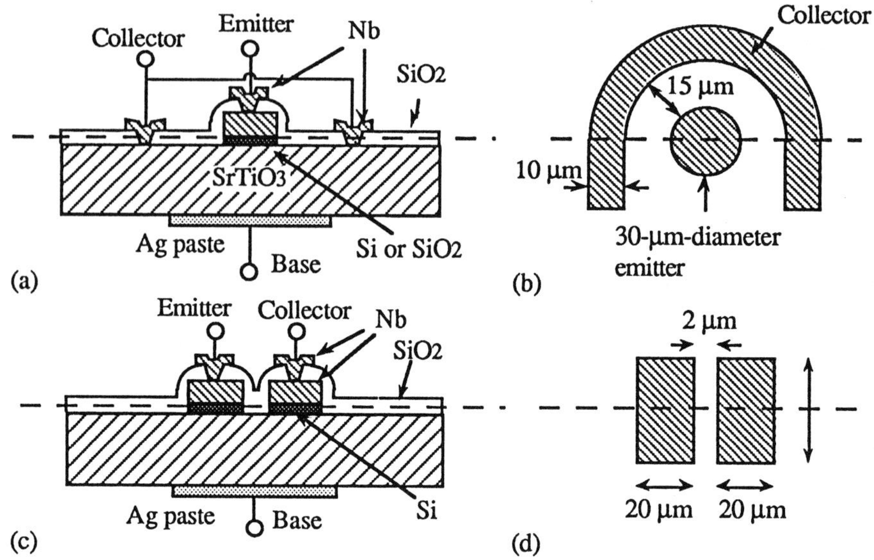

Fig. 2 Fabricated device structure. (a) Cross-sectional and (b) top views of the asymmetrical emitter-collector. (c) Cross-sectional and (d) top views of a device with a symmetrical emitter-collector.

define the emitter area. The entire wafer was then covered with a 450-nm-thick SiO_2 layer deposited by RF magnetron sputtering. Contact holes were made by RIE using CHF_3+15%O_2 gas. After 650-nm-thick Nb wiring layer was sputter-deposited, it was then patterned by RIE to form wiring layers and bonding pads.

4. Transistor Characteristics

Figures 3 (a) to (c) are collector current versus collector-emitter voltage characteristics at 4.2 K. Figure 3 (d) is collector current versus base-emitter voltage characteristics for a fixed collector-emitter voltage.

Base current I_B was less than 20 pA throughout for our experimental conditions. Because I_C, controlled by the base voltage, exceeds a few microamperes, devices had a current gain much larger than 1. The voltage gain defined by $G_V=-\partial V_{CE}/\partial V_{BE}$ was about 2.5 for a Si-barrier asymmetrical device on a 0.5-mm-thick substrate (Fig. 3 (a)). For an asymmetrical SiO_2 barrier device on a thinner substrate (0.3 mm), the voltage gain was about 4 (Fig. 3 (b)). As shown in Fig. 3 (c), the Si barrier device with a symmetrical emitter-collector arrangement also exhibited current-voltage characteristics, with voltage and current gain exceeding 1, almost the same as for asymmetrical devices. The voltage gain for this device was about 2.

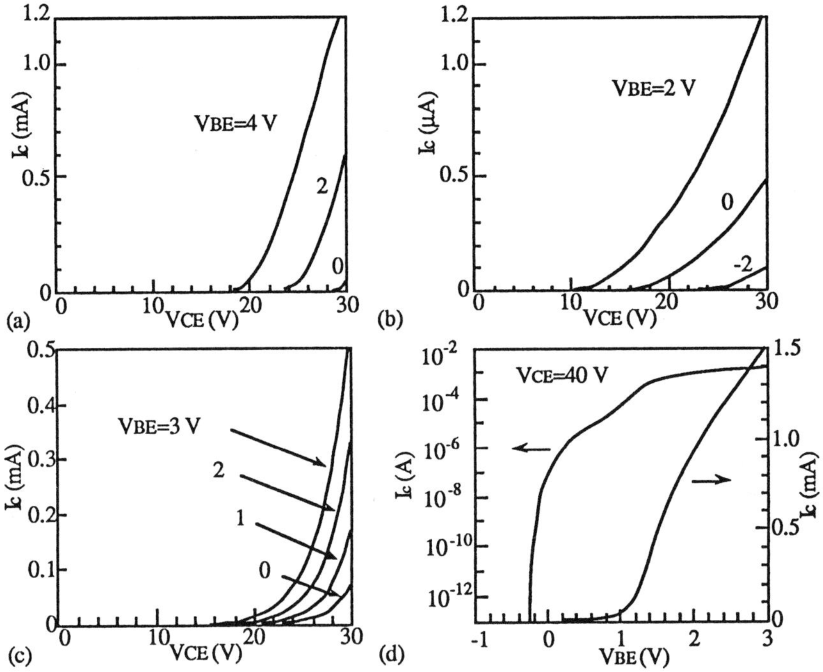

Fig. 3 I-V characteristics measured at 4.2 K. I_C-V_{CE} characteristics of (a) an asymmetrical device with a Si barrier on a 0.5-mm-thick substrate, (b) an asymmetrical device with a SiO_2 barrier on a 0.3-mm-thick substrate, and (c) a Si barrier symmetric device. (d) I_C-V_{BE} characteristics for V_{CE}=40 V for the asymmetrical device in Fig. 2 (a).

Figure 3 (d) shows emitter-grounded I_C-V_{BE} characteristics for a constant collector-emitter bias voltage of V_{CE}=40 V for the asymmetrical emitter-collector device (Fig. 2 (a)). When V_{BE} was less than about -0.2 V, the transistor was off with no measurable I_C observed. For V_{BE} exceeding -0.2 V, I_C increased rapidly with increasing V_{BE}. The maximum device transconductance for the asymmetrical Si barrier device having V_{CE} fixed at 40 V was 1.4 mS at a collector current of 0.3 mA and V_{BE} of 1.4 V.

5. Conclusion

Our results show that a transistor can be made from metallic electrodes, thin insulating barriers, and high-dielectric-constant $SrTiO_3$. Although the operating voltage of the experimental transistor was 10 to 30 V, it will probably be reduced by device downsizing and improved materials and processes. Because we observed rectifying characteristics for a voltage swing of a few tens of millivolts for Nb/(Si or SiO_2)/Nb-doped

$SrTiO_3$ diodes at 4.2 K [8], transistor operation at a bias voltage of a few tens of millivolts seems quite feasible.

One important feature of our proposed transistor is that it can be implemented with different barrier materials and electrodes. The only requirement for barrier materials is that they do not have a very high dielectric constant and that they enable carrier flow by either tunneling or emission. Conventional semiconductors such as Si and GaAs are not required to get transistor action, which helps toward making transistors out of different exotic materials. Our proposed structure is applicable, for example, to making transistors out of high-Tc superconductors, since good-quality high-Tc films and related semiconducting oxides can be grown on $SrTiO_3$ substrates.

Acknowledgments

This work was done under the management of the R&D Association for Future Electron Devices (FED) as part of the R&D of Basic Technology for Future Industries supported by the New Energy and Industrial Technology Development Organization (NEDO).

References

[1] W. J. Gallagher, *IEEE Trans. Magn.*, **MAG-21**, 709 (1985).
[2] S. M. Faris, S. I. Raider, W. J. Gallagher, and R. E. Drake, *IEEE Trans. Magn.*, **MAG-19**, 1293 (1983).
[3] D. J. Frank, M. J. Brady, and A. Davidson, *IEEE Trans. Magn.*, **MAG-21**, 721 (1985).
[4] M. Tonouchi, H. Sakai, T. Kobayashi, and K. Fujisawa, *IEEE Trans. Magn.*, **MAG-23**,1674 (1987).
[5] T. Nishino, M. Hatano, H. Hasegawa, F. Murai, T. Kure, A. Hiraiwa, K. Yagi, and U. Kawabe, *IEEE EDL*, **10**, 61 (1989).
[6] H. Takayanagi and T. Kawakami, *Phys. Rev. Lett.* **54**, 2449 (1985).
[7] A. W. Kleinsasser, T. N. Jackson, D. McInturff, F. Rammo, G. D. Pettit, and J. M. Woodall, *Appl. Phys. Lett.*, **55**, 1909 (1989).
[8] H. Tamura, A. Yoshida, S. Hasuo, and T. Van Duzer, *IEEE Trans. Magn.* **27**, 2594 (1991).

Part IV

Summary and Conclusions

Summary and Conclusions

J. Clarke

Department of Physics, University of California, and
Materials Sciences Division, Lawrence Berkeley Laboratory,
Berkeley, CA 94720, USA

Dramatic changes have occurred since SQUID '85. I think all of us visiting here have been moved by the disappearance of the Berlin Wall – earlier this week, I was fascinated to be able to walk unimpeded past the Brandenburg Gate into what used to be East Berlin. And in superconductivity, of course, the emergence of the high-T_c materials has lent an unprecedented impetus to our field that none of us dreamed about six years ago. Progress in high-T_c technology is rapid, and devices operating immersed in liquid nitrogen are already a reality. Nonetheless, the organizers of SQUID '91, wisely I believe, resisted the temptation to allow high-T_c to dominate, instead choosing to keep a wide range of topics – indeed broadening the meeting to include, for example, 1- and 2-D electron gas devices. Needless to say, it is impracticable for me to mention let alone review all 48 talks and 115 posters, and I shall simply attempt to summarize those areas that interested me most.

There were two papers that I greatly admired for their thoroughness. The first was an update by Niemeyer (PTB) of the Josephson voltage standard which involves about 20,000 Nb tunnel junctions in series. Irradiated with microwaves, the arrays exhibit a very large number of constant-voltage steps crossing the zero current axis, extending up to 10V or more. The precision of the comparison of an external voltage with the voltage of one of the steps is limited by noise in the room temperature system to a few parts in 10^{10}. Although a standard operating at 77K would obviously be very convenient, the lack of high-T_c junctions with hysteretic current-voltage (I-V) characteristics implies that an array cannot be operated in the zero-current regime, and that such a standard is unlikely soon. The second long-term experiment is Soulen's noise thermometer (NBS), which consists of a resistively shunted junction. The junction converts Nyquist voltage noise across the resistor into frequency fluctuations at about 83kHz that are measured with a counting technique to give the absolute temperature with a typical accuracy of 0.1%. Over a period of 2 months, Soulen could detect no drift in the thermometer, and over a period of *nine years*, the overall reproducibility was about 0.15%. Incidentally, there is still a good deal of innovation in these two fields. In the poster session, Kohlmann described the use of the voltage standard to compare resistors to very high precision, and Seppä discussed the use of a low-noise dc SQUID amplifier to read out a noise thermometer.

Perhaps the most exciting new physics at the conference was described in a series of talks on submicron junctions. Mooij (Delft) reminded us that effects arising from single electron tunneling (SET) go back more than 20 years; however, systematic studies have become possible only recently with the

Springer Proceedings in Physics, Vol. 64 Superconducting Devices and Their Applications
Editors: H. Koch, H. Lübbig

emergence of electron-beam lithographic techniques capable of routinely producing junctions with areas $\leq 10^{-2}$ μm^2 and capacitances $C \lesssim 1$fF. The requirements to observe charging effects are $e^2/2C >> k_BT$ and R_t (tunnel resistance) >> h/e^2. Likharev (Moscow) and his colleagues provided much of the theoretical framework for this field, both for single electron effects and also for Bloch oscillations involving Cooper pairs. Fulton and Dolan performed an early key experiment on single electron effects when they coupled two small normal state junctions in series and passed single electrons on and off the island between the two junctions. This configuration produced a striking example of the "Coulomb blockade" – the fact that the current through the junctions is significantly reduced at voltages < e/2C – and demonstrated that the blockade could be offset by means of a gate voltage applied to the island via a tiny capacitance.

Urbina described a beautiful series of experiments at Saclay in which the passage of single electrons is synchronized with radiofrequency gate voltages. These experiments, which include the 4-junction turnstile (in collaboration with Delft) and the 3-junction single electron pump, yield (nearly) constant current steps at I = ± nef and ± ef, respectively, and offer the possibility of a current-frequency standard. At present, the accuracy is limited to about 0.1% by processes that are well understood, and improvements seem likely. Delsing (Göteborg) described, amongst other devices, a 25-junction array turnstile in which constant current steps were induced by a radiofrequency signal. Kuzmin (Göteborg and Moscow) described experiments on superconducting junctions in which peaks were observed in dV/dI at currents of ± 2ef. These peaks, together with the shape of the I-V characteristic, were explained by the theory of Bloch oscillations which is based on the charging of the junction capacitance by single Cooper pairs.

There were two talks on 2-D arrays of Josephson junctions. Schön (Delft) described charge-vortex duality and introduced us to various novel properties of vortices, for example, their nonzero mass. Van der Zant (Delft) described some beautiful experiments on arrays in which vortices are generated in one region and launched through a channel into a second region. At low temperatures the vortices propagate through the second region ballistically, implying that they have a mass of about 0.01 m_e and behave as quantum mechanical objects.

The future of this field looks very exciting. There are prospects of a precise current standard. As Nazarov (Moscow) explained to us, there are intriguing possibilities of SET digital devices, although some interesting challenges have to be met to eliminate the probability of errors due to thermal activation, charge macroscopic tunneling and the "quantum wind." There will undoubtedly be new and elegant experiments, not least because, as Ingold (Essen) discussed, the effects of the environment on these tiny junctions are now well understood theoretically.

Needless to say, submicron technology has been taken up with great vigor by the semiconductor field. Von Klitzing (Stuttgart) reviewed quantum wells, wires and dots in which quantum states induced by reduced dimensionality play a dominant role in the electrical and optical properties, giving rise to such phenomena as the quantum and fractional quantum Hall effects and resonant tunneling between the states of coupled quantum wells. Van Houten (Eindhoven) described the effects of Coulomb blockade oscillations in quantum dots and wires

– for example, the conductance of disordered quantum wires is periodic in the gate voltage (Kastner, MIT). The combination of the Coulomb blockade and quantized energy levels may lead to a variety of novel effects that have yet to be discovered.

Let me now turn to superconducting devices, starting with radio frequency applications. Martens (Sandia) described a high-T_c flux flow transistor, which involves the flow of vortices across a parallel array of weak links patterned in a single layer of YBCO or TBCCO. The critical value of the current flowing through the array is determined by the onset of flux flow. The nucleation of the vortices is controlled by the magnetic field produced by the current in a nearby control-line, enabling one to modulate the voltage across the device by varying the control-current to achieve transistor-like characteristics. The device has achieved 10-12 dB of gain as an amplifier in the 2 to 10 GHz range, and a noise figure of 2 to 4 dB. Other applications include oscillators and phase shifters, raising the possibility of all-high-T_c electronic circuits. In a totally different context, Doderer (Tübingen) described his use of a low temperature scanning electron microscope to image the standing waves induced by self-resonance or external microwaves in a low-T_c Josephson junction. These beautiful images graphically illustrate various modes in both one- and two-dimensional junctions.

Naturally, much of the conference was devoted to the technology of Josephson junctions, SQUIDs, and applications of SQUIDs. Today's low-T_c junction technology is very much focused on refractory trilayers, namely Nb, NbN, NbZr or NbCN with Al_2O_3 or MgO barriers. A high degree of reproducibility in the critical current and resistance has been achieved by some groups, making the fabrication of SQUIDs (and much more complicated circuits) relatively routine. Ketchen described the IBM edge junction dc SQUIDs that have achieved a noise energy of about $2\hbar$ at 291 mK, and gave us a view of the future by discussing SQUIDs with junctions in the deep submicron regime. With 0.1 x $0.1\mu m^2$ junctions, the projected noise energy is $3\hbar$ at 4.2K. These devices are based on trilayers made by Lincoln Labs. on 5-inch wafers that will be processed on the IBM silicon line, and involve planarization with a chemical-mechanical polishing technique. About 20 posters demonstrated that the technology of low-T_c dc SQUIDs, driven by the needs of applications, is progressing steadily. A number of these posters described damping resistors to eliminate the infamous resonances arising from the parasitic capacitance between the SQUID and the input coil. Undamped, these resonances can be a major source of excess noise, but this problem has now been eliminated. The biggest group of posters, however, was concerned with new methods to read out dc SQUIDs, for example, to simplify multichannel biomagnetic systems. Thus, we saw such schemes as 2f modulation (Furukawa), on-chip digital (Fujimaki), $PI^{3/2}$ (Grönberg), relaxation oscillator (Kawai), additional positive feedback (Drung), adaptive positive feedback (Seppä) and cooled CMOS (Crocoll). The simplicity of the positive feedback schemes is particularly appealing: one uses inductive feedback to skew the voltage-flux (V-Φ) curve and operates the SQUID where $dV/d\Phi$ is large. The signal from the SQUID is coupled directly into a room temperature amplifier, followed by an integrator. On the subject of novelty, I appreciated Cabrera's laser diode switch (Stanford) that modulates the input circuit to a SQUID (as opposed to the SQUID itself), thereby reducing l/f noise and drift in both input circuit and the SQUID. This is the only method I know that can reduce l/f flux noise in a SQUID.

What of the applications? Cabrera told us about his elegant experiments to determine the Cooper pair mass, the absolute Aharonov-Bohm effect, the distribution of flux quanta trapped in a superconducting loop, the pinning force on a trapped vortex and the kinetic inductance, all using a SQUID and a laser switch. Awschalom (IBM) investigated quantum tunneling and spin dynamics in nanometer-scale magnets, and Benoit (Grenoble) described his efforts to look for persistent currents in a normal metal ring. Posters discussed nondestructive testing of steel structures and the testing of multilayer electronic cards. Other applications included SQUID-readout for a noise thermometer and a transition-edge thermometer. However, the applications were dominated by biomagnetic measurements, nicely reviewed by Foglietti (Rome). I am impressed by the progress being made with multichannel systems. No fewer than four groups – Helsinki, PTB, Twente and Rome – and four companies – BTi, Siemens, Phillips and Dornier – make systems with between 28 and 37 channels. Indeed, the Helsinki group has a 122 channel system under construction. These systems are usually operated in a magnetically and electrically shielded room, and electronic cancellation of background noise is standard. Schneider reported a magnetoencephalogram study of 20 epileptic subjects in whom the focus was localized; of these, 17 were treated with a γ-knife and 14 were cured. Both the diagnosis and treatment are entirely noninvasive. Other potential studies include brain infarction, migraines and the recovery of areas of the brain damaged by a stroke. Matlashov (Moscow) discussed his modular system with each sensor in a very small dewar, a cluster of which can be arranged as needed around the subject. Drung described the PTB system consisting of axial first-order gradiometers in which the signals from two planar multiloop magnetometers are subtracted. This arrangement offers both simplicity and flexibility, since a single SQUID can be used as a reference for an array of sensors, and one can operate a given array as both axial and planar gradiometers simultaneously. The system is usually operated inside a shielded room, but has sufficient dynamic range to operate in an unshielded environment.

The impressive performance of the biomedical systems is a double-edged sword. They have progressed to the point where serious clinical studies are now feasible, and are being undertaken. But will magnetic measurements be accepted by the medical community as a standard diagnostic tool? The pros and cons of magnetic vs. electrical measurements remain unclear, and there is the ever present threat of "cost containment" that may make it difficult to launch yet another expensive clinical instrument. I know of no other application of SQUIDs remotely approaching the commercial potential of biomedicine; success in this endeavor will provide a tremendous boost to the whole field, while failure may, I fear, lead to serious reductions in funding.

Let me turn finally to progress in HTSC technology, reviewed by Braginski (Jülich). Many groups now grow epitaxial films of YBCO, mostly with *in situ* laser or single-target sputtering processes. Photolithographic processing of these films and the fabrication of multilayer structures are progressing rapidly. Both Braginski and Okabe (Tokyo) stressed the lack of HTSC tunnel junctions with hysteretic I-V characteristics. Indeed, in the light of the experiments reported by Yoshida (Toshiba) and many others on SN interfaces and Kupriyanov's theoretical work (Moscow) on electron scattering at the surface of HTSC, the prospects of achieving such junctions seem to be fading. However, the field has made a lot of mileage from grain boundaries, although I can only echo Okabe's question "Why are natural grain boundaries so good?" A fair degree of control of parameters has been achieved with grain boundaries grown at steps, on bicrystals or with a bi-epitaxial process. Other promising structures include proximity effect junctions involving, for example, normal metal (d'Iorio, BTi) or PBCO (Rogalla, Twente) barriers making contact with the YBCO along the a-b plane.

High-T_c SQUIDs are a growth industry, with 18 papers – 13 on dc and 5 on rf devices. Koch (IBM) showed that the performance of YBCO SQUIDs grown on bicrystals is in very good agreement with computer simulations of the transfer functions and white noise, implying that the junctions were at least reasonably close to the resistively shunted junction model. As with all HTSC SQUIDs, the level of l/f noise remains high: possible sources are fluctuations in the critical current or resistance of the junctions, which can be reduced by appropriate double-modulation bias schemes, or flux noise due to the thermally activated motion of vortices pinned in the SQUID, which cannot be so reduced (except by Cabrera's scheme). Koch reported that attempts to reduce l/f noise by double modulation had sometimes been successful, but often not, for reasons still under investigation. It appears that flux noise, described by Ferrari (Berkeley), is generally lower than junction noise, except perhaps, at the highest operating temperatures. Ironically, the mechanism of flux noise is much better understood than that of junction noise, and can even be used to determine pinning energies. Wellstood (Berkeley) described SQUIDs involving bi-epitaxial junctions (made by Char *et al.*, Conductus). These junctions involve 45° grain boundaries grown at the edge of a seed layer, and can be placed at will on a chip. Several groups have achieved white noise energies comparable with or even exceeding the performance of commercially available Nb hybrid SQUIDs; the best at 77K so far is 3×10^{-31} $JHz^{-1/2}$ at 71kHz, (Kawasaki, IBM). However, the l/f noise energy, at best 10^{-28} JHz^{-1} at 1Hz, remains high, and its reduction is a major challenge for the future.

Wellstood also described the development of epitaxial YBCO-insulator-YBCO multilayers that provide the essence of an interconnect technology – insulating crossovers and superconducting vias. Both he and Koch reported flip-chip magnetometers in which a SQUID on one chip is pressed against the multiturn input coil of a flux transformer on a second chip. The best magnetic field sensitivity yet achieved at 77K and 1Hz is about $1pTHz^{-1/2}$ (Berkeley/Conductus). Koch also reported a flip-chip first-order gradiometer, and Char (Conductus/Berkeley) an integrated magnetometer, involving 7 epitaxial layers, in which the SQUID and flux transformer were grown on a single chip. Two groups showed magnetocardiograms taken with HTSC SQUIDs at 77K: Vasiliev (Dubna) with a bulk rf SQUID, and the Berkeley/Conductus group with their flip-chip magnetometer.

Thin-film high-T_c devices have made great strides in the last year or so. SQUIDs have progressed to the point where real-life applications – geophysics, nondestructive evaluation and even magnetocardiology – are only a year or two away. One might hope for commerically available SQUIDs on a similar time-scale. The whole field of superconducting and quantum devices is progressing more rapidly now than at any time in the past that I can recall, and there is every reason to believe that the next Berlin SQUID conference – maybe in 1995? – will be just as exciting as this one.

In concluding, I should like to express the thanks of us all to the sponsors and organizers of SQUID '91. The conference was supported by the European Physical Society, the Physikalish-Technische Bundesanstalt and the Dr. Wilhelm Heinrich Heraeus and Else Heraeus Foundation, to all of whom we express our grateful thanks. The superb organization followed in the tradition of Drs. H.D. Hahlbohm and H. Lübbig who originally conceived the SQUID conferences and organized the first three. They passed the helm to Dr. Hans Koch who, supported by Dr. R. Cantor, D. Drung and S.N. Erné and by a wonderfully able group of helpers, worked so hard to make the conference run smoothly and be such a resounding success. We thank you all very much.

Index of Contributors

Springer Proceedings in Physics

Managing Editor: H. K. V. Lotsch

1 *Fluctuations and Sensitivity in Nonequilibrium Systems*
Editors: W. Horsthemke and D. K. Kondepudi

2 *EXAFS and Near Edge Structure III*
Editors: K. O. Hodgson, B. Hedman, and J. E. Penner-Hahn

3 *Nonlinear Phenomena in Physics* Editor: F. Claro

4 *Time-Resolved Vibrational Spectroscopy*
Editors: A. Laubereau and M. Stockburger

5 *Physics of Finely Divided Matter*
Editors: N. Boccara and M. Daoud

6 *Aerogels* Editor: J. Fricke

7 *Nonlinear Optics: Materials and Devices*
Editors: C. Flytzanis and J. L. Oudar

8 *Optical Bistability III*
Editors: H. M. Gibbs, P. Mandel, N. Peyghambarian, and S. D. Smith

9 *Ion Formation from Organic Solids (IFOS III)*
Editor: A. Benninghoven

10 *Atomic Transport and Defects in Metals by Neutron Scattering*
Editors: C. Janot, W. Petry, D. Richter, and T. Springer

11 *Biophysical Effects of Steady Magnetic Fields*
Editors: G. Maret, J. Kiepenheuer, and N. Boccara

12 *Quantum Optics IV*
Editors: J. D. Harvey and D. F. Walls

13 *The Physics and Fabrication of Microstructures and Microdevices*
Editors: M. J. Kelly and C. Weisbuch

14 *Magnetic Properties of Low-Dimensional Systems*
Editors: L. M. Falicov and J. L. Morán-López

15 *Gas Flow and Chemical Lasers*
Editor: S. Rosenwaks

16 *Photons and Continuum States of Atoms and Molecules*
Editors: N. K. Rahman, C. Guidotti, and M. Allegrini

17 *Quantum Aspects of Molecular Motions in Solids*
Editors: A. Heidemann, A. Magerl, M. Prager, D. Richter, and T. Springer

18 *Electro-optic and Photorefractive Materials*
Editor: P. Günter

19 *Lasers and Synergetics*
Editors: R. Graham and A. Wunderlin

20 *Primary Processes in Photobiology*
Editor: T. Kobayashi

21 *Physics of Amphiphilic Layers*
Editors: J. Meunier, D. Langevin, and N. Boccara

22 *Semiconductor Interfaces: Formation and Properties*
Editors: G. Le Lay, J. Derrien, and N. Boccara

23 *Magnetic Excitations and Fluctuations II*
Editors: U. Balucani, S. W. Lovesey, M. G. Rasetti, and V. Tognetti

24 *Recent Topics in Theoretical Physics*
Editor: H. Takayama

25 *Excitons in Confined Systems*
Editors: R. Del Sole, A. D'Andrea, and A. Lapiccirella

26 *The Elementary Structure of Matter*
Editors: J.-M. Richard, E. Aslanides, and N. Boccara

27 *Competing Interactions and Microstructures: Statics and Dynamics*
Editors: R. LeSar, A. Bishop, and R. Heffner

28 *Anderson Localization*
Editors: T. Ando and H. Fukuyama

29 *Polymer Motion in Dense Systems*
Editors: D. Richter and T. Springer

30 *Short-Wavelength Lasers and Their Applications*
Editor: C. Yamanaka

31 *Quantum String Theory*
Editors: N. Kawamoto and T. Kugo

32 *Universalities in Condensed Matter*
Editors: R. Jullien, L. Peliti, R. Rammal, and N. Boccara

33 *Computer Simulation Studies in Condensed Matter Physics: Recent Developments*
Editors: D. P. Landau, K. K. Mon, and H.-B. Schüttler

34 *Amorphous and Crystalline Silicon Carbide and Related Materials*
Editors: G. L. Harris and C. Y.-W. Yang

35 *Polycrystalline Semiconductors: Grain Boundaries and Interfaces*
Editors: H. J. Möller, H. P. Strunk, and J. H. Werner

36 *Nonlinear Optics of Organics and Semiconductors*
Editors: T. Kobayashi

37 *Dynamics of Disordered Materials*
Editors: D. Richter, A. J. Dianoux, W. Petry, and J. Teixeira

38 *Electroluminescence*
Editors: S. Shionoya and H. Kobayashi

39 *Disorder and Nonlinearity*
Editors: A. R. Bishop, D. K. Campbell, and S. Pnevmatikos

40 *Static and Dynamic Properties of Liquids*
Editors: M. Davidović and A. K. Soper

41 *Quantum Optics V*
Editors: J. D. Harvey and D. F. Walls

42 *Molecular Basis of Polymer Networks*
Editors: A. Baumgärtner and C. E. Picot

43 *Amorphous and Crystalline Silicon Carbide II: Recent Developments*
Editors: M. M. Rahman, C. Y.-W. Yang, and G. L. Harris